Introduction to Trellis-Coded Modulation with Applications

Introduction to Trellis-Coded Modulation with Applications

Ezio Biglieri
Politecnico di Torino, Italy

Dariush Divsalar
Jet Propulsion Laboratory
California Institute of Technology

Peter J. McLane
Queen's University, Canada

Marvin K. Simon
Jet Propulsion Laboratory
California Institute of Technology

Macmillan Publishing Company
New York
Maxwell Macmillan Canada, Inc.
Toronto
Maxwell Macmillan International
New York Oxford Singapore Sydney

Editor: John Griffin
Text Designer: Eileen Burke
Illustrations: Publication Services

Copyright © 1991 by Macmillan Publishing Company, a division of Macmillan Inc.

Printed in the Republic of Singapore

All rights reserved. No part of this book may be reproduced or transmitted in any form or by any means, electronic or mechanical, including photocopying, recording, or any information storage and retrieval system, without permission in writing from the publisher.

Distribution rights in this edition are controlled exclusively by Maxwell Macmillan Publishing Singapore Pte. Ltd. and restricted to selected countries. This book may be sold only in the country to which it has been sold or consigned by Maxwell Macmillan Publishing Singapore Pte. Ltd. Re-export to any other country is unauthorized and a violation of the rights of the copyright proprietor.

Macmillan Publishing Company
866 Third Avenue, New York, New York 10022

Maxwell Macmillan Canada, Inc.
1200 Eglinton Avenue East
Suite 200
Don Mills, Ontario M3C 3N1

Library of Congress Cataloging in Publication Data

Introduction to trellis-coded modulation, with applications/Ezio
  Biglieri... [et al.].
  p. cm.
  Includes bibliographical references and index.
  1. Digital modulation. 2. Modulation (Electronics) 3. Coding
theory. I. Biglieri, Ezio
  TK5106.7.I58   1991   91-10991
  621.381'536—dc20   CIP
ISBN 0-02-309965-8 (Hardcover Edition)
ISBN 0-02-946542-7 (International Edition)

IE Printing: 1 2 3 4 5   Year: 2 3 4 5

ISBN 0-02-946542-7

Dariush Divsalar dedicates this book to his wife Guity, daughters Dina and Shahrzad, and to his mother and father for their support.

Marvin K. Simon dedicates this book to his wife Anita and children Brette and Jeff for their encouragement and in particular their devotion and understanding during the many hours spent away from them.

Peter J. McLane dedicates this book to his wife Colleen for her encouragement and understanding.

Preface

Ever since 1974, when Massey in the seminal paper [1] formally suggested the notion of improving system performance by looking at modulation and coding as a *combined* entity, researchers have been investigating ways of implementing this idea into reality. Prior to that time, just about all coded digital communication systems were designed by independently specifying the modulator/demodulator and the encoder/decoder portion, which led to some disappointments. To quote from [1], sometimes the communication engineer "presumes that the purpose of coding is to 'correct errors' made by the demodulator." In this case "he commonly refers to codes in general as 'error correcting codes.' He somehow believes that if the 'error probability' is unacceptably large, then coding should be able to rescue his design and he is inevitably disappointed to learn that the necessary code redundancy is so large that his overall system is quite inefficient."

Perhaps the most significant contribution toward carrying out Massey's thoughts was the invention of *trellis-coded modulation (TCM)* as described by Ungerboeck in his classic 1982 paper [2]. (Similar ideas were independently developed at about the same time by Imai and Hirakawa in a paper that does not seem to have received the attention it deserves at the time of its publication [3]). The primary advantage of TCM over modulation schemes employing traditional error-correction coding was the ability to achieve increased power efficiency without the customary expansion of bandwidth introduced by the coding process. Thus, channels that were power-limited and bandwidth-limited were an ideal application for TCM.

The purpose of this book is twofold. First, it is intended to introduce the neophyte to this timely and exciting subject. The first chapters, and in particular Chapter 3 and Chapter 4, should serve the purpose of giving the reader a grasp of the theory of TCM and the techniques needed for its analysis. Our second purpose is to describe the results of research in this area that have occurred over the last seven years since its original inception. Like Goha, who loses his money in the desert at night and searches for it in his tent because "it's dark outside, and here there is light," we have chosen to describe in more depth the results of research carried out by ourselves. However, we did our best to provide a substantial coverage of other topics that may be of some interest in TCM analysis and design.

Samuel Johnson wrote in the preface to his dictionary: "No dictionary of a living tongue ever can be perfect, since while it is hastening to publication, some words are budding and some falling away." This is true in general for the scientific literature as well, and even more so in an active research area like TCM, where the soil appears to be tropically rich and new ideas are continually brought to the forum, while many more are waiting in the wings. For this reason, and since this is indeed the first book completely devoted to TCM, the authors want to emphasize the fact that the book is really intended as an educational primer rather than a complete, encyclopedic treatise on the subject. This is why we chose to use the word "Introduction" in the title. Despite this, the book is still a highly technical detailed complete work, and the reader should come away with a thorough grasp of the subject.

The flavor of the book is both theoretical and practical. As such, it should appeal to the university environment as well as provide a valuable reference for those working in industry. This is an obvious consequence of the vast university and industrial experience of the four authors. Contained in several of the chapters are results related to applications of the theory to practical real-world problems, for example, fading channels and commercial modems.

The authors have successfully taught much of this material in portions of courses at UCLA and at Queen's University, and in three-day short courses on TCM. To that extent, the book is sufficiently compact that the basic principles can be extracted in a relatively short period of time.

Two of the authors (E. B. and P. J. McL.) had their interaction made easier by the NATO Grant No. 87/0368, which supported their travel expenses during the preparation of the manuscript. Our students and colleagues read portions of the manuscript and provided comments and suggestions. Among them, E. B. would like to thank Yow-Jong Liu, Flavio Lorenzelli, Reginaldo Palazzo, Jr., and Teresa M. Thesken.

P. J. McLane would like to thank Queen's University, the Telecommunications Research Institute of Ontario (TRIO), and the Natural Sciences and Engineering Research Council of Canada for their support. Mary Ann Simpson of TRIO is thanked for her excellent work on LATEX in preparing four chapters of the book. P. J. McLane's students, Richard Buz, Lorne Berg, Alex Fung, and Martin Turgeon did many of the TCM designs given in Chapters 5 and 8, and they are gratefully acknowledged. The digital computer programs in Appendix C were first written by Martin Turgeon and were modified by Loren Berg: both are thanked. Liang Wong is thanked for his contribution to the material in Section 11.1 and Paul Ho for general discussions on TCM. L. J. Mason of the Communication Research Centre, Ottawa, Canada, deserves special thanks for material that was used in Chapter 2 on nonbinary BCH codes, and M. Sablatash of the same organization is thanked for some material on error-correcting codes. Finally, Chun Loo, also of this organization, is thanked for material in Chapter 1 on fading models for mobile satellite communications.

The other two authors (D. D. and M. K. S) wish to thank the Jet Propulsion Laboratory, California Institute of Technology, located in Pasadena, California, for providing an academic environment within a government-supported institution sufficient to motivate them to jointly perform the research reported on in this book. To this end, what began a number of years back as two researchers collaborating in an area of common interest has evolved into a symbiotic working relationship that is largely responsible for the success of their part of this venture.

E. B.
D. D.
P. McL.
M. S.

REFERENCES

1. J. L. Massey, "Coding and modulation in digital communications," *Proc. 1974 International Zürich Seminar on Digital Communications,* Zürich, Switzerland, pp. E2(1)–E2(4), Mar. 1974.
2. G. Ungerboeck, "Channel coding with multilevel/phase signals," *IEEE Trans. Inform. Theory,* Vol. IT-28, No. 1, pp. 55–67, Jan. 1982.
3. H. Imai and S. Hirakawa, "A new multilevel coding method using error-correcting codes," *IEEE Trans. Inf. Theory,* Vol. IT-23, pp. 371–377, 1977.

Contents

CHAPTER 1
Introduction   1
- 1.1 Digital communications structure   1
  - 1.1.1 Source encoder   2
  - 1.1.2 Channel encoder   2
  - 1.1.3 Modulator   4
  - 1.1.4 The communications channel   6
  - 1.1.5 The receiver   8
- 1.2 Discrete memoryless channels   11
  - 1.2.1 Uncoded baseband communication   11
  - 1.2.2 Bit error probability: Binary signals   12
- 1.3 Entropy   15
  - 1.3.1 Discrete sources   16
  - 1.3.2 Continuous source   17
- 1.4 Channel capacity   18
  - 1.4.1 Related information theory measures   19
  - 1.4.2 Channel capacity   21
  - 1.4.3 Symmetric channels   21
  - 1.4.4 Kuhn–Tucker conditions   23
  - 1.4.5 Bandlimited Gaussian channel   25
- 1.5 Shannon's two theorems   26
- 1.6 Computational cutoff rate   27

CHAPTER 2
Error-correcting codes   33
- 2.1 Introduction   33
- 2.2 Parity check codes   34
- 2.3 Matrix description: Error-correcting codes   36
- 2.4 Algebraic concepts   40
  - 2.4.1 Primitive element   42
  - 2.4.2 Extension field $GF(q)$ 42
- 2.5 Cyclic codes   48

2.6 BCH codes   48
   2.6.1 Binary BCH codes   49
   2.6.2 Nonbinary BCH codes   52
   2.6.3 Reed–Solomon codes   54
2.7 Convolutional codes   56
   2.7.1 FSM and trellis representation   56
   2.7.2 $\mathbf{G}(D)$ and $\mathbf{H}(D)$ matrices   58
   2.7.3 Viterbi algorithm   59
   2.7.4 Error-state diagrams   61

CHAPTER 3
TCM: Combined modulation and coding   67

3.1 Introducing TCM   67
3.2 The need for TCM   69
3.3 Fundamentals of TCM   70
   3.3.1 Uncoded transmission   70
3.4 The concept of TCM   71
3.5 Trellis representation   73
3.6 Some examples of TCM schemes   74
3.7 Set partitioning   77
3.8 Representations for TCM   79
   3.8.1 Ungerboeck representation   79
   3.8.2 Analytical representation   82
3.9 Decoding TCM   87
   3.9.1 Definition of branch metric   88
   3.9.2 The Viterbi algorithm   90
3.10 Bibliographical notes   93
Appendix 3A: Orthogonal expansion of the function f 94

CHAPTER 4
Performance evaluation   99

4.1 Upper bound to error probability   99
   4.1.1 The error-state diagram   102
   4.1.2 The transfer function bound   102
   4.1.3 Consideration of different channels   113
4.2 Lower bound to error probability   114
4.3 Examples   116
4.4 Computation of d_{free} 125
   4.4.1 Using the error-state diagram   125
   4.4.2 A computational algorithm   128
   4.4.3 The product-trellis algorithm   131
4.5 Lower bounds to the achievable d_{free} 133
   4.5.1 A simple lower bound   135

4.6 Sphere-packing upper bounds 135
  4.6.1 A universal upper bound 137
4.7 Other sphere-packing bounds 138
  4.7.1 Constant-energy signal constellations 138
  4.7.2 Rectangular signal constellations 138
  4.7.3 An upper bound for PSK signals 138
  4.7.4 An asymptotic upper bound 145
4.8 Power density spectrum 145

CHAPTER 5
One- and two-dimensional modulations for TCM 149

5.1 Introduction 149
5.2 Step-by-step design procedure 149
  5.2.1 Derivation of the analytic description 150
  5.2.2 Design rules and procedure 152
5.3 One-dimensional examples 153
5.4 Two-dimensional examples 163
5.5 Trellis code performance and realization 169
5.6 Trellis coding with asymmetric modulations 174
  5.6.1 Analysis and design 175
  5.6.2 Best rate 1/2 codes combined with asymmetric 4-PSK (A4-PSK) 179
  5.6.3 Best rate 2/3 codes combined with asymmetric 8-PSK (A8-PSK) 187
  5.6.4 Best rate 3/4 codes combined with asymmetric 16-PSK (A16-PSK) 195
  5.6.5 Best rate 1/2 codes combined with asymmetric 4-AM 201
  5.6.6 Trellis-coded asymmetric 16-QAM 203

CHAPTER 6
Multidimensional modulations 207

6.1 Lattices 209
  6.1.1 Some examples of lattices 210
  6.1.2 Structural characteristics of lattices 212
  6.1.3 Example of lattice constellations for TCM 212
  6.1.4 Partition of lattices 216
  6.1.5 Calderbank–Sloane TCM schemes based on lattices 217
6.2 Group alphabets 220
  6.2.1 Set partitioning of a GA 223
6.3 Ginzburg construction 224
  6.3.1 Set partitioning 229
  6.3.2 Designing a TCM scheme 231

xiv Contents

    6.4   Wei construction   232
           6.4.1   A design example   233
    6.5   Trellis-encoded CPM   240
           6.5.1   Review of CPM   240
           6.5.2   Parameter selection   242
           6.5.3   Designing the TCM scheme   243
           6.5.4   Performance examples   246
Appendix 6A: Examples of group alphabets   246
                        6A.1   Permutation alphabets   247
                        6A.2   Cyclic-group alphabets   250
Appendix 6B: Decomposition of the CPM modulator   250
                        6B.1   Phase trellis and tilted phase trellis   251
                        6B.2   Decomposing the CPM modulator   252

CHAPTER 7
Multiple TCM   259

    7.1   Two-state MTCM   261
           7.1.1   Mapping procedure for two-state MTCM   264
           7.1.2   Evaluation of minimum squared free distance   265
    7.2   Generalized MTCM   268
           7.2.1   Set-partitioning method for generalized MTCM   269
           7.2.2   Set mapping and evaluation of squared free distance   273
    7.3   Analytical representation of MTCM   282
    7.4   Bit error probability performance   287
    7.5   Computational cutoff rate and MTCM performance   290
    7.6   Complexity considerations   292
    7.7   Concluding remarks   293

CHAPTER 8
Rotationally invariant trellis codes   295

    8.1   Introduction   295
           8.1.1   Rotational invariance   296
           8.1.2   Rotational invariant code   297
           8.1.3   Design rules   302
           8.1.4   Design procedure   303
           8.1.5   16-point examples   304
    8.2   Generation: Rotationally invariant codes   310
           8.2.1   Eight-point example   310
           8.2.2   Sixteen-point examples   313
    8.3   Multidimensional RIC   322
           8.3.1   Linear examples   323
           8.3.2   Nonlinear example   329
    8.4   Bit error rate performance   335

8.4.1 Nonlinear codes 335
8.4.2 Linear codes 336

CHAPTER 9
Analysis and performance of TCM for fading channels 343

9.1 Coherent detection of trellis-coded M-PSK on a slow-fading Rician channel 344
    9.1.1 Channel model 344
    9.1.2 System model 344
    9.1.3 Upper bound on pairwise error probability 346
    9.1.4 Upper bound on bit error probability 350
    9.1.5 Simulation results 366
9.2 Differentially coherent detection of trellis-coded M-PSK on a slow-fading Rician channel 371
    9.2.1 System model 371
    9.2.2 Analysis model 372
    9.2.3 The maximum-likelihood metric for trellis-coded M-DPSK 373
    9.2.4 Upper bound on pairwise error probability 374
    9.2.5 Upper bound on average bit error probability 378
    9.2.6 Simulation results 385
9.3 Differentially coherent detection of trellis-coded M-PSK on a fast-fading Rician channel 387
    9.3.1 Analysis model 388
    9.3.2 Upper bound on pairwise error probability 388
    9.3.3 Upper bound on average bit error probability 389
    9.3.4 Characterization of the autocorrelation and power spectral density of the fading process 390
    9.3.5 Simulation results 392
9.4 Asymptotic results 393
    9.4.1 An example 398
    9.4.2 No interleaving/deinterleaving 399
9.5 Further discussion 401
Appendix 9A: Proof that d whose square is defined in (9.19) satisfies the conditions for a distance metric 402
Appendix 9B: Derivation of the maximum-likelihood branch metric for trellis-coded M-DPSK with ideal channel state information 405

CHAPTER 10
Design of TCM for fading channels 411

10.1 Multiple trellis-code design for fading channels 412
10.2 Set partitioning for multiple trellis-coded M-PSK on the fading channel 417

- 10.2.1 The first approach 417
- 10.2.2 The second approach 427
- 10.3 Design of Ungerboeck-type codes (unit multiplicity) for fading channels 430
- 10.4 Comparison of error probability performance with computational cutoff rate 432
- 10.5 Simulational results 433
- 10.6 Further discussion 435

CHAPTER 11
Analysis and design of TCM for other practical channels 437

- 11.1 Intersymbol interference channels 437
  - 11.1.1 Model 438
  - 11.1.2 LMS equalization 444
  - 11.1.3 Trellis-code performance 447
- 11.2 Channels with phase offset 453
  - 11.2.1 Upper bound on the average bit error probability performance of TCM 454
  - 11.2.2 Carrier synchronization loop statistical model and average pairwise error probability evaluation 461
  - 11.2.3 The case of binary convolutional coded BPSK modulation 464
  - 11.2.4 A TCM example 470
  - 11.2.5 Concluding remarks 474
- 11.3 TCM over satellite channels 475
  - 11.3.1 A modem for land mobile satellite communications 476
  - 11.3.2 An SCPC modem 478
- 11.4 Trellis codes for partial response channels 479
  - 11.4.1 Trellis codes for the binary $(1 - D)$ channel 485
  - 11.4.2 Convolutional codes with precoder for $(1 - D)$ channels 487
- 11.5 Trellis coding for optical channels 490
  - 11.5.1 Signal sets with amplitude and pulse-width constraints 492
  - 11.5.2 Trellis-coded modulation for optical channels 494
- 11.6 TCM with prescribed convolutional codes 502
  - 11.6.1 Application to M-PSK modulation 503
  - 11.6.2 Application to M-AM and QAM modulations 506

APPENDIX A
Fading channel models — 511

APPENDIX B
Computational techniques for transfer functions — 521

APPENDIX C
Computer programs: Design technique — 527

Index — 541

Introduction to Trellis-Coded Modulation with Applications

CHAPTER 1

Introduction

We begin by outlining the structure of digital communication systems and end the first chapter with an outline of Shannon's information theory, which includes some aspects of elementary channel and source coding. The second chapter contains the rest of our material on traditional coding theory: a consideration of BCH codes and some material on convolutional codes. The main intent of the first two chapters is to supply the reader with the necessary background in information theory and error correction coding to understand the theory of trellis-coded modulation. Another goal is to provide the essentials of information theory and coding where the book is used for a single course on these subjects that contains a significant component on trellis-coded modulation.

1.1 Digital Communications Structure

Digital communications systems have a definite structure and knowledge of this structure is helpful in understanding the role of coding and modulation systems. The simplest structure, shown in Fig. 1.1, is for a point-to-point communication system—not that for a communications network or a point-to-multipoint system. We have a transmitter, T_x, a receiver, R_x, and a channel that links the transmitter and receiver.

The transmitter, channel, and receiver shown in Fig. 1.1 can be further subdivided. Let us begin by considering a subdivision of the transmitter structure (Fig. 1.2). We have an information source that we will take as binary, which means that its output is a sequence in which the only elements are 1s and 0s. The source is followed by a source encoder, a channel encoder, and a modulator. We now describe the function of each of these entities. Note that if the source is analog—for example, a speech or video source—we shall assume that it has been digitized.

FIGURE 1.1 Simplest digital communications structure.

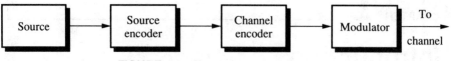

FIGURE 1.2 Transmitter block diagram.

1.1.1 Source Encoder

A good, or desirable source is random—such sources have maximum information. Clearly, if a 1 and a 0 have an equal probability of occurrence, knowledge of the source output provides a maximum amount of information. For instance, if the source nearly always outputs a 1, its output can be predicted and knowledge of the output provides very little information. Usually, sources are not random and contain significant amounts of redundancy. For example, in a video image, neighboring picture elements are usually strongly related. The role of a source encoder is to randomize the source. A measure of randomness is entropy, a concept borrowed from thermodynamics. The function of a source encoder, then, is as illustrated in Fig. 1.3.

Why do we want the source to be encoded to a disordered state? The answer lies in the utilization of one of the scarce resources of the telecommunications problem—the channel. We should not waste the scarce resources of the channel by sending predictable quantities over this link between the receiver and the transmitter. The channel should only be used to carry the unpredictable information from the source, that is, the output from the source encoder.

1.1.2 Channel Encoder

The goal of the channel encoder is to introduce an error correction capability into the source encoder output to combat channel transmission errors. To achieve this goal, some redundancy must be added to the source encoder output. This may seem confusing at first because we have just argued that all redundancy must be stripped from the source outputs for efficient channel transmission. Indeed, this book is about a technique, trellis coding, for adding redundancy to the source outputs so that the channel is utilized from a very efficient point of view. However, the first two chapters are about the traditional way of adding redundancy, through parity checks, and then transmitting the information plus parity bits across the channel in

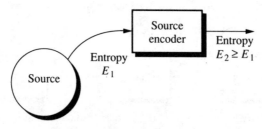

FIGURE 1.3 Function of a source encoder.

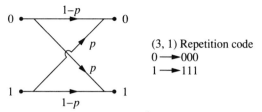

FIGURE 1.4 BSC plus (3, 1) repetition code.

a time-serial manner. Note that the same parity bits will be appended to a unique collection of source output bits called the message. In this way, the redundancy we add to the message is controlled and the receiver will have knowledge of the structure of this redundancy. This is the difference between the original redundancy in the source symbols, which is not controlled, and the redundancy added in channel coding, which is controlled. Let us consider a simple example of channel coding. The binary symmetric channel (BSC) plus a (3, 1) repetition code are illustrated in Fig. 1.4.

The transmission diagram is a summarizing diagram for transmission over the channel, illustrating the fact that transmission errors occur with a prescribed probability of p. The channel coder appends two identical parity bits to the source symbol, and the resulting 3-bit word is transmitted over the channel one bit at a time (in Section 1.1.3 we show how this could be done). We can regard channel transmission as three uses of the BSC shown in Fig. 1.4. If $(0, 1, 0)$ is the 3-bit output from three uses of the BSC, we should declare $(0, 0, 0)$ as transmitted (denoted as t_x), since if 0 was the source bit sent, we have corrected a channel transmission error in position 2. Thus our decoder for the BSC's 3-bit outputs is based on majority rule and so will always result in one channel error being corrected no matter where it appears in the 3-bit word.

The situation described above can be represented using the cube shown in Fig. 1.5. Two possible code words transmitted over the BSC are separated by a Hamming distance, d^H, of 3 and nearest-neighbor decoding results on a single error being corrected. In general, if two code words differ in their component position, we add one to their component distance, and examining all components gives the Hamming distance between the two code words. Here we have $d^H = 3$. In general, for more than two code words the greatest chance for error comes in comparing two code words of least distance, d^H_{min}. In addition, the number of errors that can be corrected by a code with the shortest Hamming distance, d^H_{min}, is $t = \lfloor (d^H_{min} - 1)/2 \rfloor$, where $\lfloor x \rfloor$ is the largest integer less than or equal to x. In the present example we have $t = 1$ correctable channel transmission errors per code word received as $d^H_{min} = 3$.

The detection of errors is also a key item in channel coding because we could always request, through a feedback channel, retransmission of a code word detected to contain errors. In the present example only one error can be detected: for instance, $(0, 1, 1)$ could be received (denoted as r_x) when $(0, 0, 0)$ was transmitted, but this error pattern is not detectable since the decoder must also consider $(1, 1, 1)$ as a candidate transmitted code word. Clearly, a single channel error can always

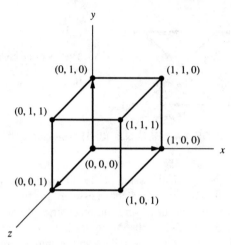

FIGURE 1.5 Decoding represented as points on a cube.

be detected for the present example. In general, $\lfloor d_{min}^H/2 \rfloor$ transmission errors can always be detected and for the present example we have only one error detected. However, if we used the code $1 \rightarrow (1, 1, 1, 1)$ and $0 \rightarrow (0, 0, 0, 0)$, we would have $d_{min}^H = 4$, and thus two errors detected but still only one error corrected. Note here that only 1 out of 4 bits sent is an information bit, and we say that the rate of the code is 1/4. The earlier code had rate 1/3 and thus less error detection capability than given in the rate 1/4 case. Our concept of error detection here is different than in most textbooks on coding theory, where the number of errors detected is taken as $d_{min}^H - 1$. In traditional coding the rate 1/3 code is transmitted by using the BSC three times for each information bit. To realize this in practice, we must either speed up the rate of symbol transmission by a factor of 3 or keep the same rate of symbol transmission and be content with one-third the information transmission rate relative to when no channel coding is used. Thus, in either case, an increased channel bandwidth is required per information bit transmitted. This book is about an alternative to this approach that involves no change in information transmission rate; rather, the number of points in the signal constellation for modulation is increased to achieve the required redundancy.

1.1.3 Modulator

In Fig. 1.2 the modulator interfaces the channel encoder to the channel. The source, source encoder, and channel encoder taken together can be viewed as a modified binary source that feeds the modulator, and as such, the modulator can be regarded as interfacing the source to the channel. Physical channels can require electrical signals, radio signals, or optical signals. The modulator takes in the source outputs and outputs waveforms that suit the physical nature of the channel and are also chosen to yield either system simplicity or optimal detection performance. A baseband binary modulator is shown in Fig. 1.6. We call this a baseband modulator because no sinusoidal carrier signal is involved. On a channel that has symmetric

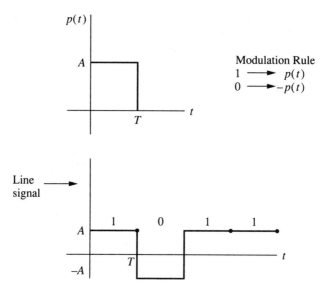

FIGURE 1.6 Baseband modulator.

interference, the signal selection in Fig. 1.6 is optimal in that it will yield for a fixed transmitted power the least number of errors in detection in the receiver. In the quaternary modulator shown in Fig. 1.7, two source bits per symbol interval T are required, whereas before, a single bit will do. In the quaternary case the symbol transmission rate is $2/T$ bits per second (bits/s). This is an example of pulse amplitude modulation; the amount of channel bandwidth such signals

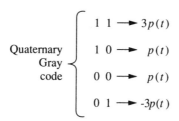

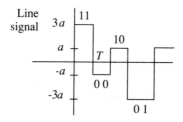

FIGURE 1.7 Quaternary modulator.

require is related only to the rate at which the modulator signals are changed, that is, to the rate $1/T$. Thus the quaternary case has twice the throughput of the binary case and clearly cannot have the same error performance, because the receiver must sort out which of four signals were transmitted for the quaternary case, whereas a signal selection over two possibilities suffices in the binary case.

To consider carrier modulation, consider the sinusoidal signal

$$s(t) = A(t)\cos[\omega_c t + \theta(t)] \quad (1.1)$$

where $A(t)$ is the amplitude; ω_c is the frequency in radians per second and equals $2\pi f_c$, with f_c the frequency in hertz; and $\theta(t)$ is the phase. In carrier modulation we can vary any or all of the parameters (A, ω_c, θ): varying A is called amplitude modulation, varying θ is called phase modulation, and varying ω_c is called frequency modulation; in all cases the variation is (hopefully) linearly related to the message to be transmitted. Some examples are given in Figs. 1.8 and 1.9. Note that binary phase modulation (called binary phase shift keying, BPSK) is equivalent to binary amplitude shift keying (BASK). The type of quadrature amplitude modulation (QAM) shown in Fig. 1.9 is called 64-QAM in that the signal constellation contains 64 points. Thus $6/T$ bits per symbol interval T can be transmitted over the channel. Inherent in the use of this modulation is the fact that two carrier signals that differ in phase by $\pm 90°$ can be separated in the receiver to recover the signals $X(t)$ and $Y(t)$, known as the in-phase and quadrature signals, respectively. This can be done in a coherent receiver, which is a receiver that must acquire and track any nonmodulation phases that exist in the received signal.

1.1.4 The Communications Channel

The simplest channel is the additive noise channel: here the signal is received with no distortion except additive noise. That is, if $r(t)$ is the received signal,

$$r(t) = s(t) + n(t) \quad (1.2)$$

where $s(t)$ is the transmitted signal and $n(t)$ is the additive noise. The classical theory of communication over the additive noise channel is given in reference [1]. A channel where the received signal is distorted, or at least can be distorted, is shown in Fig. 1.10. This phenomenon is called the intersymbol interference channel, as modulation symbols spill over into other symbol intervals, thus causing distortion. Additive noise is also present in the received signal but is not shown on the waveforms in Fig. 1.10.

Define the distribution of signal power as a function of frequency as the power spectrum of a signal. A power spectrum for a QPSK signal is displayed in Fig. 1.11. A QPSK signal involves modulation with a discontinuous phase angle. The other signal spectra in Fig. 1.11—minimum shift keying (MSK), duobinary minimum shift keying (DuMSK), and tamed frequency modulation (TFM)—involve phase modulation with increasing smoothness [2]. This smoothness produces a more compact spectrum. Call the bandwidth of a signal the set of frequencies that contain 98% of its power; that is, the area under the curve over this set of frequencies in Fig. 1.10 that contains 98% of the total area. If the 3-dB bandwidth of the linear

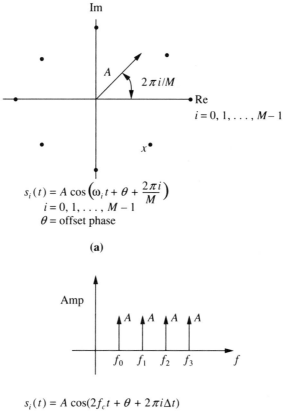

(a)

(b)

FIGURE 1.8 Digital phase and frequency modulation: (a) MPSK; (b) FSK.

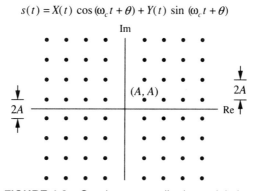

FIGURE 1.9 Quadrature amplitude modulation.

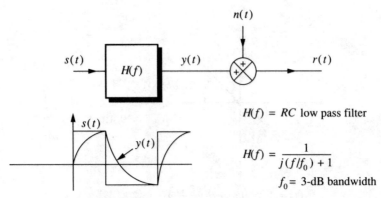

FIGURE 1.10 Intersymbol interference channel.

filter in Fig. 1.9 is significantly less than this bandwidth, intersymbol interference (ISI) results. The classical theory of communication over the ISI channel is given in [3]; a recent textbook [2] considers additive noise, ISI, and some nonlinear channels.

Much of this book is written for modulation and channel coding for the additive noise channel where the additive noise is white Gaussian noise—that is, the signal power spectrum is flat over the bandwidth of all signals sent over the channel. Very little is considered for the ISI channel because the application of trellis codes to this channel is in the early stages of research. A channel that will receive some attention is the nonfrequency selective fading channel (Fig. 1.12). Indeed, the greatest gains in performance that trellis codes have attained are for this channel. Let the input signal be $s(t)$ in equation (1.1); then the output or faded signal is

$$y(t) = G(t)A(t)\cos[\omega_c t + \theta(t) + \psi(t)]. \qquad (1.3)$$

In (1.3) the shape of $s(t)$ is not changed; only its amplitude and phase are altered. A typical fading function, $G(t)$, for the model developed in [4] for the Canadian Mobile Satellite Communications System is shown in Fig. 1.13. The classical theory of fading channels in mobile radio systems is given in Jakes's textbook [5], and a good section on fading channels appears in [6]. In addition, material of fading channel models can be found in Appendix A. It should be noted that fading channels represent an example of a multiplicative noise process rather than the additive noise case considered earlier.

In frequency-selective fading, $s(t)$ in (1.2) is distorted as well as attenuated in a time-varying manner. The channel in this case is a combination of the fading channel under consideration and the ISI channel. We do not consider such challenging channels in this book.

1.1.5 The Receiver

The receiver follows the channel in the block diagram in Fig. 1.1. Now the transmitter represents an operation on the source and the function of the transmitter

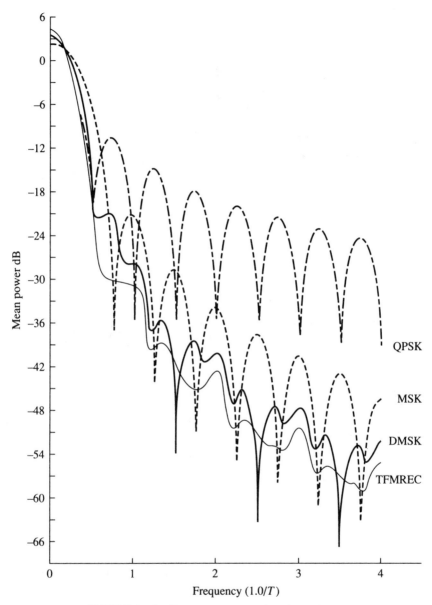

FIGURE 1.11 Power spectrum of various signals.

is to invert this operation and recover the source symbols. Figure 1.2 represents this operation by showing the transmitter in block diagram form. The inverse of this block diagram, the receiver, is shown in Fig. 1.14. Indeed, each block in Fig. 1.14 is the inverse of a corresponding block in Fig. 1.2. The demodulator is the inverse of the modulation process, the channel decoder inverts the channel encoder process, and so on. The various blocks are viewed independently, much as in

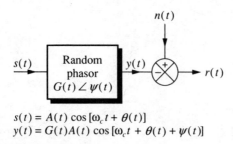

$s(t) = A(t) \cos [\omega_c t + \theta(t)]$
$y(t) = G(t)A(t) \cos [\omega_c t + \theta(t) + \psi(t)]$

FIGURE 1.12 Nonfrequency-selective fading channel.

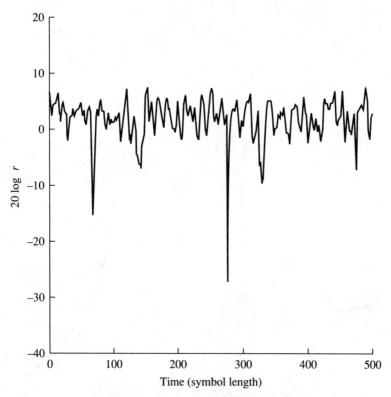

FIGURE 1.13 Amplitude fading function.

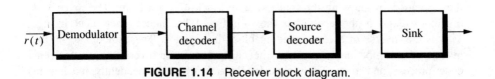

FIGURE 1.14 Receiver block diagram.

the case of multilevel data protocols. Trellis coding serves to merge the processes of modulation and coding, and recent work [7] is aimed at merging the roles of source coding, channel coding, and modulation. We consider an example of a demodulator in the next section. An example of a channel decoder was treated earlier—the majority rule or nearest-neighbor decoder discussed in relation to Fig. 1.4.

1.2 Discrete Memoryless Channels

An example of a discrete memoryless channel (DMC), the BSC, is given in Fig. 1.4. This channel has two inputs and two outputs. In general, a DMC can have a finite number of inputs and outputs. In any case, all of the channels described in Section 1.1.4 involved a continuous-time variable. In this section we show how to derive a DMC from a continuous-time channel description. The latter is a physical channel description, whereas the former is an abstract version of the channel.

1.2.1 Uncoded Baseband Communication

Consider the case of the transmitter, additive noise channel, and receiver shown in Fig. 1.15. The modulation will be as shown in Fig. 1.6, namely, $1 \to p(t)$ and $0 \to -p(t)$, where $p(t)$ is the rectangular pulse. The receiver is shown in Fig. 1.16; this is a matched filter and it is optimum in the sense of having the smallest error probability among all receivers [1]. A typical output is shown in Fig. 1.16, together with a sampler that is synchronous with the end of a pulse. These samples are quantized into two levels with a threshold at zero. If the sample is positive, a binary 1 is declared to have been transmitted; otherwise, a binary 0 is declared.

The BSC channel is shown in Fig. 1.4. This channel represents a summary of binary data transmission over the continuous-time channel shown in Fig. 1.15. The BSC is completely described by p, the probability of error per binary digit sent over the channel. Thus, to determine the BSC, we must find p.

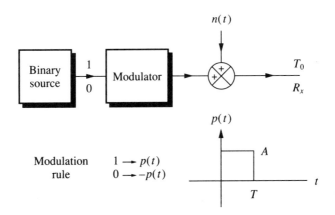

FIGURE 1.15 Baseband system with no coding.

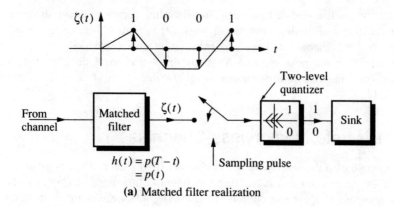

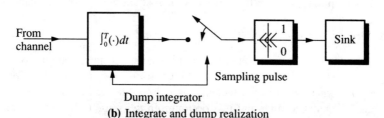

FIGURE 1.16 Matched filter receiver.

1.2.2 Bit Error Probability: Binary Signals

For the rectangular pulse shown in Fig. 1.6, the matched filter receiver [1] is shown in Fig. 1.16. This filter can be regarded as a simple integrator over the symbol interval—called an integrate-and-dump receiver, which is also shown in Fig. 1.16. Perfect symbol timing is assumed here; techniques for symbol timing are given in [8].

Let us assume first that a binary 1 was transmitted. Then the sampler output in Fig. 1.16 is

$$z = \int_0^T A\,dt + \int_0^T n(t)\,dt \qquad (1.4)$$
$$= AT + n.$$

The autocorrelation function of the additive white Gaussian noise is

$$R_n(\tau) = \mathbf{E}\{n(t+\tau)n(t)\}$$
$$= \frac{N_0}{2}\delta(\tau) \qquad (1.5)$$

where $\mathbf{E}\{X\}$ is the expectation of the random variable X and $\delta(\tau)$ is the Dirac delta function. This means that the power spectral density (PSD), the Fourier transform of $R_n(\tau)$, is $N_0/2$ for all frequencies.

1.2 / Discrete Memoryless Channels

In (1.4) and (1.5) we will need the mean and variance of the random variable n. Clearly, n is Gaussian, as the integral operation is linear and $n(t)$ is assumed to be Gaussian. Recall from [9] that the Gaussian property of a random variable is preserved under a linear transformation. Also, $\mathbf{E}\{n\} = 0$ as $\mathbf{E}\{n(t)\} = 0$ and thus $\sigma^2 = \mathbf{E}\{n^2\}$. Hence, from (1.4) and (1.5),

$$\sigma^2 = \mathbf{E}\left\{\int_0^T \int_0^T n(t)n(s)\,dt\,ds\right\}$$
$$= \int_0^T \int_0^T \mathbf{E}\{n(t)n(s)\}\,dt\,ds. \tag{1.6}$$

Now use of (1.5) in (1.6) gives a double integral involving $\delta(t-s)$ in the integrand. A property of $\delta(x)$ that we require to proceed is

$$f(x_0) = \int_a^b f(x)\delta(x - x_0)\,dx \tag{1.7}$$

as long as $a < x_0 < b$. After combining (1.5) and (1.6) using the result in (1.7), we obtain $\sigma^2 = N_0 T/2$.

Now n in (1.4) is a Gaussian random variable with zero mean and variance $N_0 T/2$; we write that n is $\mathcal{N}(0, N_0 T/2)$, which means that n has the probability density function (pdf)

$$p_n(x) = \frac{1}{\sigma\sqrt{2\pi}}e^{-x^2/2\sigma^2}. \tag{1.8}$$

When binary 1 is transmitted, the receiver makes an error when $z \leq 0$. That is, from (1.5), the error event is $\{AT + n \leq 0\}$. Thus

$$P(e\,|\,1_s) = P(n \leq -AT) \tag{1.9}$$

where 1_s means that a binary 1 is transmitted. Use of the pdf of (1.8) in (1.9) gives

$$P(e\,|\,1_s) = \frac{1}{2}\,\text{erfc}\left(\frac{AT}{\sigma\sqrt{2}}\right) \tag{1.10}$$

where $\text{erfc}(x)$ is given by

$$\text{erfc}(x) = \frac{2}{\sqrt{\pi}}\int_x^\infty e^{-t^2}\,dt \tag{1.11}$$

and $\text{erfc}(-x) = 2 - \text{erfc}(-x)$, which is needed to derive (1.10). To put (1.10) and (1.11) in a more useful form, we need to determine the signal energy for $p(t)$. In general,

$$E_s = \int_{-\infty}^\infty s^2(t)\,dt.$$

Then for $s(t)$ set equal to $p(t)$ in Fig. 1.4, that is, $s(t)$ a rectangular pulse of height A and duration T, it follows that, $E_p = A^2 T$. To compute the average energy over both binary digits, we have

$$\overline{E_b} = \frac{A^2 T}{2} + \frac{(-A)^2 T}{2} \qquad (1.12)$$
$$= A^2 T.$$

Thus, as we showed earlier that $\sigma^2 = N_0 T/2$, it follows from (1.10), (1.11), and (1.12) that

$$P(e|1_s) = \frac{1}{2} \text{erfc}\left(\sqrt{\frac{E_b}{N_0}}\right). \qquad (1.13)$$

It is a simple exercise for the reader to show that $P(e|1_s) = P(e|0_s)$. This involves repeating the argument above for a binary 0 transmitted and considering the error event $\{-AT + n \geq 0\}$. To determine the average error probability, we must average our results for a binary 1 transmitted and a binary 0 transmitted. We have

$$P(e) = P(e|0_s)P(0_s) + P(e|1_s)P(1_s) \qquad (1.14)$$

where 0_s means that a binary 0 was transmitted. Now above we have shown that $P(e|1_s) = P(e|0_s)$. Thus $P(e) = P(e|1_s)$ as $P(1_s) + P(0_s) = 1$, and from (1.14),

$$P(e) = \frac{1}{2} \text{erfc}\left(\sqrt{\frac{E_b}{N_0}}\right). \qquad (1.15)$$

This completes our derivation of the BSC for the situation in Fig. 1.15, since $P(e) = p$ for the BSC in Fig. 1.4, and thus p is equal to the right-hand side of (1.15). The plot of $P(e)$ in (1.15) shown in Fig. 1.17 is for optimal binary transmission. A useful approximation to $P(e)$ in (1.15) can be obtained by approximating $\text{erfc}(x)$ for $x \geq \sqrt{2}$ by

$$\text{erfc}(x) = \frac{e^{-x^2}}{x\sqrt{\pi}} \qquad (1.16)$$

For other values of x the simple PC-based algorithm in [10] is very useful.

More than one use of the BSC means repeated transmission over the system shown in Fig. 1.15. Since the additive white Gaussian noise is independent over nonoverlapping time intervals, this repeated use of the BSC represents independent statistical trials. This fact will be used several times in the sequel.

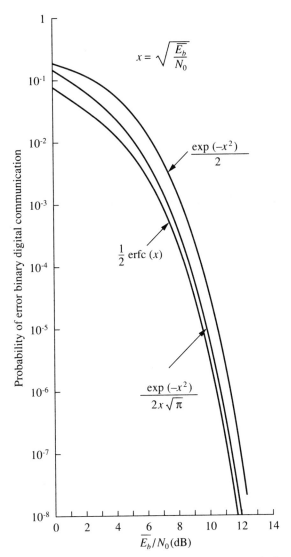

FIGURE 1.17 Error probability for binary transmission.

1.3 Entropy

A measure of information for an event A_i with probability P_i is

$$I(A_i) = -\log P_i \quad \text{bits} \tag{1.17}$$

where $\log x$ stands for $\log_2 x$. For example, consider a source consisting of $M = 2^k$ equally likely k-bit words. Then $P_i = 2^{-k}$ and in (1.17) we have $I(A_i) = k$ bits. We see that the information associated with a source outcome is the number of bits

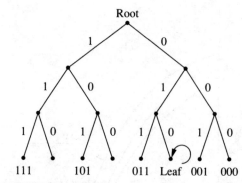

FIGURE 1.18 Binary tree for unknown source word.

required to represent the source symbol. In Fig. 1.18 we represent the source symbols as leaves in a binary tree. The effort to search the tree and recover the source symbol is proportional to $I(A_i)$, and thus there is a link between the uncertainty of information in a source symbol and the effort required to locate it in a binary search.

1.3.1 Discrete Sources

An average measure of information for a source X with M elements, each occurring with probability $Q(i)$, $i = 0, 1, \ldots, M - 1$, is the entropy of the source:

$$H(X) = -\mathbf{E}\{\log Q(i)\}$$
$$= -\sum_{i=0}^{M-1} Q(i) \log Q(i) \quad \text{bits/symbol.} \tag{1.18}$$

This represents an average measure of randomness for the source. The fact that the logarithm is used as a measure of information can be verified intuitively if we consider two independent sources and their Cartesian product. That is, any element of the combined source is one element from source X and one element from source Y. Then

$$H(X \times Y) = -\sum_{i=0}^{M-1} \sum_{j=0}^{M-1} Q(i)Q(j) \log[Q(i)Q(j)] \tag{1.19}$$
$$= H(X) + H(Y)$$

which follows from the two results

$$\log xy = \log x + \log y$$

and

$$\sum_{i=0}^{M-1} Q(i) = 1.$$

Thus the information in two combined and independent sources is the sum of their respective entropies.

As an example consider a binary source with symbols 0 and 1 occurring with probabilities $Q(0) = p$ and $Q(1) = 1 - p$. From (1.18) it follows that

$$H(X) = -p \log p - (1 - p) \log(1 - p) \qquad (1.20)$$

$$= h(p). \qquad (1.21)$$

That is, $h(p)$ is the right-hand side of (1.20); this function is a fundamental one in information theory and is sketched in Fig. 1.19. For $p = 0.50$ the source is most random, while for $p = 1$ or $p = 0$ the output is completely predictable. Hence the situation with maximum uncertainty represents that of maximum information—the information obviously occurs after we are informed what the source symbol is. In this manner information and uncertainty are really equivalent concepts.

For a discrete source with M independent elements we have $H(X) = \log M$, and this is realized in (1.18) when $Q(i) = M^{-1}$, that is, when each of the source symbols are equally likely. In general, $H(X) \leq \log M$, since no source could have more uncertainty, or randomness, than the case for equally likely symbols. If $M = 2^k$, it follows that $H(X) = k$ bits. Thus on the average it takes k bits to represent any source symbol. This is another interpretation of $H(X)$.

1.3.2 Continuous Source

Let x be the real variable and let X be a real random variable with pdf $p(x)$. Then the differential entropy of X is

$$\begin{aligned} H(X) &= -\mathbf{E}\{\ln p(x)\} \\ &= -\int_{-\infty}^{\infty} p(x) \ln p(x) \, dx \quad \text{nats/symbol} \end{aligned} \qquad (1.22)$$

where the unit of $H(X)$ is nats because the natural logarithm is used. This measure of entropy could be negative, and for this reason it is referred to in the differential sense [11].

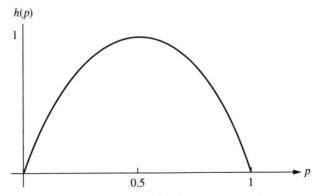

FIGURE 1.19 Binary entropy function.

For a discrete source we found in Section 1.3.1 that the maximum entropy distribution was one with equally likely source symbols. It is interesting to ask which pdf will yield maximum entropy in (1.22). We assume that the random variable should have zero mean and second moment σ^2. The optimization problem is to determine $p(x)$ such that $H(X)$ in (1.22) is a maximum subject to the constraints

$$\int_{-\infty}^{\infty} p(x)\, dx = 1 \tag{1.23}$$

$$\int_{-\infty}^{\infty} x^2 p(x)\, dx = \sigma^2. \tag{1.24}$$

This problem can be solved using the method of Lagrange multipliers in the calculus of variations [12]. One forms the functional

$$F(x) = \int_{-\infty}^{\infty} p(x)[-\ln p(x) + \lambda + \beta x^2]\, dx \tag{1.25}$$

and solves the equation

$$\frac{\partial F(x)}{\partial p} = 0$$

to get

$$p(x) = C_0 e^{\beta x^2} \tag{1.26}$$

where C_0 is a constant. Combining (1.23), (1.24), and (1.26) gives

$$p(x) = \frac{1}{\sigma\sqrt{2\pi}} e^{-x^2/2\sigma^2}. \tag{1.27}$$

The result in (1.27) shows that X is $\mathcal{N}(0, \sigma^2)$, which means that the Gaussian distribution is the maximum entropy distribution. Furthermore, for $p(x)$ in (1.27),

$$H(X)_{\max} = \frac{1}{2}\log(2\pi e\sigma^2) \quad \text{bits/source symbol}. \tag{1.28}$$

Hence the result in (1.28) is the maximum entropy for a continuous random variable with prescribed first and second moments.

The result can be generalized to the vector case. Let $\mathbf{x}$ be an N-vector with components x_i, $i = 1, \ldots, N$. Maximum entropy occurs in choosing the x_i independent and to have the pdf in (1.27). The maximum entropy is then $H(\mathbf{x})_{\max} = NH(X)_{\max}$, where $H(X)_{\max}$ is as given in (1.28). The multiplication by N occurs because the entropy of independent variates is additive, as we have shown earlier for two variates. The result just derived will be used in our discussion of Shannon's channel capacity.

1.4 Channel Capacity

We will now use the information theory concepts introduced in Section 1.3, and also some new concepts, to establish a relationship that sets the limit to communication.

1.4 / Channel Capacity

This fundamental limit, the channel capacity, was derived by Shannon [13]. Prior to Shannon's research, communication was thought to be limited by either the noise or the ISI that occurs when one tries to signal too fast (see Fig. 1.10). Shannon established a single formula that represented both effects. This formula established the limit to communication over any channel, not just those with noise or ISI. For instance, for a bandlimited channel with additive Gaussian noise, we have

$$C = B \log(1 + \text{SNR})$$

which includes both noise effects, through the signal-to-noise ratio, and ISI effects, through the bandwidth B. This formula will be derived and discussed in a later section.

1.4.1 Related Information Theory Measures

Initially, our consideration will be restricted to DMCs. An example of a DMC is the BSC of Fig. 1.4. The general DMC structure is shown in Fig. 1.20, where we have M inputs and N outputs. The input probabilities are $Q(i)$, $i = 0, 1, \ldots, M-1$; the DMC is represented by the channel transition probabilities, $P(j|i)$, $j = 0, 1, \ldots, N-1$; $i = 0, 1, \ldots, M-1$, and the output probabilities, $P(j)$, $j = 0, 1, \ldots, N-1$, which can be determined from the $Q(i)$ and the $P(j|i)$ for the range of indices given above. The uncertainty, or noise in the channel, is represented by having nonzero, nondiagonal terms in the $P(j|i)$ values.

The probability of an output symbol, $\{j\}$, is given by

$$P(j) = \sum_{i=0}^{M-1} P(j|i) Q(i). \tag{1.29}$$

Thus $H(Y)$ in (1.18) is given by

$$H(Y) = -\sum_{j=0}^{N-1} P(j) \log P(j). \tag{1.30}$$

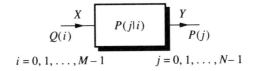

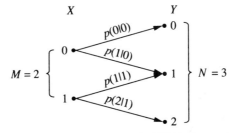

FIGURE 1.20 Structure of DMC channel.

One of the main problems in communication theory is: Given an output, $Y = j$, determine which input was transmitted. The entropy about the input given the output is

$$H(X|Y) = -\mathbf{E}\{\log P(X|Y)\}$$
$$= -\sum_{i=0}^{M-1} \sum_{j=0}^{N-1} P(i, j) \log P(i|j). \tag{1.31}$$

where $P(j, i) = P(j|i)Q(i)$. Also,

$$H(Y|X) = -\sum_{i=0}^{M-1} \sum_{j=0}^{N-1} P(j|i)Q(i) \log P(j|i). \tag{1.32}$$

The joint entropy between X and Y is

$$H(X, Y) = -\mathbf{E}\{\log P(X, Y)\}$$
$$= -\sum_{i=0}^{M-1} \sum_{j=0}^{N-1} P(i, j) \log P(i, j). \tag{1.33}$$

Manipulation of (1.18), (1.30), (1.31), and (1.32) shows that

$$H(X, Y) = H(X|Y) + H(Y)$$
$$= H(Y|X) + H(X). \tag{1.34}$$

These equations can be justified from an intuitive point of view. The joint entropy is the entropy, say, conditioned on knowledge of X plus the entropy of X. Exactly the same statement can be made for Y.

Let us work out some entropies for the BSC given in Fig. 1.4. First, the state transition matrix for the BSC channel is

$$\begin{pmatrix} 1-p & p \\ p & 1-p \end{pmatrix}. \tag{1.35}$$

We will find $H(Y|X)$ in (1.32). For a fixed value of X, say X_i, the conditional entropy is the inner sum in (1.32):

$$H(Y|X_i) = -\sum_{j=0}^{N-1} P(j|i) \log P(j|i)$$

and thus

$$H(Y|X) = -\sum_{i=0}^{M-1} Q(i) H(Y|X_i).$$

Hence, if $H(Y|X_i)$ is independent of i, $H(Y|X) = H(Y|X_i)$. This is precisely what happens in this example and, indeed, in all channels that have a symmetry property. This property will be covered later; for now we find that $H(Y|0) = H(Y|1) = h(p)$, where $h(p)$ is as defined in (1.21). Thus $H(Y|X_i)$ is independent of i and hence $H(Y|X) = h(p)$. To determine the joint entropy between X and Y, we use (1.34) and the result just derived to yield

$$H(X, Y) = h(p) + h[Q(0)]$$

where $h(p)$ is as mentioned above. Note that $H(X, Y)_{max} = 1 + h(p)$ and that this occurs for $Q(0) = Q(1) = 0.50$. Hence this choice of source probabilities yields the maximum joint information between X and Y for the BSC. It will turn out that this choice of source probabilities also leads to channel capacity, which is our next subject.

1.4.2 Channel Capacity

First, we introduce the concept of mutual information between X and Y, as in Fig. 1.20. Before transmission, the uncertainty an observer at the channel output has about the source is $H(X)$. After transmission this observer has the uncertainty $H(X|Y)$ because he or she has observed Y. The information supplied by the channel is the difference in uncertainty before and after the transmission process:

$$I(X;Y) = H(X) - H(X|Y) \qquad (1.36)$$

which is the mutual information between X and Y. An equivalent form is

$$I(X;Y) = H(Y) - H(Y|X) \qquad (1.37)$$

and this also has an interpretation. Here the observer's information is the entropy of the output $H(Y)$ less the noisy effect of the channel, which is represented by $H(Y|X)$.

Channel capacity is the maximum of $I(X;Y)$ with respect to the source probabilities, $Q(i), i = 0, 1, \ldots, M - 1$. This represents the maximum transfer of information across the channel.

1.4.3 Symmetric Channels

Channel capacity, the maximum mutual information, is easiest to find for symmetric channels. For instance, for the BSC we found earlier that $H(Y|X) = h(p)$ with $h(p)$ as in (1.21). Thus in (1.37),

$$I(X;Y) = H(Y) - h(p)$$

and hence to get channel capacity we must have $H(Y)$ as a maximum. This occurs when the outputs are all equally probable, as shown earlier for discrete sources. Due to channel symmetry this occurs when we choose the source probabilities as being equal, $Q(0) = Q(1) = 0.50$. This gives $P(0) = 0.50$ and $P(1) = 0.50$, which yields $H(Y) = 1$. We then have from the preceding equation,

$$C = 1 - h(p) \qquad (1.38)$$

and we illustrate this result in Fig. 1.21. The channel capacity is largest at the endpoints since this represents either completely perfect or completely imperfect transmission. A receiver can exploit both situations because it is assumed that p is

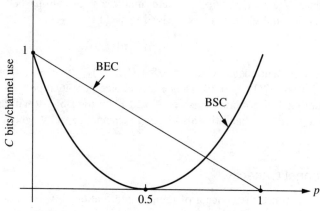

FIGURE 1.21 Channel capacity for some DMC.

known. The capacity is poorest where the channel is completely random, that is, for $p = 0.50$.

Definition 1.1 A DMC is symmetric if the outputs can be divided into subsets where the $\mathbf{P}(j|i)$ matrix for each subset has rows that are permutations of each other, and similarly for columns. □

For symmetric channels $H(Y|X)$ is independent of $Q(i)$ [14] and thus

$$C = H(Y)_{max} - H(Y|X). \tag{1.39}$$

Furthermore, $H(Y)_{max}$ occurs for $Q(i) = M^{-1}$, that is, for equally likely source symbols.

A good example of a symmetric channel is shown in Fig. 1.22. This is the binary erasure channel (BEC). An erasure occurs when the demodulator does not make a decision on a transmitted source symbol and thus decides to wait for further information about that symbol. In the example in Section 1.2, an erasure could occur if the sampled matched filter output was not large enough in absolute value, thus indicating a situation that is not a good one for which to render a bit decision. Note that the BEC in Fig. 1.20 is ideal in that binary 1 can never be transformed

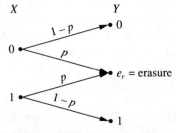

FIGURE 1.22 Binary erasure channel.

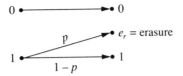

FIGURE 1.23 One-sided erasure channel.

in binary 0 and thus an error can never occur. A transmitted symbol can only be received correctly or be erased. We will find that for practical values of p the BEC has a higher capacity than that of the BSC precisely for this reason. A one-sided erasure channel is shown in Fig. 1.23. The channel matrix for the BEC in Fig. 1.22 is

$$\begin{pmatrix} 1-p & p & 0 \\ 0 & p & 1-p \end{pmatrix}$$

which can be partitioned as

$$\begin{pmatrix} 1-p & 0 \\ 0 & 1-p \end{pmatrix}$$

and

$$\begin{pmatrix} p \\ p \end{pmatrix}.$$

The first partition is for the outputs (0, 1) and the second partition is for the output e_r. These matrices in the partition satisfy the symmetry condition in the definition above. A short calculation using the elements of $\mathbf{P}(j|i)$ in (1.31) shows that $H(Y|X) = h(p)$, where $h(p)$ is as given in (1.21). We know that capacity is attained when $Q(1) = Q(0) = 0.50$. Thus, in (1.29), we find that $P(0) = P(2) = (1-p)/2$ and that $P(1) = p$. Determining $H(Y)$ from (1.30) and substituting the result into (1.39) gives

$$C = 1 - p. \tag{1.40}$$

The channel capacity for the BEC is given in (1.40) and is plotted in Fig. 1.21, where it can be compared to the capacity of the BSC. Note that for a wide range of p the capacity of the BEC exceeds that of the BSC because errors cannot be made on the BEC.

1.4.4 Kuhn–Tucker Conditions

When a channel is not symmetric, the Kuhn–Tucker conditions can sometimes be used to compute capacity. These conditions arise from the conditions for a solution to the constrained optimization problem that represents the capacity problem. Actually, few capacity problems can be solved in closed form, and computer algorithms [15] must be used to solve most problems. However, those that can be solved are treatable by the Kuhn–Tucker approach, lucidly described in the book by Gallagher [14].

To set up the optimization problem, we use (1.37) and (1.30) to write the mutual information in the form

$$I(X;Y) = \sum_{i=0}^{M-1} \sum_{j=0}^{N-1} P(j|i) Q(i) \log \frac{P(j|i)}{P(j)}. \quad (1.41)$$

The mathematical optimization problem to determine capacity is to maximize $I(X;Y)$ in (1.41) with respect to $Q(i) \geq 0$, $i = 0, 1, \ldots, M-1$, and further subject to the constraint

$$\sum_{i=0}^{M-1} Q(i) = 1.$$

The Kuhn–Tucker conditions are general necessary and sufficient conditions for a solution to constrained optimization of this type. For the capacity problem the necessary and sufficient conditions that $Q(i)$ must satisfy are

$$I(X = i; Y) = C \quad \text{if } Q(i) > 0$$

and

$$I(X = i; Y) \leq C \quad \text{if } Q(i) = 0 \quad (1.42)$$

where, as before, C is the channel capacity and $I(X = i; Y)$ is the mutual information between Y and the event $\{X = i\}$:

$$I(X = i; Y) = \sum_{i=0}^{N-1} P(j|i) \log \frac{P(j|i)}{P(j)}. \quad (1.43)$$

One can use the conditions in (1.42) to verify a guess that one might make at the $Q(i)$ that attain capacity. If the guess satisfies these conditions, capacity is realized with this guess of the $Q(i)$. To illustrate use of the conditions, consider the channel in Fig. 1.23. Note that only binary 1 can be erased and, furthermore, that this is a perfect channel. This is because if we receive 1 or e_r, we can always claim that the source symbol transmitted was binary 1. Hence the channel capacity is unity.

Let us derive this result from the Kuhn–Tucker conditions in (1.42). The channel matrix for the DMC in Fig. 1.21 is

$$\begin{pmatrix} 1 & 0 & 0 \\ 0 & p & 1-p \end{pmatrix}.$$

Use of the elements of this matrix in (1.42) and (1.43) gives

$$I(X = 0; Y) = -\log Q(0)$$

and

$$I(X = 1; Y) = -\log Q(1)$$

where we have used (1.29) to get $P(0) = Q(0)$, $P(e_r) = pQ(1)$, and $P(1) = (1-p)Q(1)$. The conditions in (1.42) are satisfied with the choice $Q(0) = Q(1) = 0.50$ and this yields $C = 1$, which is the required capacity.

The techniques given in the preceding two sections can be used to solve those capacity problems for DMCs that are solvable. Next we consider the capacity problem for a continuous channel.

1.4.5 Bandlimited Gaussian Channel

We will now derive the channel capacity for some continuous-time channels. We will use the definition of the mutual information, $I(X;Y)$, in (1.37) with $H(Y)$ in (1.22), with X replaced by Y, and

$$H(Y|X) = -E\{\ln p(y|x)\}$$
$$= -\int_{-\infty}^{\infty} p(x,y) \ln p(y|x) \, dy \, dx \text{ nats.}$$

Consider first the continuous-time sum channel described by

$$Y = X + N \qquad (1.44)$$

where X is the signal term and N is $\mathcal{N}(0, \sigma^2)$. Now from (1.44)

$$p(y|x) = \mathcal{N}(y - x, \sigma^2)$$

and thus from (1.28) in the material on the maximum entropy, continuous distribution for a source in Section 1.3.2,

$$H(Y|X) = \frac{1}{2} \log(2\pi e \sigma^2) \text{ bits/source symbol.}$$

This function is therefore independent of the source pdf and thus obeys the same property as the symmetric discrete channels of Section 1.4.4. Accordingly, the capacity follows from (1.39):

$$C = H(Y)_{max} - \frac{1}{2} \log(2\pi e \sigma^2). \qquad (1.45)$$

To get $H(Y)_{max}$, Y should be Gaussian with zero mean and variance σ_Y^2. This follows from the material in Section 1.3.2. If we choose X in (1.44) as $\mathcal{N}(0, \sigma_X^2)$, then Y in (1.44) is $\mathcal{N}(0, \sigma_Y^2)$, where $\sigma_Y^2 = \sigma_X^2 + \sigma^2$ with σ^2 the variance of the noise term in (1.44). Thus

$$H(Y)_{max} = \frac{1}{2} \log[2\pi e (\sigma_X^2 + \sigma^2)]$$

and it follows from (1.45) that

$$C = \frac{1}{2} \log(1 + \text{SNR}) \qquad (1.46)$$

where

$$\text{SNR} = \frac{\sigma_X^2}{\sigma^2}. \qquad (1.47)$$

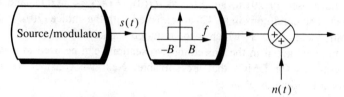

FIGURE 1.24 Bandlimited Gaussian channel.

We now extend the analysis on the sum channel to the bandlimited Gaussian channel illustrated in Fig. 1.24. The signal out of the channel filter in Fig. 1.24 is, ideally, bandlimited to B hertz. By Nyquist's sampling theorem [11], this signal can be represented by $2BT$ samples, as its time duration is T seconds. The noise can be represented by the same number of samples. All samples are spaced in time by $1/2B$ seconds, and hence the noise samples are independent. This follows as the autocorrelation function of the noise is

$$R_n(\tau) = N_0 B \operatorname{sinc} 2\tau B$$

where $\operatorname{sinc} x = \sin \pi x / \pi x$. Note that $R_n(k/2B) = 0$ for $k \neq 0$. Thus the noise samples are independent, as they are Gaussian and uncorrelated [9].

By the discussion given above we now have $2BT$ copies of the sum channel described by (1.44). Accordingly, $H(Y|X)$ and $H(Y)$ in the analysis for the sum channel are multiplied by $2BT$, as the entropies of independent quantities are additive. Since the signal duration is T seconds, the channel capacity in bits per second will be given by C/T, where C is as in (1.45) but is multiplied by $2BT$. We now have Shannon's formula for the capacity of the bandlimited Gaussian noise channel:

$$C = B \log(1 + \text{SNR}) \quad \text{bits/s} \tag{1.48}$$

where a new source symbol is sent every T seconds. The SNR is

$$\text{SNR} = \frac{2BT\sigma_s^2}{2BT\sigma^2}$$

$$= \frac{\sigma_s^2}{\sigma^2}$$

where $\sigma^2 = N_0 B$, the noise power in the channel bandwidth B.

The result in (1.48) is the capacity of the bandlimited Gaussian noise channel. As described at the beginning of this section, the formula in (1.48) combines both noise and ISI effects into a single limiting relationship for channel transmission. The limits to transmission are discussed in Section 1.5.

1.5 Shannon's Two Theorems

Shannon's two theorems on the transmission of information across a channel are the cornerstone of communication theory. They establish the ultimate limit to transmission over a channel. Furthermore, they show that transmission at rates below channel capacity can be perfect! That is, the transmission is not limited by interference of any sort: This fact is probably the most surprising of Shannon's theories.

The first theorem involves only a source encoder and a channel of capacity C; that is, no channel encoder is involved. Shannon's second theorem involves channel encoding and is more relevant to the theme of this book.

Theorem 1.1 Given a channel with capacity C, a source with entropy $H(X)$, and rate R_s symbols/sec; if $R_s H(X) < C$, then there exists a source encoder such that the source symbols can be transmitted over the channel at a rate less than C. □

Shannon's second theorem involves channel encoding and is of central importance in this book. To set the scene for this theorem, let us consider a channel encoding of source symbols at a rate of $R = k/N$. If N is increased, we increase k to keep R fixed.

Theorem 1.2 Let us view the source symbols as encoded into N-bit words or vectors and let $P_w(e)$ denote the probability of decoding a binary N-bit word in error. Then, if $R < C$, there exists a channel code $\mathscr{C}$ such that $P_w(e) \to 0$ as $N \to \infty$. □

Thus error-free transmission is possible at rates below channel capacity given that use of arbitrarily long code words is possible. The time delay to perform decoding is not a constraint here. In many applications communications must be nearly in real-time—speech transmission is an example where delays must not exceed a few hundred milliseconds, and here the transmission scheme implied in Theorem 1.1 may not be possible.

The book by Stark et al. [11] gives a simple, easy-to-follow proof of this theorem for the BSC as illustrated in Fig. 1.4. As in most of Shannon's theorems, the codes used are random and the proofs are not constructive. That is, the proofs are only of the existence type. The search continues for coding and modulation systems that meet, or closely approach, Shannon's limit. Progress has been slow because the problem is difficult. We will show that trellis coding is quite efficient in this respect for a portion of the bandwidth/transmitter power trade-off implied by Shannon's second theorem. This is discussed in [16], where it is shown that the use of trellis codes in telephone line modems has resulted in transmission rates closer to channel capacity than has occurred in other physical channels. In all cases, the proofs to Shannon's theorems involve an upper bound on $P_w(e)$ in the form

$$P_w(e) \leq C_0 2^{-N(C-R)}$$

where C_0 is a constant. Thus we see that $P_w(e)$ gets very small as $N \to \infty$ as long as $R < C$. Thus we have error-free transmission at all rates strictly less that capacity.

1.6 Computational Cutoff Rate

If coding is not used in the design of a modulation/demodulation system, the sensible criterion to use is unarguably the minimization of error probability, that is, of the probability that an information symbol emitted by the source is delivered erroneously to the user. If that same criterion is employed when a coding system is used, the purpose of coding becomes to correct the errors made in the discrete channel generated by the modulation/demodulation scheme. Error-correcting codes are used for this purpose.

Next, assume that coding is to be used over the channel. In this situation the optimum discrete channel would be the one which, *when paired to the best code that can be used on it,* gives the minimum error probability. When put in these terms, the design problem seems hardly solvable in practice. In fact, the search for the best code for use in a given discrete channel seems to be prohibitive. However, Shannon's second theorem can be used here to find a way out of this impasse. This theorem is based on a technique that is central in information theory, usually referred to as random coding. The idea on which random coding is based is the following. Instead of computing the error probability for the optimum coding scheme (which we cannot find), we compute the average error probability over the ensemble of all the possible codes used on the given channel. Obviously, at least one code will perform as well as the ensemble average; hence the ensemble average is an upper bound to the performance of the optimum code.

Shannon's second theorem (Theorem 1.2) tells us that if we transmit data over the discrete channel created by the modulation scheme, a code exists with rate $R = k/N$ for which the word error probability is bounded above by

$$P(e) \leq e^{-NE(R)} \qquad (1.49)$$

where $E(R)$, the *channel reliability function,* is a convex $\cup$, decreasing, nonnegative function of R for $0 \leq R \leq C$, where C is a quantity called *channel capacity* (see Section 1.4.2). A typical $E(R)$ function is shown in Fig. 1.25.

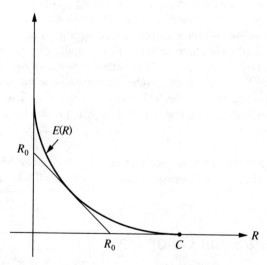

FIGURE 1.25 The function $E(R)$ of a discrete channel and its cutoff rate R_0.

Simply interpreted, inequality (1.49) tells us that we should expect lower error probabilities for codes with the same complexity (i.e., the same value of N) and the same redundancy (i.e., the same value of R) over channels having larger $E(R)$.

In conclusion, it seems that when coding is used, a meaningful comparison among discrete channels should be based on their functions $E(R)$. Now, given the fact that the actual computation of $E(R)$ may be a very demanding task, and that to make this comparison the use of a single parameter rather than a function would be far easier, how should we select this parameter? The channel capacity gives only a range of rates where reliable transmission is possible. Moreover, reliable transmission near channel capacity would be possible only for very large values of N [since $E(R)$ is very small there], and consequently for very large complexities.

It has been argued [17,18] that the sensible parameter to be used here is the *cutoff rate*, R_0, of the discrete channel. This is the rate at which the tangent to $E(R)$ of slope -1 intercepts the R axis, so that for $R \leq R_0$ we have

$$P(e) \leq e^{-N(R_0 - R)}.$$

This tangent line is shown in Fig. 1.25. The last inequality shows that while C gives only the range of rates where reliable transmission is possible, R_0 gives both a range of rates and an exponent to error probability. Thus the bigger R_0 is for a given average transmitted energy, the better the channel on which coding is to be used. Moreover, it is widely believed that R_0 is also the rate beyond which reliable transmission would become very expensive. (See [19, pp. 184–184] for a discussion of this point.) Hence R_0 can be interpreted as a *practical* upper bound to R.

Mathematically, when the signal vectors $\{s_i\}_{i=1}^M$ are transmitted over the additive white Gaussian noise channel with probabilities $\{q_i\}_{i=1}^M$, we have

$$R_0 = -\log_2 \left(\min_{\{q_i\}} \sum_{i=1}^M \sum_{j=1}^M q_i q_j e^{-|s_i - s_j|^2 / 4N_0} \right) \quad (1.50)$$

where $N_0/2$ is the variance of the components of the noise vector and the minimum is over all probability distributions on the signal vectors.

Figure 1.26 shows R_0 (in bits/waveform) versus E_b/N_0 for M-ary PSK modulation with signals used with the same probabilities [this is obtained by setting $q_i = 1/M$ for all i in (1.50)]. Inspection of this figure raises a point that is of central importance in discussion of trellis-coded modulation. Assume, for instance, that transmission of two bits per waveform is desired. A natural choice would be 4-PSK, which requires $E_b/N_0 \approx 7$ dB. However, Fig. 1.26 also shows that the same value of R_0 can be achieved *at a value of E_b/N_0 lower by about 4 dB* (i.e., with a 4-dB energy saving), if 8-PSK waveforms are used in conjunction with a coding scheme that makes them carry only two bits per waveform. This code should involve a *finite complexity*, the search for which is the main topic of this book.

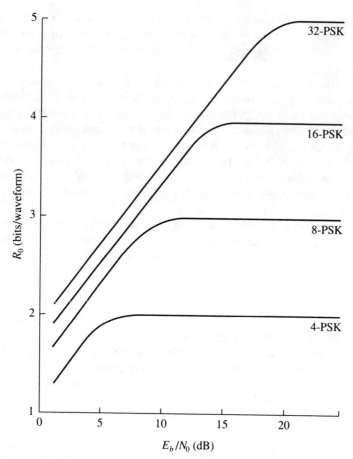

FIGURE 1.26 Cutoff rate R_0 for M-ary PSK signals used over the additive white Gaussian noise channel with the same probabilities.

REFERENCES

1. J. M. WOZENCRAFT and I. M. JACOBS, *Principles of Communication Engineering*. Wiley, New York, N.Y., 1965.
2. S. BENEDETTO, E. BIGLIERI, and V. CASTELLANI, *Digital Transmission Theory*. Prentice-Hall, Englewood Cliffs, N.J., 1987.
3. R. W. LUCKY, J. SALZ, and E. J. WELDON, *Principles of Data Communication*. McGraw-Hill, New York, N.Y., 1968.
4. C. LOO, "A statistical model for a land mobile satellite link," *IEEE Trans. Veh. Technol.*, Vol. VT-34, pp. 122–127, Aug. 1985.
5. W. C. JAKES, JR., *Microwave Mobile Communications*. Wiley, New York, N.Y., 1974.
6. J. G. PROAKIS, *Digital Communications*, 2nd ed. McGraw-Hill, New York, N.Y., 1989.

7. J. HAGENAEUR, N. SESHADRI, and C.-E. W. SUNDBERG, "Variable-rate sub-band speech coding and matched channel coding for mobile channels," *IEEE Vehicular Technology Conference*, Philadelphia, Pa., pp. 22–29, June 1988.
8. E. A. LEE and D. G. MESSERSCHMITT, *Digital Communications*. Kluwer Academic, Boston, Mass., 1988.
9. A. PAPOULIS, *Probability, Random Variables, and Stochastic Processes*, 2nd ed. McGraw-Hill, New York, N.Y., 1965.
10. N. C. BEAULIEU, "A simple series for personal computer computation of the error function $Q(\cdot)$," *IEEE Trans. Commun.*, Vol. COM-37, pp. 989–991, Sept. 1989.
11. H. STARK, F. B. TUTEUR, and J. B. ANDERSON, *Modern Electrical Communications*, 2nd ed. Prentice-Hall, Englewood Cliffs, N.J., 1988.
12. F. B. HILDEBRAND, *Methods of Applied Mathematics*, 2nd ed. Prentice-Hall, Englewood Cliffs, N.J., 1965.
13. C. E. SHANNON, "A mathematical theory of communication," *Bell Syst. Tech. J.*, Vol. 27, pp. 379–423 (Part One), pp. 623–656 (Part Two), 1948; reprinted in book form, University of Illinois Press, Urbana, Ill., 1949.
14. R. G. GALLAGER, *Information Theory and Reliable Communication*. Wiley, New York, N.Y., 1968.
15. R. E. BLAHUT, *Principles and Practice of Information Theory*. Addison-Wesley, Reading, Mass., 1987.
16. G. D. FORNEY, JR., et al., "Efficient modulation for band-limited channels," *IEEE J. Select. Areas Commun.*, Vol. SAC-2, pp. 632–647, Sept. 1984.
17. J. L. MASSEY, "Coding and modulation in digital communications," *1974 Zürich Seminar on Digital Communications*, Zürich, Switzerland, pp. E2(1)–E2(4), March 1974.
18. J. M. WOZENCRAFT and R. S. KENNEDY, "Modulation and demodulation for probabilistic coding," *IEEE Trans. Inf. Theory*, Vol. IT-12, pp. 291–297, July 1966.
19. R. E. BLAHUT, *Principles and Practice of Information Theory*. Addison-Wesley, Reading, Mass., 1987.

CHAPTER 2

Error-Correcting Codes

2.1 Introduction

This chapter is devoted to the elementary theory of error-correcting codes. The goal of this chapter is to cover the essential background material for an understanding of the channel coding concepts used in trellis-coded modulation. For instance, the introductory theory of the standard single, double, and triple random error-correcting codes is presented. Some material for Reed–Solomon codes is also presented. The general structure of the channel encoder we consider is shown in Fig. 2.1. The message is a source word composed of k binary digits, and this is formed as a k-dimensional binary vector, **m**. A parity check vector of $r = N - k$ bits is appended to each message vector **m** to form the code vector **c**. The vector **c** has dimension N. Thus the code rate is $R_c = k/N$. The code vector **c** is then t_x over the channel. In this chapter we regard this as N uses of the BSC or BEC channels given in Chapter 1. We denote such a code by $\mathcal{C}$ and say that it is a (N, k, d_{min}^H) code. Such a code can correct $\lfloor (d_{min}^H - 1)/2 \rfloor$ errors, as discussed in Chapter 1. In our brief treatment of error-correcting codes we stress the role of parity check equations as this is a central concept in trellis-coded modulation.

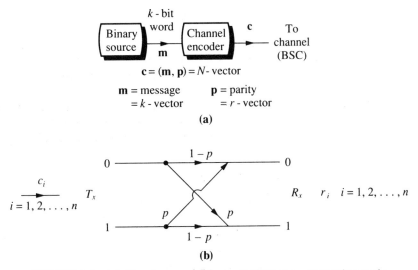

FIGURE 2.1 (a) Structure and (b) channel for error-correcting code.

2.2 Parity Check Codes

The function $a \oplus b$ is defined in Table 2.1. This reads as a EXOR b or $(a + b) \bmod 2 = a \oplus b$. Let us consider a $(3, 2, 2)$ parity check code. The parity check equation is $c_3 = c_1 \oplus c_2$, where $\mathbf{m} = (c_1, c_2)$ are information bits. The code words are (000), (011), (101), and (110). The code words are said to be of even parity as $c_1 \oplus c_2 \oplus c_3 = 0$. Thus $d_{\min}^H = 2$ and we can detect one error in three uses of a BSC. Also three channel errors will violate the parity condition; however, two errors will not. The general case is a $(N, N-1, 2)$ code, where

$$c_N = \sum_{i=1}^{N-1} \oplus c_i$$

and here any odd number of channel errors violates the parity condition, whereas any even number of errors does not. Thus, for the BSC,

$$P(e) = P(\text{decoding error}) = \sum_{i=1,\ i=\text{even}}^{N-1} \binom{N}{i} p^i (1-p)^{N-i}$$

is the probability of decoding error, where p is the error crossover probability for the BSC and

$$\binom{N}{i} = \frac{N!}{(N-i)!\, i!}.$$

Hence we see how a single parity check can give error detection. We now show how this technique can give error correction. Consider Elias's [1] iterated coding technique for the $(3, 2, 2)$ parity check code. Consider the array

$$\begin{pmatrix} i_1 & i_2 & p_1 \\ i_3 & i_4 & p_2 \\ p_3 & p_4 & p_5 \end{pmatrix}.$$

Here $p_1 = i_1 \oplus i_2$, $p_2 = i_3 \oplus i_4$, $p_3 = i_1 \oplus i_3$, $p_4 = i_2 \oplus i_4$, and $p_5 = p_1 \oplus p_2$ and (i_1, i_2, i_3, i_4) are the source bits. This array is then transmitted over the channel on a row-by-row basis. The code rate is $\frac{4}{9} = \left(\frac{2}{3}\right)^2$, that is, the code rate of the $(3, 2, 2)$ parity check code squared.

The code word is received and the blocks of nine detected bits are entered into the 3-by-3 array above. If row i and column j do not satisfy the parity check equations, then the (i, j)th element is complemented. In this manner the decoded array is

TABLE 2.1 Binary EXOR Function.

a	b	$a \oplus b$
0	0	0
1	0	1
0	1	1
1	1	0

derived. Clearly, a single error in any one of nine positions can always be corrected. This is in keeping with the result $t = \lfloor (d_{min}^H - 1)/2 \rfloor$ as $d_{min}^H = 4$. Since rows or columns will not check for any error pattern $(0, e_1, 0, 0, e_2, 0, 0, 0, 0)$ where the position of e_1 and e_2 is arbitrary, we see that two errors can be detected. This is in keeping with the relationship $\lfloor d_{min}^H/2 \rfloor$ giving the number of errors detected.

The concept of parity checks can be generalized to show how the position of errors within a code word can be determined. Consider the following (6, 3, 3) code:

$$c_4 = c_1 \oplus c_2 \tag{2.1}$$

$$c_5 = c_2 \oplus c_3 \tag{2.2}$$

$$c_6 = c_1 \oplus c_3 \tag{2.3}$$

where $\mathbf{m} = (c_1, c_2, c_3)$ are the information bits. Clearly, the parity check variables, c_4, c_5, and c_6 are independent combinations of c_1, c_2, and c_3. The parity check equations are

$$s_1 = c_1 \oplus c_2 \oplus c_4 \tag{2.4}$$

$$s_2 = c_2 \oplus c_3 \oplus c_5 \tag{2.5}$$

$$s_3 = c_1 \oplus c_3 \oplus c_6 \tag{2.6}$$

where the s_i, $i = 1, 2, 3$, account for the fact that channel errors occurring in a block of 6 bits can give nonzero values to these variables. When no errors occur, the $s_i = 0$, $i = 1, 2, 3$, because the parity checks have been chosen to satisfy (2.1) through (2.3). There are $2^3 = 8$ possible values for the s_i in (2.4) through (2.6) as $r = N - k = 3$. To see some of these values, let $\mathbf{m} = (0, 0, 0)$, giving $\mathbf{c}$ as a 6-vector with all components zero. Assume that a channel error occurs at position 2, c_2 changed from 1 to 0. Then from (2.4)–(2.6), $\mathbf{s} = (1, 1, 0)$ and no other error pattern containing a single error can give this value of $\mathbf{s}$. Hence, in the decoder, if this value of $\mathbf{s}$ occurs, one complements c_2 to get the decoded code vector. In Table 2.2 we list all values of $\mathbf{s}$ and their error patterns. The vector $\mathbf{s}$ is known as the *syndrome vector*, and use of the results in Table 2.2 for decoding is called *syndrome decoding*. It is easily shown that this is a maximum likelihood decoding rule and thus minimizes the probability of decoding error. Note that one double error is corrected by this decoder. There are other choices for the correctable double

TABLE 2.2 Error Pattern and Syndrome for (6, 3, 3) Code.

Error Pattern						Syndrome		
0	0	0	0	0	0	0	0	0
e	0	0	0	0	0	1	0	1
0	e	0	0	0	0	1	1	0
0	0	e	0	0	0	0	1	1
0	0	0	e	0	0	1	0	0
0	0	0	0	e	0	0	1	0
0	0	0	0	0	e	0	0	1
e	0	0	0	e	0	1	1	1

error, but only one choice can be made for any decoder. Here the one choice we make is when $\mathbf{s} = (1, 1, 1)$ occurs: We complement the first and fifth positions in the received code word to get the decoded code word.

Parity check equations like those in (2.1)–(2.3) form the basis of error-correcting codes. In the next section we put this in a more formal setting.

2.3 Matrix Description: Error-Correcting Codes

The parity check equations can be written in vector-matrix form: $\mathbf{c} = \mathbf{mG}$, where $\mathbf{c} = [c_1, c_2, \ldots, c_N]$, $\mathbf{m} = [i_1, i_2, \ldots, i_k]$, and $\mathbf{G} = [\mathbf{I}_k | \mathbf{P}]$. Here $\mathbf{m}$ is the message vector, $\mathbf{I}_k$ is the $k \times k$ identity matrix, and $\mathbf{P}$ is the $k \times r$ parity check array. The first k components of any code vector are the components of the message vector $\mathbf{m}$. Such a code is said to be *systematic*.

For (2.1)–(2.3), that is, the (6, 3, 3) code,

$$\mathbf{G} = \begin{bmatrix} 1 & 0 & 0 & 1 & 0 & 1 \\ 0 & 1 & 0 & 1 & 1 & 0 \\ 0 & 0 & 1 & 0 & 1 & 1 \end{bmatrix}$$

and thus

$$\mathbf{P} = \begin{bmatrix} 1 & 0 & 1 \\ 1 & 1 & 0 \\ 0 & 1 & 1 \end{bmatrix}.$$

$\mathbf{G}$ is called the *generator matrix* for the code and it is of dimension $k \times N$. Equations (2.4)–(2.6) can be expressed as $\mathbf{s} = \mathbf{yH}^T$, where we have replaced $\mathbf{c}$ by the received code word, $\mathbf{y}$. Now when $\mathbf{y} = \mathbf{c}$, $\mathbf{s} = \mathbf{0}$, as by definition, the syndrome is zero when no channel errors occur and $\mathbf{y} = \mathbf{c}$.

In terms of $\mathbf{G}$, a related matrix is $\mathbf{H}$, where $\mathbf{GH}^T = \mathbf{0}$. Thus $\mathbf{H} = [\mathbf{P}^T | \mathbf{I}_r]$, where $\mathbf{P}^T$ is the matrix transpose of $\mathbf{P}$, and $\mathbf{H}$ is an r by N matrix. $\mathbf{H}$ is known as the parity check matrix. In our example, from (2.1), (2.2), and (2.3),

$$\mathbf{H} = \begin{bmatrix} 1 & 1 & 0 & 1 & 0 & 0 \\ 0 & 1 & 1 & 0 & 1 & 0 \\ 1 & 0 & 1 & 0 & 0 & 1 \end{bmatrix}. \tag{2.7}$$

If we compute $\mathbf{s} = \mathbf{yH}^T$, we get the parity check equations (2.4)–(2.6). In our example the parity check equations are linear. Thus our general formulation, which follows from this example, represents a structure for all linear, systematic, codes. The method of decoding is to determine the syndrome $\mathbf{s} = \mathbf{yH}^T$ and then relate the syndrome to its unique error pattern. Table 2.2 gives an example.

The error-correction and error-detection capabilities can be determined from d_{min}^H, the minimum Hamming distance of the code. The Hamming distance is given by

$$d^H(\mathbf{c}_i, \mathbf{c}_j) = \sum_{k=1}^{N} (\mathbf{c}_i)_k \oplus (\mathbf{c}_j)_k$$

where $(\mathbf{c}_i)_k$ is the kth component of the vector $\mathbf{c}_i$. Then

$$d^H_{\min} = \min_{i \neq j} d^H(\mathbf{c}_i, \mathbf{c}_j).$$

Define the weight of a code word as the number of nonzero components it contains. As a linear code always has $\mathbf{c} = 0$ as a code word, corresponding to $\mathbf{m} = 0$, we see that $d^H_{\min}$ is the minimum weight of a code $\mathscr{C}$. In Table 2.3 we list the eight code words of our example. One sees that the minimum weight is 3 and thus is $d^H_{\min}$.

Another method to determine $d^H_{\min}$ is taken from the following simple theorem [2].

Theorem 2.1 Let i be the least number of columns of $\mathbf{H}$ that sums to zero (i.e., the least number that are linearly dependent). Then $d^H_{\min} = i$. □

In our example no two columns of $\mathbf{H}$ in (2.7) sum to zero. However, columns 1, 2, and 3 sum to zero. Hence $d^H_{\min} = 3$. Thus $t = 1$ errors can be corrected. It is also clear that $d^H_{\min}$ cannot exceed the number of parity equations plus one: $d^H_{\min} \leq r + 1 = 4$.

As two more examples we give the $\mathbf{G}$ and $\mathbf{H}$ matrix for the (3, 1, 3) repetition code in Fig. 2.4:

$$\mathbf{G} = [1 \quad 1 \quad 1]$$

$$\mathbf{H} = \begin{bmatrix} 1 & 1 & 0 \\ 1 & 0 & 1 \end{bmatrix} \qquad (2.8)$$

and the (3, 2, 2) parity check code of Section 2.2:

$$\mathbf{G} = \begin{bmatrix} 1 & 0 & 1 \\ 0 & 1 & 1 \end{bmatrix}$$

$$\mathbf{H} = [1 \quad 1 \quad 1].$$

Note that by Theorem 2.1 $d^H_{\min} = 3$ for the repetition code and $d^H_{\min} = 2$ for the parity check code.

TABLE 2.3 Code Words for (6, 3, 3) Code.

Message	Code Word
000	000000
100	100101
010	010110
001	001011
110	110011
101	101110
011	011101
111	111000

One of the first error-correcting codes discovered was the class of single error-correcting codes presented by Hamming [3]. They can be categorized as having **H** matrices that are all the nonzero combinations of r bits. For $r = N - k = 2$, $N = 3$, and $k = 1$, we have **H** in (2.8), that is, the (3, 1, 3) repetition code. For $r = 3$, $N = 7$, and $k = 4$,

$$\mathbf{H} = \begin{bmatrix} 1 & 0 & 1 & 1 & 1 & 0 & 0 \\ 1 & 1 & 1 & 0 & 0 & 1 & 0 \\ 0 & 1 & 1 & 1 & 0 & 0 & 1 \end{bmatrix}$$

giving

$$\mathbf{G} = \begin{bmatrix} 1 & 0 & 0 & 0 & 1 & 1 & 0 \\ 0 & 1 & 0 & 0 & 0 & 1 & 1 \\ 0 & 0 & 1 & 0 & 1 & 1 & 1 \\ 0 & 0 & 0 & 1 & 1 & 0 & 1 \end{bmatrix}.$$

The code words are listed in Fig. 2.2. The weight distribution is also shown and we note that $d_{\min}^H = 3$, thus giving $t = 1$. Displayed in Fig. 2.3 are the eight syndromes and the eight single-error vectors that can be corrected. Thus each code word is associated with a unique correctable error pattern. Hence the decoding spheres, also shown in Fig. 2.2, are nonoverlapping. The syndromes that follow from (2.7) and the equation $\mathbf{s} = \mathbf{yH}^T$ are given by the parity check equations:

$$y_1 \oplus y_3 \oplus y_4 \oplus y_5 = s_1$$
$$y_1 \oplus y_2 \oplus y_3 \oplus y_6 = s_2$$
$$y_2 \oplus y_3 \oplus y_4 \oplus y_7 = s_3.$$

Code Word	Information	Parity	w_i
c_1	0000	000	0
c_2	1000	110	3
c_3	0100	011	3
c_4	0010	111	4
c_5	0001	101	3
c_6	1100	101	4
c_7	1010	001	3
c_8	0110	100	3
c_9	1001	011	4
c_{10}	0101	110	4
c_{11}	0011	010	3
c_{12}	1110	010	4
c_{13}	1101	000	3
c_{14}	1011	100	4
c_{15}	0111	001	4
c_{16}	1111	111	7

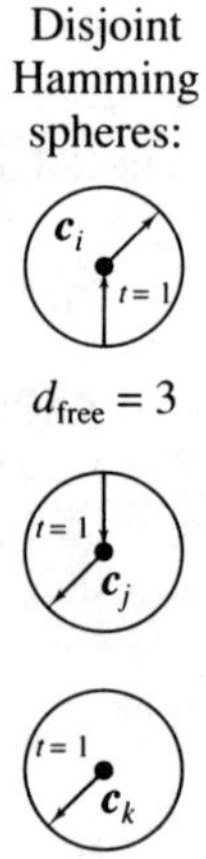

Disjoint Hamming spheres:

$d_{\text{free}} = 3$

Weight distribution:
$A_0 = 1 \quad A_3 = 7 \quad A_4 = 7 \quad A_7 = 1$
$A_k = 0$ otherwise

FIGURE 2.2 Code words of the (7, 4) Hamming code.

2.3 / Matrix Description: Error-Correcting Codes

Error Pattern	Syndrome: s_1, s_2, s_3
0000000	000
0000e00	100
00000e0	010
e000000	110
000000e	001
000e000	101
0e00000	011
00e0000	111

$$s_1 = y_1 \oplus y_3 \oplus y_4 \oplus y_5$$
$$s_2 = y_1 \oplus y_2 \oplus y_3 \oplus y_6$$
$$s_3 = y_2 \oplus y_3 \oplus y_4 \oplus y_7$$
$$\mathbf{s} = \mathbf{y}\mathbf{H}^T$$

FIGURE 2.3 Error patterns and syndromes for (7, 4) Hamming code.

These parity check equations can be represented by the Venn diagram shown in Fig. 2.4, which is constructed from the results in Fig. 2.3. There are three circles for the three entries in the syndrome vector $\mathbf{s}$. The areas are labeled with the entry of the received vector $\mathbf{y}$ that contains an error. Thus the syndrome (110) is labeled as an area common to circles 1 and 2 and is denoted as 1, since the first component of $\mathbf{y}$ is complemented to derive the decoded channel output.

The (3, 1, 3) and (7, 4, 3) Hamming codes are examples of *perfect codes*. To define a perfect code, consider the following. For an (N, k, d_{min}^H) code the number of channel outputs is 2^N. If t is the error-correcting capability of the code, we have

$$\binom{N}{0} + \binom{N}{1} + \cdots + \binom{N}{t}$$

t-correctable error patterns per code word. Thus, multiplying by the number of code words, we obtain

$$\left[\binom{N}{0} + \binom{N}{1} + \cdots + \binom{N}{t}\right] 2^k \leq 2^N$$

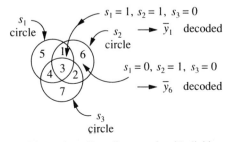

FIGURE 2.4 Venn decoding diagram for (7, 4) Hamming code.

or

$$\binom{N}{0} + \binom{N}{1} + \cdots + \binom{N}{t} \leq 2^r. \quad (2.9)$$

If we have equality, there are no channel outputs not in t-error-correcting Hamming spheres. Such codes are said to be *perfect*, as they have no capacity wasted on a partial error-correcting capability. For instance, the (6, 3, 3) code treated earlier is not a perfect code. However, the (3, 1, 3) and (7, 4, 3) Hamming codes are perfect.

There are three classes of binary perfect codes:

1. The $(N, 1, N)$ repetition codes.
2. The Hamming codes $N = 2^r - 1$, $r = N - k$.
3. The (23, 12, 7) Golay codes ($t = 3$).

Amazingly, there are no more binary, perfect codes [2].

2.4 Algebraic Concepts

To learn about standard double and triple error-correcting codes it is necessary to study some algebra. We give a brief introduction below. First, we consider the concept of a group, $\mathcal{G}$.

Definition 2.1 A group $\mathcal{G}$ is a set of elements plus an operation, $+$, such that

1. $a, b \in \mathcal{G} \Rightarrow a + b \in \mathcal{G}$.
2. $a + (b + c) = (a + b) + c$.
3. There exists an identity element, 0, such that $a + 0 = a$.
4. There exists an inverse element $(-a)$ for a such that $a + (-a) = 0$. □

The number of elements in $\mathcal{G}$ is the order of the group. An example is mod 2 addition as given in Table 2.1 for $\mathcal{G} = (0, 1)$. Another example involving ternary symbols and mod 3 addition is given in Example 2.1.

EXAMPLE 2.1

The addition rule for $\mathcal{G} = (0, 1, 2)$ is given in Table 2.4. Note from Table 2.4 that $-1 = 2$ and $-2 = 1$. □

TABLE 2.4 Addition Table for GF(3).

+	0	1	2
0	0	1	2
1	1	2	0
2	2	0	1

A further definition is that a group $\mathcal{G}$ is commutative if the plus rule satisfies the commutative property, $a + b = b + a$. This also defines an Abelian group.

Now consider the concept of a ring.

Definition 2.2 A ring $\mathcal{R}$ is a set $\mathcal{G}$ plus two operations, $+$ and $\times$, such that

- $\{\mathcal{G}, +\}$ is an Abelian group.
- $a \times (b \times c) = (a \times b) \times c$.
- $a \times (b + c) = (a \times b) + (a \times c)$. □

Examples are $\mathcal{G} = (0, 1)$ in Table 2.1 and $\mathcal{G} = (0, 1, 2)$ of Example 2.1 with multiplication performed modulo 2 and 3, respectively.

Finally, we have the concept of a field $\mathcal{F}$.

Definition 2.3 A field $\mathcal{F}$ is a set of elements, G, and two operations, $+$ and $\times$, such that

- $a, b \in \mathcal{F} \rightarrow a + b \in \mathcal{F}, a \times b \in \mathcal{F}$.
- $a + (b + c) = (a + b) + c$.
- $a \times (b \times c) = (a \times b) \times c$.
- $a + b = b + a$.
- $a \times b = b \times a$.
- $a \times (b + c) = (a \times b) + (a \times c)$.
- $a + 0 = a$.
- $a \neq 0, a \times a^{-1} = 1$. □

The set of integers $I_q = 0, 1, \ldots, q - 1$ is a field as long as q is a prime number. For instance, I_5 is a field but I_4 is not. For example, $(3 + 3) \bmod 4 = 2$ and thus $(2 \times 2) \bmod 4 = 0$ and hence 2^{-1} does not exist. When q is finite, I_q is called $GF(q)$, where GF stands for Galois field after the mathematician E. Galois. A description of $GF(3)$ is given in Example 2.2.

EXAMPLE 2.2

The addition and multiplication results for $GF(3)$ are given in Tables 2.4 and 2.5, respectively. We see from these tables that $1^{-1} = 1$ and $2^{-1} = 2$. □

A second example of a field is the $+$ rule in Table 2.1 for $GF(2) = (0, 1)$ and the $\times$ given in Table 2.6.

TABLE 2.5
Multiplication Table for GF(3).

$\times$	0	1	2
0	0	0	0
1	0	1	2
2	0	2	1

TABLE 2.6
Multiplication Table for GF(2).

×	0	1
0	0	0
1	0	1

2.4.1 Primitive Element

We will show that a finite field, $GF(q)$, can be represented in terms of any primitive element. Let $\beta^e = 1$ for the smallest nonzero integer e, where

$$\beta^e = \overbrace{(\beta \times \beta \times \beta \times \cdots \times \beta)}^{e} \bmod q,$$

then e is the *order* of β. If $e = q - 1$ for $\beta \in GF(q)$, that is, the order of β is $q - 1$, then $\beta = \alpha$ is said to be *primitive*. For instance, in $GF(3)$ in Example 2.2, $\beta = 2$ is primitive as $2^2 \pmod 3 = 1$. However, $\beta = 1$ is not primitive, as its order e is unity. We denote a primitive element of $GF(q)$ as α.

2.4.2 Extension Field $GF(q)$

Consider the finite field $GF(q)$. Form m-dimensional vectors with components from this field. There are q^m such vectors and under mod q addition these vectors form a group, with addition done on a component-by-component basis. The question is how to form a multiplication rule in order to have a field. We do this through an equivalent polynomial for the vectors. Let $p(x)$ be irreducible over $GF(q)$. That is, $p(x) \neq 0$ for all $x \in GF(q)$. Then for $\alpha \in GF(q^m)$,

$$\alpha^i \times \alpha^j = [\alpha_i(x)\alpha_j(x)] \bmod p(x).$$

This is the multiplication rule for elements of $GF(q^m)$, where $\alpha(x) \bmod p(x)$ is the remainder when $\alpha(x)$ is divided by $p(x)$. For $GF(2^2)$ we use $p(x) = x^2 + x + 1$ and note that $p(0) \neq 0$ and $p(1) \neq 0$. The subsequent polynomial and vector description of $GF(2^2)$ is given in Table 2.7. Then we get the × table given in

TABLE 2.7 Vectors and Polynomials for GF(2^2) for $p(x) = x^2 + x + 1$.

Element	Vector	Polynomial
0	00	0
1	01	1
2	10	x
3	11	$x + 1$

TABLE 2.8 Multiplication Table for $GF(2^2)$.

×	0	1	x	$x+1$
0	0	0	0	0
1	0	1	x	$x+1$
x	0	x	$x+1$	1
$x+1$	0	$x+1$	1	x

Table 2.8. As an example,

$$x^2 \div (x^2 + x + 1)$$

has the remainder $x + 1$. Other parts of the table are derived accordingly. Note from the table that $\alpha = x$ and $\alpha = x + 1$ are both primitive, as $x^3 \bmod p(x) = 1$ and $(x + 1)^3 \bmod p(x) = 1$. Note further that $p(\alpha) = 0$ for α a primitive element: for instance, $\alpha = x \Rightarrow \alpha^2 = \alpha + 1$. Letting $\alpha = x$, we get the following description of $GF(2^2)$ in Table 2.9 and the following $+$ and $\times$ tables in Tables 2.10 and 2.11, respectively, as determined by vector addition and polynomial multiplication.

For example, $\alpha + \alpha^2 = (1, 0) + (1, 1) = (0, 1) = 1$. Also, $\alpha \times \alpha^2 = \alpha^3 = 1$ as $\alpha \times \alpha^2 = \alpha \times (\alpha + 1) = \alpha^2 + \alpha = 1$. Of course, multiplication in $\alpha_i \times \alpha_j$ is by addition of exponents. Thus the $+$ and $\times$ tables for $GF(q^m)$ are not usually given. One just works out the results from Table 2.9 using exponent addition for multiplication and vector addition. For instance, from Table 2.9, $\alpha \times \alpha^2 = \alpha^3 = 1$ and $\alpha + \alpha^2 = (0, 1) = 1$ as just worked out. Tables of $GF(2^3)$, $GF(2^4)$, $GF(2^5)$, and $GF(2^6)$ follow in Tables 2.12 through 2.15, respectively. As a nonbinary case we give the table of vectors for $GF(3^4)$ in Table 2.16.

As an example of $\times$ and $+$ operations, consider $GF(3^4)$ given in Table 2.16. Clearly, $\alpha^{10} \times \alpha^{20} = \alpha^{30}$ and $\alpha^{37} \times \alpha^{63} = \alpha^{100} = \alpha^{20}$ as $\alpha^{80} = 1$. Also, $\alpha^{37} + \alpha^{63} = \alpha^{71}$ by vector addition. To see this, note that $\alpha^{37} = (2, 2, 2, 2)$ and $\alpha^{63} = (2, 0, 2, 2)$ and component addition over $GF(3)$ yields the vector $(1, 2, 1, 1)$, which, from Table 2.16, is α^{71}.

Given the procedure above, one can solve linear equations and factor polynomials using well-known techniques. Consider the following example.

TABLE 2.9
Vector Description of $GF(2^2)$.

Element	Vector	
0	0	0
1	0	1
α	1	0
α^2	1	1

TABLE 2.10 Addition Table for $GF(2^2)$.

+	0	1	α	α^2
0	0	1	α	α^2
1	1	0	α^2	α
α	α	α^2	0	1
α^2	α^2	α	1	0

TABLE 2.11 Multiplication Table for $GF(2^2)$.

×	0	1	α	α^2
0	0	0	0	0
1	0	1	α	α^2
α	0	α	α^2	1
α^2	0	α^2	1	α

TABLE 2.12 Vectors for $GF(2^3)$: $p(x) = x^3 + x + 1$.

Zero and Powers of α	Vector over GF(2)
0	000
1	001
α	010
α^2	100
α^3	011
α^4	110
α^5	111
α^6	101

Source: Adapted from [4].

TABLE 2.13 Vectors $GF(2^4)$: $p(x) = x^4 + x + 1$.

Zero and Powers of α	Vector over GF(2)
0	0000
1	0001
α	0010
α^2	0100
α^3	1000
α^4	0011
α^5	0110
α^6	1100
α^7	1011
α^8	0101
α^9	1010
α^{10}	0111
α^{11}	1110
α^{12}	1111
α^{13}	1101
α^{14}	1001

Source: Adapted from [2].

TABLE 2.14 Vectors for GF(2^5): $p(x) = x^5 + x^2 + 1$.

Zero and Powers of α	Vector over GF(2)
0	00000
α^0	00001
α^1	00010
α^2	00100
α^3	01000
α^4	10000
α^5	00101
α^6	01010
α^7	10100
α^8	01101
α^9	11010
α^{10}	10001
α^{11}	00111
α^{12}	01110
α^{13}	11100
α^{14}	11101
α^{15}	11111
α^{16}	11011
α^{17}	10011
α^{18}	00011
α^{19}	00110
α^{20}	01100
α^{21}	11000
α^{22}	10101
α^{23}	01111
α^{24}	11110
α^{25}	11001
α^{26}	10111
α^{27}	01011
α^{28}	10110
α^{29}	01001
α^{30}	10010

Source: Adapted from [2].

EXAMPLE 2.3

Consider solution of

$$X + \alpha^{41} Y = \alpha^{20}$$
$$\alpha^2 X + \alpha^{71} Y = \alpha^{78}$$

where α is a primitive element of $GF(3^4)$ (see Table 2.16). Eliminating X, we have

$$(\alpha^{71} - \alpha^{43})Y - \alpha^{78} - \alpha^{22}$$

TABLE 2.15 Vectors for $GF(2^6) : p(x) = x^6 + x + 1$.

Zero and Powers of α	Vectors over $GF(2)$	Zero and Powers of α	Vectors over $GF(2)$
0	000000	α^{31}	101001
1	100000	α^{32}	100100
α	010000	α^{33}	010010
α^2	001000	α^{34}	001001
α^3	000100	α^{35}	110100
α^4	000010	α^{36}	011010
α^5	000001	α^{37}	001101
α^6	110000	α^{38}	110110
α^7	011000	α^{39}	011011
α^8	001100	α^{40}	111101
α^9	000110	α^{41}	101110
α^{10}	000011	α^{42}	010111
α^{11}	110001	α^{43}	111011
α^{12}	101000	α^{44}	101101
α^{13}	010100	α^{45}	100110
α^{14}	001010	α^{46}	010011
α^{15}	000101	α^{47}	111001
α^{16}	110010	α^{48}	101100
α^{17}	011001	α^{49}	010110
α^{18}	111100	α^{50}	001011
α^{19}	011110	α^{51}	110101
α^{20}	001111	α^{52}	101010
α^{21}	110111	α^{53}	100101
α^{22}	101011	α^{54}	111010
α^{23}	100101	α^{55}	011101
α^{24}	100010	α^{56}	111110
α^{25}	010001	α^{57}	011111
α^{26}	111000	α^{58}	111111
α^{27}	011100	α^{59}	101111
α^{28}	001110	α^{60}	100111
α^{29}	000111	α^{61}	100011
α^{30}	110011	α^{62}	100001

Source: Adapted from [2].

where, from Table 2.16, $-\alpha^{22} = \alpha^{62}$ and $-\alpha^{43} = \alpha^3$. Now through vector addition in Table 2.16

$$\alpha^{30} Y = \alpha^{40}$$

and thus $Y = \alpha^{10}$. Solution for X gives

$$X = \alpha^{20} - \alpha^{41} Y = \alpha^{20} - \alpha^{51}$$

or as $-\alpha^{51} = 1201 = \alpha^{11}$, $X = \alpha^{20} + \alpha^{11} = \alpha^8$, which completes the example.

□

TABLE 2.16 Vectors for $GF(3^4) : p(x) = x^4 + x + 2$.

Zero and Powers of α	Vectors over $GF(3)$	Zero and Powers of α	Vectors over $GF(3)$
0	0000	α^{40}	0002
α^0	0001	α^{41}	0020
α^1	0010	α^{42}	0200
α^2	0100	α^{43}	2000
α^3	1000	α^{44}	0012
α^4	0021	α^{45}	0120
α^5	0210	α^{46}	1200
α^6	2100	α^{47}	2021
α^7	1012	α^{48}	0222
α^8	0111	α^{49}	2220
α^9	1110	α^{50}	2212
α^{10}	1121	α^{51}	2102
α^{11}	1201	α^{52}	1002
α^{12}	2001	α^{53}	0011
α^{13}	0022	α^{54}	0110
α^{14}	0220	α^{55}	1100
α^{15}	2200	α^{56}	1021
α^{16}	2012	α^{57}	0201
α^{17}	0102	α^{58}	2010
α^{18}	1020	α^{59}	0112
α^{19}	0221	α^{60}	1120
α^{20}	2210	α^{61}	1221
α^{21}	2112	α^{62}	2201
α^{22}	1102	α^{63}	2022
α^{23}	1011	α^{64}	0202
α^{24}	0101	α^{65}	2020
α^{25}	1010	α^{66}	0212
α^{26}	0121	α^{67}	2120
α^{27}	1210	α^{68}	1212
α^{28}	2121	α^{69}	2111
α^{29}	1222	α^{70}	1122
α^{30}	2211	α^{71}	1211
α^{31}	2122	α^{72}	2101
α^{32}	1202	α^{73}	1022
α^{33}	2011	α^{74}	0211
α^{34}	0122	α^{75}	2110
α^{35}	1220	α^{76}	1112
α^{36}	2221	α^{77}	1111
α^{37}	2222	α^{78}	1101
α^{38}	2202	α^{79}	1001
α^{39}	2002		

2.5 Cyclic Codes

First, BCH codes are cyclic codes. This means that if **c** is a code word, so is any cyclic shift of **c**. This follows if the code polynomial for **c** is given by $c(x) = m(x)g(x)$, where $m(x)$ is the message polynomial, and $g(x)h(x) = x^N - 1$, where $g(x)$ is called the *generator polynomial*, is of degree $r = N - k$. For instance, for $N = 7$ and $r = 3$, $g(x) = x^3 + x + 1$, giving **m** = (1110), which implies that $m(x) = 1 + x + x^2$ and futhermore, **c** = (1000110) means that $c(x) = x^5 + x^4 + 1$. In fact, this $g(x)$ generates the (7, 4) Hamming code given in Section 2.4. In the polynomial case, however, we do not have the code in systematic form.

It is useful to have the code words in systematic form. Let us let the parity bits start the code vector and the message bits appear at the end. The operation $x^r m(x)$ will shift the message polynomial to the end of the code polynomial. The procedure is as follows:

1. Form $x^r m(x)$ from $m(x)$.
2. Form $x^r m(x) \mod g(x) = b(x)$.
3. Then $c(x) = b(x) + x^r m(x)$.

For instance, for the example above, $m(x) = 1 + x + x^2$ and thus $x^3 m(x) \mod g(x) = x = b(x)$. Hence $c(x) = x^5 + x^4 + x^3 + x$ or **c** = (0101110). We note that cyclic codes are easily generated using shift register structures and sequences and this material is covered in [2]. Note that $b(x)$ represents the parity checks for the cyclic code.

2.6 BCH Codes

The BCH codes were presented in [5] and [6]. Before proceeding we must define the concept of conjugate roots. Let β be an element of $GF(q^m)$, primitive or not. If β is a root of a polynomial $g(x)$, then β^q is also a root: To get all roots, one continues raising any root to a power of q until the roots start to repeat. Also, $m_\beta(x)$ is the polynomial of least degree that has β as a root: It is called the *minimum polynomial* for β. Also, $m_\alpha(x) = p(x)$, the irreducible polynomial used to generate $GF(q^m)$ and here, as before, α is a primitive element of this finite field.

As an example, let us determine $m_\alpha(x)$ for $GF(2^4)$ in Table 2.13. We have

$$\text{roots } m_\alpha(x): \alpha, \alpha^2, \alpha^4, \alpha^8, \alpha^{16} = \alpha$$

since $\alpha^{15} = 1$. Thus

$$m_\alpha(x) = (x + \alpha)(x + \alpha^2)(x + \alpha^4)(x + \alpha^8)$$

as $-\alpha^n = \alpha^n$ over $GF(2^m)$. Expanding $m_\alpha(x)$ using the vectors for $GF(2^4)$ in Table 2.13 to, say, compute $\alpha + \alpha^2$, we find, as expected,

$$m_\alpha(x) = x^4 + x + 1 = p(x).$$

2.6 / BCH Codes

The specification of the generator polynomial for BCH codes is quite simple. Choose $d - 1$ consecutive roots

$$\beta, \beta^2, \ldots, \beta^{d-1}$$

where d is the design distance of the code. Then N is the order of β and

$$g(x) = \text{LCM}[m_\beta(x), m_{\beta^2}(x), \ldots, m_{\beta^{2t}}(x)] \quad (2.10)$$

where LCM represents "lowest common multiple"; the LCM is performed over the bracketed terms, $2t = d - 1$, and one does not have to include the minimum polynomials for the conjugate roots for β. The degree of $g(x)$ is $r = N - k$ from which k and the code rate $R_c = k/N$ follow. If $\beta = \alpha$, the code is primitive; otherwise, it is nonprimitive. Also, if we have a primitive code, $d_{\min}^H = d$, but in general, $d_{\min}^H \geq d$. The latter is a possible advantage for nonprimitive codes.

2.6.1 Binary BCH Codes

Let us first determine the $t = 2$, primitive, binary BCH code of least N. It is the code usually used in applications [7] for double error correction. It will turn out to be a (15, 7, 5) code.

Fortunately, the minimum polynomials over binary fields have been nicely given in a table by Peterson and Weldon [8]. This is universally used to find $g(x)$ from (2.10) and is given in Table 2.17. We refer to it as the *P-W table*.

EXAMPLE 2.4

As stated above, we desire a $t = 2$ (i.e., $d = 5$) binary BCH code that is primitive and based on $GF(2^4)$. Thus the consecutive roots can be

$$\alpha, \alpha^2, \alpha^3, \alpha^4.$$

To find the number of minimum polynomials needed in (2.10), we consider various $m_\alpha(x)$ and their roots (recall that α^2 is one conjugate root for α):

roots $m_\alpha(x)$: $\alpha, \alpha^2, \alpha^4, \alpha^8, \alpha^{16} = \alpha$

roots $m_{\alpha^3}(x)$: $\alpha^3, \alpha^6, \alpha^{12}, \alpha^{24} = \alpha^9, \alpha^{18} = \alpha^3$.

As $\alpha, \alpha^2, \alpha^3, \alpha^4$ are among the roots of $m_\alpha(x)$ and $m_{\alpha^3}(x)$ and also as these polynomials have no common roots, $g(x)$ in (2.10) is given by $m_\alpha(x)m_{\alpha^3}(x)$. Using the P-W table in Table 2.17, we find that $m_\alpha(x) = (23)_8$ and $m_{\alpha^3}(x) = (37)_8$. Thus

$$m_\alpha(x) = 010011$$
$$= x^4 + x + 1$$
$$= p(x)$$

and

$$m_{\alpha^3} = 011111$$
$$= x^4 + x^3 + x^2 + x + 1.$$

TABLE 2.17 Minimum Polynomials for $GF(2^m)$.

Degree	2	1	7H								
Degree	3	1	13F								
Degree	4	1	23F	3	37D	5	07				
Degree	5	1	45E	3	75G	5	67H				
Degree	6	1	103F	3	127B	5	147H	7	111A	9	015
	11	155E	21	007							
Degree	7	1	211E	3	217E	5	235E	7	367H	9	277E
	11	325G	13	203F	19	313H	21	345G			
Degree	8	1	435E	3	567B	5	763D	7	551E	9	675C
	11	747H	13	453F	15	727D	17	023	19	545E	21	613D
	23	543F	25	433B	27	477B	37	537F	43	703H	45	471A
	51	037	85	007							
Degree	9	1	1021E	3	1131E	5	1461G	7	1231A	9	1423G
	11	1055E	13	1167F	15	1541E	17	1333F	19	1605G	21	1027A
	23	1751E	25	1743H	27	1617H	29	1553H	35	1401C	37	1157F
	39	1715E	41	1563H	43	1713H	45	1175E	51	1725G	53	1225E
	55	1275E	73	0013	75	1773G	77	1511C	83	1425G	85	1267E
Degree	10	1	2011E	3	2017B	5	2415E	7	3771G	9	2257B
	11	2065A	13	2157F	15	2653B	17	3515G	19	2773F	21	3753D
	23	2033F	25	2443F	27	3573D	29	2461E	31	3043D	33	0075C
	35	3023H	37	3543F	39	2107B	41	2745E	43	2431E	45	3061C
	47	3177H	49	3525G	51	2547B	53	2617F	55	3453D	57	3121C
	59	3471G	69	2701A	71	3323H	73	3507H	75	2437B	77	2413B
	83	3623H	85	2707E	87	2311A	89	2327F	91	3265G	93	3777D
	99	0067	101	2055E	103	3575G	105	3607C	107	3171G	109	2047F
	147	2355A	149	3025G	155	2251A	165	0051	171	3315C	173	3337H
	179	3211G	341	0007							

Now $g(x) = m_\alpha(x)m_{\alpha^3}(x)$, which can be found from the following tabular array:

x^8	x^7	x^6	x^5	x^4	x^3	x^2	x^1	x^0
1	1	1	1	1				
			1	1	1	1	1	
				1	1	1	1	1
1	1	1	0	1	0	0	0	1

giving

$$g(x) = x^8 + x^7 + x^6 + x^4 + 1.$$

Note that because this is a primitive code, $d_{min}^H = 5$, $N = 5$, and $r = 8$. We have a (15, 7, 5) code. □

EXAMPLE 2.5

To consider a nonprimitive example, we consider $GF(2^6)$ in Table 2.15. Now $63 = 21 \times 3$ and thus $\beta = \alpha^3$ has order 21. Thus $N = 21$. Say we want $t = 2$ so that $d = 5$. The design consecutive roots are $\beta, \beta^2, \beta^3,$ and β^4, where $\beta = \alpha^3$.

TABLE 2.17 *(continued)*

Degree	11	1	4005E	3	4445E	5	4215E	7	4055E	9	6015G
11	7413H	13	4143F	15	4563F	17	4053F	19	5023F	21	5623F
23	4757B	25	4577F	27	6233H	29	6673H	31	7237H	33	7335G
35	4505E	37	5337F	39	5263F	41	5361E	43	5171E	45	6637H
47	7173H	49	5711E	51	5221E	53	6307H	55	6211G	57	5747F
59	4533F	61	4341E	67	6711G	69	6777D	71	7715G	73	6343H
75	6227H	77	6263H	79	5235E	81	7431G	83	6455G	85	5247F
87	5265E	89	5343B	91	4767F	93	5607F	99	4603F	101	6561G
103	7107H	105	7041G	107	4251E	109	5675E	111	4173F	113	4707F
115	7311C	117	5463F	119	5755E	137	6675G	139	7655G	141	5531E
147	7243H	149	7621G	151	7161G	153	4731E	155	4451E	157	6557H
163	7745G	165	7317H	167	5205E	169	4565E	171	6765G	173	7535G
179	4653F	181	5411E	183	5545E	185	7565G	199	6543H	201	5613F
203	6013H	205	7647H	211	6507H	213	6037H	215	7363H	217	7201G
219	7273H	293	7723H	299	4303B	301	5007F	307	7555G	309	4261E
331	6447H	333	5141E	339	7461G	341	5253F				
Degree	12	1	10123F	3	12133B	5	10115A	7	12153B	9	11765A
11	15647E	13	12513B	15	13077B	17	16533H	19	16047H	21	10065A
23	11015E	25	13377B	27	14405A	29	14127H	31	17673H	33	13311A
35	10377B	37	13565E	39	13321A	41	15341G	43	15053H	45	15173C
47	15621E	49	17703C	51	10355A	53	15321G	55	10201A	57	12331A
59	11417E	61	13505E	63	10761A	65	00141	67	13275E	69	16663C
71	11471E	73	16237E	75	16267D	77	15115C	79	12515E	81	17545C
83	12255E	85	11673B	87	17361A	89	11271E	91	10011A	93	14755C
95	17705A	97	17121G	99	17323D	101	14227H	103	12117E	105	13617A
107	14135G	109	14711G	111	15415C	113	13131E	115	13223A	117	16475C
119	14315C	121	16521E	123	13475A	133	11433B	135	10571A	137	15437G
139	12067F	141	13571A	143	12111A	145	16535C	147	17657D	149	12147F
151	14717F	153	13517B	155	14241C	157	14675G	163	10663F	165	10621A

Source: Adapted from [8].

Consider the roots of the minimum polynomials:

$$\text{roots } m_\beta(x): \beta^2, \beta^4, \beta^8, \beta^{16}, \beta^{32} = \beta^{11}, \beta^{22} = \beta$$

$$\text{roots } m_{\beta^3}(x): \beta^3, \beta^6, \beta^{12}, \beta^{24} = \beta^3$$

as $\beta^{21} = 1$. In the P-W table in Table 2.17 we have $m = degree = 6$, and recalling that $m_\beta = m_{\alpha^3}$ and $m_{\beta^3} = m_{\alpha^9}$, we have

$$m_\beta(x) = (127)_8$$
$$= 001010111$$
$$= x^6 + x^4 + x^2 + x + 1$$
$$m_{\beta^3}(x) = (015)_8$$
$$= 001101$$
$$= x^3 + x^2 + 1$$

and from (2.10), after constructing an array like that in Example 2.4, we obtain
$$g(x) = x^9 + x^8 + x^7 + x^5 + x^4 + x + 1.$$
Thus $r = 9$, and checking d_{min}^H, we find that $d_{min}^H = d = 5$. Hence we have a (21, 12, 5) code, but no increase in minimum distance over the design distance. □

We note that the design of primitive BCH codes can also be based on using $1, \alpha, \alpha^2, \alpha^3$, say, as consecutive roots when $t = 2$. This is discussed in [4].

Our next example is the usual $t = 3$ code used in applications [9]. It is the (23, 12, 7) Golay code. It is a nonprimitive, binary BCH code with a design distance of 5. However, in general, $d_{min}^H \geq d = 5$ and one finds that $d_{min}^H = 7$, giving $t = 3$. The development follows that in [4].

EXAMPLE 2.6

We choose $m = 11$ and work over $GF(2^{11})$. Now $2^{11} - 1 = 2047 = 23 \times 89$, and thus $\beta = \alpha^{89}$ has order 23. Also, a code based on β has $N = 23$. We choose β, β^2, β^3, and β^4 as consecutive roots and thus $d = 5$. Now $m_\beta(x)$ has the roots $(\beta = \alpha^{89})$

$$\beta, \quad \beta^2, \quad \beta^4, \quad \beta^8, \quad \beta^{16}, \quad \beta^{32} = \beta^9, \quad \beta^{18},$$
$$\beta^{36} = \beta^{13}, \quad \beta^{26} = \beta^3, \quad \beta^6, \quad \beta^{12}, \quad \beta^{24} = \beta$$

and thus all the consecutive roots needed are in this collection. Accordingly, by (2.10), $g(x) = m_\beta(x)$. From the P-W table in Table 2.17, we have $degree = m = 11$, for α^{89},

$$m_\beta(x) = (5343)_8$$
$$= 101011100011$$
$$= x^{11} + x^9 + x^7 + x^6 + x^5 + x + 1$$
$$= g(x).$$

Hence $r = N - k = 11$, giving $k = 12$ and a code rate, $R_c = 12/23$. By trial and error we find that $d_{min}^H = 7$, which is larger than the design distance, $d = 5$. □

It was stated earlier that the (23, 12, 7) BCH code was a perfect code. This follows from the computation

$$1 + \binom{23}{1} + \binom{23}{2} + \binom{23}{3} = 2^{11}$$

which is the $t = 3$ version of (2.9).

2.6.2 Nonbinary BCH Codes

The design of nonbinary BCH codes follows the same ideas as in the binary case. All use the construction of tables like that of Tables 2.12 through 2.16. In our second example the concept of a subgroup is used, which can result in existing tables being used for performing addition rather than always having to construct new tables.

EXAMPLE 2.7

Consider the problem of designing an $N = 80$, $t = 1$ code with ternary symbols. Thus we can use a primitive element of $GF(3^4)$ and make use of Table 2.16, as $3^4 - 1 = 80$, the order of α.

The consecutive roots are α and α^2, and the roots of $m_\alpha(x)$ and $m_{\alpha^2}(x)$ are

$$\text{roots } m_\alpha(x): \alpha, \alpha^3, \alpha^9, \alpha^{27}, \alpha^{81} = \alpha$$

$$\text{roots } m_{\alpha^2}(x): \alpha^2, \alpha^6, \alpha^{18}, \alpha^{54}, \alpha^{162} = \alpha^2.$$

Now $m_\alpha(x) = p(x)$, the irreducible polynomial used to generate Table 2.16. The problem is to find $m_{\alpha^2}(x)$, since by (2.10) and the fact that α, α^2 are the consecutive roots, $g(x) = m_\alpha(x) m_{\alpha^2}(x)$. Now, lacking a table like the P-W table, we proceed from first principles. Hence

$$m_{\alpha^2}(x) = (x - \alpha^2)(x - \alpha^6)(x - \alpha^{18})(x - \alpha^{54})$$
$$= (x^2 - (\alpha^2 + \alpha^6)x + \alpha^8)(x^2 - (\alpha^{18} + \alpha^{54})x + \alpha^{72}).$$

Now from Table 2.9 and vector addition, $\alpha^2 + \alpha^6 = \alpha^{15}$ and $\alpha^{18} + \alpha^{54} = \alpha^{55}$. Also, $-\alpha^{15} = \alpha^{55}$ and using these results in the equation directly above, we find, after a little more work like that above,

$$m_{\alpha^2}(x) = x^4 + x^2 + \alpha^{40} x + 1$$

where $GF(3) = \{0, 1, \alpha^{40}\}$ as $2 = \alpha^{40}$, the element of $GF(3^4)$ of order 2. Thus by (2.10),

$$g(x) = (x^4 + x + 2) m_{\alpha^2}(x)$$

or, again, after expansion and coefficient addition,

$$g(x) = x^8 + x^6 + x^3 + x^2 + 2x + 2.$$

Thus $r = 8$ and we have a primitive, $(80, 72, 3)$ code over $GF(3)$. The field we do the addition and multiplication in to find the code, here $GF(3^4)$, is called the *error-locator field*. The name follows, since this field is central to decoding operations [4, 10]. The symbol field is $GF(3)$. □

EXAMPLE 2.8

To consider a nonprimitive code and also the use of subfields, consider the design of an $N = 20$ code over $GF(9)$. As $80 = 4 \times 20$ we see that $\alpha^4 \in GF(3^4)$ has order 20. The error-locator field will be $GF(3^4)$ and the symbol field is $GF(9)$. If we can find an element of $GF(3^4)$ of order 8, it can serve as the primitive element of $GF(9)$. Then $GF(9)$ can be a subfield of $GF(3^4)$. Now over $GF(3^4)$, α^{10}

has order 8. Thus

$$GF(9) = \{0, 1, \alpha^{10}, \alpha^{20}, \alpha^{30}, \alpha^{40}, \alpha^{50}, \alpha^{60}, \alpha^{70}\}$$
$$= \{0, 1, 2, 3, 4, 5, 6, 7, 8\}.$$

That is, $3 = \alpha^{20}$, and so on. Now multiplication and addition for $GF(9)$ can be used via Table 2.16 for $GF(3^4)$. For instance, $\alpha^{10} + \alpha^{50} = 2 + 7 = 0$. Finally, as $9^2 - 1 = 80$, $m = 2$, and the degree of all minimum polynomials are at most 2.

The consecutive roots are β, β^2 as $t = 1$. Now the roots over $GF(9)$ of $m_\beta(x)$ and $m_{\beta^2}(x)$ are

$$\text{roots } m_\beta(x): \beta, \beta^9, \beta^{81} = \beta$$
$$\text{roots } m_{\beta^2}(x): \beta^2, \beta^{18}, \beta^{162} = \beta^2.$$

Thus $g(x) = m_\beta(x) m_{\beta^2}(x)$. To find $m_\beta(x)$, we have

$$m_\beta(x) = (x - \beta)(x - \beta^9)$$
$$= (x - \alpha^4)(x - \alpha^{36})$$
$$= x^2 - (\alpha^4 + \alpha^{36})x + \alpha^{40}$$
$$= x^2 + \alpha^{10}x + \alpha^{40}$$

as from Table 2.16, $\alpha^4 + \alpha^{36} = \alpha^{50}$ and $\alpha^{10} + \alpha^{50} = 0$. Similarly,

$$m_{\beta^2}(x) = (x - \beta^2)(x - \beta^{18})$$
$$= x^2 + \alpha^{10}x + 1.$$

Then use of (2.10) and the same rules as above gives

$$g(x) = m_\beta(x) m_{\beta^2}(x)$$
$$= x^4 + \alpha^{50}x^3 + \alpha^{20}x^2 + \alpha^{40}$$
$$= x^4 + 6x^3 + 3x^2 + 5.$$

Thus $r = 4$ and we have a (20, 16, 3) code over $GF(9)$. □

The underlying theory for constructing nonbinary BCH codes can be found in [10]. We note that the $GF(2^6)$ table in Table 2.14 would be used in designing an $N = 63$ code over $GF(8)$ as $8^2 - 1 = 63$ and $2^6 - 1 = 63$. Thus $m = 2$ and $GF(8)$ can be generated using $\alpha^9 \in GF(2^6)$, as it has order 7. Other challenging cases can be found in [10], especially in the problems at the end of Chapters 5 and 7.

2.6.3 Reed–Solomon Codes

The best known nonbinary BCH codes are the *Reed–Solomon* (RS) *codes*. They are a popular class of codes for application in burst error environments. They are simple nonbinary BCH codes, as the error-locator and symbol fields are the same. Thus $m = 1$ and all minimum polynomials have degree 1.

As stated above, the RS codes are nonbinary BCH codes, where the symbol field and error-locator field are the same. Accordingly, all minimum polynomials have degree 1 and by (2.10),

$$g(x) = (x - \alpha)(x - \alpha^2) \cdots (x - \alpha^{d-1})$$

as $\alpha, \alpha^2, \ldots, \alpha^{d-1}$ are the consecutive roots. If we have $GF(q)$ as the field for the RS code, $N = q - 1$, and if the degree of $g(x)$ is r, then $r = N - k = d - 1$. Thus $d = r + 1$ and as $d_{min}^H \leq r + 1$, we have $d_{min}^H = r + 1$, where r is the degree of the RS polynomial.

We present two examples on finding $g(x)$ for RS codes.

EXAMPLE 2.9

Let the field be $GF(4) = (0, 1, \alpha, \alpha^2)$, where α is a primitive element of $GF(2^2) = GF(4)$. Let us assume that we want a code with $N = 3$ and $r = 2$. Thus $d_{min}^H = r + 1 = 3$, giving $t = 1$ as the error-correcting capability of the code. Thus the consecutive roots can be α, α^2, and by (2.10),

$$g(x) = (x - \alpha)(x - \alpha^2).$$

Use of Table 2.10 gives $\alpha + \alpha^2 = 1$ and thus upon expanding $g(x)$ above,

$$g(x) = x^2 + x + 1.$$

There are four code words when the message is in $GF(4)$: $(0, 0, 0)$, $(1, 1, 1)$, (α, α, α), and $(\alpha^2, \alpha^2, \alpha^2)$. Hence we have a (3, 1, 3) repetition code over $GF(4)$.   □

EXAMPLE 2.10

In our second example we require a *nibble*-oriented code, where a nibble is one-half a byte. Thus we seek 4-bit code words and hence there are 16 code words and the symbol field is $GF(16) = GF(2^4)$. Let us assume that the required block length is 5. Now as $2^4 - 1 = 15 = 5 \times 3$ we set $\beta = \alpha^3$, as α^3 has order 5. Thus the code we determine is nonprimitive. To keep the analysis simple, let us further assume that the code must correct one error so that $t = 1$. Thus we need β, β^2 as consecutive roots and by (2.10),

$$g(x) = (x - \beta)(x - \beta^2)$$

where $\beta = \alpha^3 \in GF(2^4)$. Expansion of $g(x)$ above and use of the vectors in Table 2.13 gives

$$g(x) = x^2 + \alpha^2 x + \alpha^9$$

where α^i is an element of the symbol field $GF(2^4)$. Thus we have a (5, 3) code as $r = 2$. The code rate in bits is 12/20. Suppose that the input message is

(0110, 0001, 1000) or $(\alpha^5, 1, \alpha^3)$. Then

$$m(x) = \alpha^5 x^2 + x + \alpha^3$$

and $c(x) = m(x)g(x)$ gives the code vector $\mathbf{c} = (\alpha^5, \alpha^9, \alpha^8, \alpha^6, \alpha^2)$. □

2.7 Convolutional Codes

For convolutional codes the code words are generated through linear operations on source bits. Typically, the systems are linear over $GF(2)$ and have memory. An example with two input source bits and three output bits is given in Fig. 2.5. This code has rate $R_c = 2/3$. The number of storage elements is two and as the inputs are binary, this is called a *four-state code*. Convolutional codes play a central role in this book as they are commonly used to generate trellis codes. We cover the essential theory that is required for trellis-coded modulation. The example in Fig. 2.5, which is the best four-state, rate 2/3 code [11], will be used as a general example.

2.7.1 FSM and Trellis Representation

There are both a finite-state machine (FSM) and a trellis representation for trellis codes. We give the former in Fig. 2.6. The state variables are $a = 00$, $b = 10$, $c = 01$, and $d = 11$, where $\alpha_{i-1}^{(1)}$ and $\alpha_{i-1}^{(2)}$ are the states with the $(\alpha_i^{(1)}, \alpha_i^{(2)})$ inputs. The FSM represents all that can occur and the branch variables are $\alpha_i^{(1)} \alpha_i^{(2)} / \beta_i^{(1)} \beta_i^{(2)} \beta_i^{(3)}$, with the latter being outputs. The branch labels are computed from the input–output equations for the rate 2/3 code in Fig. 2.5:

$$\beta_i^{(1)} = \alpha_i^{(1)} \oplus \alpha_i^{(2)} \oplus \alpha_{i-i}^{(1)} \oplus \alpha_{i-1}^{(2)} \tag{2.11}$$

$$\beta_i^{(2)} = \alpha_{i-1}^{(1)} \oplus \alpha_i^{(2)} \tag{2.12}$$

$$\beta_i^{(3)} = \alpha_i^{(1)} \oplus \alpha_i^{(2)} \oplus \alpha_{i-1}^{(2)}. \tag{2.13}$$

From the FSM we can derive the trellis diagram. This is shown in Fig. 2.7. Again there are four states, and as there are two input bits, we have four transitions per

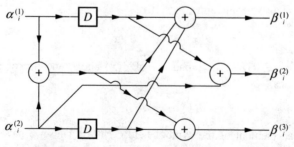

FIGURE 2.5 Best four-state, rate 2/3 code.

2.7 / Convolutional Codes   57

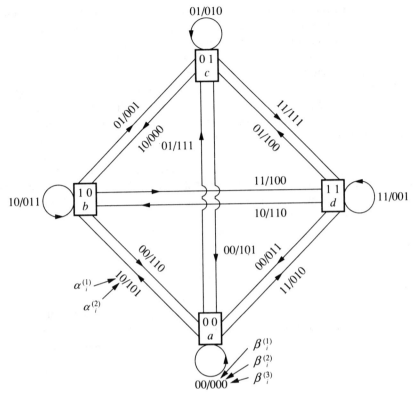

FIGURE 2.6 Finite-state machine for code in Fig. 2.5.

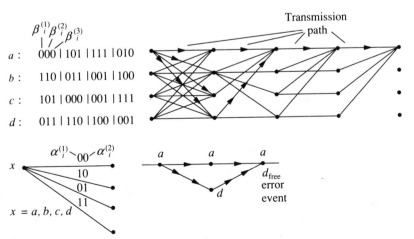

FIGURE 2.7 Trellis diagram and error events for code of Fig. 2.5.

state variable. The state variables can also be labeled according to the output triples used per state. This labeling is also given in the trellis diagram in Fig. 2.7. It is this labeling that is often used in trellis-coded modulation systems.

Let us consider transmission over a BSC, and thus we will have three channel uses per pair of input bits. A sequence of input bits will give rise to two paths through the trellis: the transmitted path, which is error-free, shown as the all-zeros path in Fig. 2.7, and the path observed from the BSC. When we have no errors, these two paths agree. The most likely error path is the one closest to the transmitted path in Hamming distance. As shown in Fig. 2.7, the transmitted path is a–a–a and the most likely error path is a–d–a and the minimum Hamming distance is 3. Thus $d_{\text{free}}^H = 3$ for this rate 2/3 code. These two paths constitute what is called an *error event*. At low crossover probability, p, this error event dominates error performance, as the next-most-likely error path has distance 4. Note that the minimum-distance error event contains two bit errors. Also, the error events are invariant relative to using the all-zeros path as the reference transmitted path. This is because the code in Fig. 2.5 is linear.

2.7.2 G(D) and H(D) Matrices

Let D be the unit delay operator. Then we can give a representation of convolutional code that is analogous to the equations $\mathbf{c} = \mathbf{mG}$, $\mathbf{s} = \mathbf{yH}^T$, and $\mathbf{GH}^T = 0$ used for linear, block codes. Let $\mathbf{m} = (\alpha_i^{(1)}, \alpha_i^{(2)})$. Then from equations (2.10) to (2.12),

$$\mathbf{c} = \mathbf{mG}(D)$$

where $\mathbf{c} = (\beta_i^{(1)}, \beta_i^{(2)}, \beta_i^{(3)})$ and

$$\mathbf{G}(D) = \begin{bmatrix} 1+D & D & 1 \\ 1+D & 1 & 1+D \end{bmatrix}.$$

Note that $\mathbf{G}(D)$ is not in systematic form. To find $\mathbf{H}(D)$ in $\mathbf{G}(D)\mathbf{H}(D)^T = 0$, it is easiest first to put $\mathbf{G}(D)$ in this form. By elementary operations we have

$$\mathbf{G}'(D) = \begin{bmatrix} 1 & 0 & (1+D)^{-1} + D^2(1+D)^{-2} \\ 0 & 1 & D(1+D)^{-1} \end{bmatrix}.$$

Thus $\mathbf{G}'$ is in the form $[\mathbf{I}_2 | \mathbf{P}]$, and thus $\mathbf{H}(D) = [\mathbf{P} | \mathbf{I}_1]$, or

$$\mathbf{H}'(D) = [(1+D)^{-1} + D^2(1+D)^{-2}, D(1+D)^{-1}, 1].$$

Clearing fractions yields

$$\mathbf{H}(D) = [1 + D + D^2, D(1+D), (1+D^2)] \tag{2.14}$$

where $(1+D)^2 = 1 + D^2$ over $GF(2)$.

The parity check equation for this can be derived from $\mathbf{H}(D)$ in (2.14). Thus, as $\mathbf{c} = (\beta_i^{(1)}, \beta_i^{(2)}, \beta_i^{(3)})$ is a code word, $\mathbf{cH}(D)^T = 0$, or from (2.11)–(2.13),

$$\beta_i^{(1)} \oplus \beta_{i-1}^{(1)} \oplus \beta_{i-2}^{(1)} \oplus \beta_{i-1}^{(2)} \oplus \beta_{i-2}^{(2)} \oplus \beta_i^{(3)} \oplus \beta_{i-2}^{(3)} = 0.$$

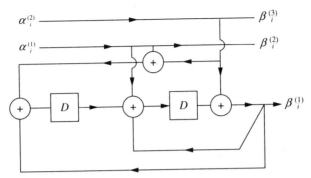

FIGURE 2.8 Feedback form of the code of Fig. 2.5.

From this equation we can derive the feedback form for generating this code, which is given in Fig. 2.8. This is just a realization of this parity check equation as a difference equation over $GF(2)$. Both the feedback form in Fig. 2.8 and the feedforward form in Fig. 2.5 are used in representing trellis codes.

2.7.3 Viterbi Algorithm

There are at least three popular methods to decode convolutional codes: the feedback decoder, the sequential decoder, and Viterbi decoder. We consider only the Viterbi decoder, as it is the decoder that is usually used in trellis-coded modulation. The other decoders are covered in [12].

The decoding problem is to determine the transmitted path through the trellis. For instance, if the all-zeros path, a–a–a–..., is sent, one would hope that this is the decoded path. Now the channel output, that is, the result of sending the transmitted path through the BSC, represents a perturbation of the transmitted path.

Thus the decoding problem is: Given $(r_i^{(1)}, r_i^{(2)}, r_i^{(3)})$ per trellis interval, where $r_i^{(j)} = 0, 1$, determine the most likely transmitted path through the trellis. If we assume that the uses of the BSC are independent (i.e., we have random errors), the problem reduces to minimizing the Hamming distance between the $\{r_i\}$ and our estimate of the $\{\beta_i\}$, denoted as $\{\hat{\beta}_i\}$:

$$e(\hat{\boldsymbol{\beta}}) = \sum_{i=0}^{\infty} \sum_{j=1}^{3} \hat{\beta}_i^{(j)} \oplus r_i^{(j)}. \tag{2.15}$$

Sample values of $\hat{\beta}_i$ are the output bits used on the trellis transition given in Fig. 2.7. For instance, for the transition $a \to a$, $(\hat{\beta}_i^{(1)}, \hat{\beta}_i^{(2)}, \hat{\beta}_i^{(3)}) = (0, 0, 0)$.

The optimization problem that given the $\{r_i\}$, find the minimum in (2.15) subject to the $\hat{\beta}_i^{(j)}$, $j = 1, 2, 3$, satisfying (2.11) through (2.13), can be solved using Bellman's dynamic programming [13]. The solution of this problem, which in various forms occurs throughout digital communications, is termed the *Viterbi algorithm*. We can invoke a dynamic programming solution because (1) the optimization function in (2.15) is additive, (2) the difference equation in (2.11) through (2.13) is Markovian, and (3) the outputs of the BSC per trellis interval are memoryless.

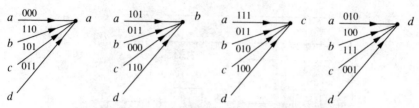

FIGURE 2.9 Transition diagram for Viterbi algorithm.

A list of all the transitions per state and their values for $(\hat{\beta}_i^{(1)}, \hat{\beta}_i^{(2)}, \hat{\beta}_i^{(3)})$ are given in Fig. 2.9. For each trellis transition we do the following. Let $M(x_i)$, $x_i = a, b, c, d$ be the accumulated state metric, that is, the sum in (2.15) up to trellis interval i; the input for the discrete time transition $i \to i+1$ is $(r_{i+1}^{(1)}, r_{i+1}^{(2)}, r_{i+1}^{(3)})$. For each state, x_{i+1}:

1. Compute

$$M(x_{i+1}) = \min_{(\hat{\beta}_{i+1}^{(1)}, \hat{\beta}_{i+1}^{(2)}, \hat{\beta}_{i+1}^{(3)})} [M(x_i) + \sum_{j=1}^{3} r_{i+1}^{(j)} \oplus \hat{\beta}_{i+1}^{(j)}].$$

2. Call the best $(\hat{\beta}_{i+1}^{(1)}, \hat{\beta}_{i+1}^{(2)}, \hat{\beta}_{i+1}^{(3)})$ the winning transition and store it as well as $M(x_{i+1})$, the state metric for the winning transition.

When these steps are implemented, we are left with one single transition path per state per trellis interval. The collection of these winning transition paths over time are called *survivor paths* and are illustrated in Fig. 2.10. For instance, the survivor for state a after three trellis intervals from a cold start is the path (a, a, a). The decoded path is then the minimum of $M(x_{i+1})$ over all survivor paths. In reality, one should wait until survivor paths merge, that is, their initial segments coincide. In practice, one stores the result for D_d trellis intervals and then makes the choice of the best survivor path; this is the path with smallest $m(\cdot)$. An example for $D_d = 4$ is shown in Fig. 2.11: A merge is shown over the first two intervals of *all* survivor paths. Note that single trellis transition decisions, that is, input symbol decisions (see Fig. 2.7), are made with a delay of D_d trellis intervals. The variable D_d is called the *decision depth* of the decoder. Large D_d is best, but it is subject to the principle of diminishing returns as D_d gets large. A rule of thumb used in

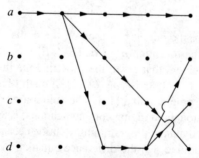

FIGURE 2.10 Survivor paths for four-state code.

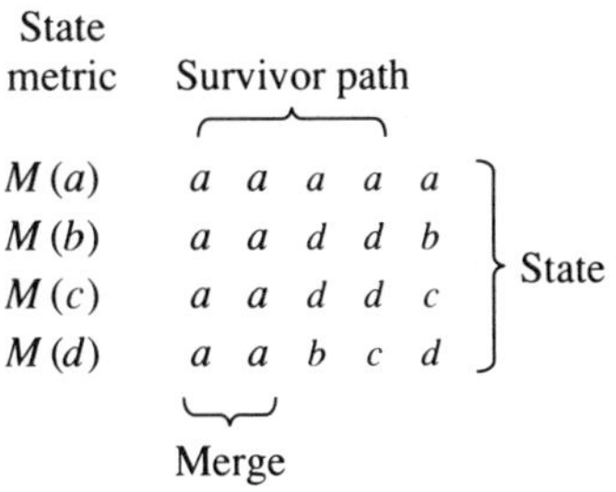

```
                State
                metric      Survivor path

                M (a)       a  a  a  a  a
                M (b)       a  a  d  d  b  ⎤
                M (c)       a  a  d  d  c  ⎬ State
                M (d)       a  a  b  c  d  ⎦
                               Merge
```

FIGURE 2.11 Decision depth for four-state trellis and paths in Fig. 2.10.

trellis codes is to use three times the number of state variables; that is, $D_d = 12$ for the present example.

2.7.4 Error-State Diagrams

In Fig. 2.7 regard the all-zeros sequence as the transmitted bit stream. All possible sequences that diverge and then later merge with this sequence can be summarized in a single diagram having the same number of states as the code: in the present example in Fig. 2.5, four states. In Fig. 2.7 some error events have been noted and they can be identified with the states (b, c, d). Also, a beginning state and ending state, which are both (a), are needed. Note that the error events are all relative to a single instance of diverging and merging with the transmitted, all-zeros path. Multiple occurrences of these events are not part of the theory. A probabilistic analysis of the single instance of diverging and merging is possible. Assuming stationarity and long-term average error probabilities, we find that analysis of this single class of error events represents the performance of the system [12].

The error-state diagram for the code under consideration is shown in Fig. 2.12. Let us consider the label on the branch $a \to b$ to represent the general picture. The label is D^2LI, where

1. D^2 means that the Hamming distance between $a \to a$ and $a \to b$ in Fig. 2.7 is 2.
2. The power of L represents the length of this error branch.
3. The power of I tells us that there is one error in transmitted bits on this branch; that is, 00 versus 10, as shown in Fig. 2.7.

The other labels, called *branch gains*, are made up in the same manner.

Let us extract from the error-state diagram the error path, among all error paths, that go from a to a and have the smallest power of D. There is one such path, $a \to d \to a$, and it has gain, that is, the product of branch gains, equal to $D^3L^2I^2$. This tells us that the minimum Hamming distance error path is of distance $d_{\min}^H = 3$, of length 2, and contains two bit errors. Referring back to Fig. 2.7, we see that this is a representation of transmitted path $a \to a \to a$ versus error path $a \to d \to a$. We study two paths in the error-state diagram in Fig. 2.12 at Hamming distance 4. They are $a \to b \to a$ and $a \to d \to d \to a$. The first has gain D^4L^2I

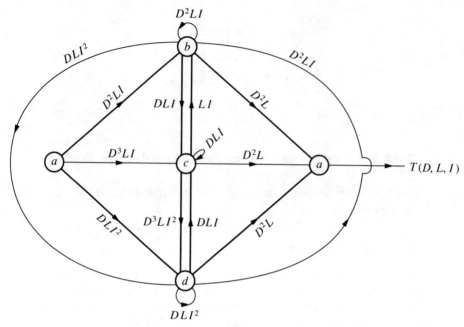

FIGURE 2.12 Error-state diagram for code of Fig. 2.5.

and the second, $D^4L^3I^4$. Now higher powers of D require more and more errors in the BSC. Thus, at small p, or high SNR, the sum of all gains from $a \to a$, $T(D, L, I)$, is approximately (not all D^4 terms are included)

$$T(D, L, I) = D^3L^2I^2 + D^4L^2I(1 + LI^3). \tag{2.16}$$

$T(D, L, I)$ is called the *transfer function* for the error-state diagram in Fig. 2.12.

To consider (2.16) from another point of view, let us assume that we had worked the transfer function for small p for Fig. 2.12. Let us further assume that the result is given by (2.16). Then we can say that the dominating error event at small p has Hamming distance 3, is of length 2, and contains two bit errors. A similar statement can be made for the next-higher power of D in (2.16).

To determine the effect of all error events, we must solve for $T(D, L, I)$ in Fig. 2.12. The nodal transfer coefficients are $(\epsilon_a, \epsilon_b, \epsilon_c, \epsilon_d)$, with $\epsilon_a = 1$. Thus, from Fig. 2.12,

$$T(D, L, I) = D^2L\epsilon_T \tag{2.17}$$

where

$$\epsilon_T = \epsilon_b + \epsilon_c + \epsilon_d. \tag{2.18}$$

Now from the equations at nodes $(\epsilon_b, \epsilon_c, \epsilon_d)$, we have

$$\epsilon_b = D^2LI + D^2LI\epsilon_b + LI\epsilon_c + D^2LI\epsilon_d$$

$$\epsilon_c = D^3LI + DLI\epsilon_T$$

$$\epsilon_d = DLI^2 + DLI^2\epsilon_d + DLI^2\epsilon_b + D^3LI^2\epsilon_c.$$

2.7 / Convolutional Codes

Now from (2.17) it suffices to solve these three equations for ϵ_T in (2.18). Adding these three equations and simplifying gives

$$\epsilon_T = \frac{\eta_1 + \eta_2 D^3 LI}{1 - (DLI + \eta_3 + \eta_2 DLI)}$$

where

$$\eta_1 = DLI(D^2 + D + I) \tag{2.19}$$

$$\eta_2 = LI(1 + D^3 I - D^2 - DI) \tag{2.20}$$

$$\eta_3 = DLI(D + I). \tag{2.21}$$

Then, by (2.17),

$$T(D, L, I) = \frac{D^2 L(\eta_1 + \eta_2 D^3 LI)}{1 - (DL + \eta_3 + \eta_2 DLI)}. \tag{2.22}$$

Expanding the denominator in (2.22) in a power series in D, we find that the term $D^4 I^3 L^3 (1 + LI)$ must be added to the right-hand side of (2.16) to approximate $T(D, L, I)$ up to D^4.

Reference [12] shows that the probability of an error event is upper bounded by

$$P(\{E\}) \le T(D, 1, 1) \tag{2.23}$$

where the probability of error for any path in the error-state diagram can be bounded by $P_d \le D^d$, where d is the Hamming distance for this path, that is, the power of D for any path from ϵ_a to $T(D, L, I)$ in Fig. 2.12. Also, for the BSC in Fig. 2.4, with crossover probability p, [12] shows that $D = \sqrt{4p(1-p)}$. Thus, for some given p, $P(E)$ in (2.23) is obtained from (2.22) and (2.19) through (2.21) for the computed value of D.

Often, the information bit error probability is of interest. Note from (2.16) that differentiating with respect to I will weight each error event by the number of bit errors if I is set to unity. If b is the number of input bits per trellis interval, and as we have performed an analysis on a per trellis interval basis, it follows that [12]

$$P_b \le \frac{1}{b} \frac{\partial}{\partial I} T(D, I, 1) \Big|_{I=1}. \tag{2.24}$$

Instead of differentiating, one can [12] further bound the result in (2.24) with

$$\frac{\partial T(D, I, 1)}{\partial I} < \frac{T(D, 1, 1+\epsilon) - T(D, 1, 1)}{\epsilon} \tag{2.25}$$

and one can choose a small ϵ like 10^{-6} to approximate the derivative. Here $T(D, L, I)$ is given in (2.22). Except for a few simple examples, the approximation in (2.25) leads to fewer manipulative errors.

To determine either $P(E)$ in (2.23) or P_b in (2.24) and (2.25), we use a numerical approach that starts with the equations just after (2.18) in matrix form:

$$\mathbf{x} = \mathbf{Ax} + \mathbf{b} \tag{2.26}$$

$$T(D, L, I) = D^2 L \mathbf{1}^T \mathbf{x} \tag{2.27}$$

where

$$\mathbf{x} = (\epsilon_b, \epsilon_c, \epsilon_d), \quad \mathbf{1} = (1, 1, 1)$$

$$\mathbf{A} = \begin{pmatrix} D^2 LI & LI & D^2 LI \\ DLI & DLI & DLI \\ DLI^2 & D^3 LI^2 & DLI^2 \end{pmatrix}$$

and

$$\mathbf{b} = DLI \begin{pmatrix} D \\ D^2 \\ I \end{pmatrix}.$$

Then the solution to (2.27) is

$$\mathbf{x} = (\mathbf{I} - \mathbf{A})^{-1} \mathbf{b}$$

and thus in (2.27),

$$T(D, L, I) = D^2 L \mathbf{1}^T (\mathbf{I} - \mathbf{A})^{-1} \mathbf{b}. \tag{2.28}$$

Then the error probabilities in (2.23), (2.24), and (2.25) are computed by setting $D = \sqrt{4p(1-p)}$, $L = 1$, and $I = 1$ in (2.28). No numerical instabilities have been observed in this method when used for trellis codes [14, 15].

This completes our introduction to error-correcting codes. As will be evident in later chapters, trellis coding is strongly related to the theory of convolutional codes. Block coding concepts have been applied in trellis coding; for instance, [16], [17], and [18] treat the block-coded case.

REFERENCES

1. F. M. REZA, *An Introduction to Information Theory*. McGraw-Hill, New York, N.Y., 1961.
2. S. LIN and D. J. COSTELLO, JR., *Error Control Coding: Fundamentals and Applications*. Prentice-Hall, Englewood Cliffs, N.J., 1983.
3. R. W. HAMMING, "Error detecting and error correcting codes," *Bell Syst. Tech. J.*, Vol. 42, pp. 79–94, Apr. 1950.
4. A. M. MICHELSON and A. H. LEVESQUE, *Error-Control Techniques for Digital Communications*. Wiley, New York, N.Y., 1985.
5. R. C. BOSE and D. K. RAY-CHAUDHURI, "On a class of error-correcting binary group codes," *Inf. Control*, Vol. 3, pp. 68–79, Mar. 1960.
6. A. HOCQUENGHEM, *Codes Correcteurs d'Erreurs*. Chiffres, France, 1959, pp. 147–156.
7. M. KAVEHRAD and P. J. MCLANE, "Performance of low complexity channel coding and diversity for spread spectrum in indoor wireless communication," *AT&T Tech. J.*, Vol. 64, pp. 1927–1967, Oct. 1985.
8. W. W. PETERSON and E. J. WELDON, JR., *Error-Correcting Codes*, 2nd ed. MIT Press, Cambridge, Mass., 1972.

9. L. B. MILSTEIN, R. L. PICKHOLTZ, and D. L. SCHILLING, "Optimization of the processing gain of an FSK-FH system," *IEEE Trans. Commun.*, Vol. COM-28, pp. 1062–1079, July 1960.
10. R. E. BLAHUT, *The Theory and Practice of Error Control Codes*. Addison-Wesley, Reading, Mass., 1983.
11. S. BENEDETTO, E. BIGLIERI, and V. CASTELLANI, *Digital Transmission Theory*. Prentice-Hall, Englewood Cliffs, N.J., 1987.
12. A. J. VITERBI and J. K. OMURA, *Principles of Digital Communication and Coding*. McGraw-Hill, New York, N.Y., 1979.
13. B. GLUSS, *An Elementary Introduction to Dynamic Programming*. Allyn & Bacon, Boston, Mass., 1972.
14. R. MCKAY, P. J. MCLANE, and E. BIGLIERI, "Analytical performance bounds on average bit error probability for trellis coded PSK transmitted over fading channels," *Proc. International Conference on Communications*, Boston, Mass., June 11–14, 1989; *IEEE Trans. Commun.*, to appear.
15. R. BUZ and P. J. MCLANE, "Error bounds for multi-dimensional TCM," *Proc. International Conference on Communications*, Boston, Mass., June 11–14, 1989.
16. S. I. SAYEGH, "A class of optimum block codes in signal space," *IEEE Trans. Commun.*, Vol. COM-34, pp. 1043–1045, Oct. 1986.
17. M. ISAKSSON, "A class of block codes with expanded signal-sets for PSK-modulation," *Report TRITA-TTT-8806*, Royal Institute of Technology, Stockholm, Sweden, Aug. 1988.
18. G. R. FORNEY, JR., et al., "Efficient modulation for band-limited channels," *IEEE J. Select. Areas Commun.*, Vol. SAC-2, pp. 632–647, 1984.

CHAPTER 3

TCM: Combined Modulation and Coding

In this chapter we introduce trellis-coded modulation (TCM), used for data communication with the purpose of gaining noise immunity over uncoded transmission without altering the data rate. TCM is a combined coding and modulation scheme for improving the reliability of a digital transmission system without increasing the transmitted power or the required bandwidth. In a power-limited environment, the desired system performance should be achieved with the smallest possible power. One solution is the use of error-correcting codes, which increase the power efficiency by adding extra bits to the transmitted symbol sequence. This procedure requires the modulator to operate at a higher data rate and hence requires a larger bandwidth. In a bandwidth-limited environment, increased efficiency in frequency utilization can be obtained by choosing higher-order modulation schemes (e.g., 8-PSK instead of 4-PSK), but a larger signal power would be needed to maintain the same signal separation and hence the same error probability. *The trellis code solution combines the choice of a higher-order modulation scheme with that of a convolutional code, while the receiver, instead of performing demodulation and decoding in two separate steps, combines the two operations into one.*

3.1 Introducing TCM

An example will introduce the concept of TCM. Consider a digital communication scheme to transmit data from a source emitting two information bits every T seconds. Several solutions are possible (Fig. 3.1).

(a) Use no coding and 4-PSK modulation, with one signal every T seconds. In this situation, every signal carries two information bits.

(b) Use a convolutional code with rate 2/3 and 4-PSK modulation. Since every signal now carries 4/3 information bits, it must have a duration of $2T/3$ to match the information rate of the source. This implies that with respect to the uncoded scheme, the bandwidth increases by a factor of 3/2.

(c) Use a convolutional code with rate 2/3 and 8-PSK modulation to avoid reducing the signal duration. Each signal carries two information bits, and hence no bandwidth expansion is incurred because 8-PSK and 4-PSK occupy the same bandwidth.

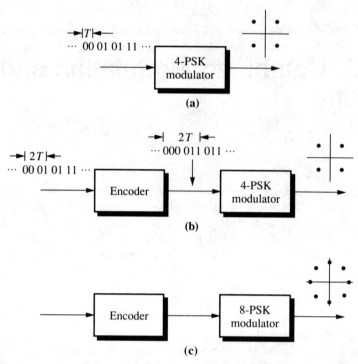

FIGURE 3.1 Three digital communication schemes transmitting 2 bits every T seconds: (a) uncoded transmission with 4-PSK; (b) 4-PSK with a rate 2/3 convolutional encoder and bandwidth expansion; (c) 8-PSK with a rate 2/3 convolutional encoder and no bandwidth expansion.

We see that with solution (c) we can use coding with no bandwidth expansion. One might expect that the use of a higher-order signal constellation would involve a power penalty with respect to 4-PSK; thus the coding gain achieved by the rate 2/3 convolutional code should offset this penalty, the net result being some coding gain at no price in bandwidth.

This idea is indeed not new, since multilevel modulation of convolutionally encoded symbols was a known concept before the introduction of TCM. The innovative aspect of TCM is the concept that convolutional encoding and modulation should not be treated as separate entities, but rather, as a unique operation. As a conclusion, the received signal, instead of being first demodulated *and then* decoded, is processed by a receiver that combines demodulation and decoding in a single step. The consequence of this is that the parameter governing the performance of the transmission system is not the *free Hamming distance* of the convolutional code, but rather, over the additive white Gaussian noise channel, the *free Euclidean distance* between transmitted signal sequences. Thus the optimization of the TCM design will be based on Euclidean distances rather than on Hamming distances, so that the choice of the code and of the signal constellation will not be performed separately. Finally, the detection process will involve *soft* rather than *hard* decisions (i.e., received signals are processed *before* making decisions as to which transmitted symbol they correspond).

3.2 The Need for TCM

In general, the selection of modulation and coding formats for transmission on the AWGN channel is based on the following parameters:

- $\mathcal{R}$, the information rate (i.e., the number of bits per second that we want to transmit over the channel).
- B, the transmission bandwidth available (in hertz).
- The error probability to be achieved at a given signal-to-noise ratio.

It should be observed that the value of B depends on the definition of bandwidth that has been accepted for the specific application. In general, we can say that

$$B = \frac{\alpha}{T}$$

where T is the duration of one of the waveforms used by the modulator. For example, uncoded PSK has a power density spectrum proportional to

$$\frac{\sin^2 \pi(f - f_0)T}{[\pi(f - f_0)T]^2}$$

where f_0 is the carrier frequency. For null-to-null bandwidth, we have $B = 2/T$ (i.e., $\alpha = 2$). For 3-dB bandwidth, $\alpha = 0.88$. For equivalent noise bandwidth, we have $\alpha = 1$.

The ratio $r = \mathcal{R}/B$, in bits/s/Hz, is the *throughput*, the number of bits per second that we can transmit in a bandwidth of 1 Hz. Hence the number of bits per signal (or *bits per channel use*) is given by

$$R = T\mathcal{R}$$
$$= TBr.$$

When $\alpha = 1$, we have $R = r$; that is, the number of bits per channel use is equal to the throughput. For example, 9600 bits/s can be transmitted over a 2400-Hz channel by using $9600/2400 = 4$ bits/channel use (i.e., a 16-ary uncoded modulation scheme).

Consider, for example, the design of a digital radio communication system, in which we need a constant-envelope transmitted signal with the following parameters:

- $B = 4800$ Hz, with B defined as null-to-null.
- $\mathcal{R} = 9600$ bits/s.

In this case we get

$$r = \frac{9.6}{4.8} = 2 \text{ bits/s/Hz}.$$

If we use PSK, we need a bandwidth $B = 2/T$; hence

$$T\mathcal{R} = TBr = 4 \text{ bits/signal}.$$

A candidate system is then 16-PSK, which provides an adequate solution if it satisfies the requirement for error probability. For instance, if $P(e) \leq 10^{-5}$ is required, uncoded 16-PSK is adequate if the available signal-to-noise ratio E_b/N_0

is higher than 18 dB. But if the signal-to-noise ratio is, say, 15 dB, we may want to introduce a TCM scheme that provides a coding gain in excess of 3 dB without altering the bandwidth occupancy of the transmitted signal.

3.3 Fundamentals of TCM

We assume here a discrete-time, continuous-amplitude model for the transmission of data on the additive white Gaussian noise channel. In this communication model, introduced independently and at about the same time by the Russian scholar Kotel'nikov and by C. E. Shannon, the messages to be delivered to the user are represented by points, or vectors, in an N-dimensional Euclidean space $\mathbf{R}^N$, called the *signal space*. When the vector x is transmitted, the received signal is represented by the vector

$$z = x + \nu$$

where ν is a noise vector whose components are independent Gaussian random variables with mean zero and the same variance $N_0/2$. The vector x is chosen from a set Ω' consisting of M' signal vectors, the *signal constellation*. The average square length

$$E' = \frac{1}{M'} \sum_{x \in \Omega'} \|x\|^2$$

will be referred to as the average signal *energy*.

Consider now the transmission of a sequence $\{x_i\}_{i=0}^{K-1}$ of K signals, where the subscript i denotes discrete time. The receiver that minimizes the average error probability over the sequence operates as follows. It first observes the received sequence $y_0, \ldots, y_{K-1}$, then it decides that $X_0, \ldots, X_{K-1}$ was transmitted if the squared Euclidean distance

$$d^2 = \sum_{i=0}^{K} \|y_i - x_i\|^2$$

is minimized for $x_i = X_i, i = 0, \ldots, K - 1$, or, in words, if the sequence $X_0, \ldots, X_{K-1}$ is closer to the received sequence than to any other allowable signal vector sequence. The resulting sequence error probability, as well as the symbol error probability, is upper bounded, at least for high signal-to-noise ratios, by a decreasing function of the ratio $d_{\min}^2/N_0$, where $d_{\min}^2$ is the minimum squared Euclidean distance between two allowable signal vector sequences.

3.3.1 Uncoded Transmission

An important special case occurs when the signals form an *independent* sequence. In this case, the allowable signal sequences are all the elements of Ω'^K, and hence d^2 is minimized by minimizing separately the individual terms $\|y - x\|^2$ for $x \in \Omega'$. The performance of this "symbol-by-symbol receiver" will then depend on the minimum distance

$$d_{min}^2 = \min_{x' \neq x''} \|x' - x''\|^2$$

as x', x'' run through Ω'. In fact, the symbol error probability is upper bounded (and for high signal-to-noise ratios, closely approximated) by

$$P(e) \leq \frac{M' - 1}{2} \text{erfc}\left(\frac{d_{min}}{2\sqrt{N_0}}\right). \quad (3.1)$$

With this model, the problem of designing a good communication system is that of choosing a set of vector signals such that the minimum distance between any two of them is a maximum, once the quantities M', N, and E' are given.

It is convenient to define two quantities that are useful in comparing different constellations: their *information rate*,

$$R = \frac{\log_2 M'}{N} \quad (3.2)$$

and their *normalized squared minimum distance*,

$$\delta^2 = \frac{d_{min}^2}{E'} \log_2 M'. \quad (3.3)$$

The first parameter is also referred to as the *bandwidth efficiency* of the signal set, while the second is its *energy efficiency*. The former is the ratio between the number of information bits carried by a single signal in the constellation and the number of dimensions. To justify the latter definition, observe that the upper bound (3.1) can be rewritten in the form

$$P(e) \leq \frac{M' - 1}{2} \text{erfc}\left(\frac{\delta}{2}\sqrt{\frac{E_b}{N_0}}\right) \quad (3.4)$$

where

$$E_b = \frac{E'}{\log_2 M'}$$

represents the average energy per bit. It is seen from (3.4) that the same $P(e)$ can be achieved with a smaller signal-to-noise ratio E_b/N_0 if δ is larger.

3.4 The Concept of TCM

One way of improving the system performance is that of removing the assumption that the signals are independent. This can be done by restricting the transmitted sequences to a subset of Ω'^K. Now, to do this, the transmission rate will also be reduced. To avoid this unwanted reduction, we may choose to increase the size of Ω'. For example, if we change Ω' into $\Omega \supset \Omega'$ and M' into $M > M'$, and we select M'^K sequences as a subset of Ω^K, we can have sequences that are less tightly packed and hence increase the minimum distance between them.

In conclusion, we obtain a minimum distance d_{free} between two possible sequences that turns out to be greater than the minimum distance $d_{\min}$ between signals in Ω', the constellation from which they were drawn. Hence use of maximum-likelihood sequence detection will yield a *distance gain* $d_{\text{free}}^2/d_{\min}^2$.

On the other hand, to avoid a reduction of the value of the transmission rate, the constellation is expanded from Ω' to Ω. This may entail an increase in the average energy expenditure from E' to E, and hence an "energy loss" E/E'. Thus we define the *asymptotic coding gain* of a TCM scheme as

$$\gamma = \frac{d_{\text{free}}^2/E}{d_{\min}^2/E'}$$

where E' and E are the average energies spent to transmit with uncoded and coded transmission, respectively.

The introduction of interdependencies among the signals and the expansion of the signal set are two of the basic ideas underlying trellis-coded modulation (another is *set partitioning*, described later).

We assume that the signal x_n transmitted at discrete time n depends not only on the source symbol a_n transmitted at the same time instant (as it would be with memoryless modulation), but also on a finite number of previous source symbols:

$$x_n = f(a_n, a_{n-1}, \ldots, a_{n-L}). \tag{3.5}$$

By defining

$$\sigma_n = (a_{n-1}, \ldots, a_{n-L}) \tag{3.6}$$

as the *state* of the encoder at time n, we can rewrite (3.5) in the more compact form

$$x_n = f(a_n, \sigma_n) \tag{3.7}$$

$$\sigma_{n+1} = g(a_n, \sigma_n). \tag{3.8}$$

Equations (3.7) and (3.8) can be interpreted as follows. The function $f(\cdot, \cdot)$ describes the fact that each channel symbol depends not only on the corresponding source symbol, but also on the parameter σ_n. In other words, at any time instant

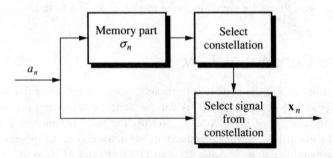

FIGURE 3.2 General model for TCM.

the transmitted symbol is chosen from a constellation that is selected by the value of σ_n. The function $g(\cdot, \cdot)$ describes the *memory part* of the encoder and shows the evolution of the modulator states (Fig. 3.2). Here we shall assume that the functions f and g are time invariant, although it is possible to consider time-varying TCM schemes as well.

3.5 Trellis Representation

For a graphical representation of the functions f and g it is convenient to use a *trellis*. The values that can be taken by σ_n, the encoder state at time n, are the *nodes* of the trellis. With each source symbol we associate a branch that stems from each modulator state at time n and reaches the encoder state at time $n + 1$. The branch is labeled by the corresponding value of f. The trellis structure is determined by the function g, while f describes how channel symbols are associated with each branch along the trellis.

If the source symbols are M'-ary, each node must have M' branches stemming from it (one per each source symbol). This implies that in some cases two or more branches connect the same pair of nodes; when this occurs, we say that *parallel transitions* take place.[1]

Figure 3.3 shows an example of this representation. It is assumed that the encoder has four states, the source emits binary symbols, and a constellation with four signals $x_1, \ldots, x_4$ is used. The distance properties of a TCM scheme can be studied through its trellis diagram. Observe first that optimum decoding is the search of the most likely path through the trellis once the received sequence has been observed at the channel output. This search is best done using the Viterbi algorithm (see, e.g., [1]). Because of the noise, the path chosen may not coincide with the correct path (i.e., the one traced by the sequence of source symbols), but will occasionally diverge from it (at time n, say) and remerge at a later time, $n + L$. When this happens, we say that an *error event* of length L has taken place. Thus the free distance of a TCM scheme is the minimum Euclidean distance between a pair of paths forming an error event.

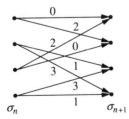

FIGURE 3.3 Example of a trellis describing a TCM scheme with four states and four channel symbols used to transmit from a binary source.

---

[1] Note that for uncoded transmission, the trellis degenerates to a single state, and thus *all transitions* are parallel transitions.

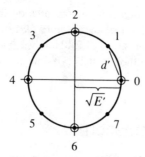

FIGURE 3.4 Two quaternary constellations used in a TCM scheme.

3.6 Some Examples of TCM Schemes

Here we describe a few examples of TCM schemes based on their trellis diagrams. We first consider the transmission of quaternary source symbols (i.e., 2 bits per symbol). With uncoded transmission a channel alphabet with $M' = 4$ would be adequate. We shall examine TCM schemes with $M = 2M' = 8$. We consider in particular coherent PSK transmission. With $M' = 4$ we obtain

$$\frac{d^2_{min}}{E'} = 2$$

a figure that will be used as a reference to compute the coding gain of PSK-based TCM. We use TCM schemes based on two quaternary constellations, $\{0, 2, 4, 6\}$ and $\{1, 3, 5, 7\}$ shown in Fig. 3.4. We have

$$E' = \frac{d'^2}{4\sin^2(\pi/8)}.$$

Consider first a scheme with two states, as shown in Fig. 3.5. If the encoder is in state S_1, alphabet $\{0, 2, 4, 6\}$ is used. If it is in state S_2, alphabet $\{1, 3, 5, 7\}$ is used instead. Let us compute the free distance of this TCM scheme. This can be done by choosing the smallest of the distances between signals associated with parallel transitions and the distances associated with a pair of paths in the trellis that originate from a common node and merge into a single node at a later time. The pair of paths giving the free distance is shown in Fig. 3.5, and we have, by

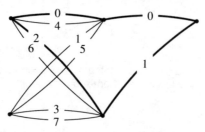

FIGURE 3.5 TCM scheme based on a two-state trellis, $M' = 4$, and $M = 8$.

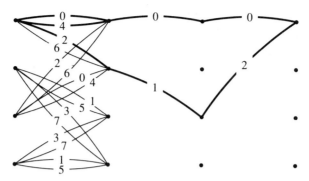

FIGURE 3.6 TCM scheme based on a four-state trellis, $M' = 4$, and $M = 8$. Two error events are shown. The free distance is determined by the error event with $L = 1$.

denoting the Euclidean distance between signals i and j by $d(i, j)$:

$$\frac{d^2_{\text{free}}}{E} = \frac{1}{E}[d^2(0, 2) + d^2(0, 1)] = 2 + 4\sin^2\frac{\pi}{8} = 2.586.$$

Hence we obtain a coding gain

$$\gamma = \frac{2.856}{2} = 1.293 \Rightarrow 1.1 \text{ dB}.$$

Let us now use a TCM scheme with a more complex structure, in order to increase the coding gain. With the same constellation of Fig. 3.4, take a trellis with four states as in Fig. 3.6. We associate the alphabet $\{0, 2, 4, 6\}$ with states S_1 and S_3, and $\{1, 3, 5, 7\}$ with S_2 and S_4. The error event leading to d_{free} has length 1 (a parallel transition) and is shown in Fig. 3.6. We get

$$\frac{d^2_{\text{free}}}{E} = d^2(0, 4) = 4$$

and hence

$$\gamma = \frac{4}{2} = 2 \Rightarrow 3 \text{ dB}.$$

A further step can be taken by choosing a trellis with eight states as shown in Fig. 3.7. For simplicity, the four symbols associated with the branches emanating from each node are used as node labels. The first symbol in each node label is associated with the uppermost transition from the node, the second symbol with the transition immediately below it, and so on. The error event leading to d_{free} is also shown. It yields

$$\frac{d^2_{\text{free}}}{E} = \frac{1}{E}[d^2(0, 6) + d^2(0, 7) + d^2(0, 6)] = 2 + 4\sin^2\frac{\pi}{8} + 2 = 4.586$$

and hence

$$\gamma = \frac{4.586}{2} = 2.293 \Rightarrow 3.6 \text{ dB}.$$

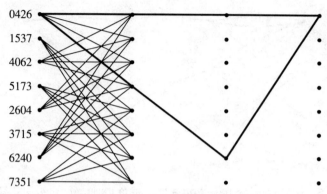

FIGURE 3.7 TCM scheme based on an eight-state trellis, $M' = 4$, and $M = 8$.

Consider now the transmission of 3 bits per symbol and AM–PM schemes. The octonary alphabet of Fig. 3.8 will be used as a reference uncoded scheme. It yields

$$\frac{d_{min}^2}{E'} = 0.8.$$

A TCM scheme with eight states and based on this alphabet is shown in Fig. 3.9. The subconstellations used are

$$\{0, 2, 5, 7, 8, 10, 13, 15\} \quad \text{and} \quad \{1, 3, 4, 6, 9, 11, 12, 14\}.$$

We have

$$E = 2.5d^2$$

and the free distance is obtained from

$$\frac{d_{free}^2}{E} = \frac{1}{E}[d^2(10, 13) + d^2(0, 1) + d^2(0, 5)]$$

$$= \frac{1}{E}[0.8E + 0.4E + 0.8E]$$

$$= 2$$

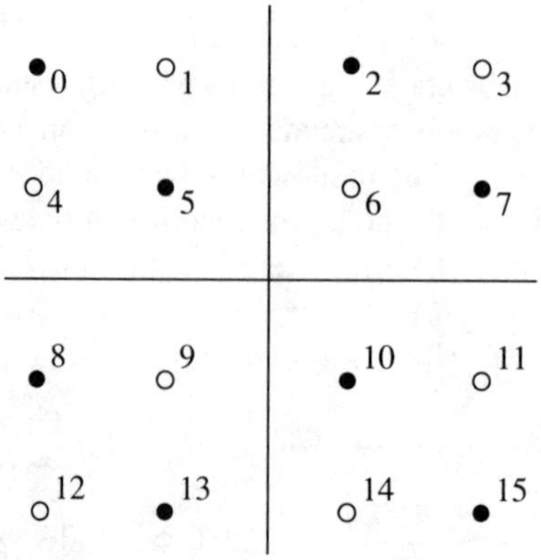

FIGURE 3.8 Octonary AM–PM alphabet.

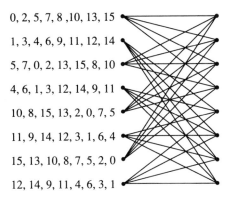

0, 2, 5, 7, 8, 10, 13, 15
1, 3, 4, 6, 9, 11, 12, 14
5, 7, 0, 2, 13, 15, 8, 10
4, 6, 1, 3, 12, 14, 9, 11
10, 8, 15, 13, 2, 0, 7, 5
11, 9, 14, 12, 3, 1, 6, 4
15, 13, 10, 8, 7, 5, 2, 0
12, 14, 9, 11, 4, 6, 3, 1

FIGURE 3.9 TCM scheme based on an eight-state trellis, $M' = 8$, and $M = 16$.

so that

$$\gamma = \frac{2}{0.8} = 2.5 \Rightarrow 3.98 \text{ dB}.$$

3.7 Set Partitioning

Consider the determination of d_{free}. This is the distance between the signals associated with a pair of paths that originate from an initial split and, after L time instants, merge into a single node as shown in Fig. 3.10. Assume first that the free distance is determined by parallel transitions (i.e., $L = 1$). If A denotes the set of signals associated with the branches emanating from a given node, then d_{free} equals the minimum distance among signals in A.

Consider next $L > 1$. With A, B, C, D denoting subsets of signals associated with each branch, and $d(X,Y)$ denoting the minimum Euclidean distance between one signal in X and one in Y, d_{free}^2 will have the expression

$$d_{\text{free}}^2 = d^2(A, B) + \cdots + d^2(C, D).$$

This implies that in a good code, the subsets assigned to the same originating state or to the same terminating state ("adjacent transitions") must have the largest possible distance. To implement these rules, Ungerboeck [2] suggested the following technique, called *set partitioning*. Set partitioning has been described as "the key

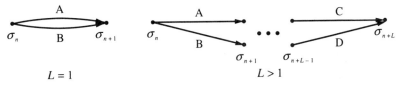

FIGURE 3.10 Pair of splitting and remerging paths for $L = 1$ (parallel transitions) and $L > 1$.

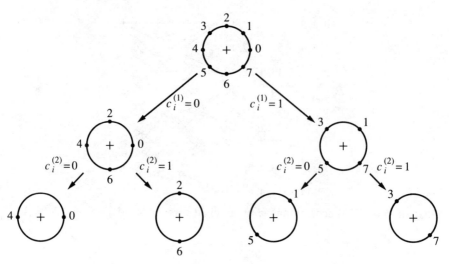

FIGURE 3.11 Set partition of an 8-PSK constellation.

that cracked the problem of constructing efficient coded modulation techniques for bandlimited channels."

The M-ary constellation is successively partitioned into $2, 4, 8, \ldots$ subsets with size $M/2, M/4, M/8, \ldots$, having progressively larger minimum distances $d_{\min}^{(1)}, d_{\min}^{(2)}, d_{\min}^{(3)}, \ldots$ (Figs. 3.11 and 3.12). Then

- U1. To parallel transitions are assigned members of the same partition; and
- U2. To adjacent transitions are assigned members of the next larger partition.

These two rules, in conjunction with the symmetry requirement:

- U3. All the signals are used equally often,

are conjectured to give rise to the best TCM schemes and will be referred to hereafter as the three *Ungerboeck rules*.

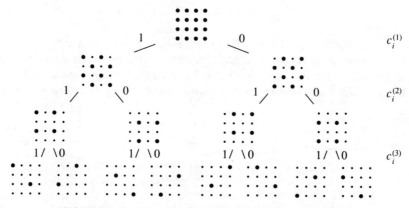

FIGURE 3.12 Set partition of a 16-QAM constellation.

3.8 Representations for TCM

In this section we examine two representations of a TCM encoder. The first one, due to Ungerboeck [2, 3], consists of modeling the memory part of the TCM encoder through a convolutional encoder. The second one, due to Calderbank and Mazo [4], is based on an analytical expression for the mapping of source symbols into signals.

3.8.1 Ungerboeck Representation

If a_i, the source symbol at time i, can take on 2^k values, we can represent it as a sequence of k binary digits $b_i^{(1)}, b_i^{(2)}, \ldots, b_i^{(k)}$ that are presented simultaneously to the encoder. In general, x_i, the encoder output at time i, will depend on the previous $v_j \geq 0$ bits of the jth binary input stream, $j = 1, \ldots, k$:

$$x_i = f(b_i^{(1)}, b_{i-1}^{(1)}, \ldots, b_{i-v_1}^{(1)}, \ldots, b_i^{(k)}, \ldots, b_{i-v_k}^{(k)}) \tag{3.9}$$

(see Fig. 3.13). From this model we can represent the trellis encoder as being composed of two parts:

- A *binary convolutional encoder*, which has k binary inputs $b_i^{(1)}, \ldots, b_i^{(k)}$ and n binary outputs $c_i^{(1)}, \ldots, c_i^{(n)}$. The sum

$$v = \sum_{i=1}^{k} v_i$$

is the *memory* of the encoder, and it determines the number of states, which is 2^v.
- A *modulator* part, which is memoryless and associates with the binary n-tuple at the output of the convolutional encoder a signal $x_i \in \Omega$.

Ungerboeck codes

Based on this representation, we shall now describe an exceedingly important family of TCM codes that we call *Ungerboeck codes*. They satisfy the following requirements:

1. $M = 2M'$; that is, M signals are used to transmit $\log_2 M - 1$ bits per signal.
2. The convolutional code in Fig. 3.13 is *linear*.

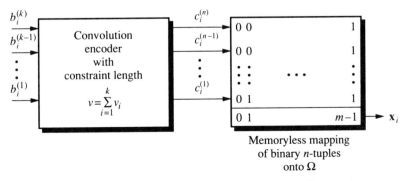

FIGURE 3.13 Ungerboeck representation of a trellis encoder.

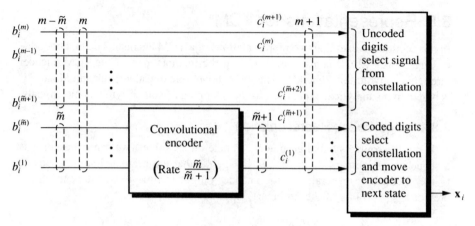

FIGURE 3.14 Block diagram of an Ungerboeck code ($m = \log_2 M$).

It is convenient to represent an Ungerboeck code by the scheme of Fig. 3.14. At every time i, the rate $\tilde{m}/(\tilde{m} + 1)$ convolutional encoder receives $\tilde{m}$ input bits and generates $\tilde{m} + 1$ coded digits. These, in turn, determine the subconstellation from which the transmitted signal has to be chosen. In the figure, the $m - \tilde{m}$ source digits left uncoded are explicitly shown. The presence of uncoded digits causes parallel transitions; an edge in the trellis diagram of the code is associated with $2^{m-\tilde{m}}$ signals. The correspondence between the encoded digits and the subconstellations obtained from set partitioning is shown in Figs. 3.11 and 3.12.

Constructing the trellis

We shall now see through an example how we can get the trellis description of a TCM scheme from its Ungerboeck representation. The example we use to illustrate the trellis construction is based on a TCM scheme that has now been adopted as the international standard for the coded 9.6-kbit/s two-wire full-duplex voiceband modem [5]. The encoder for this scheme is represented in Fig. 3.15.

It is apparent from the scheme that there are

- $m = 4$ binary input symbols (by disregarding the differential encoder, these are denoted by $Y1_n, Y2_n, Q3_n, Q4_n$), corresponding to $m + 1 = 5$ binary symbols entering the mapper. The mapper has $2^{m+1} = 32$ signal elements available, as depicted in Fig. 3.16.
- A convolutional encoder with $\tilde{m} = 2$ inputs and $\tilde{m} + 1 = 3$ outputs, that is, with rate 2/3, and $v = 3$ delay elements.

By referring to the previous discussion, in particular to Fig. 3.14, we observe that for the present TCM scheme the number of states in the trellis is $2^v = 8$, with a subconstellation of $2^{m-\tilde{m}} = 4$ signals (parallel transitions) associated with each branch of the trellis.

The trellis structure is then obtained as follows. With reference to the notations of Fig. 3.15, let us denote by $W1_n W2_n W3_n$ the current state, so that the next state will correspond to $W1_{n+1} W2_{n+1} W3_{n+1}$. We must evaluate which states can be

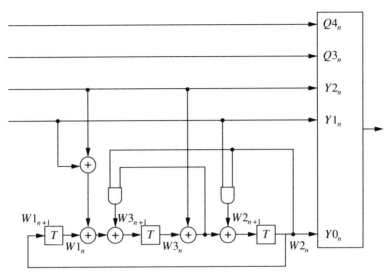

FIGURE 3.15 TCM scheme for the coded 9.6 kbit/s two-wire full-duplex voiceband modem.

reached from the current state. From Fig. 3.15 we have

$$W1_{n+1} = W2_n$$
$$W2_{n+1} = (W3_n \oplus Y2_n) \oplus (W2_n \odot Y1_n)$$
$$W3_{n+1} = W1_n \oplus Y1_n \oplus Y2_n \oplus [(W3_n \oplus Y2_n) \odot W2_n].$$

Thus if, for instance, $W1_n W2_n W3_n = 000$, the states that can be reached are 000 (for $Y1_n Y2_n = 00$; in this case, since $Y0_n = W2_n$, we have $Y0_n Y1_n Y2_n = 000$, so that the subconstellation of signals denoted a in Fig. 3.16 is associated with this transition), 001 (for $Y1_n Y2_n = 10$, and hence $Y0_n Y1_n Y2_n = 010$, corresponding to signals c), 011 (for $Y1_n Y2_n = 01$, and hence $Y0_n Y1_n Y2_n = 001$,

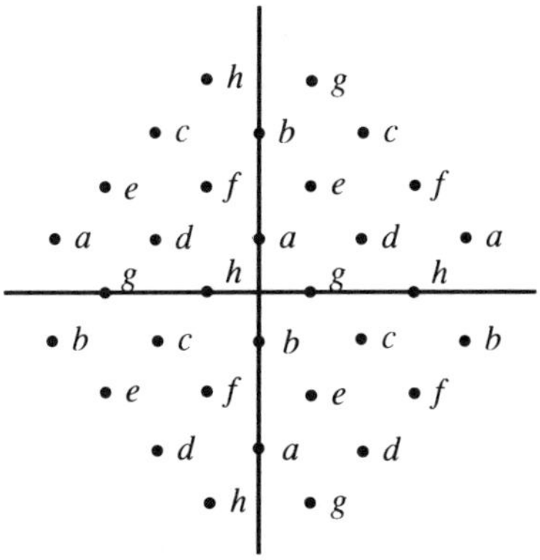

	$Y0_n$	$Y1_n$	$Y2_n$
a	0	0	0
b	0	0	1
c	0	1	0
d	0	1	1
e	1	0	0
f	1	0	1
g	1	1	0
h	1	1	1

FIGURE 3.16 32-signal constellation for the CCITT TCM scheme.

corresponding to signals d). Similarly, the states that can be reached from $W1_n W2_n W3_n = 001$ are 010 (for $Y1_n Y2_n = 00$), 011 (for $Y1_n Y2_n = 10$), 001 (for $Y1_n Y2_n = 01$), and 000 (for $Y1_n Y2_n = 11$). The situation is summarized in Fig. 3.17, where $A, B, \ldots, H$ denote the subconstellations formed by signals labeled $a, b, \ldots, h$, respectively, in Fig. 3.16.

3.8.2 Analytical Representation

Consider again the information source. It outputs a sequence (a_n) of symbols taken from an alphabet with 2^m elements. The effect of the trellis encoder is to transform this sequence of symbols into a sequence of signals taken from Ω. We have

$$x_i = f(a_i, a_{i-1}, \ldots, a_{i-L}). \tag{3.10}$$

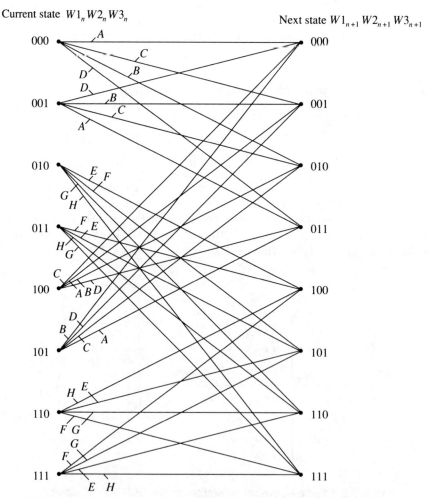

FIGURE 3.17 One stage of the trellis diagram for the CCITT TCM scheme.

Equation (3.10) provides the basis for a different description of the trellis encoder. Since it can be viewed as the input–output relationship of a system with memory, any analytical description of this transfer function will provide a representation for the encoder.

The function f must be nonlinear

It should be observed that the function $f(\cdot)$ must be *nonlinear* in order to achieve a coding gain. In fact, a linear input–output relationship corresponds to the effects of intersymbol interference, which cannot improve the performance of a system. This can be seen through a simple example. Consider the binary symbols $a_n \in \{-1, +1\}$ and

$$f(a_1, \ldots, a_n) = h_1 a_1 + \cdots + h_n a_n.$$

We can obtain an upper bound to the free distance of the code by observing that d_{free} is certainly less than or equal to the distance pertaining to any specified pair of paths. By choosing one path corresponding to the sequence of repeated +1's and the second path corresponding to the same sequence with only one +1 changed into a −1, their square distance is given by

$$\begin{aligned}
d^2 &= \|(h_1 + h_2 + \cdots + h_n) - (-h_1 + h_2 + \cdots + h_n)\|^2 \\
&\quad + \|(h_1 + h_2 + \cdots + h_n) - (h_1 - h_2 + \cdots + h_n)\|^2 + \cdots \\
&\quad + \|(h_1 + h_2 + \cdots + h_n) - (h_1 + h_2 + \cdots - h_n)\|^2 \\
&= 4(h_1^2 + h_2^2 + \cdots + h_n^2)
\end{aligned}$$

so that

$$\frac{d_{\text{free}}^2}{E} \leq 4 \frac{h_1^2 + h_2^2 + \cdots + h_n^2}{h_1^2 + h_2^2 + \cdots + h_n^2} = 4$$

while in the absence of coding (i.e., with $h_1 = 1$ and $h_i = 0, i \neq 1$), we have

$$\frac{d_{\min}^2}{E'} = \frac{4}{1} = 4$$

which shows that this form of linear coding does not increase the separation between signals.

On the contrary, if we consider, for example, the nonlinear f-function

$$f(a_1, a_2, a_3) = a_2 - \beta a_1 a_3$$

we could show that its minimum normalized square distance is given by

$$\frac{d_{\text{free}}^2}{E} = 4 \frac{\min\{1 + 2\beta^2, 4\beta^2 - 4\beta + 2\}}{1 + \beta^2}$$

that is, the coding gain afforded by the corresponding code is

$$\gamma = \frac{\min\{1 + ?\beta^2, 4\beta^2 - 4\beta + 2\}}{1 + \beta^2}$$

a quantity greater than 1 if $\beta > 1$. It should also be observed that the coded symbols can take on values $\pm 1 \pm \beta$; hence the choice $\beta = 2$ gives rise to equally spaced ("symmetric") symbols. However, this choice does not provide the largest gain, which is obtained instead as $\beta \to \infty$. (This issue will be raised again in Chapter 5.)

Expanding the function f

In this section we consider the expansion of the function f introduced above. In particular, we shall see that this function can always be written as a *finite* sum of products of its arguments. For simplicity's sake, here we restrict our attention to the case in which the arguments of f are allowed to take on only *two* values. The most general setting is considered in some detail in Appendix 3A at the end of this chapter.

Let $a_1, \ldots, a_n$ take on values in the set $\{0, 1\}$. Then any function of these variables may be written as a sum of the products of the a_i's, namely,

$$f(a_1, a_2, \ldots, a_n) = k^{(0)} + \sum_i k_i^{(1)} a_i + \sum_{j>i} k_{ij}^{(2)} a_i a_j + \sum_{h>j>i} k_{ijh}^{(3)} a_i a_j a_h$$
$$+ \cdots + k_{1\ldots n}^{(n)} a_1 a_2 \cdots a_n$$

(3.11)

where the coefficients $k^{(\cdot)}$ are a set of constants that may be determined iteratively from the 2^n values that the function f can assume. The procedure for their computation is as follows.

First note that there are $\binom{n}{1}$ $k^{(1)}$'s, $\binom{n}{2}$ $k^{(2)}$'s, ..., or a total of

$$\binom{n}{0} + \binom{n}{1} + \binom{n}{2} + \cdots + \binom{n}{n} = 2^n$$

coefficients (i.e., as many as the values taken by f). By setting $a_i = 0$ for all i, we obtain

$$k^{(0)} = f(0, \ldots, 0).$$

If now we set $a_j = 1$ and $a_i = 0$ for $i \neq j$, we determine $k_j^{(1)}$ for $j = 1, \ldots, n$. Continuing in this fashion, setting two a_i's equal to unity and the rest of the a_i's equal to zero determines the coefficients $k^{(2)}$. Finally, setting all $a_i = 1$ gives $k_{1\ldots n}^{(n)}$.

When dealing with TCM, it is more convenient to consider variables a_i taking values in the set $\{-1, +1\}$ rather than in $\{0, 1\}$. If we make the linear transformation

$$b_i = 1 - 2a_i$$

we get ± 1-valued variables, and consequently,

$$f(b_1, b_2, \ldots, b_n) = d^{(0)} + \sum_i d_i^{(1)} b_i + \sum_{j>i} d_{ij}^{(2)} b_i b_j$$
$$+ \sum_{h>j>i} d_{ijh}^{(3)} b_i b_j b_h + \cdots + d_{1\ldots n}^{(n)} b_1 b_2 \cdots b_n.$$

(3.12)

Unfortunately, the ± 1 values of the b_i's do not lend themselves to an iterative solution for the coefficients of the expansion. Nevertheless, these coefficients can be derived simply as follows. Let **f** denote a column vector whose 2^n components are all the values that f can take. Next, let **d** denote the vector of the (unknown) coefficients. Finally, let **B** be a $2^n \times 2^n$ matrix where each row represents the 2^n values taken by all the products of the b_i's called for in (3.12) for each n-tuple $b_1, \ldots, b_n$. Then (3.12) can be given the matrix form

$$\mathbf{f} = \mathbf{Bd}.$$

Now we note that **B** has orthogonal rows (and columns), and

$$\mathbf{BB}^T = 2^n \mathbf{I}$$

where the superscript T denotes transpose and **I** is the $2^n \times 2^n$ identity matrix (actually, **B** is a Hadamard matrix). Thus

$$\mathbf{B}^{-1} = \frac{1}{2^n} \mathbf{B}^T \qquad (3.13)$$

and consequently the vector of the unknown coefficients can be computed simply from

$$\mathbf{d} = \frac{1}{2^n} \mathbf{B}^T \mathbf{f}.$$

The preceding equation can be interpreted by saying that **d** is the Hadamard transform of the vector **f**.

EXAMPLE 3.1

As a simple illustration of the foregoing theory, consider the case of a rate 1/2 TCM scheme with a 4-AM constellation and memory $v = 2$ [4]. A four-state trellis diagram that characterizes this scheme is shown in Fig. 3.18, where the states have been labeled with ± 1. The values along the transitions between states represent the output symbols $f(b_1, b_2, b_3)$, that is, the value determined by (3.12) in going from the present state (b_2, b_3) to the next state (b_1, b_2) in response to the input symbol b_1. For $n = 3$, by choosing $d^{(0)} = 0$ (which does not entail any loss of generality in the present case), (3.12) can be written as

$$f(b_1, b_2, b_3) = d_1^{(1)} b_1 + d_2^{(1)} b_2 + d_3^{(1)} b_3 + d_{12}^{(2)} b_1 b_2 + d_{13}^{(2)} b_1 b_3 + d_{23}^{(2)} b_2 b_3.$$

From the trellis diagram of Fig. 3.18, we have

$$f(1, 1, 1) = -1$$
$$f(-1, 1, 1) = 3$$
$$f(1, -1, 1) = -3$$
$$f(-1, -1, 1) = 1$$
$$f(1, 1, -1) = 3$$
$$f(-1, 1, -1) = -1$$
$$f(-1, -1, -1) = -3$$
$$f(1, -1, -1) = 1.$$

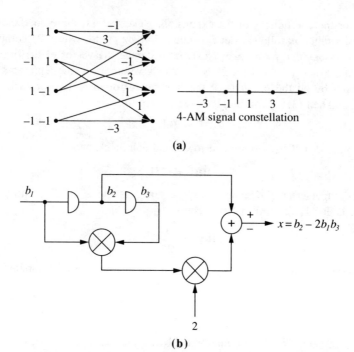

FIGURE 3.18 Rate 1/2 TCM scheme with 4-AM and four states: (a) trellis diagram; (b) scheme of the transmitter based on the analytical representation.

The coefficients d can be obtained by computing (3.13), where we have

$$\mathbf{f}^T = [-1 \quad 3 \quad -3 \quad 1 \quad 3 \quad -1 \quad -3 \quad 1]$$

$$\mathbf{d} = \begin{bmatrix} d_1^{(1)} & d_2^{(1)} & d_3^{(1)} & d_{12}^{(2)} & d_{13}^{(2)} & d_{23}^{(2)} & d_{123}^{(3)} \end{bmatrix}$$

and

$$B = \begin{bmatrix} 1 & 1 & 1 & 1 & 1 & 1 & 1 \\ -1 & 1 & 1 & -1 & -1 & 1 & -1 \\ 1 & -1 & 1 & -1 & 1 & -1 & -1 \\ -1 & -1 & 1 & 1 & -1 & -1 & 1 \\ 1 & 1 & -1 & 1 & -1 & -1 & -1 \\ -1 & 1 & -1 & -1 & 1 & -1 & 1 \\ -1 & -1 & -1 & 1 & 1 & 1 & -1 \\ 1 & -1 & -1 & -1 & -1 & 1 & 1 \end{bmatrix}.$$

A simple computation yields the expression

$$f(b_1, b_2, b_3) = b_2 - 2b_1 b_3.$$

A simple implementation of the last equation as a transmitter is illustrated in Fig. 3.18. Note that this figure represents the *combined* modulation/coding process with

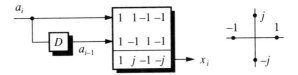

FIGURE 3.19 TCM scheme.

out the necessity of separating it into its component parts, that is, a convolutional code and a memoryless mapper, as in Ungerboeck's representation. □

EXAMPLE 3.2

Consider the TCM scheme represented in Fig. 3.19. We can write

$$f(b_1, b_2) = A + Bb_1 + Cb_2 + Db_1b_2.$$

The expansion coefficients A, B, C, D can be obtained by solving the system of linear equations

$$A + B + C + D = 1$$
$$A + B - C - D = j$$
$$A - B + C - D = -1$$
$$A - B - C + D = -j$$

which gives $A = C = 0, B = (1 + j)/2, D = (1 - j)/2$. Hence

$$f(b_1, b_2) = \frac{1+j}{2} b_1 + \frac{1-j}{2} b_1 b_2.$$ □

3.9 Decoding TCM

In this section we consider the Viterbi algorithm as applied to decoding trellis-coded modulation (TCM) signals. If the TCM signal is described by using a trellis, whose branches are associated with transitions between encoder states and with signals transmitted over the channel, the task of the TCM decoder is to estimate the path that the encoded signal sequence traverses through the trellis. This is done by associating with each branch of the trellis a number, called the *branch metric,* and looking for the path whose total metric is minimum. This path corresponds to the transmitted signal sequence. Thus the decoding problem can be split in two parts:

- Definition of a branch metric, and its computation based on the observed values of the received signal.
- Evaluation of the minimum-metric path.

In the following we consider these two problems separately.

3.9.1 Definition of Branch Metric

Maximum-likelihood demodulation

Consider a sequence of M-ary symbols to be transmitted, say,

$$\mathbf{x} = (x_0, x_1, \ldots, x_{K-1})$$

where each x_i can take on M values. These symbols are used to modulate a signal $s(t)$, which is sent through the channel. Thus the transmitted signal can be given the form

$$v(t) = \sum_{k=0}^{K-1} x_k s(t - kT) \tag{3.14}$$

where, as usual, T denotes the symbol period (equivalently, $1/T$ denotes the baud rate). If the channel effect is to add the random noise $n(t)$ to the transmitted signal, we can write the received signal in the form

$$y(t) = v(t) + n(t)$$
$$= \sum_{k=0}^{K-1} x_k s(t - kT) + n(t).$$

The task of the demodulator is to process the observed signal $y(t)$ in order to produce an estimate $\hat{\mathbf{x}}$ of the transmitted symbol sequence $\mathbf{x}$,

$$\hat{\mathbf{x}} = (\hat{x}_0, \hat{x}_1, \ldots, \hat{x}_{K-1}).$$

We are obviously interested in *optimum* processing, where by "optimum" we mean "resulting in a transmission scheme whose error probability is minimized."

If we make the assumption that all the information symbols are equally likely (i.e., are transmitted with equal probabilities), then the probability of erroneously demodulating the sequence is minimized by processing $y(t)$ according to the so-called *maximum-likelihood* procedure. This procedure compares the conditional probabilities of the received signal given each possible transmitted sequence and chooses the transmitted sequence corresponding to the largest among them. That is, the decoder chooses $\hat{\mathbf{x}}$ if

$$P[y(t)|\hat{\mathbf{x}}] = \max_{\text{all } \mathbf{x}} P[y(t)|\mathbf{x}].$$

The actual form of these conditional probabilities (the *likelihood functions*) depends on the channel. For the additive white Gaussian noise channel, by rewriting $y(t)$ as a sum of signal "chips," each with duration T seconds:

$$y(t) = \sum_{k=0}^{K-1} Y(t - kT)$$

we have

$$\log P[y(t)|\mathbf{x}] = \sum_{k=0}^{K-1} \log P[Y(t-kT)|x_i] = C - \sum_{k=0}^{K-1} \|Y(t-kT) - x_k s(t-kT)\|^2$$

where we define

$$\|f(t)\|^2 = \int |f(t)|^2 \, dt$$

and C is a constant that can be disregarded in the maximization. In conclusion, maximum-likelihood demodulation of a sequence of signals $x_k, k = 0, \ldots, K-1$, is equivalent to the minimization, with respect to all the possible sequences, of the quantity, called the *metric* associated with $y(t)$:

$$m[y(t), \mathbf{x}] = \sum_{k=0}^{K-1} \|Y(t - kT) - x_k s(t - kT)\|^2. \quad (3.15)$$

Symbol-by-symbol demodulation

Consider first the special case of *uncoded* symbols x_k. In this situation all the terms $\|Y(t - kT) - x_k s(t - kT)\|^2$ in (3.15) turn out to be *independent*, so that

$$\min_{\mathbf{x}} m[y(t), \mathbf{x}] = \sum_{k=0}^{K-1} \min_{x_k} \|Y(t - kT) - x_k s(t - kT)\|^2 \quad (3.16)$$

that is, the minimum of the sum turns out to be equal to the sum of the minima. We are in a situation called *symbol-by-symbol demodulation*, in which separate demodulation of the symbol transmitted in each interval is indeed optimum. The demodulation is done by using the *minimum-distance rule*, that is, by looking for the value of x_k such that the *Euclidean distance*

$$\|Y(t - kT) - x_k s(t - kT)\|^2$$

between the observed signal $Y(t - kT)$ and the candidate signal $x_k s(t - kT)$ is a minimum.

In practice, we may want to consider only *signal samples* instead of continuous signals. The resulting simplicity is, of course, paid for in terms of a (possibly slight) loss of optimality if the channel is not ideal. In this case we have to minimize

$$\|y_k - x_k s_k\|^2$$

or if $s(t)$ is such that $s_k = 1$, simply

$$\|y_k - x_k\|^2.$$

The last quantity is the *metric* associated with the observed sample y_k and the transmitted symbol x_k. The task of the demodulator is to observe y_k and to compute the distance between it and each candidate x_k. The closest x_k is the demodulator's choice.

Notice also that we can write, for complex (i.e., two-dimensional) signals:

$$\|y_k - x_k\|^2 = \|y_k\|^2 - 2\mathbf{R}[y_k x_k^*] + \|x_k\|^2$$

where $\mathbf{R}$ denotes real part and the superscript * denotes complex conjugate. Since we must minimize this with respect to x_k, we can disregard $\|y_k\|^2$, and consider the metric

$$-2\mathbf{R}[y_k x_k^*] + \|x_k\|^2.$$

Optimum demodulation of correlated symbols

With TCM, the symbols x_k, instead of forming a sequence of independent random variables, are *interrelated*. An immediate result of this interrelation is that (3.16) does not hold anymore, forcing us to give up symbol-by-symbol demodulation. The reason for the lack of optimality of symbol-by-symbol demodulation of correlated symbols can be understood as follows. If the constraints created by the encoder on the symbol sequence are not taken into account, the demodulated sequence $\hat{x}_0, \ldots, \hat{x}_{K-1}$ *may not be a valid one*.

Now, how can we find the symbol sequence minimizing the metric if symbol-by-symbol demodulation is not allowed? A brute-force approach is the following. We observe that the number of sequences of length K that can possibly be transmitted is *finite* and does not exceed M^K (M is the number of values taken on by each symbol x_k). Thus we may just *compute the metric for all possible symbol sequences and choose the symbol sequence giving the minimum value*. This procedure has the following drawbacks:

- *Delay.* To demodulate the symbol sequence, we must wait until the whole signal $y(t)$ associated with it has been observed. This entails a delay of at least KT seconds, which may not be acceptable. In fact, usually K is very large, corresponding to the entire time that communication takes place (minutes, hours, ...).
- *Storage.* The size of the memory required to store all the possible transmitted sequences may be huge (up to M^K sequences must be stored).
- *Complexity.* To select the minimum metric, a very large number of comparisons must be performed.

In conclusion, the brute-force approach is not practical. As we shall see in the next chapter, the solution to our minimization problem with encoded symbols can be obtained by using the *Viterbi algorithm*, a technique whereby *the number of computations necessary for the selection of the optimum symbol sequence, as well as the memory size, grow only linearly with the sequence length K*, rather than exponentially. However, the Viterbi algorithm does not automatically solve the delay problem: This is due to the fact that to minimize a function we should wait until all of it has been observed. Thus, to solve the delay problem, we shall resort to a *suboptimum* version of the Viterbi algorithm, called the *truncated Viterbi algorithm*.

3.9.2 The Viterbi Algorithm

The Viterbi algorithm was proposed in 1967 for decoding convolutional codes. Shortly after its invention, it was observed that the algorithm is based on the principles of *dynamic programming*, a general technique for solving minimization problems (see, e.g., [6] or [1, App. F and pp. 342–347]).

Our application of the Viterbi algorithm to decode TCM is based on the trellis diagram representing the TCM encoder. For the moment, assume for simplicity that there is only one signal associated with each branch of the trellis (i.e., no parallel transitions). In this situation we have a one-to-one correspondence between the sequence of source symbols and the path through the trellis. Also, assume sampled signals with $s_k = 1$ and real or complex signals. The symbol x_k, or, equivalently, a branch of the trellis, is transmitted every T seconds.

The Viterbi algorithm finds the path with the minimum path metric by sequentially moving through the trellis stage by stage. In every stage [i.e., in every T-second interval, say, from kT to $(k + 1)T$], the receiver observes the sample y_k received in that interval and computes the metric $\|y_k - x_k\|^2$ associated with all the branches. For example, if an eight-state TCM scheme transmits 2 bits per channel use, there are eight nodes and four branches emanating from each node, for a total of 32 branch metrics. However, x_k takes on only eight values, so that only *eight* metrics must be computed.

The path metric for a particular path through a given node is the sum of the *partial path metrics* for the portion of the path to the left of the node and the portion to the right of the node. Among the possible paths to the left, the demodulator will prefer the one with the smallest partial path metric, called the *survivor path* for that node. *We remove from consideration all partial paths to the left other than the survivor path, and for each node we store the survivor path and its partial metric.*

At each stage of the trellis we do not know which node the optimal path passes through, so we keep one *survivor* path for each and every node. To get this survivor path for a given node at stage k, we look at all the branches leading to that node from stage $k - 1$, and at the partial path metrics of the nodes from which they stem. We determine the partial path metric for each of those paths by summing the partial path metric of the survivor at time $k - 1$ and the branch metric. The survivor path at that node is chosen as the path with the smallest path metric (Fig. 3.20).

We must store, for each node at stage k, the survivor path and the associated path metric. Then the algorithm proceeds to time $k + 1$ by computing the new branch metrics, and so on.

EXAMPLE 3.3

We illustrate the Viterbi algorithm by using a simple example. Consider the trellis shown in Fig. 3.21. The figure also shows the branch metrics, the survivor paths at each node, and the partial path metric of each surviving path. □

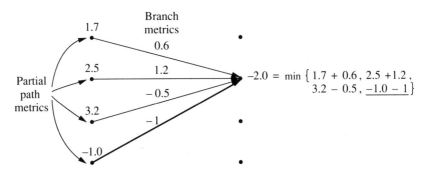

FIGURE 3.20 One step of the Viterbi algorithm.

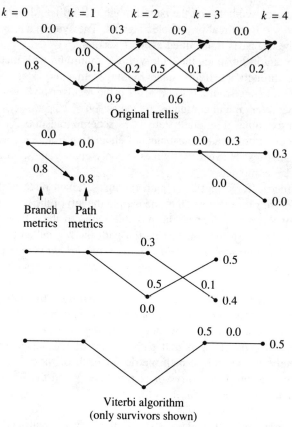

FIGURE 3.21 Two-state trellis with the branch metrics of the transitions marked and the Viterbi algorithm illustrated. The Viterbi algorithm iteratively finds the path with the minimum path metric.

In the presence of parallel transitions, that is, when more than one signal (a "subconstellation") is associated with the transition from one trellis state to the next, a minor modification should be made in the algorithm. In this situation, the branch metric computation is based on the preliminary detection of the symbol x_k in the branch subconstellation that lies closest to the received signal y_k. Its metric can be used thereafter for that branch, and the Viterbi algorithm can proceed conventionally.

The computational complexity of the Viterbi algorithm is the same at each stage, and hence *its computational complexity and storage requirements are proportional to the length of the transmitted sequence K*. One practical problem remains unsolved, however: the delay problem. In fact, the algorithm does not reach a conclusion on the entire symbol sequence until the end of the transmission. For very long symbol sequences we cannot afford either the resulting long delay in making decisions or the very large memory that would be required.

For these reasons, it is usual to make a modification of the algorithm to render it practically implementable. In practice, decisions must be made *before* the entire sequence y_k, $k = 0, \ldots, K - 1$, has been observed. In other words, it is necessary to *force* decisions at stage k on all paths prior to stage $k - d$, for some *truncation depth d*. The usual approach is to compare all the partial path metrics for the partial paths at stage k and note which is smallest. The demodulated input symbol x_{k-d} is then the symbol associated with the branch of this path at stage $k - d$. If d is chosen to be large enough, analysis and computer simulations have shown that the degradation in error probability due to these premature decisions becomes irrelevant.

The advantage of this *truncated Viterbi algorithm* is twofold:

- It solves the delay problem. Instead of waiting until the entire sequence has been received to make a decision, it introduces a fixed delay, and one demodulated symbol is output every T seconds.
- It solves the memory problem. At stage k, only the last d branches of the survivor paths must be kept in the memory.

The price to be paid is, of course, a loss of optimality, resulting in an increase of error probability and the fact that the sequence of demodulated symbols $\hat{x}_k$ is not necessarily a code sequence.

It should be observed here that this delay can create problems if an adaptive equalizer is used. In fact, the decisions that would be used in the adaptation algorithm are returned to the equalizer after a number of signaling intervals. To avoid this delay, a possible solution is to use *preliminary* (i.e., symbol-by-symbol) decisions to adapt the equalizer. Although these decisions are less reliable than the final decisions taken by the Viterbi processor, it is expected that a large enough percentage of them will be correct, so as to take the equalizer to a correct steady state.

3.10 Bibliographical Notes

In a seminal 1974 paper, Massey [7] formally suggested that the performance of a digital communication system could be improved by looking at modulation and coding as a *combined* entity rather than as two separate operations. The basic principles of trellis-coded modulation were described by Ungerboeck in his classic 1982 paper [2], although the idea of TCM was presented in an earlier conference [8] (see also [3] and [9]). At about the same time, Imai and Hirakawa published a paper [10] in which block and convolutional codes were combined with modulation.

TCM experienced a fast transition from research to practical applications. In 1984, a generation of modems using TCM became available, achieving reliable transmission at speeds of 14.4 kbits/s on private-line modems and 9.6 kbits/s on switched-network modems. A TCM scheme with an asymptotic coding gain of 4 dB [5] was adopted in 1984 by the CCITT (International Telegraph and Telephone Consultative Committee) for use in voiceband modems [11, 12] (see also [13]). In late 1985, a TCM-based modem operating at 19.2 kbits/s was introduced [14].

APPENDIX 3A: ORTHOGONAL EXPANSION OF THE FUNCTION f

In this appendix we describe a general representation for functions that depend on a finite number of arguments, which in turn can take on a finite number of values. This representation, which turns out to be a generalization of the one described by Agazzi et al. in [15] and used by Calderbank and Mazo in [4], allows us to provide a parametric representation for the function $f(\cdot)$ introduced in Section 3.8.2, and generalizes the discussion contained there. The expansion studied here is in the form of an orthogonal *Volterra series*.

Let $\{a_i\}_{i=1}^{n}$ be a set of independent, identically distributed random variables taking on K real values from the set $\mathcal{A}$. Let $a = (a_i, \ldots, a_n)$. Any function $f(a)$ can be expanded in a series of orthogonal polynomials as follows:

$$f(a) = k^{(0)}Q^{(0)} + \sum_{i=0}^{n} k_i^{(1)} Q_i^{(1)}(a) + \sum_{i=0}^{n}\sum_{j=0}^{m} k_{ij}^{(2)} Q_{ij}^{(2)}(a) + \cdots \quad (3A.1)$$

where $Q_{i_1 \ldots i_\ell}^{(\ell)}(a)$ is an ℓ-degree polynomial in the variables $a_{i_1}, \ldots, a_{i_\ell}$ which satisfies the following orthogonality conditions:

$$E\left[Q_{i_1 \ldots i_\ell}^{(\ell)}(a) Q_{j_1 \ldots j_m}^{(m)}(a)\right] = 0 \quad (3A.2)$$

if $\ell \neq m$, or $\ell = m$ but $j_1, \ldots, j_m$ is not a permutation of $i_1, \ldots, i_\ell$. Here E denotes expectation taken with respect to the random vector a. (By normalizing the Q-polynomials, it is obviously possible to have an *orthonormal* expansion. In the following, unless the contrary is explicitly stated, we shall assume that the Q-polynomials are orthonormal.)

Now, recall that the random variables $a_i, i = 1, \ldots, n$ take on a *finite* number K of values from the set $\mathcal{A}$. The following theorem holds.

Theorem 3.A.1 The right-hand side of (3A.1) includes no more than K^n terms.

Proof. The proof of the theorem is constructive and is based on the derivation of the expansion (3A.1). We first show how to construct the orthogonal polynomials $Q_{i_1 \ldots i_\ell}^{(\ell)}(a)$; then, we show that there are no more than K^n of them.

Denote by $j_1, \ldots, j_s$, $s \leq \ell$, the distinct indices obtained from $i_1, \ldots, i_\ell$ after eliminating repetitions. For $k = 1, \ldots, s$, then let ν_k denote the number of i-indices equal to j_k. Hence let

$$Q_{i_1 \ldots i_\ell}^{(\ell)}(a) = P_{\nu_1}(a_1) \cdots P_{\nu_s}(a_s) \quad (3A.3)$$

where $P_{\nu_i}(\cdot), i = 1, \ldots, s$, are polynomials of degree ν_i orthogonal with respect to the random variables a_m:

$$E[P_{\nu_i}(a_m) P_{\nu_k}(a_m)] = 0, \quad \nu_i \neq \nu_k, \quad 1 \leq m \leq n. \quad (3A.4)$$

For example, we have

$$Q_{112}^{(3)}(a) = P_2(a_1) P_1(a_2)$$

and
$$Q^{(4)}_{1123}(\mathbf{a}) = P_2(a_1)P_1(a_2)P_1(a_3).$$

By substituting (3A.3) into (3A.2), one can see that the orthogonality condition is satisfied.

Consider then the derivation of the polynomials $P(\cdot)$. This can be obtained by exploiting an observation of Zadeh [16]. Our starting point will be a sequence of linearly independent monomials $f_0(a), f_1(a), \ldots$ Then we have explicitly, through a process equivalent to Gram–Schmidt orthogonalization,

$$P_0(a) = f_0(a)$$

and

$$P_k(a) = \det \begin{bmatrix} f_k(a) & f_{k-1}(a) & \cdots & f_0(a) \\ E[f_{k-1}(a)f_k(a)] & E[f^2_{k-1}(a)] & \cdots & E[f_{k-1}(a)f_0(a)] \\ & & \cdots & \\ E[f_0(a)f_k(a)] & E[f_0(a)f_{k-1}(a)] & \cdots & E[f^2_0(a)] \end{bmatrix}$$

for $k = 1, 2, \ldots$.

Since a can take on K values, the monomials

$$1, a, a^2, \ldots, a^{K-1}$$

are linearly independent. In fact, if $P(a)$ is an arbitrary real polynomial, the equality $E[P(a)]^2 = 0$ is possible only if $P(a)$ vanishes at the K points corresponding to the values taken by a. But if $P(a)$ has degree $K - 1$, this is possible only if it vanishes identically.

The number of orthonormal P-polynomials cannot exceed K (see [17, p. 23]). Thus the Q-polynomials are obtained by forming products of P-polynomials, chosen by combining no more than one element in each one of the following lines:

$$1$$
$$P_1(a_1), P_2(a_1), \ldots, P_{K-1}(a_1)$$
$$P_1(a_2), P_2(a_2), \ldots, P_{K-1}(a_2)$$
$$\cdots$$
$$P_1(a_n), P_2(a_n), \ldots, P_{K-1}(a_n).$$

Now, a single element can be chosen in $1 + n(K - 1)$ ways. Two elements can be chosen from the last n lines in $\binom{n}{2}(K - 1)^2$ ways, ..., n elements can be chosen from the last n lines in $\binom{n}{n}(K - 1)^2$ ways. By using Newton's formula, we see that the total number of choices, and hence the total number of Q-polynomials, is indeed K^n. □

Corollary 3A.1 It is seen from the expression for $P_k(a)$ that with the choice $f_i(a) = a^i$ the computation of the P-polynomials, and hence of the Q-polynomials, involves only a finite number of *moments* of the random variable a. □

Corollary 3A.2 With the choice $f_i(a) = a^i$ all the P-polynomials have zero mean, with the only exception of $P_0(a) = 1$. In fact, for orthogonality we must have, if $\nu \neq 0$,

$$E[P_\nu(a)] = E[P_\nu(a)P_0(a)] = 0. \qquad \square$$

Corollary 3A.3 The coefficients $k_{i_1\ldots i_\ell}^{(\ell)}$ in expansion (3A.1) can be evaluated using the following formula, which derives from the orthonormality of the Q-polynomials:

$$k_{i_1\ldots i_\ell}^{(\ell)} = E[f(\mathbf{a})Q_{i_1\ldots i_\ell}^{(\ell)}(\mathbf{a})]. \tag{3A.5}$$

$\square$

EXAMPLE 3A.1

Let $\mathcal{A} = \{0, 1\}$. Take $f_i(a) = a^i$, and observe that since for $n > 1$ we have $a^n = a$, only $f_0(a)$ and $f_1(a)$ are linearly independent. Thus the only P-polynomials are $P_0(a) = 1$ and $P_1(a) = a - p$, where p denotes the expectation of a. Hence for any n we have

$$f(a_1, \ldots, a_n) = k^{(0)} + \sum_{i=1}^n k_i^{(1)}(a_i - p) + \sum_{j=1}^n \sum_{\substack{i=1 \\ i \neq j}}^n k_{ij}^{(2)}(a_i - p)(a_j - p) + \cdots$$

$$+ k_{12\ldots n}^{(n)}(a_1 - p)(a_2 - p)\cdots(a_n - p). \qquad \square$$

EXAMPLE 3A.2

Let $\mathcal{A} = \{-1, 0, 1\}$. Since we have $a^3 = a$, the only P-polynomials are $P_0 = 1$, $P_1(a) = a$, and $P_2(a) = (a^2 - \mu_2)$, where $\mu_2 = E[a^2] = 2/3$. $\square$

EXAMPLE 3A.3

Let $\mathcal{A} = \{-3, -1, 1, 3\}$. There are four P-polynomials: $P_0 = 1$, $P_1(a) = a$, $P_2(a) = a^2 - \mu_2$, and $P_3(a) = a^3 - (\mu_4/\mu_2)a$. Here $\mu_4 = E[a^4] = 41$ and $\mu_2 = E[a^2] = 5$. $\square$

If the K^n values taken on by the function f are arranged in a column vector $\mathbf{f}$, the K^n k-coefficients are arranged in a similar vector $\mathbf{k}$, and the K^{2n} values taken on by the Q-polynomials are arranged in a $K^n \times K^n$ matrix $\mathbf{Q}$, expansion (3A.1) can be given the matrix form

$$\mathbf{f} = \mathbf{Q}\mathbf{k}. \tag{3A.6}$$

Assume that the values taken by the random variables a_i, $i = 1, \ldots, n$, are equally likely. Then from the orthonormality of the Q-polynomials it follows that the matrix $\mathbf{Q}$ has the property

$$\mathbf{Q}^T\mathbf{Q} = K^n\mathbf{I} \tag{3A.7}$$

where **I** denotes the identity matrix. Hence, by premultiplying (3A.6) by $\mathbf{Q}^T$, we obtain

$$\mathbf{k} = K^{-n}\mathbf{Q}^T\mathbf{f} \tag{3A.8}$$

which is the matrix equivalent of (3A.5). As observed in [15], the last equation can be interpreted by viewing the vector of coefficients on the right-hand side as a discrete transform of the vector **f**. The type of transformation depends on the matrix **Q**, that is, on the distribution of the random variable a on which the orthogonal polynomials ultimately depend.

REFERENCES

1. S. BENEDETTO, E. BIGLIERI, and V. CASTELLANI, *Digital Transmission Theory.* Prentice-Hall, Englewood Cliffs, N.J., 1987.
2. G. UNGERBOECK, "Channel coding with multilevel/phase signals," *IEEE Trans. Inf. Theory*, Vol. IT-28, pp. 56–67, Jan. 1982.
3. G. UNGERBOECK, "Trellis-coded modulation with redundant signal sets. Part I: Introduction; Part II: State of the art," *IEEE Commun. Mag.*, Vol. 25, No. 2, pp. 5–21, Feb. 1987.
4. A. R. CALDERBANK and J. E. MAZO, "A new description of trellis codes," *IEEE Trans. Inf. Theory*, Vol. IT-30, pp. 784–791, Nov. 1984.
5. L. F. WEI, "Rotationally invariant convolutional channel coding with expanded signal space. Part II: Nonlinear coding," *IEEE J. Select. Areas Commun.*, Vol. SAC-2, pp. 672–686, 1984.
6. G. D. FORNEY, JR., "The Viterbi algorithm," *IEEE Proc.*, Vol. 61, pp. 268–278, 1973.
7. J. L. MASSEY, "Coding and modulation in digital communications," *1974 International Zurich Seminar on Digital Communications*, Zurich, Switzerland, Mar. 1974.
8. G. UNGERBOECK and I. CSAJKA, "On improving data-link performance by increasing the channel alphabet and introducing sequence coding," *1976 International Symposium on Information Theory*, Ronneby, Sweden, June 1976.
9. G. D. FORNEY, JR., et al., "Efficient modulation for band-limited channels," *IEEE J. Select. Areas Commun.*, Vol. SAC-2, No. 5, pp. 632–647, Sept. 1984.
10. H. IMAI and S. HIRAKAWA, "A new multilevel coding method using error-correcting codes," *IEEE Trans. Inf. Theory*, Vol. IT-23, No. 3, pp. 371–377, May 1977.
11. CCITT, "A family of 2-wire, duplex modems operating at data signalling rates of up to 9600 bits/s for use on the general switched telephone network and on leased telephone-type circuits," *Recommendation V.32*, Málaga-Torremolinos, Spain, 1984.
12. CCITT STUDY GROUP XVII, "Recommendation V.32 for a family of 2-wire, duplex modems operating on the general switched telephone network and on leased telephone-type circuits," *Document AP VIII-43-E*, May 1984.

13. CCITT Study Group XVII, "Draft recommendation V.33 for 14400 bits per second modem standardized for use in point-to-point 4-wire leased telephone-type circuits," *Circular 12, COM XVII/YS,* Geneva, Switzerland, May 17, 1985.
14. E. von Taube and K. W. Seitz, "Modem design boosts data processing throughput," *Comput. Des.,* pp. 71–74, Apr. 15, 1986.
15. O. Agazzi, D. G. Messerschmitt, and D. A. Hodges, "Nonlinear echo cancellation of data signals," *IEEE Trans. Commun.,* Vol. COM-30, No. 11, pp. 2421–2433, Nov. 1982.
16. L. A. Zadeh, "On the representation of nonlinear operators," *IRE Westcon Conv. Rec.,* Pt. 2, pp. 105–113, 1957.
17. G. Szegö, *Orthogonal Polynomials,* 3rd ed. Colloquium Publications, 23, American Mathematical Society, Providence, R.I., 1967.

CHAPTER 4

Performance Evaluation

In this chapter we consider the error probability, the free distance, and the power density spectrum of TCM schemes. The most common parameter used for the evaluation of the quality of a digital communication system is error probability, so our presentation will be centered on its evaluation. We shall discover that, asymptotically, the error probability is upper- and lower-bounded by a function that decreases monotonically when d_{free} increases. This fact shows that the free Euclidean distance is the most significant single parameter useful for comparing TCM schemes employed for transmission over the additive white Gaussian noise channel when the signal-to-noise ratio is large enough. This also explains why the increase from minimum distance to free distance caused by the introduction of TCM was called *asymptotic coding gain*.

Since there appears to be no general technique for choosing an optimum TCM scheme, the selection of any such scheme is typically based on a search among a wide subclass. Thus it is extremely important that computationally efficient algorithms for the computation of free distance and error probability be available. Interest in TCM schemes with a large number of states fosters the investigation of sufficient conditions for the performance of a specific TCM scheme to be evaluated using algorithms that are as simple as possible. We shall describe some simple conditions that allow a considerable simplification in the algorithms for the computation of both error probability and free distance.

4.1 Upper Bound to Error Probability

Consider again the Ungerboeck model for rate $m/(m+1)$ TCM schemes, as shown in Fig. 4.1. Here a linear, rate $m/m + 1$ convolutional code accepts binary source symbols m at a time and transforms them into blocks $\mathbf{c}_i$ of $m + 1$ binary symbols that are fed into a memoryless mapper $f(\cdot)$. This mapper outputs channel symbols x_i. The binary $(m + 1)$-tuples $\mathbf{c}_i$ will often be referred to as the *labels* of the signals x_i.

Since there is a one-to-one correspondence between x_i and its label $\mathbf{c}_i$, an *error event of length L* can be equivalently described by giving two L-tuples of coded symbols or two L-tuples of labels

$$\mathbf{c}_k, \mathbf{c}_{k+1}, \ldots, \mathbf{c}_{k+L-1}$$

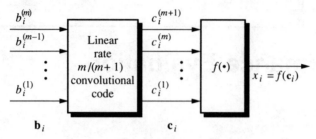

FIGURE 4.1 Model of a TCM encoder.

and

$$\mathbf{c}'_k = \mathbf{c}_k \oplus \mathbf{e}_k, \quad \mathbf{c}'_{k+1} = \mathbf{c}_{k+1} \oplus \mathbf{e}_{k+1}, \quad \ldots, \quad \mathbf{c}'_{k+L-1} = \mathbf{c}_{k-L+1} \oplus \mathbf{e}_{k+L-1}$$

where $\mathbf{e}_i$, $i = k, \ldots, k+L-1$, form a sequence of binary error vectors.

Now, let $\mathbf{X}_L$ and $\mathbf{X}'_L$ denote two sequences of x-symbols with length L. An *error event* of length L occurs when the demodulator chooses, instead of the transmitted sequence $\mathbf{X}_L$, another sequence $\mathbf{X}'_L$ of channel symbols corresponding to a trellis path that splits from the correct path at a given time, and remerges exactly L discrete times later. The error probability is then obtained by summing over L, $L = 1, 2, \ldots$, the probabilities of error events of length L (i.e., the joint probabilities that $\mathbf{X}_L$ be transmitted and $\mathbf{X}'_L$ detected).

The union bound provides the following inequality for the probability of an error event:[1]

$$P(e) \leq \sum_{L=1}^{\infty} \sum_{\mathbf{X}_L} \sum_{\mathbf{X}'_L \neq \mathbf{X}_L} P[\mathbf{X}_L] P[\mathbf{X}_L \to \mathbf{X}'_L] \tag{4.1}$$

where $P[\mathbf{X}_L \to \mathbf{X}'_L]$ denotes the "pairwise" error probability (i.e., the probability that when $\mathbf{X}_L$ is transmitted $\mathbf{X}'_L$ be preferred to $\mathbf{X}_L$ by the demodulator).

Since we assume a one-to-one correspondence between output symbols and labels, by letting $\mathbf{C}_L$ denote an L-sequence of labels $\mathbf{c}_i$ and $\mathbf{E}_L$ an L-sequence of error vectors $\mathbf{e}_i$, we can rewrite (4.1) in the form

$$P(e) \leq \sum_{L=1}^{\infty} \sum_{\mathbf{C}_L} P[\mathbf{C}_L] \sum_{\mathbf{C}'_L \neq \mathbf{C}_L} P[\mathbf{C}_L \to \mathbf{C}'_L]$$

$$= \sum_{L=1}^{\infty} \sum_{\mathbf{C}_L} P[\mathbf{C}_L] \sum_{\mathbf{E}_L \neq 0} P[\mathbf{C}_L \to \mathbf{C}_L \oplus \mathbf{E}_L] \tag{4.2}$$

$$= \sum_{L=1}^{\infty} \sum_{\mathbf{E}_L \neq 0} P[\mathbf{E}_L]$$

---

[1] Strictly speaking, unless the transmission is assumed to start at time $-\infty$ and end at time $+\infty$, this probability is a function of the discrete time k at which the error event starts. We assume that the transmission is long enough, and disregard this difficulty.

where

$$P[\mathbf{E}_L] = \sum_{\mathbf{C}_L} P[\mathbf{C}_L] P[\mathbf{C}_L \to \mathbf{C}_L \oplus \mathbf{E}_L]. \tag{4.3}$$

Equation (4.3) expresses the average pairwise probability of an error event of length L caused by the error sequence $\mathbf{E}_L$.[2]

Now recall that the transmission takes place over the additive white Gaussian noise channel with noise power spectral density $N_0/2$, and the detection is maximum likelihood. With this channel model we can upper bound $P[\mathbf{C}_L \to \mathbf{C}'_L]$ by using the *Bhattacharyya bound*: We have

$$P[\mathbf{C}_L \to \mathbf{C}'_L] \le \exp\left\{-\frac{1}{4N_0}\|f(\mathbf{C}_L) - f(\mathbf{C}'_L)\|^2\right\} \tag{4.4}$$

$$= \exp\left\{-\frac{1}{4N_0}\sum_{n=1}^{L}\|f(\mathbf{c}_n) - f(\mathbf{c}'_n)\|^2\right\}.$$

Now define the function

$$W(\mathbf{E}_L) = \sum_{\mathbf{C}_L} P[\mathbf{C}_L] Z^{\|f(\mathbf{C}_L) - f(\mathbf{C}_L \oplus \mathbf{E}_L)\|^2} \tag{4.5}$$

where $Z = \exp(-1/4N_0)$ (this is the Bhattacharyya bound to $P[\mathbf{E}_L]$). By observing that $P[\mathbf{C}_L] = P[\mathbf{X}_L]$, (4.1) can be rewritten in the form

$$P(e) \le \sum_{L=1}^{\infty}\sum_{\mathbf{E}_L \neq \mathbf{0}} W(\mathbf{E}_L). \tag{4.6}$$

Equation (4.6) shows that $P(e)$ is upper-bounded by a sum, over the possible error events, of functions of the vectors $\mathbf{E}_L$ causing the error events. Our next task toward the evaluation of $P(e)$ will be to enumerate these vectors. We shall do this by using the transfer function of an *error-state diagram*, which is a graph whose branches have matrix labels. Specifically, recall that under our assumptions the source symbols have equal probabilities $1/M = 2^{-m}$, and define the $N \times N$ *error weight matrices* $\mathbf{G}(\mathbf{e}_i)$ as follows. The entry p, q of $\mathbf{G}(\mathbf{e}_i)$ is *zero* if no transition from the code trellis state p to the state q is possible. Otherwise, it is given by

$$[\mathbf{G}(\mathbf{e}_i)]_{p,q} = \frac{1}{M}\sum_{\mathbf{c}_{p\to q}} D^{\|f(\mathbf{c}_{p\to q}) - f(\mathbf{c}_{p\to q}\oplus \mathbf{e}_i)\|^2} \tag{4.7}$$

where D is an indeterminate and $\mathbf{c}_{p\to q}$ are the label vectors generated by the transition from p to q. (The sum accounts for the possible parallel transitions between states.)

---

[2] Observe that in the error analysis of linear convolutional codes all the terms in (4.3) contribute equally to $P[\mathbf{E}_L]$. For this reason it is customary to assume, without any loss of generality, that the all-zero encoded symbol sequence was transmitted over the channel. Since TCM schemes include the *nonlinear* mapper $f(\cdot)$, it may not be possible to make this assumption here, at least without a careful analysis of the scheme. Later on we shall study conditions for this simplification to be valid for TCM.

With these notations, to any sequence $\mathbf{E}_L = \mathbf{e}_1, \ldots, \mathbf{e}_L$ there corresponds a sequence of L error weight matrices $\mathbf{G}(\mathbf{e}_1), \ldots, \mathbf{G}(\mathbf{e}_L)$, and we have

$$W(\mathbf{E}_L) = \frac{1}{N}\mathbf{1}^T \prod_{n=1}^{L} \mathbf{G}(\mathbf{e}_n)\mathbf{1} \tag{4.8}$$

where $\mathbf{1}$ is the column N-vector all of whose elements are 1 (consequently, for any $N \times N$ matrix $\mathbf{A}$, $\mathbf{1}^T\mathbf{A}\mathbf{1}$ is the sum of all the entries of $\mathbf{A}$). It should be apparent that the element p, q of the matrix $\prod_{n=1}^{L} \mathbf{G}(\mathbf{e}_n)$ enumerates the Euclidean distances involved in the transition from state p to state q in exactly L steps. Thus what we need next to compute $P(e)$ is to sum $W(\mathbf{E}_L)$ over the possible error sequences $\mathbf{E}_L$, according to (4.6).

4.1.1 The Error-State Diagram

We observe that the error vectors in the sequence $\mathbf{e}_1, \ldots, \mathbf{e}_L$ are not independent. Thus to compute (4.6), we need a model accounting for their dependence. This task will be accomplished by describing the connections between error vectors $\mathbf{e}_n$ through a state diagram. This turns out to be a copy of the one describing the code; we shall call it the *error-state diagram*, as distinguished from the code-state diagram.

The error-state diagram has a structure determined only by the linear convolutional code of our model and differs from the code-state diagram only in the denomination of its states and in the branch labels, which are the matrices $\mathbf{G}(\mathbf{e}_i)$. To prove this, observe that due to the linearity of the convolutional code, the vectors $\mathbf{c}$ at its output form an additive commutative group $\mathscr{C}$. Also, the set of possible vectors $\mathbf{c}$ corresponding to the all-zero state forms a *subgroup,* and the vectors $\mathbf{c}$ corresponding to any other state form a coset of that group (see Lemma 4.1, p. 107). Now the error vectors $\mathbf{e}$ are differences of vectors $\mathbf{c}$, so they are elements of $\mathscr{C}$. Consequently, the connections among vectors $\mathbf{e}_i$ are the same as the connections among vectors $\mathbf{c}_i$.

4.1.2 The Transfer Function Bound

From (4.8) and (4.6) we have

$$P(e) \leq T(D)\big|_{D=e^{-1/4N_0}} \tag{4.9}$$

where

$$T(D) = \frac{1}{N}\mathbf{1}^T\mathbf{G}\mathbf{1} \tag{4.10}$$

and the matrix

$$\mathbf{G} = \sum_{L=1}^{\infty} \sum_{\mathbf{E}_L \neq 0} \prod_{n=1}^{L} \mathbf{G}(\mathbf{e}_n) \tag{4.11}$$

is the matrix transfer function of the error-state diagram. $T(D)$ will be called the (scalar) *transfer function* of the error-state diagram.

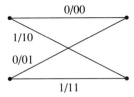

FIGURE 4.2 Trellis diagram for a two-state, $m = 1$ TCM scheme. The branch labels denote the source symbol and the components of the **c**-vector associated with it.

EXAMPLE 4.1

Consider the TCM scheme one section of whose trellis diagram is shown in Fig. 4.2. Here $m = 1$ and $M = 4$ (binary source, quaternary channel symbols). The error-state diagram is shown in Fig. 4.3 with the corresponding labels. If we denote the error vector as $\mathbf{e} = (e_2 e_1)$ and we let $\bar{e} = 1 \oplus e$ (i.e., $\bar{e}$ denotes the complement of e), then we can write the general form of the matrix $\mathbf{G}(\mathbf{e})$ as

$$\mathbf{G}(e_2 e_1) = \frac{1}{2} \begin{bmatrix} D^{\|f(00)-f(e_2 e_1)\|^2} & D^{\|f(10)-f(\bar{e}_2 e_1)\|^2} \\ D^{\|f(01)-f(e_2 \bar{e}_1)\|^2} & D^{\|f(11)-f(\bar{e}_2 \bar{e}_1)\|^2} \end{bmatrix}. \qquad (4.12)$$

The transfer function of the error-state diagram is

$$\mathbf{G} = \mathbf{G}(10)[\mathbf{I}_2 - \mathbf{G}(11)]^{-1} \mathbf{G}(01) \qquad (4.13)$$

where $\mathbf{I}_2$ denotes the 2×2 identity matrix.

Now (4.12) and (4.13) could be written without specifying the signals used in the TCM scheme. Actually, to give the signal constellation corresponds to specifying the four values taken on by the function $f(\cdot)$. These will provide the values of the entries of $\mathbf{G}(e_2 e_1)$ from which the transfer function $T(D)$ is computed.

Consider first quaternary PAM (Fig. 4.4). In this case we have

$$f(00) = 3, \quad f(01) = 1, \quad f(10) = -1, \quad f(11) = -3$$

and consequently,

$$\mathbf{G}(01) = \frac{1}{2} \begin{bmatrix} D^4 & D^4 \\ D^4 & D^4 \end{bmatrix}$$

$$\mathbf{G}(10) = \frac{1}{2} \begin{bmatrix} D^{16} & D^{16} \\ D^{16} & D^{16} \end{bmatrix}$$

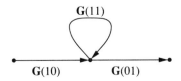

FIGURE 4.3 Error-state diagram for a two-state, $m = 1$ TCM scheme.

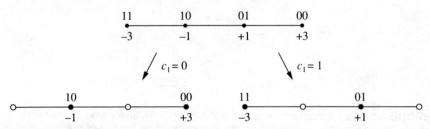

FIGURE 4.4 Signal constellation and set partition for 4-AM.

and
$$G(11) = \frac{1}{2}\begin{bmatrix} D^{36} & D^4 \\ D^4 & D^{36} \end{bmatrix}$$

so that from (4.13) we obtain
$$\mathbf{G} = \frac{1}{2}\frac{D^{20}}{1 - \frac{1}{2}(D^4 + D^{36})}\begin{bmatrix} 1 & 1 \\ 1 & 1 \end{bmatrix}. \tag{4.14}$$

In conclusion, we get the transfer function
$$T(D) = \frac{1}{2}\mathbf{1}^T \mathbf{G}\mathbf{1} = \frac{D^{20}}{1 - \frac{1}{2}(D^4 + D^{36})}. \tag{4.15}$$

If we consider a 4-PSK constellation with unit energy (Fig. 4.5), we have
$$f(00) = 1, \quad f(01) = j, \quad f(10) = -1, \quad f(11) = -j$$

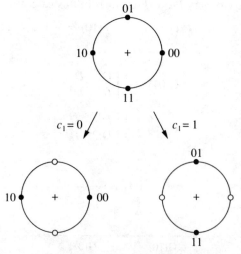

FIGURE 4.5 Signal constellation and set partition for 4-PSK.

so that

$$\mathbf{G}(01) = \frac{1}{2}\begin{bmatrix} D^2 & D^2 \\ D^2 & D^2 \end{bmatrix}$$

$$\mathbf{G}(10) = \frac{1}{2}\begin{bmatrix} D^4 & D^4 \\ D^4 & D^4 \end{bmatrix}$$

and

$$\mathbf{G}(11) = \frac{1}{2}\begin{bmatrix} D^2 & D^2 \\ D^2 & D^2 \end{bmatrix}.$$

In conclusion

$$\mathbf{G} = \frac{1}{2}\frac{D^6}{1-D^2}\begin{bmatrix} 1 & 1 \\ 1 & 1 \end{bmatrix} \qquad (4.16)$$

which gives the transfer function

$$T(D) = \frac{1}{2}\mathbf{1}^T\mathbf{G}\mathbf{1} = \frac{D^6}{1-D^2}. \qquad (4.17)$$

By substituting $D = e^{-1/4N_0}$ in (4.15) or in (4.17), we get an upper bound to symbol error probability. □

Interpretation and symmetry considerations

From (4.9) and its derivation, let us interpret the meaning of the matrix $\mathbf{G}$. We may observe that the entry p, q of $\mathbf{G}$ provides an upper bound to the probability that an error event occurs starting from node p and ending at node q. Similarly, $(1/N)\mathbf{G}\mathbf{1}$ is a vector whose pth entry is a bound to the probability of any error event starting from node p, and $(1/N)\mathbf{1}'\mathbf{G}$ is a vector whose qth entry is a bound to the probability of any error event ending at node q.

Inspection of the matrix $\mathbf{G}$ leads to the consideration of different degrees of symmetry implied in a TCM scheme. In some cases the matrix $\mathbf{G}$ has equal entries; this is the case of the 4-PSK example above. This fact can be interpreted by saying that *all the paths* in the trellis are on an equal footing (i.e., they contribute equally to the error probability).[3] Hence, in the analysis of the TCM scheme, we may take any single path as a reference and compute error probabilities by assuming any given transmitted sequence. A simple sufficient condition for this symmetry to hold is that all the matrices $\mathbf{G}(\mathbf{e})$ have equal entries. However, this condition is not necessary, as proved by examining the 4-PAM example above: It has $\mathbf{G}$ with equal entries, although the entries of $\mathbf{G}(11)$ are unequal.

If all the matrices $\mathbf{G}(\mathbf{e})$ have equal entries, then in the computation of the transfer function bound, the branches of the error-state diagram can simply be labeled by the common entries of these matrices, thus leading to a *scalar* transfer function.

---

[3] More precisely, we should say that they contribute equally to the *upper bound to* error probability. However, here and in the following we avoid making this distinction.

However, for the computations to be done in terms of scalars, it is not necessary to require such a high degree of symmetry. All that is needed is the looser symmetry that arises when *the sum of all the elements in any row (or column) of* **G** *does not depend on the row (or column) itself.* This symmetry corresponds to having *all the states* on an equal footing and allows consideration of *a single reference state rather than all the pairs of states* for the computation of error probabilities. More specifically, only the error events leaving from a fixed node (in the case of equal row sums) or reaching a fixed node (in the case of equal column sums) need be considered. This point is discussed and made more precise in the next paragraph.

Algebraic conditions for scalar transfer functions. We now state some simple conditions that, if satisfied, will make it possible to compute a transfer function bound based on *scalar,* rather than matrix, branch labels. Given a square matrix **A**, if **1** is an eigenvector of its transpose $\mathbf{A}^T$,

$$\mathbf{1}^T \mathbf{A} = \alpha \mathbf{1}^T$$

where α is a constant, the sum of its elements in a column does not depend on the column order. We call **A** *column-uniform.* Similarly, if **1** is an eigenvector of the square matrix **B**,

$$\mathbf{B1} = \beta \mathbf{1}$$

where β is a constant, the sum of its elements in a row does not depend on the row order. In this case we call **B** *row-uniform.*

Now, the product and the sum of two column- (row)-uniform matrices is itself column- (row)-uniform. For example, if $\mathbf{B}_1$ and $\mathbf{B}_2$ are row-uniform with eigenvalues β_1 and β_2, respectively, and we define $\mathbf{B}_3 = \mathbf{B}_1 + \mathbf{B}_2$ and $\mathbf{B}_4 = \mathbf{B}_1 \mathbf{B}_2$, we have

$$\mathbf{B}_3 \mathbf{1} = (\beta_1 + \beta_2)\mathbf{1}$$

and

$$\mathbf{B}_4 \mathbf{1} = \beta_1 \beta_2 \mathbf{1}$$

which show that $\mathbf{B}_3$ and $\mathbf{B}_4$ are also row-uniform, with eigenvalues $\beta_1 + \beta_2$ and $\beta_1 \beta_2$, respectively. Also, for an $N \times N$ matrix **A** that is either row- or column-uniform, we have

$$\mathbf{1}^T \mathbf{A} \mathbf{1} = N\alpha.$$

From the above it follows that if all the matrices **G(e)** are either row-uniform or column-uniform, the transfer function [which is a sum of products of error matrices, as seen explicitly in (4.11)] can be computed by using scalar labels on the branches of the error-state diagram. These labels are the sums of the elements in a row (column). In this case we say that the TCM scheme is *uniform.* By recalling the definition of the matrices **G(e)**, we have that **G(e)** is row-uniform if the transitions *stemming from* any node of the trellis carry the same set of labels, irrespective of the order of the transitions. It is column uniform if the transitions *leading* to any node of the trellis carry the same set of labels.

Conditions for uniformity

Simple conditions for uniformity, that is, for scalar labels in the error-state diagram, will now be made explicit.

We start with a lemma [1] we shall need later on.

Lemma 4.1 Denote by $\mathscr{C}_0$ the set of 2^m vectors **c** corresponding to the all-zero state of the convolutional encoder. This set forms a commutative group. Every state in the trellis has either $\mathscr{C}_0$ or the coset $\mathscr{C}_0 + \tilde{\mathbf{c}}$ associated with it. (Only one coset exists.)

Proof. If the encoder is in the all-zero state, each of its $m + 1$ outputs is the modulo-2 sum of some subset of the m inputs. Since the code is linear, the sum of any two input m-tuples produces the corresponding sum of the output $(m + 1)$-tuples, so the closure property of groups is satisfied. Since each $(m + 1)$-tuple is its own inverse, the outputs **c** form a subgroup $\mathscr{C}_0$ of the group of all the binary $(m + 1)$-tuples. If the encoder is in a nonzero state, its outputs are found by adding a fixed $(m + 1)$-tuple, say $\tilde{\mathbf{c}}$, to each of the $(m + 1)$-tuples in $\mathscr{C}_0$. Thus this new set is either $\mathscr{C}_0$ itself or a coset of $\mathscr{C}_0$. Only one such coset exists, since the intersection of $\mathscr{C}_0$ and $\mathscr{C}_0 + \tilde{\mathbf{c}}$ is the empty set and their union has cardinality $2^{(m+1)}$, that is, is the set of all the binary $(m + 1)$-tuples. □

From the lemma we see that two types of states exist. States of the first type generate outputs **c** chosen from $\mathscr{C}_0$, while states of the second type produce outputs from $\mathscr{C}_0 + \tilde{\mathbf{c}}$. Now recall that each row of the matrices **G(e)** corresponds to a particular state. Some of them correspond to $\mathscr{C}_0$, while the others correspond to $\mathscr{C}_0 + \tilde{\mathbf{c}}$. Recalling (4.7), this implies that all the rows of the error matrices will have equal sums if, for any **e**,

$$\sum_{\mathbf{c} \in \mathscr{C}_0} D^{\|f(\mathbf{c}) - f(\mathbf{c} \oplus \mathbf{e})\|^2} = \sum_{\mathbf{c} \in \mathscr{C}_0} D^{\|f(\mathbf{c} \oplus \tilde{\mathbf{c}}) - f(\mathbf{c} \oplus \tilde{\mathbf{c}} \oplus \mathbf{e})\|^2}. \quad (4.18)$$

This is a sufficient condition for scalar labels. For a simpler sufficient condition, consider the transformation induced in the signal set $\{f(\mathbf{c}), \mathbf{c} \in \mathscr{C}_0\}$ by the addition to the label vectors of one and the same vector $\tilde{\mathbf{c}}$. As **c** runs in $\mathscr{C}_0$ we get the one-to-one correspondence

$$f(\mathbf{c}) \longrightarrow f(\mathbf{c} \oplus \tilde{\mathbf{c}}), \quad \mathbf{c} \in \mathscr{C}_0. \quad (4.19)$$

A sufficient condition for uniformity is that (4.19) be an *isometry*, that is, a one-to-one mapping that preserves the scalar products (and, consequently, the distances). To prove that this condition is sufficient for uniformity, observe that because of the isometry, the distance of $f(\mathbf{c})$ from $f(\mathbf{c} \oplus \mathbf{e})$ will be the same as the distance of $f(\mathbf{c} \oplus \tilde{\mathbf{c}})$ from $f(\mathbf{c} \oplus \tilde{\mathbf{c}} \oplus \mathbf{e})$ for any **c** and for any **e**, which is what we wanted to prove.

Figure 4.6 shows an example of a constellation and its labeling giving rise to a uniform TCM scheme. It refers to an 8-PSK constellation with "natural" labeling. In this case the mapping

$$f(\mathbf{c}) \longrightarrow f(\mathbf{c} \oplus (011))$$

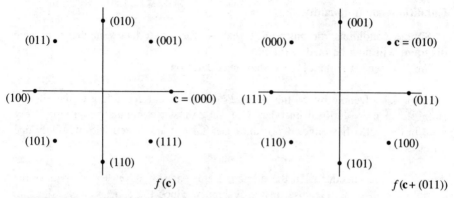

FIGURE 4.6 8-PSK constellation with labeling giving rise to an isometric TCM scheme.

is the combination of a reflection along the vertical axis and of a rotation by $-\pi/4$. Since the composition of two isometries is itself an isometry, the mapping above is an isometry. Thus a TCM scheme that associates with the $\mathscr{C}_0$-states the symbols $f(000)$, $f(010)$, $f(100)$, and $f(110)$ and with the $(\mathscr{C}_0 \oplus \tilde{\mathbf{c}})$-states the symbols $f(011)$, $f(001)$, $f(111)$, $f(101)$ is uniform. Another example is shown in Fig. 4.7, which refers to a 16-QAM constellation. Here the mapping

$$f(\mathbf{c}) \longrightarrow f(\mathbf{c} \oplus (0001))$$

is a specular reflection along the vertical axis, an isometry again.

If the uniformity condition holds, the error matrices have equal row sums, and consequently we can associate with the branches of the error-state diagrams labels that are *scalars*. Specifically, we define the *weight profile* $W(\mathbf{e})$ as the common value of the sum of the elements of each row of $\mathbf{G}(\mathbf{e})$:

$$W(\mathbf{e}) = \frac{1}{M} \sum_{\mathbf{c} \in \mathscr{C}_0} D^{\|f(\mathbf{c}) - f(\mathbf{c} \oplus \mathbf{e})\|^2} \tag{4.20}$$

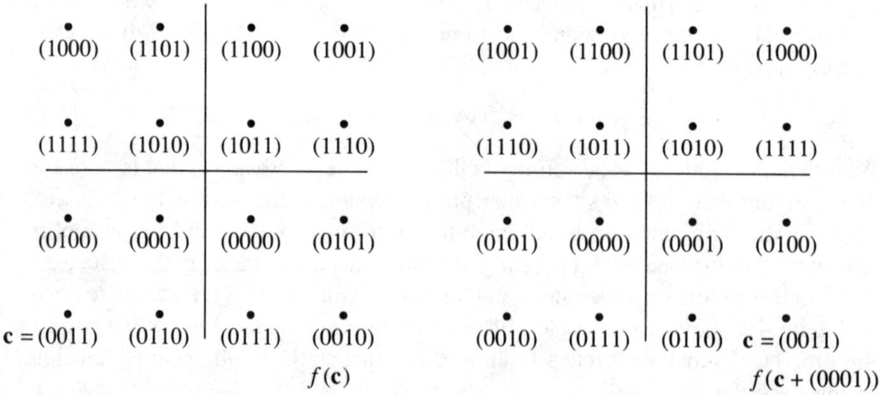

FIGURE 4.7 16-QAM constellation with labeling giving rise to an isometric TCM scheme.

and use them as labels in the error-state diagram. We get

$$T(D) = \sum_{\mathbf{e}} W(\mathbf{e})$$

which, in conjunction with (4.9), provides the transfer-function upper bound to $P(e)$.

Observe further that if we restrict $\mathbf{c}$ in (4.19) to run in $\mathscr{C}_0$, (4.19) defines a correspondence between the two subsets of signals that are obtained in the first set-partition level, each of them being associated with one-half of the trellis states. Thus our condition for uniformity implies that these two subsets must be related by an isometry (e.g., a rotation or a reflection). However, the existence of an isometry between these subsets is not sufficient for uniformity. For a counterexample, consider the TCM scheme of Fig. 4.8. The first partition generates two subconstellations that can be obtained one from the other by a reflection along the vertical axis. Thus these two subconstellations have the same set of mutual distances. However, in this case it is not possible to find a vector $\tilde{\mathbf{c}}$ such that (4.19) holds. A simple direct computation shows that the weight profiles for this TCM scheme are different.

The weight profile of the error vector $\mathbf{e}$ can also be given the form

$$W(\mathbf{e}) = \frac{1}{M} \sum_{\mathbf{c}(\nu)} D^{\|f[\mathbf{c}(\nu)] - f[\mathbf{c}(\nu) \oplus \mathbf{e}]\|^2} \quad (4.21)$$

where ν denotes the order of the component of $\mathbf{c}$ that varies in the isometry (4.19), and $\mathbf{c}(\nu)$ denotes the vector $\mathbf{c}$ with its νth component chosen as equal to either 1 or zero, arbitrarily.

Asymptotic considerations

The entry p, q of the matrix $\mathbf{G}$ is a power series in the indeterminate D. We denote the general term in the series as $\nu_{pq}(d_\ell) D^{d_\ell^2}$, where

$$\nu_{pq}(d_\ell) = \frac{1}{M^{L_1}} n_1 + \frac{1}{M^{L_2}} n_2 + \cdots$$

and n_h, $h = 1, 2, \ldots$, is the number of error paths starting from node p at time 0 (say), remerging L_h time instants later at node q, and whose associated distance is d_l. Since $1/M^{L_h}$ is the probability of a sequence of symbols of length L_h, $\nu_{pq}(d_l)$ can be interpreted as the average number of competing paths at distance d_ℓ associated with any path in the code trellis starting at node p and ending at node q. Consequently, the quantity

$$\nu_p(d_\ell) = \sum_q \nu_{pq}(d_\ell)$$

can be interpreted as the average number of competing paths at distance d_ℓ associated with any path in the code trellis leaving from node p and ending at any node. Similarly,

$$N(d_\ell) = \frac{1}{N} \sum_{p,q} \nu_{pq}(d_\ell)$$

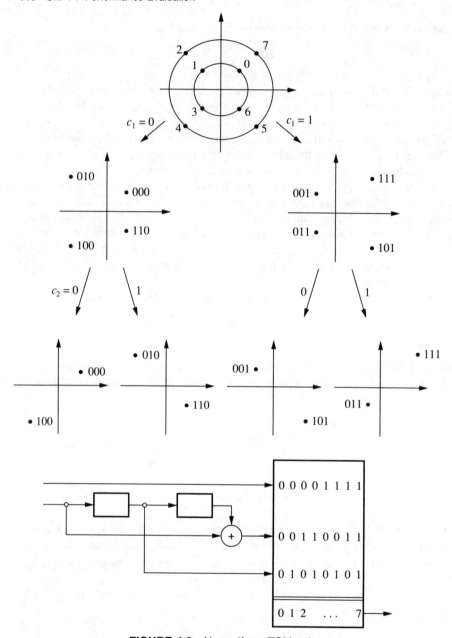

FIGURE 4.8 Nonuniform TCM scheme.

is the average number of competing paths at distance d_ℓ associated with a path in the code trellis.

For large signal-to-noise ratios, that is, as $N_0 \to 0$, the only terms in the entries of the matrix **G** that provide a contribution to the error probability significantly different from zero will be of the type $\nu_{pq}(d_{\text{free}})D^{d^2_{\text{free}}}$. Thus, asymptotically, $P(e)$ will approach $N(d_{\text{free}})e^{-d^2_{\text{free}}/4N_0}$.

An improved upper bound

An upper bound on $P(e)$ tighter than (4.9) can be obtained by substituting for the Bhattacharyya bound in (4.4) a tighter expression [2, p. 247]. Specifically, let us observe that we have, exactly,

$$P[\mathbf{C}_L \to \mathbf{C}'_L] = \frac{1}{2} \operatorname{erfc}\left[\frac{\|f(\mathbf{C}_L) - f(\mathbf{C}'_L)\|}{2\sqrt{N_0}}\right] \quad (4.22)$$

where $\frac{1}{2} \operatorname{erfc}(\cdot)$ is the Gaussian integral function.

Since the minimum value taken by $\|f(\mathbf{C}_L) - f(\mathbf{C}'_L)\|$ is what we have called d_{free}, use of the inequality

$$\operatorname{erfc}\left(\sqrt{\frac{x+y}{2}}\right) \leq \operatorname{erfc}\left(\sqrt{\frac{x}{2}}\right) e^{y/2}, \qquad x \geq 0, \quad y \geq 0 \quad (4.23)$$

leads to the bound

$$P[\mathbf{C}_L \to \mathbf{C}'_L] \leq \frac{1}{2} \operatorname{erfc}\left(\frac{d_{\text{free}}}{2\sqrt{N_0}}\right) e^{d_{\text{free}}^2/4N_0} \exp\left\{-\frac{1}{4N_0}\|f(\mathbf{C}_L) - f(\mathbf{C}'_L)\|^2\right\}. \quad (4.24)$$

In conclusion, we have the bound on error probability:

$$P(e) \leq \frac{1}{2} \operatorname{erfc}\left(\frac{d_{\text{free}}}{2\sqrt{N_0}}\right) e^{d_{\text{free}}^2/4N_0} T(D) \bigg|_{D = e^{-1/4N_0}}. \quad (4.25)$$

We also have, approximately for high signal-to-noise ratios,

$$P(e) \simeq N(d_{\text{free}}) \frac{1}{2} \operatorname{erfc}\left(\frac{d_{\text{free}}}{2\sqrt{N_0}}\right). \quad (4.26)$$

Bit error probability

A bound on bit error probability can also be obtained by following the footsteps of a similar derivation for convolutional codes [2, pp. 245–246]. All that is needed here is a change in the error matrices. Explicitly, the entry p, q of the matrix $\mathbf{G}(\mathbf{e})$ must be multiplied by the factor I^ϵ, where ϵ is the number of incorrect input bits associated with the error vector $\mathbf{e}$ and the transition from state p to state q.

With this new definition of the error matrices, the entry p, q of the matrix $\mathbf{G}$ can now be expressed in a power series in the two indeterminates D and I. The general term of the series will be $\mu_{pq}(d_l, \epsilon_h) D^{d_l} I^{\epsilon_h}$, where $\mu_{pq}(d_l, \epsilon_h)$ can be interpreted as the average number of paths having distance d_l and ϵ_h bit errors with respect to any path in the trellis starting at node p and ending at node q. If we take the derivative of these terms with respect to I and set $I = 1$, each of them will provide the expected number of bit errors per branch generated by the incorrect paths from p to q. If we further divide these quantities by m, the number of source bits per trellis transition, and we sum the series, we obtain the following upper bound on bit error probability:

$$P_b \leq \frac{1}{m} \frac{\partial}{\partial I} T(D, I) \bigg|_{I=1, D=e^{-1/4N_0}} \quad (4.27)$$

Another upper bound (TUB) can be obtained by substituting for the Bhattacharyya bound in (4.4) the tighter inequality derived above. We get

$$P_b \leq \frac{1}{2m} \operatorname{erfc}\left(\frac{d_{\text{free}}}{2\sqrt{N_0}}\right) \exp\left(\frac{d_{\text{free}}^2}{4N_0}\right) \frac{\partial}{\partial I} T(D, I) \bigg|_{I=1, D=Z} \quad (4.28)$$

Computation of the transfer function

The computation of the matrix transfer function may be performed in two separate steps. First, the transfer function of the error-state diagram with formal labels (proper attention should be paid to the noncommutativity of matrix operations if the TCM scheme is nonuniform). This can be accomplished by using graph-reduction techniques, or methods based on the solution of a system of linear equations [2, pp. 241 ff.]. Then the appropriate labels (the error matrices, or the weight profiles) are substituted for the formal labels, and the matrix **G** computed.

Here we consider, for illustration's sake, a simple example of application of the linear-equations method to the computation of the transfer function $T(D)$. We assume scalar labels here (the reader may want to generalize this example to the matrix-label case). Consider Fig. 4.9. The labels of each edge are shown. Let ξ_i denote the transfer function from node ⓪ to node ⓘ. By observing that $T(D) = \xi_4$, and that node ④ can only be reached from node ③ through the edge labeled g, we have

$$T(D) = g\xi_3. \quad (4.29)$$

Further, observe that node ③ can be reached either from node ② (through the edge labeled f) or from node ① (through the edge labeled d). Thus we may write

$$\xi_3 = f\xi_2 + c\xi_1.$$

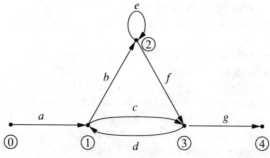

FIGURE 4.9 Computation of the transfer function of a state diagram.

Similarly, we have the equations

$$\xi_2 = e\xi_2 + b\xi_1$$

$$\xi_1 = a + d\xi_3.$$

By solving the last three equations, we get

$$\xi_3 = a\frac{c - ce + bf}{1 - e - cd + cde - bdf}$$

which, in conjunction with (4.29), provides the desired expression of $T(D)$.

Convergence considerations. Catastrophic codes

In our previous analysis we have not considered the issue of the convergence of the series providing the transfer function $T(D)$ or $T(D, I)$. Actually, we assume implicitly that $T(D)$, or the derivative of $T(D, I)$, converges for large enough values of the signal-to-noise ratio.

Now there are situations in which the transfer function does not converge due to the fact that one or more of its coefficients take value infinity. We shall see in later chapters that this situation may actually occur for certain TCM schemes in which two encoded sequences with a finite Euclidean distance correspond to source symbol sequences with infinite Hamming distance [3]. The scheme is called *catastrophic* in this case (see [2, pp. 250–251]) because a finite number of channel errors may cause an infinite number of source symbol errors.

4.1.3 Consideration of Different Channels

Here we briefly consider a generalization of the results presented so far to transmission over a channel that is not additive white Gaussian, with a detection rule that may not be maximum likelihood. In later chapters we examine in detail a number of channel models for which these concepts apply.

Let $\mathbf{R}_L$ be the received sample sequence of length L, $\mathbf{R}_L = (r_1, \ldots, r_L)$. Also, let $m(\mathbf{R}_L, \mathbf{X}_L)$ denote the metric used by the demodulator. This means that $\mathbf{X}'_L$ is the demodulator choice whenever

$$m(\mathbf{R}_L, \mathbf{X}'_L) \geq m(\mathbf{R}_L, \mathbf{X}_L) \qquad (4.30)$$

for all $\mathbf{X}_L \neq \mathbf{X}'_L$. We assume that the metric is *additive*, that is,

$$m(\mathbf{R}_L, \mathbf{X}_L) = \sum_{n=1}^{L} m(r_n, x_n). \qquad (4.31)$$

Let us bound $P[\mathbf{E}_L]$ from above. For any $\lambda > 0$ we have

$$P[\mathbf{C}_L \to \mathbf{C}_L \oplus \mathbf{E}_L] \qquad (4.32)$$

$$\leq E\left\{e^{\lambda \sum_{n=1}^{L}[m(r_n,f(\mathbf{c}_n\oplus\mathbf{e}_n))-m(r_n,f(\mathbf{c}_n))]}\bigg|\mathbf{X}_L\right\} \quad (4.33)$$

$$= \prod_{n=1}^{L} E\{e^{\lambda[m(r_n,f(\mathbf{c}_n\oplus\mathbf{e}_n))-m(r_n,f(\mathbf{c}_n))]}|x_n\} \quad (4.34)$$

where the inequality (4.33) follows from the Chernoff bound, while equality (4.34) follows from the assumption that the channel is memoryless.

Let us define the quantity $\Delta_\lambda(f(\mathbf{c}_n), f(\mathbf{c}_n \oplus \mathbf{e}_n))$ as follows:

$$e^{-\Delta_\lambda(f(\mathbf{c}_n),f(\mathbf{c}_n\oplus\mathbf{e}_n))} = \mathbf{E}\{e^{\lambda[m(r_n,f(\mathbf{c}_n\oplus\mathbf{e}_n))-m(r_n,f(\mathbf{c}_n))]}|\mathbf{c}_n\}. \quad (4.35)$$

Here the Chernoff parameter λ should be chosen so as to minimize the upper bound (4.34).[4] Thus, from (4.1), we get

$$P(e) \leq \sum_{L=1}^{\infty}\sum_{\mathbf{E}_L\neq 0}\sum_{\mathbf{C}_L} P[\mathbf{C}_L]e^{-\Delta_\lambda[f(\mathbf{C}_L),f(\mathbf{C}_L\oplus\mathbf{E}_L)]}. \quad (4.36)$$

By defining the function

$$W(\mathbf{E}_L) = \sum_{\mathbf{C}_L} P[\mathbf{C}_L]e^{-\Delta_\lambda[f(\mathbf{C}_L),f(\mathbf{C}_L\oplus\mathbf{E}_L)]} \quad (4.37)$$

we can rewrite (4.36) in the form (4.6), and the previous results hold with the new definition of the error weight matrices:

$$[\mathbf{G}(\mathbf{e}_i)]_{p,q} = \frac{1}{M}\sum_{\mathbf{c}_{p\to q}} e^{-\Delta_\lambda[f(\mathbf{c}_{p\to q}),f(\mathbf{c}_{p\to q}\oplus\mathbf{e}_i)]}. \quad (4.38)$$

Comparison of (4.38) with (4.7) shows that the function Δ_λ is for the general channel model what the squared Euclidean distance is for the additive white Gaussian channel model.

4.2 Lower Bound to Error Probability

In this section we consider a lower bound to error probability for transmission over the AWGN channel. The bound is based on the fact that the error probability of any real-life receiver is larger than that of a receiver that makes use of side information provided by, say, a genie [4].

The genie-aided receiver operates as follows. The genie observes a long sequence of transmitted symbols, or, equivalently, the sequence

$$\mathbf{C} = (\mathbf{c}_i)_{i=0}^{K-1}$$

---

[4] Notice that λ cannot depend on the index n: If the optimum Chernoff parameter turns out to be a function of n, then a suboptimum λ must be chosen.

of labels, and tells the receiver that the transmitted sequence was either **C** *or* the sequence

$$\mathbf{C}' = (\mathbf{c}'_i)_{i=0}^{K-1}$$

where $\mathbf{C}'$ is picked at random among the possibly transmitted sequences having the smallest Euclidean distance from $\mathbf{C}$ (not necessarily d_{free}, because $\mathbf{C}$ may not have any sequence $\mathbf{C}'$ at free distance).

The error probability for this genie-aided receiver is

$$P_G(e \mid \mathbf{C}) = \frac{1}{2} \operatorname{erfc}\left(\frac{\|f(\mathbf{C}) - f(\mathbf{C}')\|}{2\sqrt{N_0}}\right). \tag{4.39}$$

Now consider the unconditional probability $P_G(e)$. We have

$$P_G(e) = \frac{1}{2} \sum_{\mathbf{C}} P(\mathbf{C}) \operatorname{erfc}\left(\frac{\|f(\mathbf{C}) - f(\mathbf{C}')\|}{2\sqrt{N_0}}\right)$$

$$\geq \frac{1}{2} \sum_{\mathbf{C}} I(\mathbf{C}) P(\mathbf{C}) \operatorname{erfc}\left(\frac{d_{\text{free}}}{2\sqrt{N_0}}\right) \tag{4.40}$$

where $I(\mathbf{C}) = 1$ if

$$\min_{\mathbf{C}'} d[f(\mathbf{C}), f(\mathbf{C}')] = d_{\text{free}}$$

and $I(\mathbf{C}) = 0$ otherwise. In conclusion,

$$P(e) \geq \psi \frac{1}{2} \operatorname{erfc}\left(\frac{d_{\text{free}}}{2\sqrt{N_0}}\right)$$

where

$$\psi = \sum_{\mathbf{C}} P(\mathbf{C}) I(\mathbf{C}) \tag{4.41}$$

represents the probability that at any given time a code trellis path chosen at random has another path splitting from it at that time, and remerging later, such that the Euclidean distance between them is d_{free}. If all the sequences have this property, we get the lower bound

$$P(e) \geq \frac{1}{2} \operatorname{erfc}\left(\frac{d_{\text{free}}}{2\sqrt{N_0}}\right) \tag{4.42}$$

but this is not valid in general. For (4.42) to hold, it is sufficient that all the trellis paths be on an equal footing, so that in particular all of them have a path at d_{free}. This is obtained if each one of the error matrices $\mathbf{G}(\mathbf{e})$ has equal entries.

More generally, for a given TCM scheme, let Λ denote the minimum length of the error events that achieve d_{free}, and $N_\Lambda(d_{\text{free}})$ the number of paths of length Λ

that have a competing path at d_{free}. If m denotes the number of information bits at any node, there are $N \cdot 2^{m\Lambda}$ paths of length Λ starting from any node. Thus the ratio $N_\Lambda(d_{\text{free}})/N2^{m\Lambda}$ is the probability that at any given time a trellis path chosen at random has another path splitting from it at that time, and remerging later, such that the Euclidean distance between them is d_{free} *and* their length is exactly Λ. We have

$$\psi \geq \frac{N_\Lambda(d_{\text{free}})}{N 2^{m\Lambda}} \geq \frac{1}{N 2^{m\Lambda - 1}} \qquad (4.43)$$

where the last inequality follows from the fact that necessarily $N_\Lambda(d_{\text{free}}) \geq 2$.

Finally, we may get a lower bound to the *bit* error probability by observing that the average fraction of erroneous information bits in the splitting branch of an error event cannot be lower than $1/m$. Thus

$$P_b \geq \frac{\psi}{m} \frac{1}{2} \operatorname{erfc}\left(\frac{d_{\text{free}}}{2\sqrt{N_0}}\right).$$

4.3 Examples

In this section we present some examples of computation of error probabilities for uniform TCM schemes, that is, schemes that satisfy the uniformity conditions [5]. From the theory developed before, it is seen that this computation involves two steps. The first one is the evaluation, for each error vector **e**, of the weight profile $W(\mathbf{e})$. Then each weight profile is multiplied by I^ϵ, where ϵ is the number of bit errors caused by **e**. The second step is the computation of the transfer function $T(D, I)$ of the error-state diagram, whose branches are labeled by the weights above.

$$\boxed{m = 1}$$

Let the memoryless mapper of the Ungerboeck model be the natural binary mapper. For $m = 1$ the channel symbols are 0, 1, 2, 3. Consider again the trellis shown in Fig. 4.2 for a two-state code. Figure 4.10 shows the corresponding error-state diagram, whose branch labels are labeled t_1, t_2, and t_3. Let α denote the transfer function from the starting node to the single-error state. From $\alpha = t_1 + \alpha t_2$ and $T(D, I) = t_3 \alpha$ we obtain

$$T(D, I) = \frac{t_1 t_3}{1 - t_2}. \qquad (4.44)$$

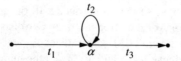

FIGURE 4.10 Error-state diagram for a two-state, $m = 1$ TCM scheme.

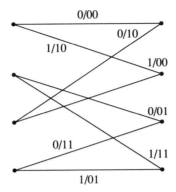

FIGURE 4.11 Trellis diagram for a four-state, $m = 1$ TCM scheme.

A four-state code is shown in Fig. 4.11, while the corresponding error-state diagram is shown in Fig. 4.12. Each node label in the latter figure is the transfer function from the starting node to that node. The relevant equations for this graph are

$$\alpha = t_1 + t_5\gamma$$
$$\beta = t_6\beta + t_2\alpha$$
$$\gamma = t_3\beta + t_4\alpha$$
$$T(D, I) = t_7\gamma.$$

Solution yields

$$\beta = \frac{t_1 t_2}{(1 - t_4 t_5)(1 - t_6) - t_2 t_3 t_5}$$

so that

$$T(D, I) = \frac{1}{1 - t_4 t_5}\left[\frac{t_1 t_2 t_3 t_7}{(1 - t_4 t_5)(1 - t_6) - t_2 t_3 t_5} + t_1 t_4 t_7\right]. \tag{4.45}$$

As a special case, consider the TCM scheme with $m = 1$ and quaternary amplitude modulation (4-AM). The signal constellation is shown in Fig. 4.4. To

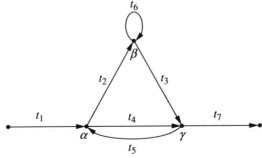

FIGURE 4.12 Error-state diagram for a four-state, $m = 1$ TCM scheme.

derive the values of t_i we use (4.21), in the form

$$W(e_2 e_1) = \frac{1}{2} \sum_{c_2 = 0, 1} D^{\|f[\mathbf{c}(1)], f[\mathbf{c}(1) \oplus \mathbf{e}]\|^2}$$

where $\mathbf{c}(1) = (c_2, 0)$ and $\mathbf{e} = (e_2, e_1)$. For $W(10)$, $c_2 = 0$ gives D^{16}, since we compare signals 0 and 2 in Fig. 4.4. Similarly, for $c_2 = 1$ we get the same exponent of D through comparison of signals 2 and 0. For an error in the parity bit, $c_1 = 0$, we get D^4 for $c_2 = 0, 1$. Thus $W(01) = D^4$. Finally, for errors in c_2 and c_1 we get D^{36} for $c_2 = 0$, but D^4 for $c_2 = 1$. In conclusion, the weight profiles are

$$W(00) = 1$$
$$W(01) = D^4$$
$$W(10) = D^{16}$$
$$W(11) = \tfrac{1}{2}(D^4 + D^{36}).$$

The situation for 4-PSK is shown in Fig. 4.5. The smallest distance squared is two, and the largest is four. The weight profiles are

$$W(00) = 1$$
$$W(01) = D^2$$
$$W(10) = D^4$$
$$W(11) = D^2.$$

For two-state, 4-AM TCM schemes, we have, from Figs. 4.3 and 4.4, and (4.44),

$$T(D, I) = \frac{D^{20} I}{1 - \tfrac{1}{2}(D^4 + D^{36})}$$

The bit error probability upper bound (4.28) derived above gives

$$P_b \leq \frac{\tfrac{1}{2} \operatorname{erfc}\left(\sqrt{5 E_s / N_0}\right)}{\left[1 - \tfrac{1}{2} \exp(-9 E_s / N_0) - \tfrac{1}{2} \exp(-E_s / N_0)\right]^2}$$

where E_s is the smallest energy in the signal constellation of Fig. 4.4, which is unity. Thus, defining the average energy per symbol $\overline{E_s}$ and the average energy per bit $\overline{E_b}$, we have $\overline{E_s} = \overline{E_b} = 5 E_s$, and

$$P_b \leq \frac{\tfrac{1}{2} \operatorname{erfc}\left(\sqrt{\overline{E_b} / N_0}\right)}{\left[1 - \tfrac{1}{2} \exp(-9 \overline{E_b} / 5 N_0) - \tfrac{1}{2} \exp(-\overline{E_b} / 5 N_0)\right]^2}.$$

The preceding equation shows that the asymptotic coding gain is 0 dB in this case.

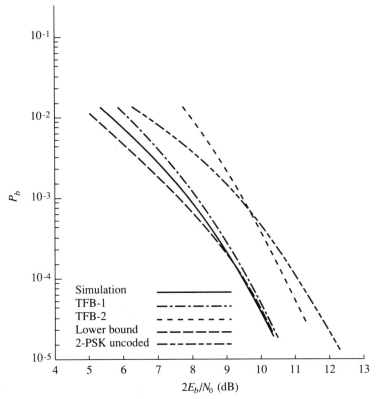

FIGURE 4.13 Upper and lower bounds to bit error rate for a two-state, 4-PSK TCM scheme.

For 4-PSK, Figs. 4.3 and 4.5 and (4.44) together give

$$T(D, I) = \frac{D^6 I}{1 - D^2 I}$$

and (4.28) gives

$$P_b \leq \frac{\frac{1}{2} \operatorname{erfc}\left(\sqrt{3E_b/2N_0}\right)}{(1 - e^{-E_b/2N_0})^2}$$

as $\overline{E_b} = E_b$ for the signal constellation on the unit circle. The asymptotic coding gain here is $10 \log_{10} 3/2 = 1.76$ dB.

A lower bound to P_b is $\frac{1}{2} \operatorname{erfc}\left(\sqrt{3E_b/2N_0}\right)$. Both bounds, plus a digital computer simulation result, are shown in Fig. 4.13. In Fig. 4.13, as in the performance curves that follow, TFB-1 is based on the bound (4.27). TFB-2 is based on the weaker bound

$$P_b \leq \frac{1}{m} \left. \frac{\partial T(D, I)}{\partial I} \right|_{I=1, D=e^{-1/4N_0}} \quad (4.46)$$

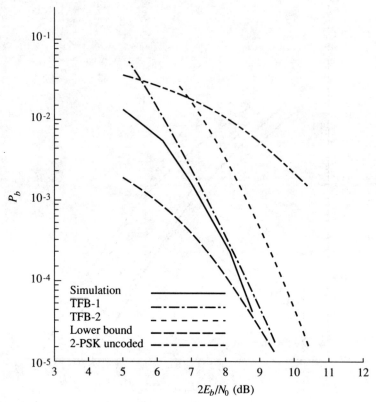

FIGURE 4.14 Upper and lower bounds to bit error rate for a four-state, 4-PSK TCM scheme.

which is the same as replacing $\frac{1}{2}\text{erfc}(\sqrt{x/2})$ by $e^{-x/2}$ in TBF-1. The bounds are quite accurate in estimating P_b for $P_b \leq 5 \times 10^{-3}$. For four-state 4-PSK we obtain, from (4.45) and Figs. 4.12 and 4.5,

$$T(D, I) = \frac{ID^{10}}{1 - 2D^2 I}. \tag{4.47}$$

The upper bound from (4.27) is

$$P_b \leq \frac{\frac{1}{2}\text{erfc}\left(\sqrt{5E_b/2N_0}\right)}{\left(1 - 2e^{-E_b/N_0}\right)^2}$$

and the lower bound is

$$P_b \geq \frac{1}{2}\text{erfc}\left(\sqrt{\frac{5E_b}{2N_0}}\right).$$

The asymptotic coding gain is $10\log_{10} 5/2 \approx 4$ dB. Our upper and lower bounds, as well as a digital computer simulation result, are given in Fig. 4.14.

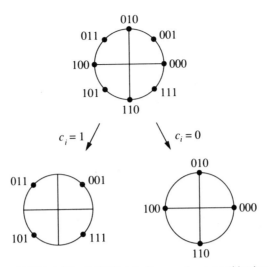

FIGURE 4.15 8-PSK: labeling and set partitioning.

$\boxed{m = 2}$

We now consider the case $m = 2$, corresponding to three bits at the output of the first block of Fig. 4.1. The error weight distribution is given by

$$W(e_3 e_2 e_1) = \frac{1}{4} \sum_{c_2 \in \{0,1\}, c_3 \in \{0,1\}} D^{\|f[\mathbf{c}(1)] - f[\mathbf{c}(1) \oplus \mathbf{e}]\|^2}.$$

The error weight profile is shown in Fig. 4.15. To consider a typical computation, consider $W(011)$. As before, set $c_1 = 0$. Then the sequences shown in Table 4.1 represent all the error sequences. Thus

$$W(011) = \frac{D^{0.586} + D^{3.414}}{2}$$

as given in Fig. 4.15.

TABLE 4.1 Computation of $W(011)$.

$\mathbf{c}(1)$	$\mathbf{c}(1) \oplus \mathbf{e}$	$d^2[f(\mathbf{c}(1)), f(\mathbf{c}(1) \oplus \mathbf{e})]$
000	011	3.414
010	001	0.586
100	111	3.414
110	101	0.586

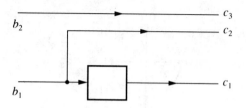

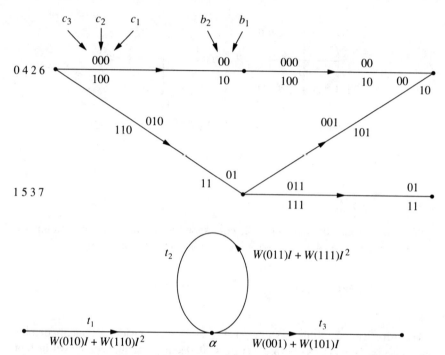

FIGURE 4.16 Code structure, trellis diagram, and error-state diagram.

A two-state code is shown in Fig. 4.16. The same figure also shows the trellis and the error-state diagram. By using in (4.44) the 8-PSK error weights of Fig. 4.15, we get

$$P(e) \leq D^{2.586} \frac{I(1+I)(1+bI)}{1 - aI(1+I)}$$

where

$$a = \tfrac{1}{2}(D^{0.586} + D^{3.414}) \tag{4.48}$$

and

$$b = D^{2.858}. \tag{4.49}$$

Our bounds, plus simulation results, are given in Fig. 4.17.

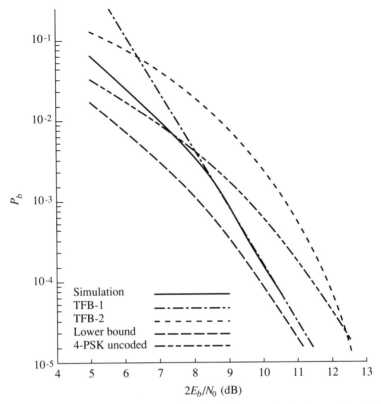

FIGURE 4.17 Upper and lower bound to the bit error rate for two-state, 8-PSK TCM scheme.

We now consider the four-state, 8-PSK scheme with the uncoded modulation as 4-PSK. The trellis diagram is given in Fig. 4.18. The error-state diagram is as in Fig. 4.12, but the branch gains must be for the weight profile of 8-PSK. This is provided in Fig. 4.15.

From Figs. 4.18 and 4.15, the branch labels in Fig. 4.12 are

$$t_1 = W(010)I + W(110)I^2 = D^2I(I+1)$$
$$t_2 = W(011)I + W(110)I^2 = \tfrac{1}{2}(D^{0.586} + D^{3.414})I(I+1)$$
$$t_3 = W(011)I + W(111)I^2 = \tfrac{1}{2}(D^{0.586} + D^{3.414})(I+1)$$
$$t_4 = W(001) + W(101)I = D^{0.586} + ID^{3.414}$$
$$t_5 = W(000)I + W(100)I^2 = I + D^4I^2$$
$$t_6 = W(001)I + W(101)I^2 = D^{0.586}I + D^{3.414}I^2$$
$$t_7 = W(010) + W(110)I = D^2(I+1)$$

Substitution of these t_i, $i = 1, \ldots, 7$, into (4.45) provides $T(D, I)$ for the error

124 Ch. 4 / Performance Evaluation

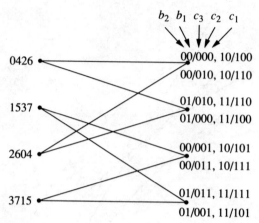

FIGURE 4.18 Trellis diagram for an 8-PSK TCM scheme.

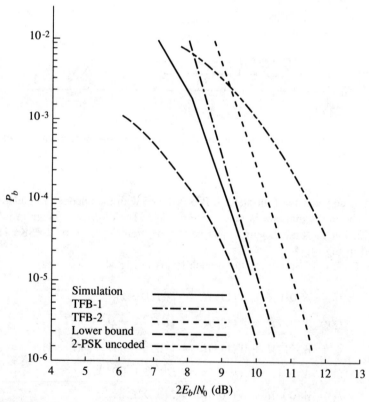

FIGURE 4.19 Upper and lower bound to the bit error rate for a four-state, 8-PSK TCM scheme.

events in the trellis of Fig. 4.18 that do not involve parallel transitions. After considerable algebra it can be shown that

$$\left.\frac{\partial T}{\partial D}\right|_{I=1} = 4D^{d_c^2} \frac{\hat{c}_3(2 - c_3 - c_3^2) + (1 - 2c_3 + 2c_3^2)D^{2.828}}{(1 - 2c_3 - c_3 D^4)^2}$$

where $d_c^2 = 4.586$, $c_3 = D^{0.586} + D^{3.414}$, and $\hat{c}_3 = c_3 D^{-0.586}$. Here d_c^2 is the minimum distance when parallel transitions are not considered. The bit error probability for parallel transitions is $\frac{1}{4}\text{erfc}\left(\sqrt{2E_b/N_0}\right)$. Thus the upper bound on P_b is given by

$$P_b \leq \frac{1}{4}\text{erfc}\left(\sqrt{\frac{2E_b}{N_0}}\right) + \frac{1}{2}D^{-d_c^2}\left.\frac{\partial T(D, I)}{\partial I}\right|_{I=1, D=e^{-E_b/2N_0}}. \quad (4.50)$$

A lower bound in this case is provided by the first term in the right-hand side of (4.50). Upper and lower bounds, along with simulation results for 4-PSK, are given in Fig. 4.19.

4.4 Computation of d_{free}

Earlier we saw how the Euclidean free distance of a TCM scheme defines the asymptotic coding gain of the scheme. In this chapter the results on upper and lower bounds on the error probability over the AWGN show that d_{free} plays a central role in determining the performance. Consequently, if a single parameter is to be used to assess the quality of a TCM scheme, the sensible one is d_{free}. Finally, the tighter upper bound (4.25) requires for its calculation the previous knowledge of d_{free}. Consequently, it makes sense to look for an algorithm to compute this parameter independently of $P(e)$.

4.4.1 Using the Error-State Diagram

The first technique we describe for the computation of d_{free} is based on the error-state diagram that was described in the context of the evaluation of an upper bound to error probability. We have already observed that asymptotically $P(e)$, and in particular its upper bound based on Bhattacharyya inequality, approaches $N(d_{\text{free}})D^{d_{\text{free}}^2}$. Thus the transfer function of the error-state diagram with branch labels (4.7) includes information about d_{free}. Obviously, if the transfer function $T(D)$ is obtained in a closed form, the value of d_{free}^2 follows immediately from the expansion of that function in a power series: The smallest exponent of D in that series is d_{free}^2. However, in most cases a closed form for $T(D)$ will not be available, so we must get this exponent by using numerical techniques based on our ability of computing $T(D)$ for any value of D.

We shall soon describe one such technique, but before that we want to comment further on a simplification that takes place in the branch labels of the error-state diagram when we want to use it to the only purpose of evaluating d_{free}^2.

Consider the error matrix (4.7) again. In general, its elements are sums of terms generated by parallel transitions. Now, for the computation of d_{free}^2 it is not necessary to keep track of all the possible distances, since only the smallest among them is relevant. Thus the entries of the error matrices include *only one term*, the one given by the summand in the right-hand side of (4.7) that has the smallest exponent.

Getting d_{free}^2 from $T(D)$

Observe that the transfer function can be written in the form

$$T(D) = N(d_{\text{free}})D^{d_{\text{free}}^2} + N(d_{\text{next}})D^{d_{\text{next}}^2} + \cdots \quad (4.51)$$

where d_{next} is the second smallest Euclidean distance. In [6, p. 127] it is shown that the function $\phi_1(D) = \ln[T(eD)/T(D)]$ decreases monotonically to the limit d_{free}^2 as $D \to 0$. Consequently, for any $D > 0$, $\phi_1(D)$ will provide an upper bound to d_{free}^2. A lower bound is obtained by considering the function

$$\phi_2(D) = \frac{\ln T(D)}{\ln D}.$$

In fact, take the logarithm of both sides of (4.51). We have

$$d_{\text{free}}^2 \ln D = \ln T(D) - \ln N(d_{\text{free}}) - \ln\left[1 + \frac{N(d_{\text{free}})}{N(d_{\text{next}})} D^{d_{\text{next}}^2 - d_{\text{free}}^2} + \cdots\right].$$

By taking a small enough $D > 0$, we get

$$\frac{\ln T(D)}{\ln D} = d_{\text{free}}^2 - \epsilon(D) \quad (4.52)$$

where $\epsilon(D)$ is a function that is positive and tends to zero monotonically as $D \to 0$. Thus, if we take a decreasing sequence of values of D, from $\phi_1(D)$ and $\phi_2(D)$, we get two sequences that are increasingly closer to each other, and hence to d_{free}^2. A similar technique could be used to evaluate $N(d_{\text{free}}^2)$ [3].

Exploiting uniformity

As for the computation of error probability, a good deal of simplification arises when the symmetries of the TCM scheme allow the use of *scalar* branch labels in the error-state diagram. $T(D)$ can be computed by using scalar branch labels if the uniformity conditions are satisfied. In this case the labels are the sums of the elements in any row of the error matrices. As before, since only the smallest distance is of concern to us, we can keep *only one term* of the summation—the one with the smallest exponent.

Since the elements in a row of the error matrix correspond to the channel symbols that can be transmitted from a single state of the trellis, we must pick the minimum-exponent term among

$$D^{\|f(\mathbf{c}) - f(\mathbf{c} \oplus \mathbf{e})\|^2} \qquad \mathbf{c} \in \mathscr{C}_0$$

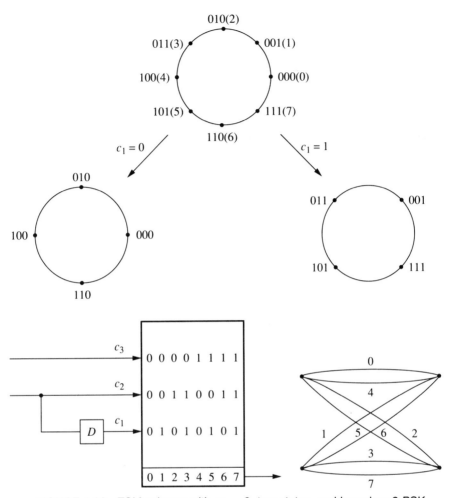

FIGURE 4.20 TCM scheme with $m = 2$, two states, and based on 8-PSK.

where as before $\mathscr{C}_0$ is the set of 2^m vectors **c** corresponding to the all-zero state of the convolutional code. (Of course, if we choose any other state, the result will not change since we assume that the uniformity conditions are satisfied.) Thus we can define the function, depending only on the error vector **e**:

$$w(\mathbf{e}) = \min_{\mathbf{c} \in \mathscr{C}_0} \|f(\mathbf{c}) - f(\mathbf{c} \oplus \mathbf{e})\|^2$$

(this is called the *Euclidean weight* of **e**), and label the nodes of the error-state diagram with the values $D^{w^2(\mathbf{e})}$.

EXAMPLE 4.2

Consider the TCM scheme of Fig. 4.20. First, observe that the uniformity conditions are satisfied by this scheme. The parallel transitions give a squared Euclidean

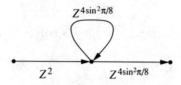

FIGURE 4.21 Error-state diagram for Example 4.2.

distance equal to 4, while the Euclidean weights are

$$w(000) = 0$$
$$w(001) = 2\sin\frac{\pi}{8}$$
$$w(010) = \sqrt{2}$$
$$w(011) = 2\sin\frac{\pi}{8}$$
$$w(100) = 2$$
$$w(101) = 2\sin 3\frac{\pi}{8}$$
$$w(110) = \sqrt{2}$$
$$w(111) = 2\sin\frac{\pi}{8}.$$

Figure 4.21 shows the error-state diagram with the branch labels. By inspection, it is seen that

$$d_{\text{free}}^2 = 2 + 4\sin^2\frac{\pi}{8} = 2.856. \qquad \square$$

4.4.2 A Computational Algorithm

We now describe an algorithm for the computation of d_{free}, as first derived by Saxena [7] and Mulligan and Wilson [8]. Our presentation here follows [9, pp. 561 ff.]. Consider the trellis describing the TCM scheme. Every pair of branches in a section of the trellis defines one distance between the signals labeling the branches. If there are parallel transitions, every branch will be associated with an entire subconstellation. In this case, only the minimum distance between any two signals extracted from the pair of subconstellations will be used. The squared distance between the signal sequences associated with two paths in the trellis is obtained by summing the individual squared distances. The algorithm is based on the update of the entries of a matrix $\mathbf{D}^{(n)} = (\delta_{pq}^{(n)})$, which are the minimum squared distances between all pairs of paths diverging from any initial state and reaching the states p and q at discrete time n. Two such pairs of paths are shown in Fig. 4.22. Notice that the matrix $\mathbf{D}^{(n)}$ is symmetric, and that its elements on the main diagonal are the distances between remerged paths (the "error events"). The algorithm goes as follows.

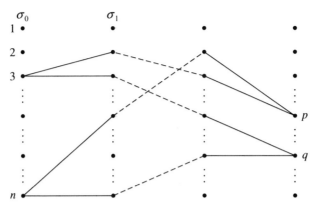

FIGURE 4.22 Two pairs of paths diverging at time $n = 0$ and reaching the states p, q at the same time.

Step 1. For each state p, find the M states (the "predecessors") from which a transition to p is possible, and store them in a table. Set $\delta_{pq} = -1$ for all p and $q \geq p$.

Step 2. For each pair of states (p, q), find the minimum squared Euclidean distance between pairs of paths diverging from the same initial states (whatever they are) and reaching p, q in one time unit. Two such pairs are shown in Fig. 4.23. This distance is $\delta_{pq}^{(1)}$.

Step 3. For both states in the pair (p, q) find in the table of step 1 the M predecessors $p_1, \ldots, p_M$ and $q_1, \ldots, q_M$ (Fig. 4.24). In general there are M^2 possible paths at time $n - 1$ that pass through p and q at time n. They pass through the pairs

$$(p_1, q_1), (p_1, q_2), \ldots, (p_1, q_M)$$
$$\ldots$$
$$(p_M, q_1), (p_M, q_2), \ldots, (p_M, q_M).$$

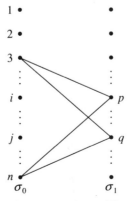

FIGURE 4.23 Two pairs of paths starting from different states and reaching the same pair of states in one time instant.

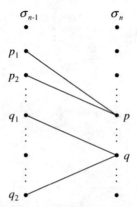

FIGURE 4.24 Predecessors of states p, q.

The minimum distance among all the paths passing through (p, q) at time n is

$$\delta_{pq}^{(n)} = \min \Big\{ \delta_{p_1 q_1}^{(n-1)} + d^2(p_1 \to p, q_1 \to q),$$
$$\delta_{p_1 q_2}^{(n-1)} + d^2(p_1 \to p, q_2 \to q),$$
$$\ldots \qquad (4.53)$$
$$\delta_{p_1 q_M}^{(n-1)} + d^2(p_1 \to p, q_M \to q),$$
$$\ldots$$
$$\delta_{p_M q_M}^{(n-1)} + d^2(p_M \to p, q_M \to q) \Big\}.$$

In (4.53), the distances $\delta^{(n-1)}$ come from the calculations of step 2, while, for example, $d(p_1 \to p, q_1 \to q)$ denotes the Euclidean distance between the two signals associated with the transitions $p_1 \to p$ and $q_1 \to q$. These can be computed once and for all. When one of the previous distances $d_{\ell m}^{(n-1)}$ is equal to -1, the corresponding term on the right-hand side of (4.53) disappears. In fact, $\delta_{\ell m}^{(n-1)} = -1$ means that no pair of paths can pass through the states ℓ and m at time $n - 1$. When $p = q$, $\delta_{pp}^{(n)}$ represents the squared distance between two paths remerging at the nth step on the state p. This is an error event. Thus, if $\delta_{pp}^{(n)} < \delta_{pp}^{(n-1)}$, then $\delta_{pp}^{(n)}$ will take the place of $\delta_{pp}^{(n-1)}$ in matrix $\mathbf{D}^{(n)}$.

Step 4. If

$$\delta_{pq} < \min_p \delta_{pp}^{(n)} \qquad (4.54)$$

for at least one pair (p, q), then set $n = n + 1$ and go back to step 3. Otherwise, stop iterating and set

$$d_{\text{free}}^2 = \min_p \delta_{pp}^{(n)}.$$

Condition (4.54) verifies that all the paths still open at time n have distances not less than the minimum distance of an error event, and guarantees that the latter is actually d_free.

4.4.3 The Product-Trellis Algorithm

We now describe a general algorithm for the evaluation of the free distance based on the consideration of a *product trellis*. This algorithm, which is valid for any TCM scheme, uses a state diagram with *scalar* labels, but the diagram has N^2 states.[5] This is the product of two trellises, each providing one path in the pair whose distance has to be computed. In this product trellis, *supernodes* correspond to pairs of nodes in the original trellis, while *superbranches* correspond to pairs of branches. With each superbranch we can associate the Euclidean distance between the two sets of signals associated with each branch.

We classify the superstates as follows. A superstate (i, j) is called *good* if $i = j$, *bad* if $j \neq j$. A *superpath* in the product trellis joining only good superstates corresponds to two identical paths in the original trellis, while a superpath, including a bad superstate, denotes the presence of two different paths. The computation of the free distance can be done by using the product trellis: A pair of paths that originate from the same node and eventually remerge into a single node correspond, in the product trellis, to a superpath originating from a good superstate and reaching a good superstate after passing through at least one bad superstate.

EXAMPLE 4.3

Consider the TCM scheme of Fig. 4.20. The state diagram corresponding to its product trellis is shown in Fig. 4.25.

There are two good superstates, $(1, 1)$ and $(2, 2)$, and two bad superstates, $(1, 2)$ and $(2, 1)$. It is seen by inspection that

$$d^2_\text{free} = 2 + 4\sin^2\frac{\pi}{8} = 2.586$$

corresponding to the superpath

$$(1, 1) \to (1, 2) \to (1, 1).$$

For a more systematic analysis, let us consider matrices whose rows and columns are labeled by the superstates. The entry corresponding to row (i, j) and column (m, n) is D^{δ^2}, where δ is the smallest Euclidean distance between the two subconstellations associated with the transitions $i \to m$ and $j \to n$. In particular, let $\mathbf{S}_{GB}(D)$ account for the transitions from a good superstate to a bad superstate, $\mathbf{S}_{BB}(D)$ account for the transitions from a bad superstate to a bad superstate,

---

[5] A generalization of this algorithm can also be used to compute error probabilities, although we shall not describe it here (see, e.g., [10]).

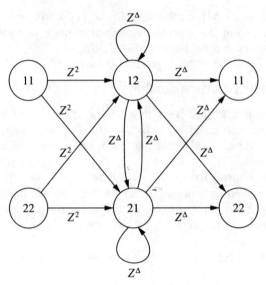

FIGURE 4.25 Superstate diagram corresponding to the product trellis.

and $S_{BG}(D)$ for the transitions from a bad superstate to a good superstate. The state diagram corresponding to the product trellis can be concisely represented in the form of Fig. 4.26. The transfer functions from a good superstate to a good superstate in this diagram are the elements of the $N \times N$ matrix (N is the number of states in the original trellis diagram, that is, the number of good superstates):

$$\mathbf{T}(D) = \mathbf{S}_{GB}(D)\mathbf{S}_{BG}(D) + \mathbf{S}_{GB}(D)\mathbf{S}_{BB}(D)\mathbf{S}_{BG}(D) + \mathbf{S}_{GB}\mathbf{S}_{BB}^2\mathbf{S}_{BG} + \cdots$$
$$= \mathbf{S}_{GB}(D)[\mathbf{I} - \mathbf{S}_{BB}(D)]^{-1}\mathbf{S}_{BG}(D)$$

where $\mathbf{I}$ denotes the identity matrix, and we assume that the inverse of $\mathbf{I} - \mathbf{S}_{BB}(D)$ exists. The transfer function from any good state to any good state is obtained by summing up the elements of the matrix $\mathbf{T}(D)$: We get

$$T(D) = a_1 D^{d_{\text{free}}^2} + \text{higher-order terms} = \mathbf{1}^T \mathbf{T}(D) \mathbf{1} \qquad (4.55)$$

where $\mathbf{1}$ denotes the vector all of whose elements are 1.

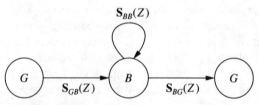

FIGURE 4.26 Concise representation of the state diagram corresponding to a product trellis for the computation of d_{free}.

By defining $\Delta = 4\sin^2(\pi/8)$, we get

$$S_{GB}(D) = \begin{bmatrix} D^2 & D^2 \\ D^2 & D^2 \end{bmatrix}$$

$$S_{BB}(D) = \begin{bmatrix} D^\Delta & D^\Delta \\ D^\Delta & D^\Delta \end{bmatrix}$$

and

$$S_{BG}(D) = \begin{bmatrix} D^\Delta & D^\Delta \\ D^\Delta & D^\Delta \end{bmatrix}.$$

In conclusion,

$$T(D) = 8\frac{D^{2+\Delta}}{1 - 2D^\Delta}$$

which shows again that $d_{\text{free}}^2 = 2 + 4\sin^2(\pi/8)$. □

A word of caution

While d_{free} provides the best single-parameter description of the quality of a TCM scheme, some caution should be exercised when it is used to compare schemes to be operated at low-to-intermediate signal-to-noise ratios. In fact, besides d_{free}, it may be advisable to consider two other parameters as influential over the performance of the TCM scheme:

1. The *error coefficient* $N(d_{\text{free}})$. As a rule of thumb [11, p. 1142], a factor-of-2 increase in this error coefficient reduces the coding gain by about 0.2 dB at error rates on the order of 10^{-6}.
2. The *next distance* d_{next}, that is, the second smallest Euclidean distance between two paths forming an error event. If this is very close to d_{free}, the signal-to-noise ratio necessary for a good approximation to $P(e)$ based on (4.26) may become very large.

Systematic search for good TCM schemes

Once an algorithm for the evaluation of d_{free} is available, it can be used in the search for good TCM schemes. A systematic search procedure is described in [12] and [13].

4.5 Lower Bounds to the Achievable d_{free}

As we have seen previously, d_{free} is the single most important parameter establishing the quality of a TCM scheme operating over a white Gaussian noise channel at large signal-to-noise ratios. For this reason, it is interesting to determine the range of free distances that may be obtained for a given signal constellation and a given code complexity (i.e., a given number of encoder states).

Let us first consider the values of d_{free} achieved by actual designs based on two-dimensional modulations. (An explicit description of the corresponding TCM schemes is provided in Chapter 5.) Figure 4.27 shows d_{free} for selected schemes. The average signal energy of all the signal sets is normalized to unity. Free distances are expressed in decibels relative to the value $d_{min}^2 = 2$ of uncoded 4-PSK. The free distances of various schemes are shown as a function of the parameter r. This is the throughput in bits/s/Hz under the assumption that the signal bandwidth

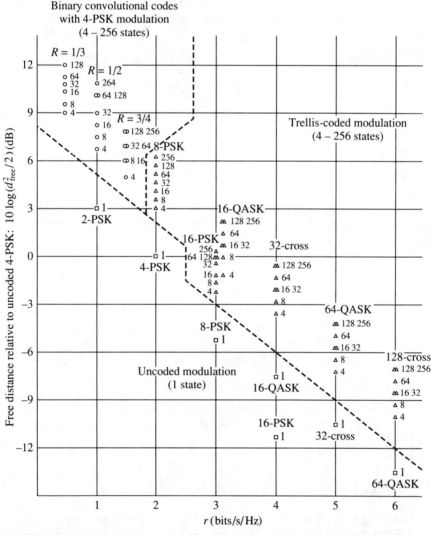

FIGURE 4.27 Free distance of selected TCM schemes based on two-dimensional modulation. For the sake of comparison, the mininum distance of uncoded modulation schemes and the free distance of the combination of binary convolutional codes with 4-PSK modulation is also shown. Reprinted with permission from [14].

is $B = 1/T$, where T is the signaling interval (see Chapter 3). As observed in [14], this figure shows how TCM schemes can achieve significant coding gains for throughput values for which efficient combinations of coding and modulation were not previously known, that is, beyond 2 bits/s/Hz. It should also be stressed that significant coding gains are achieved by TCM schemes having as few as 4, 8, and 16 states. A rule of thumb suggested in [14] is that 3 dB can be gained with 4 states, 4 dB with 8 states, nearly 5 dB with 16 states, and up to 6 dB with 128 or more states. With higher numbers of states the returns are diminishing.

4.5.1 A Simple Lower Bound

We shall now describe a lower bound to the minimum distance of a TCM scheme. This bound is constructive, and it assumes the Ungerboeck model for the scheme. Consider Fig. 4.1. The signal constellation has 2^{m+1} signals, which are set-partitioned. The uncoded digit $c_i^{(m+1)}$ chooses a subconstellation from a signal set with minimum Euclidean distance δ_{m+1}, $c_i^{(m)}$ chooses a subconstellation from a signal set with minimum distance $\delta_m < \delta_{m+1}, \ldots, c_i^{(\tilde{m}+2)}$ chooses a subconstellation from a signal set with minimum distance $\delta_{\tilde{m}+2} < \cdots < \delta_{m+1}$. Finally, the encoded digits $c_i^{(\tilde{m}+1)}, \ldots, c_i^{(1)}$ select one signal from a subconstellation with $2^{\tilde{m}+1}$ signals with minimum Euclidean distance δ_1.

Consider now an error event in the decoding procedure. It will involve at least one of the $c_i^{(j)}$'s, so that we have

$$d_{\text{free}} \geq \delta_{\tilde{m}+2}$$

if $j \geq \tilde{m} + 2$, and

$$d_{\text{free}} \geq d_{\text{free}}^{\text{H}} \delta_1$$

otherwise (here $d_{\text{free}}^{\text{H}}$ denotes the free Hamming distance of the convolutional code). In conclusion, we get

$$d_{\text{free}} \geq \min\{\delta_1, d_{\text{free}}^{\text{H}} \delta_{\tilde{m}+2}\}. \tag{4.56}$$

The bound (4.56) does not reflect much of the structure of the trellis code, and hence it may be rather loose. However, it may be instructive. For example, it can be observed how it shows separately the roles, played in determining the minimum distance, of the convolutional code (through $d_{\text{free}}^{\text{H}}$) and of the signals used by the modulator (through the Euclidean distances δ_j).

4.6 Sphere-Packing Upper Bounds

In the following analysis, we denote by 2^v the number of states in the trellis, by m the number of bits per channel use, and by N the dimensionality of the signals transmitted through the channel. As we have considered before, the free distance of a TCM scheme is the smallest among the Euclidean distances of pairs of signal sequences arising from an error event. Each trellis path in an error event of length L involves L signals with dimensionality N, or, equivalently, one point in a space whose dimensionality does not exceed NL. In this geometric framework, d_{free} is the smallest Euclidean distance between two such points for $L = 1, 2, \ldots$. If we think of these points as surrounded by NL-dimensional spheres, we can see that an efficient TCM scheme (i.e., one maximizing the least separation of points) corresponds to an efficient sphere packing in a space with NL dimensions. Thus the problem of finding upper bounds to d_{free} is equivalent to a sphere-packing problem in a Euclidean space.

The following procedure may be followed to obtain sphere-packing upper bounds [15]:

1. Calculate the number of points, say $M(L)$, and the dimensionality, say $N(L)$, of the packings for each value of L.
2. Upper bound the distance $d(L)$ between sphere centers given $M(L)$ and $N(L)$ for each L.
3. Choose the minimum $d(L)$ over all L as the upper bound on d_{free}.

Consider first the computation of $M(L)$. With reference to (3.9), let the encoded signal x_i depend on the source symbols

$$b_i^{(j)}, b_{i-1}^{(j)}, \ldots, b_{i-v_j}^{(j)}, \qquad j = 1, \ldots, m.$$

Alternatively, v_j is the number of memory elements in the jth branch of the convolutional encoder. For a rate $m/(m+1)$ linear convolutional encoder, $M(L)$ is given by [16]

$$M(L) = \prod_{j=1}^{m} \max\{2^{(L-v_j)}, 1\}. \tag{4.57}$$

To prove this, let S_A and S_B be two trellis states that are connected by a path of length L through the trellis. Without loss of generality we count the number of such paths starting at time 0 and ending at time L. To do so, consider the contents of the memory elements on the jth input stream. The initial contents $b_{-1}^{(j)}, b_{-2}^{(j)}, \ldots, b_{-v_j}^{(j)}$ are determined by S_A, the final contents $b_{L-1}^{(j)}, b_{L-2}^{(j)}, \ldots, b_{L-v_j}^{(j)}$ are determined by S_B, and the other values $b_0^{(j)}, \ldots, b_{L-v_j-1}^{(j)}$ can be chosen freely. This holds for all j, so the number of paths from S_A to S_B is given by the right-hand side of (4.57). This number will be the same for all pairs that can be connected by a path of length L, and if the encoder is assumed to be *linear*, the number of paths that split from a reference path and remerge in it later does not depend on the reference path.

Every path of length L determines a sequence $\mathbf{x} = (x_1, \ldots, x_L)$ of encoder outputs, which label the edges of the path in the code trellis. If $\mathbf{x}$ and $\mathbf{x}'$ denote two distinct (not necessarily disjoint) paths connecting S_A and S_B, then

$$d_{\text{free}}^2 \leq \|\mathbf{x} - \mathbf{x}'\|^2. \tag{4.58}$$

If $\overline{E}$ is the average energy of these $M(L)$ vectors, then

$$\overline{E} = \frac{1}{M(L)} \sum \|\mathbf{x}\|^2 \tag{4.59}$$

and (4.58) implies that

$$d_{\text{free}}^2 \leq r^*(M(L); NL)\overline{E} \tag{4.60}$$

where $r^*(M, s)$ is the maximum $d_{\text{free}}^2/\overline{E}$ that can be obtained by M vectors in the Euclidean s-dimensional space. Since every edge appears in the same number of paths, the average of $\overline{E}/L$ over all pairs of states connected by paths of length L is just E, the average signal energy. (See [17] for a rigorous proof of this fact.)

In conclusion, for some pair of states we have $\overline{E} \leq LE$, and consequently the bound:

$$\frac{d_{\text{free}}^2}{E} \leq \min_L \{r^*(M(L); NL)\}. \tag{4.61}$$

Any exact evaluation of $r^*(M, s)$ or of an upper bound to this quantity yields, through (4.61), an upper bound to the normalized free distance.

4.6.1 A Universal Upper Bound

A simple upper bound to d_{free} can be obtained by using a simplex bound on $r^*(M; s)$ [15, 16]. We observe that the most efficient way of packing M spheres in a space of dimensionality greater than or equal to $M - 1$ is in a regular simplex [9, p. 191]. Conversely, M spheres in a space of less than $M - 1$ dimensions may not be packed as efficiently as the regular simplex in $M - 1$ dimensions. This bound, which is independent of the dimension s (and therefore weak), states that

$$r^*(M; s) \leq 2\frac{M}{M-1}. \tag{4.62}$$

Proof. For M vectors $\mathbf{v}_i$, $i = 1, \ldots, M$, we have

$$\min_{i \neq j} \|\mathbf{v}_i - \mathbf{v}_j\|^2 \leq \frac{1}{M(M-1)} \sum_{i \neq j} \|\mathbf{v}_i - \mathbf{v}_j\|^2$$

$$= \frac{1}{M(M-1)} \sum_{i \neq j} \left[\|\mathbf{v}_i\|^2 + \|\mathbf{v}_j\|^2 - 2(\mathbf{v}_i, \mathbf{v}_j) \right]$$

$$= \frac{1}{M(M-1)} \left[2M^2\overline{E} - 2\|\sum_i \mathbf{v}_i\|^2 \right]$$

$$\leq \frac{2M}{M-1}\overline{E}. \qquad \square$$

Hence the normalized minimum distance of any TCM scheme satisfies

$$\frac{d_{\text{free}}^2}{E} \leq \min_L \frac{2M(L)}{M(L)-1} L. \tag{4.63}$$

The quantity $M(L)$ is given by (4.57). In particular, if $L \geq \max_i v_i$, then $M(L) = 2^{mL-v}$ whatever the trellis structure. If, in addition, m divides v, error events of length $L = (v/m) + t$, $t = 1, 2, \ldots$, can occur. This implies that [16]

$$\frac{d_{\text{free}}^2}{E} \leq \min_{t \geq 1} \left[2\frac{2^{tm}}{2^{tm}-1}\left(\frac{v}{m} + t\right) \right].$$

4.7 Other Sphere-Packing Bounds

The derivation of an upper bound valid for all TCM schemes would require a general solution to the sphere-packing problem. However, this turns out to be an open problem (see, e.g.,[18]). To derive bounds to the free distance of TCM schemes, one should rather resort to available results involving infinite packings or packings of a special type. In this section we see how these bounds can be obtained.

4.7.1 Constant-Energy Signal Constellations [15]

Consider TCM schemes based on constant-energy signals, that is, constellations of code vectors lying on the surface of a hypersphere. In this situation bounds are available for the packing of spherical caps on the surface of an s-dimensional sphere. On the hypersphere, define the angular radius of its spherical caps as the angle subtended at the center. Let M denote the maximum number of spherical caps of angular radius $\phi/2$ that may be placed on an s-dimensional unit sphere. A bound on M for all the values of ϕ was derived by Levenshtein in [19], and used by Pottie and Taylor in [15] to derive bounds on d_{free}. Some of these results are tabulated in Tables 4.2 and 4.3.

4.7.2 Rectangular Signal Constellations [15]

In [15], Pottie and Taylor derive bounds on the free distance of TCM schemes using rectangular constellations. These bounds are based on the Levenshtein bound and the modified Johnson bound [20, pp. 523 ff.]. Tables 4.4 and 4.5 summarize the results.

4.7.3 An Upper Bound for PSK Signals [21]

From (4.57) we have, for $L \geq v$,

$$M(L) = 2^{Lm-v}. \tag{4.64}$$

TABLE 4.2 Bounds to the Normalized Square Free Distance d_{free}^2 of Rate 2/3 TCM Schemes Based on 8-PSK with 2^v States.

v	Lower Bound[a]	Simplex Bound	PT Bound[b]
2	4.0	5.3	5.3
3	4.6	6.9	6.7
4	5.2	8.0	8.0
5	5.8	9.1	9.0
6	6.3	10.7	10.1
7	6.3	11.4	11.2
8	6.9	12.8	12.3

[a] From actual constructions.
[b] From [15].

TABLE 4.3 Bounds to the Normalized Square Free Distance d_{free}^2 of Rate 3/4 TCM Schemes Based on 16-PSK with 2^v States.

v	Lower Bound[a]	Simplex Bound	PT Bound[b]
2	1.3	4.0	2.6
3	1.5	4.6	3.3
4	1.6	5.3	3.9
5	1.9	6.4	4.5
6	2.0	6.9	5.1
7		8.0	5.6
8		8.5	6.2

[a] From actual constructions.
[b] From [15].

TABLE 4.4 Bounds to the Normalized Square Free Distance d_{free}^2 of Rate 2/3 TCM Schemes Based on 8-QAM with 2^v States.

v	Lower Bound[a]	Simplex Bound	PT Bound[b]
2	3.2	5.3	4.0
3	4.0	6.9	4.8
4	4.8	8.0	6.4
5	4.8	9.1	7.2
6	5.6	10.7	8.0
7	6.4	11.4	8.8
8	6.4	12.8	9.6

[a] From actual constructions.
[b] From [15].

TABLE 4.5 Bounds to the Normalized Square Free Distance d_{free}^2 of Rate 4/5 TCM Schemes Based on 32-QAM with 2^v States.

v	Lower Bound[a]	Simplex Bound	PT Bound[b]
2	0.76	2.7	0.95
3	0.95	4.0	1.14
4	1.14	4.3	1.52
5	1.14	4.3	1.71
6	1.33	5.3	1.90
7	1.52	6.1	2.10
8	1.52	6.4	2.29

[a] From actual constructions.
[b] From [15].

Let a vector

$$\mathbf{x} = (x_1, \ldots, x_L)$$

of signals of the PSK type correspond to every path of length L in the trellis. We assume, without loss of generality, that these signals have unit energy, that is, they are such that $x_j \in \mathbf{R}^2$, and $\|x_j\| = 1$, $j = 1, \ldots, L$. Thus the $M(L)$ paths connecting the states S_A and S_B correspond to $M(L)$ signal vectors, which we denote as

$$\mathbf{x}_i = (x_i^{(1)}, \ldots, x_i^{(L)}), \quad x_i^{(j)} \in \mathbf{R}^2, \quad \|x_i^{(j)}\| = 1; \quad i = 1, \ldots, M(L); \quad j = 1, \ldots, L. \tag{4.65}$$

If we represent each signal $x_i^{(j)}$ as a point on the unit circle, then the signal is completely determined by the angle $\theta_i^{(j)}$, $0 \leq \theta_i^{(j)} < 2\pi$. This in turn implies that each vector $\mathbf{x}_i$ is completely determined by the vector

$$\boldsymbol{\theta}_i = (\theta_i^{(1)}, \ldots, \theta_i^{(L)}).$$

The squared Euclidean distance between two vectors $\mathbf{x}_1$ and $\mathbf{x}_2$ is equal to

$$d_{1,2}^2(L) = 4 \sum_{j=1}^{L} \sin^2 \frac{\theta_1^{(j)} - \theta_2^{(j)}}{2} \tag{4.66}$$

We shall now proceed to bound from above the minimum Euclidean distance $d_{\text{free}}^2(L)$ among $M(L)$ vectors $\mathbf{x}_i$. Consider first the set of vectors $\boldsymbol{\theta}_i$. By introducing in it the mod-2π addition $\oplus$, they form an L-dimensional hypercube, which we denote by $C_L(2\pi) = ([0, 2\pi])^L$.

For our derivation we first need a lemma relating the minimum Euclidean distance in the signal space to the minimum Euclidean distance in $C_L(2\pi)$. The latter is defined as

$$\Delta_{\text{free}}(L) = \min_{i \neq j} \|\boldsymbol{\theta}_i - \boldsymbol{\theta}_j\| = \min_{i \neq j} \sqrt{\sum_{k=1}^{L} \left(\theta_i^{(k)} - \theta_j^{(k)}\right)^2} \tag{4.67}$$

Lemma 4.2 For any set $\{\boldsymbol{\theta}_1, \boldsymbol{\theta}_2, \ldots, \boldsymbol{\theta}_M\}$, $\boldsymbol{\theta}_i \in C_L(2\pi)$, we have

$$\Delta_{\text{free}}(L) \geq 2\sqrt{L} \arcsin \frac{d_{\text{free}}(L)}{2\sqrt{L}} = \Delta_0. \tag{4.68}$$

Proof. Observe that for $0 < z < \pi^2/4$, the function $f(z) = \sin^2 \sqrt{z}$ is concave $\cap$. Hence

$$\sin^2 \sqrt{\frac{1}{L} \sum_{k=1}^{L} y_k^2} \geq \frac{1}{L} \sum_{k=1}^{L} \sin^2 y_k. \tag{4.69}$$

Then let
$$\phi_k = \frac{\theta_m^{(k)} - \theta_n^{(k)}}{2}$$

so that $|\phi_k| < \pi$. For $0 < \phi_k < \pi$, there are two values of ϕ_k that give the same value of $\sin \phi_k$. Call y_k the smallest, so that $0 < y_k^2 < \pi^2/4$. Then

$$\sin^2 \sqrt{\frac{1}{4L} \sum_{k=1}^{L} \left(\theta_m^{(k)} - \theta_n^{(k)}\right)^2} \geq \frac{1}{L} \sum_{k=1}^{L} \sin^2 \frac{\theta_m^{(k)} - \theta_n^{(k)}}{2}$$

or

$$\sin^2 \frac{1}{2\sqrt{L}} \|\boldsymbol{\theta}_m - \boldsymbol{\theta}_n\| \geq \frac{1}{4L} d_{\text{free}}^2(L)$$

and finally,

$$\sin \frac{1}{2\sqrt{L}} \Delta_{\text{free}}(L) \geq \frac{1}{2\sqrt{L}} d_{\text{free}}(L)$$

which is equivalent to (4.68). □

Now take a body A, that is, a set of points $\mathbf{a} = (a_1, \ldots, a_L)$ in $C_L(2\pi)$. For any $\boldsymbol{\theta} \in C_L(2\pi)$, we define the shifted version of A as

$$A \oplus \boldsymbol{\theta} = \{\mathbf{a} \oplus \boldsymbol{\theta} \mid \mathbf{a} \in A\}. \quad (4.70)$$

Consider then a set $\Theta = \{\boldsymbol{\theta}_0 = \mathbf{0}, \boldsymbol{\theta}_1, \ldots, \boldsymbol{\theta}_{M-1}\}$ of M elements $\boldsymbol{\theta} \in C_L(2\pi)$ such that the sets $A \oplus \boldsymbol{\theta}_i$ are disjoint:

$$(A \oplus \boldsymbol{\theta}_i) \cap (A \oplus \boldsymbol{\theta}_j) = \emptyset, \quad \text{for } j \neq i. \quad (4.71)$$

With any pair (A, Θ), called an *A-packing*, we associate the corresponding packing density

$$\rho^*(A, \Theta) = \frac{M|A|}{|C_L(2\pi)|} \quad (4.72)$$

where $|\cdot|$ denotes the volume of its argument. We also define

$$\rho^*(A) = \max_{\boldsymbol{\theta}_1, \ldots, \boldsymbol{\theta}_{M-1}} \rho^*(A, \Theta). \quad (4.73)$$

In a similar way, we can introduce the "usual" maximum packing density $\rho(A)$ of the whole space $\mathbf{R}^L$, as packed by using shifted versions of the body A defined through the usual vector sum instead of (4.70). We have the following lemma.

Lemma 4.3

$$\rho^*(A) \leq \rho(A). \quad (4.74)$$

Proof. The hypercube $C_L(2\pi)$ tessellates the Euclidean space $\mathbf{R}^L$. Thus any A-packing of $C_L(2\pi)$ extends to an A-packing of $\mathbf{R}^L$ by replicating the packing in each copy of $C_L(2\pi)$. □

Consider then the L-dimensional sphere with radius a whose center is coincident with the center of $C_L(2\pi)$. Denote by $S_L(a)$ the intersection of this sphere with $C_L(2\pi)$, and consider the packing $(S_L(a), \Theta)$. For $a \leq \pi$ the packing is made by real spheres—otherwise, a portion of them has to be cut out. We have, because of Lemma 4.3,

$$\rho^*\left[S_L\left(\frac{\Delta_0}{2}\right), \Theta\right] \leq \rho\left[S_L\left(\frac{\Delta_0}{2}\right)\right]. \tag{4.75}$$

For $\Delta_0 \leq 2\pi$, $S_L(\Delta_0/2)$ is a real sphere, and hence the left-hand side of (4.75) can be computed explicitly. By using (4.72), we get

$$\frac{M(\Delta_0/2)^L V_L}{(2\pi)^L} \leq \rho_L \tag{4.76}$$

where V_L is the volume of the L-dimensional sphere with unit radius:

$$V_L = \frac{\pi^{L/2}}{\Gamma(L/2 + 1)}$$

and ρ_L is the maximum packing density of $\mathbf{R}^L$ by congruent spheres. Denote now by δ_ℓ the maximum *center packing density* of $\mathbf{R}^\ell$ by congruent spheres [23] (i.e., $\delta_L = \rho_L/V_L$). We have, from (4.76),

$$\frac{M\Delta_0^L}{(4\pi)^L} \leq \delta_L \tag{4.77}$$

and finally,

$$\Delta_0 \leq 4\pi\left(\frac{\delta_L}{M}\right)^{1/L}. \tag{4.78}$$

As the volume of $S_L(\Delta_0/2)$ is an increasing function of Δ_0, assume for simplicity that inequality (4.76) does not hold for $\Delta_0 = 2\pi$. This is equivalent to saying that the following is true:

$$\frac{M\pi^L V_L}{(2\pi)^L} \geq \rho_L \tag{4.79}$$

that is,

$$M \geq 2^\ell \delta_\ell. \tag{4.80}$$

By combining (4.68) and (4.78), we get

$$\arcsin\frac{d_{\text{free}}(L)}{2\sqrt{L}} \leq \frac{2\pi}{\sqrt{L}}\left(\frac{\delta_L}{M}\right)^{1/L}. \tag{4.81}$$

If the right-hand side of (4.81) does not exceed $\pi/2$, that is, if the following inequality holds,

$$M \geq \delta_L \left(\frac{4}{\sqrt{L}}\right)^L \tag{4.82}$$

then (4.81) is equivalent to

$$d_{\text{free}}(L) \leq 2\sqrt{L} \sin\left[\frac{2\pi}{\sqrt{L}}\left(\frac{\delta_L}{M}\right)^{1/L}\right]. \tag{4.83}$$

This upper bound to the value of $d_{\text{free}}(L)$ holds provided that the inequalities (4.80) and (4.82) are satisfied, which implies that

$$M \geq \delta_L 2^L \max\left\{1, \left(\frac{2}{\sqrt{L}}\right)^L\right\} \tag{4.84}$$

or if $L \geq 4$ is assumed,

$$M \geq \delta_L 2^L. \tag{4.85}$$

We can summarize our findings through the following lemma.

Lemma 4.4 Consider the M vectors

$$\mathbf{x}_i = \left(x_i^{(1)}, \ldots, x_i^{(L)}\right), \quad x_i^{(j)} \in \mathbf{R}^2, \quad \|x_i^{(j)}\| = 1; \quad i = 1, \ldots, M; \quad j = 1, \ldots, L. \tag{4.86}$$

If M satisfies inequality (4.84), the minimum Euclidean distance among these vectors is upper bounded by (4.83). □

Finally, observe that if $L \geq v$, we have

$$M(L) = 2^{Lm-v}.$$

Thus, with $L \geq \max\{4, v\}$, and using the Blichfeldt bound for ρ_L [18]:

$$\rho_L \leq (L+2)2^{-1-L/2} \tag{4.87}$$

we get the upper bound we were looking for.

Theorem 4.1 The free distance of any TCM scheme with 2^v states, rate $m/(m+1)$, and unit-energy PSK signals satisfies the inequality

$$d_{\text{free}} \leq 2 \min_{L \in L(m,v)} \left\{\sqrt{L}\sin\left[\frac{\pi}{\sqrt{L2^{m-1}}}(\delta_L 2^v)^{1/L}\right]\right\} \tag{4.88}$$

where

$$L(m, v) = \{L \geq \max\{4, v\} : [\pi 2^{(2m-1)}]^{L/2} \geq 2^{v-1}(L+2)\Gamma(L/2+1)\}.$$

(4.89)

□

Numerical values of the upper bound (4.88) are shown in Table 4.6, along with lower bounds obtained from actual designs of trellis codes based on PSK signals. The upper bound is also plotted in Fig. 4.28 in the form of a coding gain, that is, of the ratio between the maximum value of d_{free}^2 and the squared minimum distance afforded by 2^m-ary symmetric PSK modulation. All the computations were

TABLE 4.6 Bounds to the Normalized Square Free Distance of TCM Schemes with 2^{m+1} Signals and 2^v States.[a]

m	v	LB1	LB2	UB1	UB2	UB3
1	2	7.2		10.7		
1	3	8.0		13.3		
1	4	8.8		16.0		
1	5	10.4		18.3		
1	6	11.2		20.6		
1	7	12.8		22.9		
1	8	13.6		25.1		
1	9	14.5		27.4		
1	10	15.2		29.7		
1	11	16.0		32.0		
2	1	2.6	4.0		4.6	
2	2	4.0	4.0	5.3	6.1	5.3
2	3	4.6	4.6	6.9	7.6	6.7
2	4	5.2	5.2	8.0	8.9	8.0
2	5	5.8			10.2	9.0
2	6	6.3		10.7	11.5	10.1
3	1	0.7	1.2	4.0	1.2	
3	2	1.3	1.4		1.7	2.6
3	3	1.5	1.6	4.6	2.0	3.3
3	4	1.6	1.6	5.3	2.4	3.9
3	5	1.9		6.4	2.7	4.5
3	6	2.0		6.9	3.1	5.1
4	1				0.3	
4	2				0.4	
4	3				0.5	
4	4			4.3	0.6	
4	5				0.7	
4	6				0.8	

Source: The results concerning symmetric signals (i.e., signals equally spaced around the circumference) are taken from [3]. Those concerning asymmetric PSK are taken from [22].

[a] The lower bounds LB1–LB2 are from actual constructions (LB2 refers to asymmetric PSK signals, described in Chapter 5), UB1 is the simplex bound, UB2 is the bound derived in this section, and UB3 is the asymptotic bound derived in Section 4.7.4. It is observed that UB2 improves on UB1 for $m > 2$.

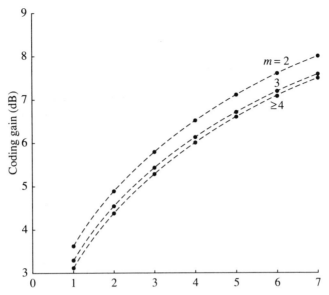

FIGURE 4.28 Maximum coding gain of TCM with PSK signals versus the logarithm of the number of states.

performed by substituting for δ_L in (4.88) its best known upper bound, as tabulated in [23].

4.7.4 An Asymptotic Upper Bound

An upper bound on the normalized minimum distance of a TCM code valid for sufficiently large memory v has been derived in [16]. The method is to reduce the problem to that of bounding the maximum number of points on an N'-dimensional unit sphere for a suitably chosen $N' > N$ with a certain minimum separating angle. The upper bounds derived in [24] are then applied. The normalized minimum distance of any trellis code satisfies, as $v \to \infty$, the following bound:

$$\frac{d_{\text{free}}^2}{E} \leq \begin{cases} \dfrac{6.57v}{N \cdot 4^{m/N}} & \text{if } \dfrac{m}{N} \geq \dfrac{1}{\ln 4} = 0.721 \\ \dfrac{1.74v}{m} & \text{if } \dfrac{m}{N} < \dfrac{1}{\ln 4} \end{cases} \qquad (4.90)$$

4.8 Power Density Spectrum

In this section we consider the power density spectrum of the digital signal at the output of the TCM modulator. In particular, we derive simple sufficient conditions for the resulting spectrum to be equal to the spectrum of an uncoded signal [25].

For simplicity, we only consider *linear* and one- or two-dimensional modulations here; that is, we shall assume that the signal transmitted over the channel has the form

$$y(t) = \sum_{n=-\infty}^{\infty} x_n s(t - nT) \qquad (4.91)$$

where $s(t)$ is a waveform defined in the time interval $(0, T)$ with Fourier transform $S(f)$, and (x_n) is a sequence of complex random variables representing the TCM encoder outputs. If we assume that the source symbols are independent and equally likely, from the regular time-invariant structure of the code trellis it follows that the sequence (x_n) is stationary. Under these conditions, from [9, pp. 32 ff.] we have the following result. If

$$E[x_n] = \mu$$

and

$$E[x_\ell x_m^*] = \sigma_x^2 \rho_{\ell-m} + |\mu|^2$$

so that $\rho_0 = 1$ and $\rho_\infty = 0$, then the power density spectrum of $y(t)$ is

$$\mathcal{G}(f) = \mathcal{G}^{(c)}(f) + \mathcal{G}^{(d)}(f) \qquad (4.92)$$

where $\mathcal{G}^{(c)}(f)$, the continuous part of the spectrum, is given by

$$\mathcal{G}^{(c)}(f) = \frac{\sigma_x^2}{T} |S(f)|^2 \sum_{\ell=-\infty}^{\infty} \rho_k e^{-j2\pi f \ell/T} \qquad (4.93)$$

and $\mathcal{G}^{(d)}(f)$, the discrete part of the spectrum (or *line spectrum*), is given by

$$\mathcal{G}^{(d)}(f) = \frac{|\mu|^2}{T^2} |S(f)|^2 \sum_{\ell=-\infty}^{\infty} \delta\left(f - \frac{\ell}{T}\right). \qquad (4.94)$$

When the random variables x_n are uncorrelated (i.e., $\rho_\ell = \delta_{0,\ell}$, δ the Kronecker symbol), we have the special case

$$\mathcal{G}^{(c)}(f) = \frac{\sigma_x^2}{T} |S(f)|^2. \qquad (4.95)$$

This is the power spectral density that we would obtain without TCM from a modulator using the same waveform $s(t)$ for signaling. In the balance of this section we investigate the conditions for TCM not to shape the signal spectrum (i.e., to give the same power density spectrum as for an uncoded signal).

We first assume that $\mu = 0$, so that no line spectrum appears. Without loss of generality we also assume that $\sigma_x^2 = 1$. Let σ_n denote the encoder state when the symbol x_n is transmitted, and σ_{n+1} the successive state. The correlations ρ_ℓ can be expressed as

$$\rho_\ell = \sum_{x_k} \sum_{\sigma_k} \sum_{\sigma_1} \sum_{x_0} x_k^* x_0 P[x_k, \sigma_k, \sigma_1, x_0].$$

With

$$P[x_k, \sigma_k, \sigma_1, x_0] = P[x_k | \sigma_k] \cdot P[\sigma_k, \sigma_1] \cdot P[x_0 | \sigma_1]$$

this becomes

$$\rho_k = \sum_{\sigma_k} \sum_{\sigma_1} E[x_k \mid \sigma_k] \cdot P[\sigma_k, \sigma_1] \cdot E[x_0 \mid \sigma_1].$$

Hence sufficient conditions for $\rho_\ell = 0$, $\ell \neq 0$, are that $E[x_n \mid \sigma_n] = 0$ or $E[x_{n-1} \mid \sigma_n] = 0$, for each σ_n individually. The first condition is equivalent to stating that for all the encoder states the symbols available to the encoder have zero mean. The second condition is equivalent to stating that for each encoder state, the average of the symbols forcing the encoder to that state is zero. This is the case for a good many TCM schemes.

Finally, consider the line spectrum. A sufficient condition for it to be zero is that $\mu = 0$, that is, that the average of symbols at the encoder output be zero.

REFERENCES

1. E. ZEHAVI and J. K. WOLF, "On the performance evaluation of trellis codes," *IEEE Trans. Inf. Theory,* Vol. IT-33, No. 2, pp. 196–201, Mar. 1987.
2. A. J. VITERBI and J. K. OMURA, *Principles of Digital Communication and Coding.* McGraw-Hill, New York, N.Y., 1979.
3. S. BENEDETTO, M. AJMONE MARSAN, G. ALBERTENGO, and E. GIACHIN, "Combined coding and modulation: Theory and applications," *IEEE Trans. Inf. Theory,* Vol. IT-34, No. 2, pp. 223–236, Mar. 1988.
4. G. D. FORNEY, JR., "Lower bounds on error probability in the presence of large intersymbol interference," *IEEE Trans. Commun.,* Vol. COM-20, pp. 76–77, 1972.
5. E. BIGLIERI and P. J. MCLANE, "Uniform distance and error probability properties of TCM schemes," *ICC'89,* Boston, Mass., June 11–14, 1989.
6. D. DIVSALAR, *Performance of Mismatched Receivers on Bandlimited Channels.* Ph.D. dissertation, University of California, Los Angeles, Calif., 1978.
7. R. C. P. SAXENA, "Optimum encoding in finite state coded modulation," *Report TR83-2,* Department of Electrical, Computer and System Engineering, Rensselaer Polytechnic Institute, Troy, N.Y., 1983.
8. M. M. MULLIGAN and S. G. WILSON, "An improved algorithm for evaluating trellis phase codes," *IEEE Trans. Inf. Theory,* Vol. IT-30, No. 6, pp. 846–851, Nov. 1984.
9. S. BENEDETTO, E. BIGLIERI, and V. CASTELLANI, *Digital Transmission Theory.* Prentice-Hall, Englewood Cliffs, N.J., 1987.
10. E. BIGLIERI, "High-level modulation and coding for nonlinear satellite channels," *IEEE Trans. Commun.,* Vol. COM-32, No. 5, pp. 616–626, May 1984.
11. G. D. FORNEY, JR., "Coset codes—Part I: Introduction and geometrical classification," *IEEE Trans. Inf. Theory,* Vol. 34, No. 5, pp. 1123–1151, Sept. 1988.
12. G. UNGERBOECK, "Channel coding with multilevel/phase signals," *IEEE Trans. Inf. Theory,* Vol. IT-28, pp. 56–67, Jan. 1982.
13. S. S. PIETROBON, R. H. DENG, A. LAFANECHÈRE, G. UNGERBOECK, and D. J. COSTELLO, JR., "Trellis-coded multi-dimensional phase modulation," *IEEE Trans. Inf. Theory,* Vol. 36, No. 1, pp. 63–89, Jan. 1990.

14. G. Ungerboeck, "Trellis-coded modulation with redundant signal sets. Part I: Introduction; Part II: State of the art," *IEEE Commun. Mag.*, Vol. 25, No. 2, pp. 5–21, Feb. 1987.
15. G. J. Pottie and D. P. Taylor, "Sphere-packing upper bounds on the free distance of trellis codes," *IEEE Trans. Inf. Theory*, Vol. 34, No. 3, pp. 435–447, May 1988.
16. A. R. Calderbank, J. E. Mazo, and V. K. Wei, "Asymptotic upper bounds on the minimum distance of trellis codes," *IEEE Trans. Commun.*, Vol. COM-33, pp. 305–309, Apr. 1985.
17. A. R. Calderbank, J. E. Mazo, and H. M. Shapiro, "Upper bounds on the minimum distance of trellis codes," *Bell Syst. Tech. J.*, Part I, Vol. 62, pp. 2617–2646, Oct. 1983.
18. C. A. Rogers, *Packing and Covering*. Cambridge University Press, Cambridge, England, 1964.
19. V. I. Levenshtein, "On bounds for packings in n-dimensional Euclidean space," *Sov. Math. Doklady*, Vol. 20, pp. 417–421, 1979.
20. F. J. MacWilliams and N. J. A. Sloane, *The Theory of Error-Correcting Codes*. North-Holland, New York, N.Y., 1978.
21. M. Burnashev and E. Biglieri, "Bounds on the minimum distance of trellis codes," *IEEE Trans. Inf. Theory*, Vol. 35, No. 3, pp. 659–662, May 1989.
22. D. Divsalar, M. K. Simon, and J. H. Yuen, "Trellis coding with asymmetric modulations," *IEEE Trans. Commun.*, Vol. COM-35, No. 2, pp. 130–141, Feb. 1987.
23. J. Leech and N. J. A. Sloane, "Sphere packing and error-correcting codes," *Canadian Journal of Mathematics*, Vol. 23, No. 4, pp. 718–745, 1971.
24. G. A. Kabatianskii and V. I. Levenshtein, "O granicakh dlya upakovok na sfere i v prostranstve" (in Russian), *Problemy Peredachi Informacii*, Vol. 14, No. 1, pp. 3–25, 1978. English translation: "Bounds for packings on a sphere and in space," *Problems of Information Transmission*, Vol. 14, No. 1, pp. 1–17, 1978.
25. E. Biglieri, "Ungerboeck codes do not shape the signal power spectrum," *IEEE Trans. Inf. Theory*, Vol. IT-32, No. 4, pp. 595–596, July 1986.

CHAPTER 5

One- and Two-Dimensional Modulations for TCM

5.1 Introduction

The concept of TCM was explained in Chapter 3. There are basically two types of realizations for TCM. First, we have the Ungerboeck form [1–3] in Fig. 5.1, where the trellis code is generated by a rate $\tilde{m}/(\tilde{m} + 1)$ code, followed by a natural bit mapper. Also, the Calderbank–Mazo form [4] in Fig. 5.2 can be used to generate the trellis code. The Calderbank–Mazo form of a TCM code is quite easy to derive and gives the trellis code in the form of a function of an input, a sliding block, of source bits. Finally, the Ungerboeck form in Fig. 5.1 is easily derived from the Calderbank–Mazo formula. This is the basis of our step-by-step design technique, presented in [5] and repeated here.

The purpose of this chapter is twofold. First, we display some of the modulation techniques that can be used in trellis coding. Also, the coding gain for the modulation is given. With each case we present a realization of the code in both Ungerboeck and Calderbank–Mazo forms. We also show how to convert codes given in Ungerboeck's tables [1–3] into realizations. The chapter concludes with some TCM codes in which the signal constellation is asymmetric.

5.2 Step-by-Step Design Procedure

In this section we describe a design procedure for feedforward linear trellis codes. Only rate $m/m + 1$ codes are considered. The input data for the design procedure are the number of code states and the signal constellation. The Ungerboeck's rules [1–3] are first used to decide on the transition connectivity of states in the trellis. A minimal complexity, algebraic or analytic, description for the trellis code is then obtained. The final step is the derivation of the underlying convolutional encoder to generate the trellis code. One- and two-dimensional codes are discussed; nonlinear trellis code design is treated in Chapter 8. Examples are given for all cases. In summary, our goal is to determine the Ungerboeck representation in Fig. 5.1, which is preferable in implementation, by first finding the analytic representation in Fig. 5.2 and then converting it to Ungerboeck form.

150　Ch. 5 / One- and Two-Dimensional Modulations for TCM

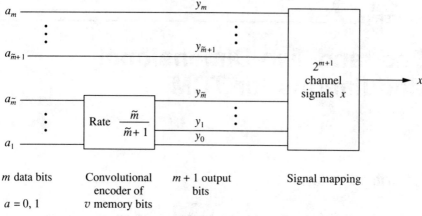

FIGURE 5.1　Ungerboeck's encoder/modulator.

5.2.1 Derivation of the Analytic Description [4]

The general theory of the Calderbank–Mazo [4] formulation of trellis codes was given in Section 3.8. An example was given in Example 3.1. Here we give a presentation in terms of a sliding block of input source bits, as this is a central concept of our step-by-step design procedure.

As described by Calderbank and Mazo, a channel signal depends on m current data bits $(a_1, a_2, \ldots, a_m)$ and on v previous input bits $(a_{m+1}, \ldots, a_{m+v})$. Thus the channel signal x is a function of the $n = m + v$ binary input bits, called the sliding block of input bits. This function can be described as follows:

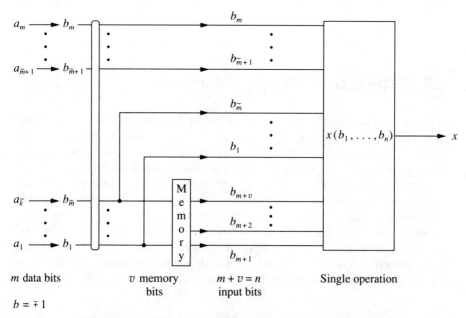

FIGURE 5.2　General structure of the analytic description.

$$x(a_1, a_2, \ldots, a_n) = c_0 + \sum_{i=1}^{n} c_i a_i + \sum_{\substack{i,j=1 \\ j>i}}^{n} c_{ij} a_i a_j$$
$$+ \sum_{\substack{i,j,\ell=1 \\ \ell>j>i}}^{n} c_{ij\ell} a_i a_j a_\ell + \cdots + c_{12\cdots n} a_1 a_2 \cdots a_n. \quad (5.1)$$

Values of ± 1 for the input bits instead of 0 and 1 are more convenient to use when obtaining the analytic description of a trellis code. Therefore, the following conversion is necessary:

$$b_i = 1 - 2a_i \qquad i = 1, 2, 3, \ldots, n. \quad (5.2)$$

With this conversion, 0 and 1 are converted to 1 and -1, respectively. An equivalent expression to (5.1) is

$$x(b_1, b_2, \ldots, b_n) = d_0 + \sum_{i=1}^{n} d_i b_i + \sum_{\substack{i,j=1 \\ j>i}}^{n} d_{ij} b_i b_j$$
$$+ \sum_{\substack{i,j,\ell=1 \\ \ell>j>i}}^{n} d_{ij\ell} b_i b_j b_\ell + \cdots + d_{12\cdots n} b_1 b_2 \cdots b_n \quad (5.3)$$

which is the same as equation (3.12).

The equation in (5.3) can be written in the vector-matrix form

$$\mathbf{X} = \mathbf{BD} \quad (5.4)$$

where

$$\mathbf{X} = \begin{bmatrix} x(1, 1, \ldots, 1) \\ x(-1, 1, \ldots, 1) \\ \vdots \\ x(-1, -1, \ldots, -1) \end{bmatrix}$$

$$\mathbf{B}_i = \begin{bmatrix} 1 & b_1 & b_2 \cdots b_n, & b_1 b_2, b_2 b_3, \ldots, & b_1 b_2 \cdots b_n \end{bmatrix}$$

and

$$\mathbf{D}^T = [d_0, d_1, \ldots, d_{12n}].$$

The ith row of $\mathbf{B}$, $\mathbf{B}_i$, takes its b_i's from the argument of the ith row of $\mathbf{X}$. For D-dimensional modulation, $\mathbf{X}$ is $2^n \times D$, $\mathbf{D}$ is the same, and $\mathbf{B}$ is always $2^n \times 2^n$. Here $n = k + v$, where k is the number of input bits and v the number of memory elements for the trellis code.

Calderbank and Mazo [4] show that $\mathbf{B}$ is an orthogonal matrix. Thus the solution to (5.3) is

$$\mathbf{D} = \frac{\mathbf{B}^T \mathbf{X}}{2^n}. \quad (5.5)$$

In practice, a computer program using simple matrix algebra can be used to calculate the coefficients. Thus, for a trellis description of a code an analytic description can be derived by knowing the input bits and the corresponding signals in the modulation alphabet. A step-by-step PC-based program, with three examples, for both solving for $\mathbf{D}$ in (5.4) and finding d_{free} is given in Appendix C. These are the two main tasks in the design procedure that must be carried out.

5.2.2 Design Rules and Procedure

To obtain analytic code descriptions of minimal complexity and optimum performance, the rules outlined by Ungerboeck [1–3] and Turgeon [5] for signal assignment to trellis transitions should be followed. An analytic code description is of minimal complexity if the formula in (5.3) contains the least number of nonzero d coefficients, and each bit in the input sliding block appears only once in the code formula.

The first step in the design procedure is to determine the number of input bits, $\tilde{m}$, in the sliding block that will pass through the encoder in either Fig. 5.1 or 5.2. The choice of $\tilde{m}$ determines the best value of the Euclidean distance for the TCM scheme and also establishes the connectivity of trellis state variables. To do this, we use Ungerboeck's three design rules, given below as rules U1, U2, and U3. The number of state variables in the code is taken as prescribed.

Ungerboeck stated three rules:

- U1. All signals should be equiprobable, as good codes should be of regular structure.
- U2. Parallel transitions should be assigned signals from the subset with the greatest intrasubset distance.
- U3. The 2^m transitions that diverge from a common state or remerge into a same state must be assigned with signals from one subset at the first level of set partitioning.

The second step of the design process is to determine the state description of the code using the two design rules T1 and T2 below, which are adapted from [5]. In the state-variable specification using U1, U2, and U3 there is considerable freedom in the assignment of code-state variables. However, some state assignments lead to complicated forms of the formula in (5.3). Using rules T1 and T2 has always resulted in the simplest formula. In these rules the signal value is the numerical value associated with a modulation symbol. The signal difference is the numerical difference between two modulation symbols when a single bit in the input sliding block is changed.

Turgeon stated two rules that will ensure minimal complexity for the analytic description of the code. In these rules the maximum signal value (MSV) is the maximum numerical value of a signal. If the signal is a signal vector, the MSV is the maximum sum of the vector coordinates.

- T1. To establish the signal sequence for state 1, determine the maximum signal value at level m of set partitioning. The assignment starts at level m with the two-signal subset containing the MSV. One then moves up one level of set partitioning and repeats the process until the first state is fully defined.
- T2. Within any given dimension, each specific input bit should be associated to a unique signal difference. Note that the first rule accounts for the first $m + 1$ signal differences. A *signal difference* [5] simply represents the difference between any two channel signals when only a single bit changes in the input sliding block of source bits. The *absolute signal difference* is the absolute value of the signal difference. To give a more concrete definition of a signal difference, let $x(b_1, \ldots, b_k, b_{k+1}, \ldots, b_n)$ be the modulation signal

for the ± 1 sliding block, $(b_1, b_2, \ldots, b_k, b_{k+1}, \ldots, b_n)$. Then the ith signal difference is

$$\delta_i = x(b_1, b_2, \ldots, b_i, \ldots, b_k, b_{k+1}, \ldots, b_n)$$
$$- x(b_1, b_2, \ldots, -b_i, \ldots, b_k, b_{k+1}, \ldots, b_n).$$

As stated earlier, the MSV is the maximum value of a signal. If the signal is a vector, the MSV is the maximum sum of the vector coordinates. In one-dimensional cases one can also use the minimum signal label (MSL) instead of the MSV in design rule T1, where the MSL is the minimum numerical value of the symbols used to label the signals in the transmitter signal constellation. Our second example, to follow, illustrates this point. If the MSV is used, the d coefficients are all positive and each bit from the sliding block of source bits appears only once in each term of the code formula. The MSL concept does not have these properties, but it does produce codes with similar structure to that given in [1] to [3]. As long as the signal differences between parallel transitions are kept the same, the state assignment for the first two states, without requiring that the d coefficients all be positive, can be deduced from U1 to U3, but we will use either the MSV or MSL in using rule T1.

To determine the Ungerboeck encoder from the Calderbank–Mazo form, we follow the procedure from Turgeon's thesis [5]. First, a mapping rule must be determined that maps the output bits $z_0, z_2, \ldots, z_m$ [for rate $m/(m+1)$ code] to the signal constellation. The next step involves expressing the output signal x in terms of the output bits. One method for accomplishing this is to find successively for z_i, $i = 0, \ldots, m$, the signal difference in each dimension between $z_i = 1$ and $z_i = -1$. The z's are then assigned to a coefficient half their corresponding signal difference.

Finally, once expressions for each dimension of x, the output signal, are determined in terms of both the input bits $b_1, b_2, \ldots, b_n$ and the output bits $z_0, z_1, \ldots, z_m$, the output bits can be related to the input bits. These relationships will yield the structure of the convolutional encoder. Since the input–output bits relationship is in terms of 1 and -1, the corresponding relationships for 0 and 1 must be determined. This process is simple enough for linear codes, for a multiplication in the ± 1 convention is equivalent to XOR in the 0 and 1 convention. Thus the output 0 and 1 bits $y_0, y_1, \ldots, y_m$ are expressed in terms of the input bits $a_1, a_2, \ldots, a_n$. For the nonlinear case, more complicated equations must be solved. This is treated in Chapter 8.

The procedure just described above assumes that $\tilde{m}$, the number of encoded input bits, is known. The rest of the input bits pass directly to the mapper and represent parallel transitions in the code trellis. The first step in the design procedure is to find $\tilde{m}$ using Ungerboeck rules U1 to U3. After this step, the design proceeds as described above.

5.3 One-Dimensional Examples

We first consider AM modulation. The modulation symbols will be $-(M-1)$, $-(M-3), \ldots, -1, 1, \ldots, (M-3), (M-1)$. Our first example concerns $M=4$.

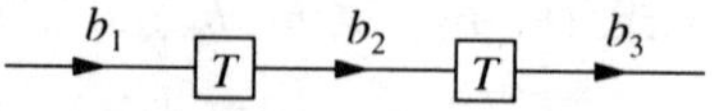

FIGURE 5.3 Input bit relationship for four-state encoder, rate 1/2.

EXAMPLE 5.1   4-AM, FOUR-STATE

In this trellis modulation scheme there are two memory bits ($v = 2$). Since this is a rate 1/2 code, $m = 1$. There are no parallel transitions, so that $\tilde{m} = 1$ as well (see Fig. 5.1). Figure 5.3 describes the relationship between the data bit and the memory units (shift registers). The general structure is given in Fig. 5.2. The delay of the shift registers is T seconds, where T is the modulation interval. Input bits b_2 and b_3 define the state of the trellis. The signal set and its partitioning is shown in Fig. 5.4. Signal assignments to the trellis transitions are done following rules T1 to T2 and U1 to U3. For the signal constellation in Fig. 5.4, the MSV is 3. The trellis structure is shown in Fig. 5.5. Rule T1 is simply applied in this case as $m = 1$ and there is only one level of set partitioning. Also shown in Fig. 5.5 are the signal differences $\delta_1 = 4$, $\delta_2 = 2$, and $\delta_3 = 4$. Note that the signal differences are minimal and are maintained throughout the trellis in Fig. 5.5. This is the essence of rule T2. The coding gain relative to 2-AM modulation is 2.55 dB. For the trellis structure in Fig. 5.5 and the definitions of (b_1, b_2, b_3) given, we have the following case of (5.3):

$$\begin{bmatrix} x(1, & 1, & 1) \\ x(-1, & 1, & 1) \\ x(1, & -1, & 1) \\ x(-1, & -1, & 1) \\ x(1, & 1, & -1) \\ x(-1, & 1, & -1) \\ x(1, & -1, & -1) \\ x(-1, & -1, & -1) \end{bmatrix} = \begin{bmatrix} 3 \\ -1 \\ 1 \\ -3 \\ -1 \\ 3 \\ -3 \\ 1 \end{bmatrix} = \begin{bmatrix} b_1 & b_2 & b_3 & b_1b_2 & b_1b_3 & b_2b_3 & b_1b_2b_3 \\ 1 & 1 & 1 & 1 & 1 & 1 & 1 \\ -1 & 1 & 1 & -1 & -1 & 1 & -1 \\ 1 & -1 & 1 & -1 & 1 & -1 & -1 \\ -1 & -1 & 1 & 1 & -1 & -1 & 1 \\ 1 & 1 & -1 & 1 & -1 & -1 & -1 \\ -1 & 1 & -1 & -1 & 1 & -1 & 1 \\ 1 & -1 & -1 & -1 & -1 & 1 & 1 \\ -1 & -1 & -1 & 1 & 1 & 1 & -1 \end{bmatrix} \begin{bmatrix} d_1 \\ d_2 \\ d_3 \\ d_{12} \\ d_{13} \\ d_{23} \\ d_{123} \end{bmatrix}$$

Note that we dropped d_0 and its associated column of 1's. Now solution for the d coefficients via (5.4) gives $d_2 = 1$ and $d_{13} = 2$, with all other components of D equal to zero. Therefore, the analytic description of the trellis code is

$$x = b_2 + 2b_1b_3. \tag{5.6}$$

The transmitter implementation for the analytic description is shown in Fig. 5.6.

The formula in (5.6) can also be derived from the *half-signal-difference rule*. This rule states that the coefficient of b_i in (5.6) is $\delta_i/2$ [5]. The three signal differences are shown in Fig. 5.5. Thus the coefficient in (5.6) for b_2 is 1, and for

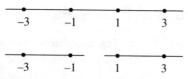

FIGURE 5.4   Set partitioning, 4-AM.

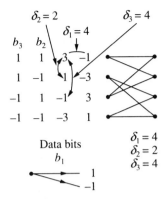

FIGURE 5.5 Trellis structure, four-state 4-AM.

b_1 and b_3 it is 2. Thus b_1 and b_3 occur as a product, as they have the same signal difference. This procedure is really practical only for simple examples. Otherwise, the computer program in Appendix C should be used to solve (5.3).

To obtain the underlying convolutional encoder, we follow the procedure given at the end of Section 5.1.2. One must first determine how the output bits relate to the channel signal. For this example, the natural mapping as shown in Fig. 5.7 will be used. Table 5.1 can be drawn up to give the signal differences between $z_i = 1$ and $z_i = -1$ for $i = 0, 1$. From Table 5.1 the relationship between the channel signal and the output bits is apparent:

$$x = 2z_1 + z_0. \tag{5.7}$$

This formula can be derived using the half-signal-difference rule described above. The signal differences are shown in Table 5.1. By comparing (5.6) and (5.7), it is clear that

$$z_1 = b_1 b_3$$
$$z_0 = b_2.$$

The corresponding relations for 0 and 1 variables are

$$y_1 = a_1 \oplus a_3$$
$$y_0 = a_2.$$

The underlying convolutional encoder and mapper is shown in Fig. 5.8, and this completes the design. □

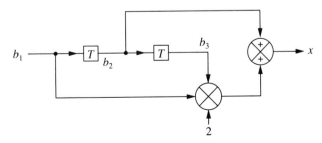

FIGURE 5.6 Analytic description transmitter implementation.

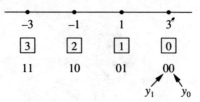

FIGURE 5.7 Signal constellation and mapping, 4-AM.

TABLE 5.1 Signal Mapping and Signal Differences for the 4-AM Signal Set of Fig. 5.4.

Output MSB	z_1	1	1	−1	−1
Bits LSB	z_0	1	−1	1	−1
Channel signal	x	3	1	−1	−3

Effect on x due to z_i:
z_1: $|4| = \delta_1$
z_0: $|2| = \delta_0$

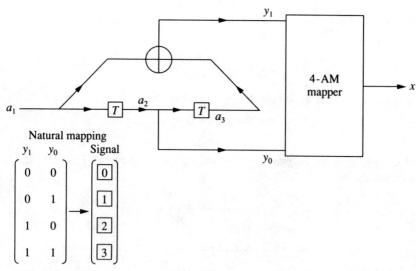

FIGURE 5.8 Four-state, 4-AM convolutional encoder and mapper.

EXAMPLE 5.2 8-PSK, EIGHT-STATE

This trellis code modulation will be generated using the signal, $e^{j(x-1)\pi/8}$, where x is an AM signal as in Example 5.1 [6]. Thus we see that a two-dimensional PSK signal can be generated from a one-dimensional signal constellation. The 8-PSK signal set is shown in Fig. 5.9. The natural mapper is used. The values in the rectangle are the values of x in the argument of the phase-modulated signal, $e^{j(x-1)\pi/8}$. The set partitioning is also shown in Fig. 5.9.

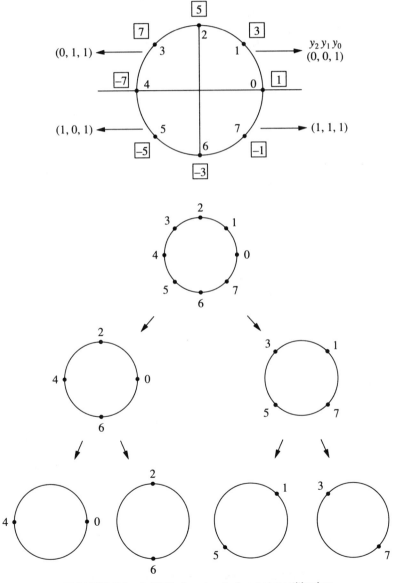

FIGURE 5.9 8-PSK signal set plus set partitioning.

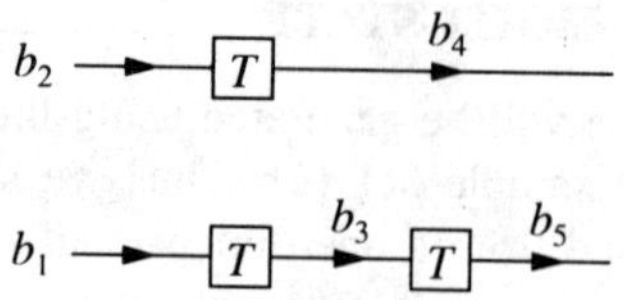

FIGURE 5.10 Inputs and state variables for eight-state code.

We give the code structure in Fig. 5.10. This defines the sliding block of source variables (b_1, b_2, b_3, b_4, b_5). Use of the U1 to U3 rules shows us that $\tilde{m} = 2$ is best and yields $d_{\text{free}}^2 = 4.586$. This gives a coding gain of 3.6 dB relative to 4-PSK, where $d_{\min}^2 = 2.0$. Rules T1 and T2, given earlier, yield the trellis diagram in Fig. 5.11. In using rule T1, we use the concept of an MSL. This gives a signal assignment that follows the signal assignment used in [1] to [3]. As $\tilde{m} = 2$, $(1, -7)$ are the first two signals of state 1 in rule T1 as the MSL $= 0$ and $(1, -7)$ are the associated two signals at the lowest level of the set partitioning in Fig. 5.9. By moving up one level, the MSL at level 1 is 2 and thus $(5, -3)$ completes the signal assignment for state 1. State 2 is defined by repeating this procedure for the other side of the set partitioning shown in Fig. 5.9.

The trellis in Fig. 5.11 has been labeled with the variables x and the values of the sliding block, ($b_1, b_2, \ldots, b_5$). Solution of (5.4) for the trellis in Fig. 5.11 gives $d_3 = -1$, $d_{14} = 4$, and $d_{25} = -2$, with all the rest of the **D** vector equal to zero. Thus

$$x = 4b_1 b_4 - 2b_2 b_5 - b_3. \tag{5.8}$$

The generation of the 8-AM signal set is given in Table 5.2. From the signal differences in Table 5.2, we have

$$x = 4z_2 - 2z_1 - z_0. \tag{5.9}$$

b_5	b_4	b_3	(b_2, b_1)			
			(1, 1)	(1, -1)	(-1, 1)	(-1, -1)
1	1	1	1	-7	5	-3
1	1	-1	3	-5	7	-1
1	-1	1	-7	1	-3	5
1	-1	-1	-5	3	-1	7
-1	1	1	5	-3	1	-7
-1	1	-1	7	-1	3	-5
-1	-1	1	-3	5	-7	1
-1	-1	-1	-1	7	-5	3

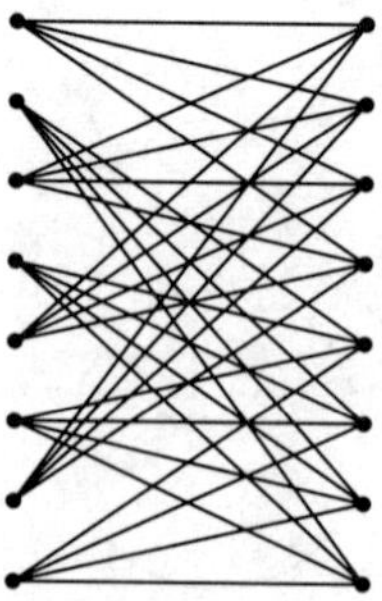

FIGURE 5.11 Trellis for 8-PSK, eight-state code.

TABLE 5.2 Generation of an 8-PSK signal set.

Output bits	z_2	1	1	1	1	−1	−1	−1	−1
	z_1	1	1	−1	−1	1	1	−1	−1
	z_0	1	−1	1	−1	1	−1	1	−1
Output signal	x	1	3	5	7	−7	−5	−3	−1

Effect on x due to z_i:
  z_2: $|8|$
  z_1: $|4|$
  z_0: $|2|$

The coefficient for z_i is the absolute signal difference and the sign of this coefficient is the sign of the signal difference as z_i changes from 1 to −1. Comparing (5.8) and (5.9), we have

$$z_0 = b_3$$
$$z_1 = b_2 b_5$$
$$z_2 = b_1 b_5.$$

Now converting to $(a_1, a_2, a_3, a_4, a_5)$, $a_i = 0, 1$, we have

$$y_0 = a_3$$
$$y_1 = a_2 \oplus a_5$$
$$y_2 = a_1 \oplus a_4$$

and the Ungerboeck form is given in Fig. 5.12. □

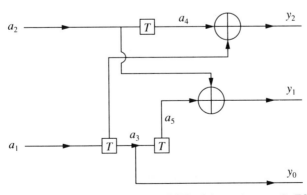

FIGURE 5.12 Ungerboeck encoder for 8-PSK, eight-state, $v = 3$, TCM scheme.

EXAMPLE 5.3   16-PSK, EIGHT-STATE

In this trellis modulation scheme there are three memory bits ($v = 3$). Since this is a rate 3/4 code, $m = 3$. There is one parallel transition, so that $\tilde{m} = 2$ (see Fig. 5.1). Figure 5.13 describes the relationship between the data bit and the memory units (shift registers). Input bits b_4, b_5, and b_6 define the state of the trellis. The signal set and its partitioning are shown in Fig. 5.14. Signal assignments to the trellis transitions are done following rules T1 to T2 and U1 to U3. The trellis structure is shown in Fig. 5.15. In setting the signals for state 1 we used the MSL, which is 0. Thus (0, 8) are the first two signals for state 1. Moving up one level, the MSL among unused signals is 2, and thus (2, 10) is the next signal pair in state 1. Moving up one level, we get (4, 12). Going back down to the lowest level (6, 14) completes the definition of state 1. State 2 follows by considering the unused signals in the other side of the set partitioning in Fig. 5.14. The minimum squared Euclidean distance between coded signal sequences is $d_{\text{free}}^2 = 1.32$, which results in a coding gain of 3.54 dB over uncoded 8-PSK modulation. In solving the matrix equation (5.4), the d coefficients are determined. The results are

$$d_1 = 8$$

$$d_4 = -1$$

$$d_{25} = -2$$

$$d_{36} = -4.$$

Therefore, the analytic description of the trellis code is

$$x = 8b_1 - 4b_3b_6 - 2b_2b_5 - b_4. \qquad (5.10)$$

The transmitter implementation using the analytic expression is shown in Fig. 5.16. Note also that the signal transmitted is $e^{j(x-1)\pi/16}$.

To obtain the underlying convolutional encoder, one must determine how the output bits relate to the channel signal. For this example the natural mapper shown

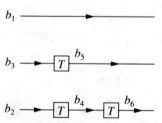

FIGURE 5.13   Input bit relationship for eight-state code, rate 3/4.

5.3 / One-Dimensional Examples 161

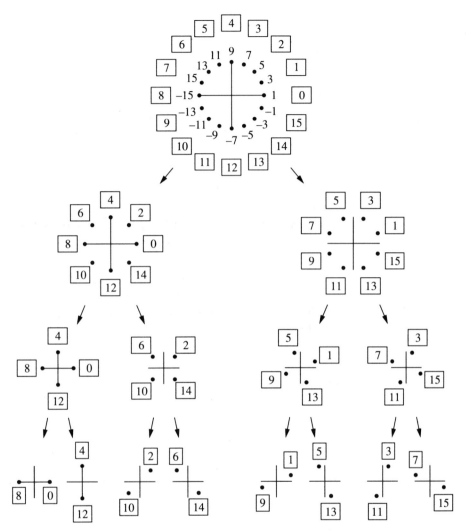

FIGURE 5.14 Set partitioning for 16-PSK.

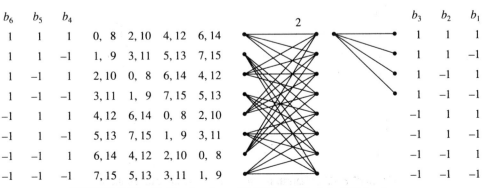

FIGURE 5.15 Trellis structure, 16-PSK, eight-state TCM scheme.

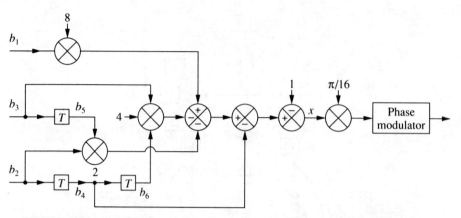

FIGURE 5.16 Analytic description transmitter, 16-PSK, eight-state TCM scheme.

in Fig. 5.17 will be used. Table 5.3 can be drawn up to ascertain the signal differences between $b_i = 1$ and $b_i = -1$ for $i = 0, 1, \ldots, m$.

From Table 5.3 the relationship between the channel signal and the output bits follows from the half-signal-difference rule and is

$$x = 8z_3 - 4z_2 - 2z_1 - z_0. \tag{5.11}$$

By comparing (5.10) and (5.11), it is clear that

$$z_3 = b_1$$

$$z_2 = b_3 b_6$$

$$z_1 = b_2 b_5$$

$$z_0 = b_4.$$

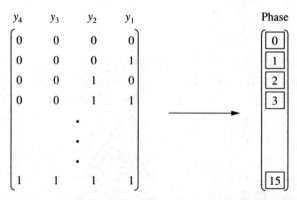

FIGURE 5.17 Natural mapper for 16-point constellation.

TABLE 5.3 Generation of a 16-PSK Signal Set.

	z_3	1	1	1	1	1	1	1	1	-1	-1	-1	-1	-1	-1	-1	-1
Output	z_2	1	1	1	1	-1	-1	-1	-1	1	1	1	1	-1	-1	-1	-1
bits	z_1	1	1	-1	-1	1	1	-1	-1	1	1	-1	-1	1	1	-1	-1
	z_0	1	-1	1	-1	1	-1	1	-1	1	-1	1	-1	1	-1	1	-1
Output signal	x	1	3	5	7	9	11	13	15	-15	-13	-11	-9	-7	-5	-3	-1

Effect on x due to z_i:
  z_3: $|16|$
  z_2: $|8|$
  z_1: $|4|$
  z_0: $|2|$

The corresponding relations for 0 and 1 variables are

$$y_3 = a_1$$
$$y_2 = a_3 \oplus b_6$$
$$y_1 = a_2 \oplus a_5$$
$$y_0 = a_4.$$

The underlying convolutional encoder and mapper are shown in Fig. 5.18. □

5.4 Two-Dimensional Examples

For our two-dimensional examples we consider first the use of the 16-QAM signal set in Fig. 5.19. We finish up the section with a 2/4-PSK code. Here 2-PSK is used in the first modulation interval and 4-PSK is used in the second interval, as two modulation intervals comprise one trellis interval. This is a two-dimensional modulation, as MPSK can be treated as e^{jx}, where x is AM.

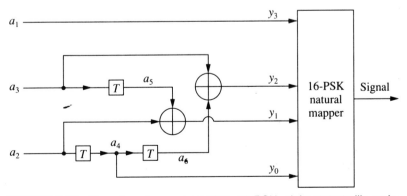

FIGURE 5.18 Ungerboeck representation, 16-PSK, eight-state trellis code.

164 Ch. 5 / One- and Two-Dimensional Modulations for TCM

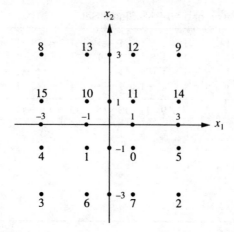

FIGURE 5.19 16-QAM signal set with signal labeling.

EXAMPLE 5.4 16-QAM, EIGHT-STATE

Let us list the important steps that must be carried out. We use $m = 3$ and choose $\tilde{m} = 2$ giving one parallel transition. This must be found by trial and error using rules U1 to U3. The coding gain is 5.33 dB.

- The signal constellation is given in Fig. 5.19.
- We are considering an eight-state code, so that $v = 3$.
- The set partitioning is carried out and is displayed in Fig. 5.20. The signal labels are given in Fig. 5.19.

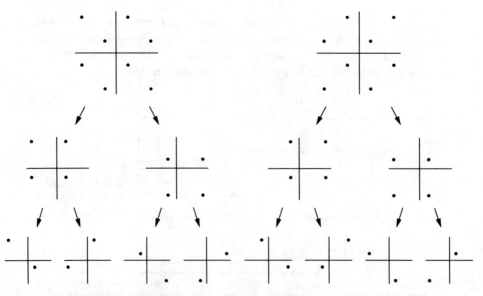

FIGURE 5.20 Set partitioning and signal notation of 16-QAM signal set.

- The trellis in Fig. 5.21 follows from rules U1 to U3 and T1 and T2. In rule T1 we have used the MSL concept. The signal assignment for states 1 and 2 follow exactly the ideas in Examples 5.2 and 5.3. A design of the code under consideration that uses the MSV concept in rule T1 is given in [5].
- From the trellis in Fig. 5.21 we construct (5.4). Here x_1 will be the in-phase component of the complex signal $x = x_1 + jx_2$.
- Solving for **D** in (5.4) gives

$$x_1 = 2b_1 b_2 b_4 b_5 - b_3 b_4 b_6$$

$$x_2 = b_3 b_6 - 2b_1.$$

- The signal set in Fig. 5.19 can be generated as in Table 5.4. Signal differences are also shown and using the half-signal-difference rule (z_0 is chosen to give

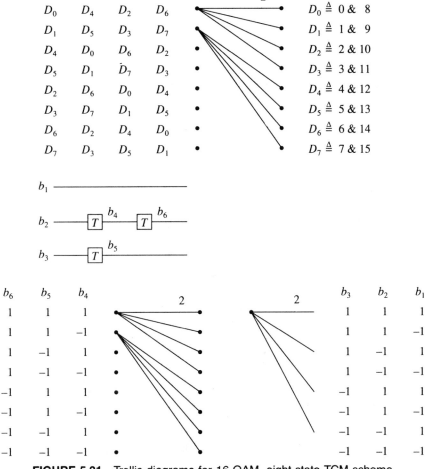

FIGURE 5.21 Trellis diagrams for 16-QAM, eight-state TCM scheme.

TABLE 5.4 Generation of the 16-QAM Signal Set in Fig. 5.19.

Output bits	z_3	1	1	1	1	1	1	1	1	-1	-1	-1	-1	-1	-1	-1	-1
	z_2	1	1	1	1	-1	-1	-1	-1	1	1	1	1	-1	-1	-1	-1
	z_1	1	1	-1	-1	1	1	-1	-1	1	1	-1	-1	1	1	-1	-1
	z_0	1	-1	1	-1	1	-1	1	-1	1	-1	1	-1	1	-1	1	-1
Output signals	x_1	1	-1	3	-3	-3	3	-1	1	-3	3	-1	1	1	-1	3	-3
	x_2	-1	-1	-3	-3	-1	-1	-3	-3	3	3	1	1	3	3	1	1

Effect on x_i due to z_i:

	x_1	x_2
z_0:	$+/-$	No effect
z_1:	$|2|$	$|2|$
z_2:	$|4|$	No effect
z_3:	$|4|$	$|4|$

the proper sign to x_1),

$$x_1 = 2z_0 z_2 z_3 - z_0 z_1$$

$$x_2 = z_1 - 2z_3.$$

Thus $z_0 = b_4$, $z_1 = b_3 b_6$, $z_2 = b_2 b_5$, and $z_3 = b_1$. Converting to (0, 1) variables, we obtain

$$y_0 = a_4$$

$$y_1 = a_3 \oplus a_6$$

$$y_2 = a_2 \oplus a_5$$

$$y_3 = a_1$$

□

which is realized in Fig. 5.22.

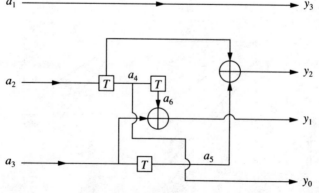

FIGURE 5.22 Ungerboeck realization of 16-QAM TCM scheme.

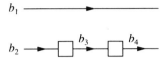

FIGURE 5.23 Input bit relationship for rate 2/3, four-state TCM scheme.

EXAMPLE 5.5 2/4-PSK, FOUR-STATE

Again this code can be viewed as a two-dimensional code with the phases of the two signals representing each dimension. In this trellis modulation scheme there are two memory bits ($v = 2$). Since this is a rate 2/3 code, $m = 2$. There is a parallel transition, so $\tilde{m} = 1$. Figure 5.23 depicts the connection between data bits and the two memory bits. Figure 5.24 illustrates the two signals in the 2/4-PSK constellations with natural mapping. The 2/4-PSK signal set is partitioned twice, resulting in four subsets, $S_0 = \{(3, 1), (-1, -1)\}$, $S_1 = \{(-1, 1), (3, -1)\}$, $S_2 = \{(1, 1), (-3, 1)\}$, and $S_3 = \{(-3, 1), (1, -1)\}$. The trellis structure and signal assignment are shown in Fig. 5.25, and from this it is determined that $d_{\text{free}}^2 = 8$. In rule T1 we have used the MSV concept. Here we have the MSV of a two-dimensional vector. In general, the MSV of a signal vector is the maximum sum of its coordinate numerical values. The code derived transmits one bit per symbol and exhibits a gain of 3.01 dB over BPSK. x_1 will be the first channel signal sent, and x_2 will be the second channel signal sent per trellis interval. Thus

First signal: $e^{j(x_1)\pi/4}$.

Second signal: $e^{j(x_2-1)\pi/2}$.

By solving (5.4) for each dimension, the following coefficients were obtained:

$$\text{For } x_1: \quad \text{For } x_2:$$
$$d_3 = 1 \quad d_1 = 1.$$
$$d_{124} = 2.$$

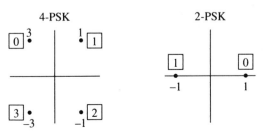

FIGURE 5.24 Signal constellation 2/4-PSK.

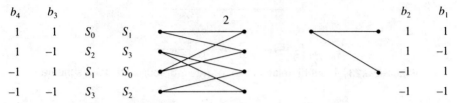

FIGURE 5.25 Trellis for 2/4-PSK, four-state.

The resulting analytic description is

$$x_1 = b_3 + 2b_1 b_2 b_4 \tag{5.12}$$

$$x_2 = b_1. \tag{5.13}$$

What has yet to be determined is which output bits represent which signal. For simplicity sake output bits z_2 and z_1 are to represent the first channel signal, x_1, and z_0 is to represent the second channel signal, x_2, where z_2 is the most significant bit for the first signal.

As in the previous example, a table will be constructed to determine the appropriate signal differences. These differences are given in Table 5.5. From the table the relationship between the channel signals and the output bits is known;

$$x_1 = 2z_2 + z_1 \tag{5.14}$$

$$x_2 = z_0. \tag{5.15}$$

As before, compare (5.14) and (5.15) to (5.12) and (5.13), respectively. The following relationships are found:

$$z_2 = b_1 b_2 b_4 \implies y_2 = a_1 \oplus a_2 \oplus a_4$$
$$z_1 = b_3 \implies y_1 = a_3$$
$$z_0 = b_1 \implies y_0 = a_1.$$

The underlying convolutional encoder and mapper are shown in Fig. 5.26. □

TABLE 5.5 Generation of a 2/4-PSK Signal Set.

Output bits	z_2	1	1	1	1	−1	−1	−1	−1
	z_1	1	1	−1	−1	1	1	−1	−1
	z_0	1	−1	1	−1	1	−1	1	−1
Output signals	x_1	3	3	1	1	−1	−1	−3	−3
	x_2	1	−1	1	−1	1	−1	1	−1

Effect on x_i due to z_j:

	x_1	x_2
z_2:	|4|	No effect
z_1:	|2|	No effect
z_0:	No effect	|2|

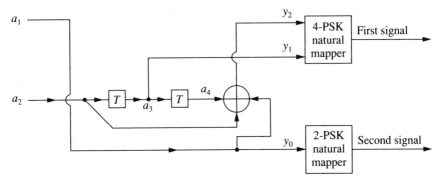

FIGURE 5.26 Ungerboeck representation for 2/4-PSK, four-state TCM scheme.

The example we have just completed is in the spirit of applying our design rule to multidimensional signals. Such signals are the subject of the next chapter. Our design technique for multidimensional signals is given later in Section 8.3.

5.5 Trellis-Code Performance and Realization

In the earlier sections of this chapter a step-by-step design procedure has been given for one- and two-dimensional trellis codes. If a design of a code is tabulated, there is no need for design. Thus most examples were given for tutorial purposes. Extensive tables on the performance of a wide class of codes have been given by Ungerboeck [1–3] and, in addition, a search procedure is presented, which is not covered in this book. In this section we present some of Ungerboeck's tabular results and show how to realize his codes in both feedback and feedforward form. These forms of realizing convolutional codes were studied in Section 2.6.2.

Coding gain performance [3] is presented in Table 5.6 for AM modulation, in Table 5.7 for PSK modulation, and in Table 5.8 for AM–PM modulation. The

TABLE 5.6 Codes for Amplitude Modulation Adapted from [3].

Number of States 2^ν	$\tilde{m}$	Parity Check Coefficients		$d^2_{\text{free}}/\Delta^2_0$	Asymptotic Coding Gain (dB)	
		h^1	h^0		$G_{4\text{AM}/2\text{AM}}$ ($m = 1$)	$G_{8\text{AM}/4\text{AM}}$ ($m = 2$)
4	1	2	5	9.0	2.55	3.31
8	1	04	13	10.0	3.01	3.77
16	1	04	23	11.0	3.42	4.18
32	1	10	45	13.0	4.15	4.91
64	1	024	103	14.0	4.47	5.23
128	1	126	235	16.0	5.05	5.81
256	1	362	515	16.0*	—	5.81
256	1	362	515	17.0	5.30	—

*The asterisk here means that the code search was not complete $\Delta_0 = d_{min}$ for coded signal set.

TABLE 5.7 Codes for Phase Modulation Adapted from [3].

Number of States 2^ν	$\tilde{m}$	Parity Check Coefficients h^2	h^1	h^0	$d^2_{\text{free}}/\Delta_0^2$	Asymptotic Coding Gain (dB) $G_{\text{8PSK/4PSK}}$ ($m = 3$)	$G_{\text{16PSK/8PSK}}$ ($m = 4$)
4	1	—	2	5	4.000*	3.01	—
8	2	04	02	11	4.586	3.60	—
16	2	16	04	23	5.172	4.13	—
32	2	34	16	45	5.758	4.59	—
64	2	066	030	103	6.343	5.01	—
128	2	122	054	277	6.586	5.17	—
256	2	130	072	435	7.515	5.75	—
4	1	—	2	5	1.324	—	3.54
8	1	—	04	13	1.476	—	4.01
16	1	—	04	23	1.628	—	4.44
32	1	—	10	45	1.910	—	5.13
64	1	—	024	103	2.000*	—	5.33
128	1	—	024	203	2.000*	—	5.33
256	2	374	176	427	2.085	—	5.51

* The asterisk means that the code search was not complete. $\Delta_0 = d_{\min}$ for coded signal set.

coding gain is presented as well as the coefficients for the parity check polynomial of the underlying convolutional code. See Section 2.6.2 for material on parity check polynomials. The structure of the Ungerboeck encoder is given in Fig. 5.1. We use $c_i = y_i$ as the encoder output to follow Ungerboeck's notation. Fundamental theory on the $\mathbf{G}(D)$ and $\mathbf{H}(D)$ matrices for convolutional codes can also be found in Section 2.6.

As a simple example, consider the four-state AM code in Table 5.6. The coefficients in octal of the two components of the parity check matrix, $\mathbf{H}(D) = (H_0(D), H_1(D))$ are $\mathbf{H}(D) = (5, 2)_8 = (101, 010)_2$. Thus

$$\mathbf{H}(D) = (1 + D^2, D). \tag{5.16}$$

The parity check polynomial is

$$\left(c_i^{(0)}, c_i^{(1)}\right)\mathbf{H}^T(D) = c_i^{(0)} \oplus c_{i-2}^{(0)} \oplus c_{i-1}^{(1)} = 0.$$

Writing this equation as a difference equation gives

$$c_i^{(0)} = c_{i-2}^{(0)} \oplus c_{i-1}^{(1)}.$$

This equation is realized in Fig. 5.27, and this represents the feedback form of the encoder. Note that the code is in systematic form. All that is required to realize this form is to write the parity check equation as a difference equation and then realize this difference equation in a diagram. This technique can be used to convert

TABLE 5.8 Codes for Two-Dimensional Modulation with Two-Dimensional Signals. (The 32-CR Signal Constellation Can Be Found in [3].)

| Number of States 2^ν | $\tilde{m}$ | Parity Check Coefficients | | | $d_{\text{free}}^2/\Delta_0^2$ | Asymptotic Coding Gain (dB) | | |
		h^2	h^1	h^0		$G_{\text{16QAM/8PSK}}$ ($m=3$)	$G_{\text{32CR/16QAM}}$ ($m=4$)	$G_{\text{64QAM/32CR}}$ ($m=5$)
4	1	—	2	5	4.0*	4.36	3.01	2.80
8	2	04	02	11	5.0	5.33	3.98	3.77
16	2	16	04	23	6.0	6.12	4.77	4.56
32	2	10	06	41	6.0	6.12	4.77	4.56
64	2	064	016	101	7.0	6.79	5.44	5.23
128	2	042	014	203	8.0	7.37	6.02	5.81
256	2	304	056	401	8.0	7.37	6.02	5.81
512	2	0510	0346	1001	8.0*	7.37	6.02	5.81

* The asterisk here means that the code search was not complete. $\Delta_0 = d_{\min}$ for coded signal set.

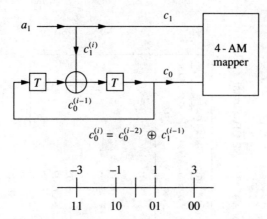

FIGURE 5.27 Realization in feedback form of 4-AM, four-state trellis code with natural mapper.

all of the trellis codes in this chapter from feedforward to feedback form. This technique is probably limited to simple codes. See [7] for algorithms to treat more complicated situations.

We now show how to realize Ungerboeck's codes in feedforward form. To determine this form requires the generator matrix $\mathbf{G}(D)$. Given the $\mathbf{H}(D)$ matrix in (5.16), $\mathbf{G}(D)$ is the solution to the equation $\mathbf{G}(D)\mathbf{H}^T(D) = \mathbf{0}$, where the superscript T represents a matrix transpose. Let $\mathbf{G}(D) = (G_0(D), G_1(D))$, and then for $\mathbf{H}(D)$ in (5.16) to get $\mathbf{G}(D)\mathbf{H}^T(D) = 0$, we have

$$G_0(D)(1 + D^2) + G_1(D)D = 0.$$

The desired solution is $G_0(D) = D$ and $G_1(D) = 1 + D^2$. Thus

$$\mathbf{G}(D) = (D, 1 + D^2).$$

The coefficient matrices are $\mathbf{G}(D) = (010, 101)_2 = (2, 5)_8$ and the encoder is shown in Fig. 5.28.

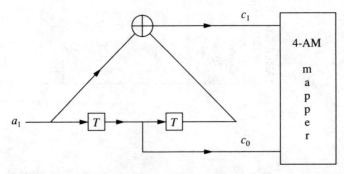

FIGURE 5.28 The code in Fig. 5.27 realized in feedforward form.

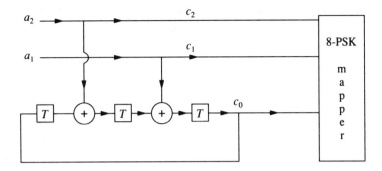

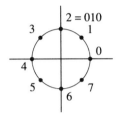

FIGURE 5.29 Feedback realization of 8-PSK, eight-state trellis code.

As a final example consider the eight-state, 8-PSK code in Table 5.7. Now $\mathbf{H}(D) = (11, 2, 4)_8$, or

$$\mathbf{H}(D) = (D^3 + 1, D, D^2) \tag{5.17}$$

which gives the parity check equation, $(c_0, c_1, c_2)\mathbf{H}^T(D)$, or

$$c_i^{(0)} = c_{i-2}^{(3)} \oplus c_{i-1}^{(2)} \oplus c_{i-3}^{(0)}.$$

Realizing this parity check equation gives the realization of the 8-PSK, eight-state code in Fig. 5.29.

To find the feedforward form, let us determine

$$\mathbf{G}(D) = \begin{pmatrix} G_{11}(D) & G_{12}(D) & G_{13}(D) \\ G_{21}(D) & G_{22}(D) & G_{23}(D) \end{pmatrix}. \tag{5.18}$$

Without loss of generality let us assume the structure in Fig. 5.30. Here a_2 will only be stored in one element, and the highest power of any polynomial in row 2 of $G(D)$ is unity. From the result $\mathbf{G}(D)\mathbf{H}^T(D) = \mathbf{0}$, we have, from (5.17) and (5.18),

$$G_{21}(D)(D^3 + 1) \oplus G_{22}(D)D \oplus G_{23}(D)D^2 = 0.$$

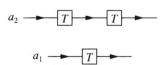

FIGURE 5.30 Assumed structure for $\mathbf{G}(D)$, 8-PSK, eight-state trellis code.

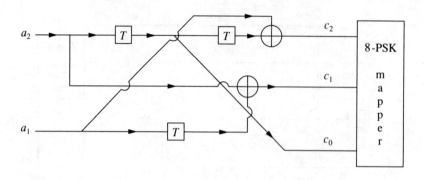

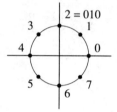

FIGURE 5.31 Feedforward realization of 8-PSK, eight-state trellis code.

A solution with degree 1 polynomials is $G_{21} = 0, G_{22} = D$, and $G_{23} = 1$. For the first row of $\mathbf{G}(D)$ in (5.18), the equation $\mathbf{G}(D)\mathbf{H}^T(D) = \mathbf{0}$ yields

$$G_{11}(D)(D^3 + 1) \oplus G_{12}(D)D \oplus G_{13}(D)D^2 = 0.$$

A solution is

$$G_{11} = D, \quad G_{12} = 1, \quad G_{13} = D^2$$

and thus

$$\mathbf{G}(D) = \begin{pmatrix} D & 1 & D^2 \\ 0 & D & 1 \end{pmatrix}.$$

This generator matrix can be realized as shown in Fig. 5.31, which gives the feedforward form of the 8-PSK, eight-state code. This code is equivalent to that generated earlier in Fig. 5.12, except that for that code

$$\mathbf{H}(D) = (D^3 + 1, \ D^2, \ D).$$

5.6 Trellis Coding with Asymmetric Modulations

In this section we show that symmetric signal sets (i.e., those with uniformly spaced signal points) that are optimum in uncoded AWGN environments are not necessarily optimum for coded systems (optimum in the sense of minimum bit error probability or maximum free Euclidean distance). Traditionally, symmetric signal constellations have been used for both uncoded and coded systems. In previous chapters we have shown how to optimize the TCM scheme for a given signal constellation. Here we also consider optimization of the constellation. Therefore, by designing the signal constellations to be asymmetric (signal points in the constellation are nonuniformly spaced), one can, in many instances, obtain a performance gain over TCM schemes based on traditional (symmetric) constellations. For example, by designing asymmetric M-PSK signal constellations and combining them with optimized trellis coding, one can further improve the performance of coded systems without increasing average or peak power or changing the bandwidth constraints imposed on the system. In fact, such properties of the signal set as constant envelope and number of dimensions are not changed by the asymmetry.

The first look at the idea of using other than conventional symmetric signal sets in coded systems appears in the work of Divsalar and Yuen [8, 9], who restricted themselves to trellis-coded M-PSK modulation. The results in [9] have been expanded by Simon and Divsalar [10] to higher-level PSK asymmetric signal sets combined with the optimum (in the sense of maximum d_free) TCM with 2, 4, 8, and 16 states. Later, the concept was extended to other one- and two-dimensional modulation types (e.g., M-AM and QAM) in [11] and [12]. In this section we summarize the results found in the foregoing references. The signal design method described in this section has also been applied to CPM modulation in [13].

Here it is shown that for signal sets with one degree of freedom (e.g., M-PSK and M-AM), the optimum asymmetric 2^{m+1}-point constellation to be used with rate $m/(m+1)$ trellis coding is the augmentation of a symmetric 2^m-point constellation with a *phase-rotated* (for M-PSK) or *amplitude-translated* (for M-AM) version of itself. For some signal sets with 2 degrees of freedom, in particular, QAM, it appears that the optimum 2^{m+1}-point asymmetric structure is achieved by augmenting the optimum 2^m two-dimensional AM–PM structure with a *rotated* version of itself. Certainly, symmetric 2^{m+1}-point QAM can be thought of as 2^m-point AM–PM augmented with itself rotated $90°$. We conjecture that, in general, for arbitrary two-dimensional structures, the optimum asymmetric set is achieved by combining the optimum symmetric set containing half the number of points with a translated and rotated version of itself. Although in this section we restrict ourselves to two-dimensional TCMs, the concept and approach can be extended to multidimensional TCM schemes.

5.6.1 Analysis and Design

Typical symmetric and asymmetric signal sets are shown in Fig. 5.32. As discussed, the asymmetric $M = 2^{m+1}$-point signal set is created by adding together the symmetrical $M/2$-point set with a translated or rotated version of itself. Another way of looking at the M-point asymmetric construction, which is more in keeping with set-partitioning technique, is to imagine partitioning the symmetric M-point constellation into two $M/2$-point constellations with maximally separated signals and then performing an appropriate rotation (M-PSK), translation (M-AM), or combination of rotation and translation (QAM) of one subset with respect to the other. Upon optimization of the amount of translation, rotation, or the combination of the two, the resulting two subsets can be used as the first level of set partitioning. Next, we discuss this procedure briefly and illustrate its application.

The approach of assigning signals to transitions of the trellis code is based on the mapping by set-partitioning rule described in Chapter 3. Each subset (including

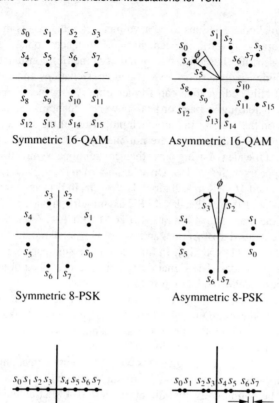

FIGURE 5.32 Symmetric and asymmetric signal sets.

the original set) is partitioned into two subsets with an equal number of signals and with the largest minimum distance among signals within the subset. Figure 5.33 demonstrates the set-partitioning method as applied to asymmetric 8-PSK. What remains is to optimize the rotation angle ϕ in asymmetric M-PSK, or translation Δ in M-AM, or both, in QAM and generalized two-dimensional signal sets. In most of the examples to be considered, for simplicity of analysis, the criterion of optimization will be to maximize the free Euclidean distance (or its square) of the TCM scheme. Next, we review the relation of this performance measure, and the average bit error probability of the overall coded system, to the transition structure of the trellis diagram.

As usual, we assume that for every m information bits, the rate $m/(m+1)$ trellis encoder produces $m+1$ output-coded symbols. These symbols are assigned to a unique member of the asymmetric 2^{m+1} signal set in accordance with the foregoing mapping procedure. Thus each transmitted complex signal x_k at time k is a nonlinear function $f(\cdot, \cdot)$ of the state σ_k of the encoder at discrete time k and the source output symbol denoted by u_k,

$$x_k = f(u_k, \sigma_k). \qquad (5.19)$$

5.6 / Trellis Coding with Asymmetric Modulations

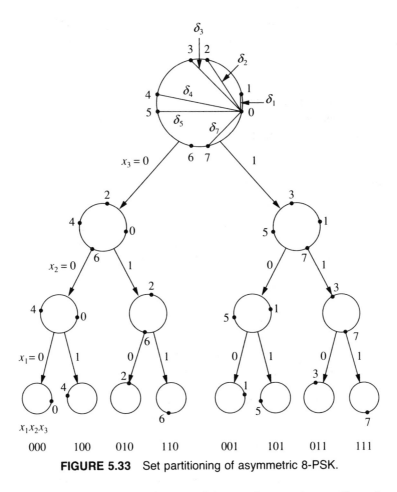

FIGURE 5.33 Set partitioning of asymmetric 8-PSK.

As discussed in Chapter 3, the next state of the encoder σ_{k+1} is a nonlinear function $g(\cdot, \cdot)$ of the present state and the input u_k. In mathematical terms,

$$\sigma_{k+1} = g(u_k, \sigma_k). \tag{5.20}$$

The received signal sample at time k is

$$r_k = x_k + n_k \tag{5.21}$$

where n_k is a sample of a zero-mean complex Gaussian noise process with variance σ^2.

To find the average bit error probability performance of the Viterbi decoder, we use here the product-trellis algorithm mentioned in Chapter 4. By using (4.4) and observing that there is a one-to-one correspondence between the vectors $\mathbf{c}_L$ of the signal labels and the vectors $\mathbf{x}_L$ of the received signal sequences, we have

$$P(\mathbf{x}_L \to \hat{\mathbf{x}}_L) \leq Z^{\sum_{k=1}^{L} \delta^2(U_k, S_k)} \tag{5.22}$$

where

$$\delta^2(U_k, S_k) = |f(u_k, \sigma_k) \quad f(\hat{u}_k, \hat{\sigma}_k)|^2 \tag{5.23}$$

with $\hat{\sigma}_k$ and $\hat{u}_k$ and the estimates of the state of the decoder and the source symbol, respectively. Also, Z is the Bhattacharyya parameter, which in this case is given by

$$Z = e^{-1/8\sigma^2}. \tag{5.24}$$

The parameter D of (5.24) can be related to the system bit energy-to-noise ratio E_b/N_0 by normalizing the signal energy to unity. Then, since the energy per symbol E_s equals n times the energy per bit E_b, we have

$$Z = e^{-mE_b/4N_0}. \tag{5.25}$$

The superstate S_k and the supersymbol U_k are defined as

$$S_k = (\sigma_k, \hat{\sigma}_k) \qquad U_k = (u_k, \hat{u}_k). \tag{5.26}$$

In (5.26), when $\sigma_k = \hat{\sigma}_k$, then S_k is called a good (correct) superstate, and when $\sigma_k \neq \hat{\sigma}_k$, S_k is called a bad (incorrect) superstate. In terms of the definitions above, the upper bound on the bit error rate is given by (4.27)

$$P_b \leq \frac{1}{m} \frac{\partial}{\partial I} T(D, I) \Big|_{I=1, D=Z} \tag{5.27}$$

and the even tighter bound (4.28)

$$P_b \leq \frac{1}{2m} \text{erfc}\left(\sqrt{\frac{mE_b}{N_0} \frac{d_{\text{free}}^2}{4}}\right) D^{-d_{\text{free}}^2} \frac{\partial}{\partial I} T(D, I) \Big|_{I=1, D=Z} \tag{5.28}$$

where $T(D, I)$ is the transfer function of the superstate diagram of the TCM scheme. It can be computed from the superstate diagram with superbranch gains given by

$$a(S_k, S_{k+1}) = \sum_{U_k} \frac{1}{2^m} I^{w(U_k)} D^{\delta^2(U_k, S_k)} \tag{5.29}$$

where $w(U_k)$ is the Hamming distance of binary sequences representing u_k and $\hat{u}_k$, and the summation is over those values of U_k that result in allowable superstate transitions. For the computation of transfer functions and that of free Euclidean distance using transfer functions, see Appendix B.

Asymptotically for large SNR values, maximizing d_{free} is synonymous with minimizing the average bit error probability. This relation is true provided that the distances between individual points in the signal set do not become too small. As we shall see, in some cases (e.g., the two-state codes), optimization of the asymmetry condition produces signal sets wherein the signal points tend to merge together in the limit as SNR approaches infinity. This results in a catastrophic trellis code, one where more and more long paths appear that have a squared distance equal to or slightly larger than the squared free distance. When this occurs, the union bound on the event error probability is no longer well approximated by its first term, or equivalently, the free distance alone is no longer sufficient for determining the coding gain. In these instances, the performance advantage achieved in terms of improvement in d_{free} does not translate directly into improvement in the required SNR.

Aside from the catastrophe that occurs in the trellis code when the signal points merge together, there are other practical reasons for not allowing this to occur. For example, with M-PSK, the use of close phase spacing in conjunction with soft decision inputs to a maximum-likelihood decoder makes the system more sensitive to phase jitter, due to imperfect carrier synchronization. Also, as the signal points come together, the distance buildup along some of the error paths can be considerably slowed, thus requiring more trellis memory (larger buffer size and decoding delay) for decoding. Therefore, it is always preferable to use bit error probability as a performance measure for optimization of the asymmetric signal set.

Based on the discussion above, the procedure for designing good trellis codes, combined with an optimum asymmetric signal constellation, can be summarized by the following steps:

Step 1
Use the mapping by set-partitioning method to partition the signal constellation, as in the example of Fig. 5.33.

Step 2
Assign signals from either of the two partitions (each containing 2^m signals) generated at the first level of partitioning in step 1 to transitions diverging from a given state. Similarly, assign signals from the other partition to transitions remerging to a given state. These assignments should be made so that the minimum distance between diverged transitions and the minimum distance between remerged transitions are as large as possible.

Step 3
Find the free Euclidean distance of the code or the bit error probability using (5.27) or (5.28).

Step 4
Maximize the free Euclidean distance or minimize the bit error probability of step 3 with respect to the rotation angle ϕ, or the translation Δ, or both. These values of ϕ and Δ then define the optimum asymmetric signal constellation.

In the following examples only the minimum-distance paths with respect to the all-zeros transmitted path are considered. However, the results have been checked against all possible transmitted paths.

5.6.2 Best Rate 1/2 Codes Combined with Asymmetric 4-PSK (A4-PSK)

The signal partitioning for trellis-coded A4-PSK is as in Fig. 5.34. For a rate 1/2 code, there will be two transitions leaving each state. We begin by considering the signal point assignment for the simplest case of two states.

Two-state trellis

For a two-state trellis, one has only two choices for the transition assignment. Either there exist multiple (two) transitions between like states or the two transitions leaving a given state go to different states. In the case of the former, the shortest error event path will be of length 1 (parallel transition); hence the maximum value of d_{free}^2 is limited by the Euclidean distance between a pair of signal points. For the set partitioning of Fig. 5.34, this corresponds to the squared distance between points 0 and 2 (or 1 and 3). This has a value of 4.0 under our assumption of

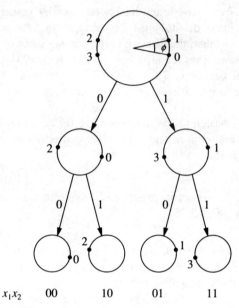

FIGURE 5.34 Set partitioning of asymmetric 4-PSK.

signal constellations with unit energy. If, on the other hand, the latter assignment is chosen as illustrated in Fig. 5.35, then the error event yielding the minimum distance is of length 2. This path, corresponding to the error event of choosing signal 2 followed by signal 1, when, in fact, signal 0 and signal 0 were successively transmitted, clearly has a larger value of d_{free}^2 than 4.0, since the squared distance of the first branch of this path is by itself 4.0. Thus this assignment is obviously the better choice.

We shall define a state transition matrix **T**, which describes the possible transitions between states corresponding to successive discrete time instants separated by a channel symbol. The ijth entry in the matrix represents the output M-PSK symbol assigned to the transition from state i to j. The absence of an entry implies that a transition between those states is not possible. Thus for the trellis of Fig. 5.35, we have

$$\mathbf{T} = \begin{bmatrix} 0 & 2 \\ 1 & 3 \end{bmatrix}. \tag{5.30}$$

We note that the signal point constellation of Fig. 5.34 can be regarded as a special case of an unbalanced Q-PSK (UQ-PSK), where the data rates on the two

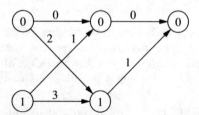

FIGURE 5.35 Trellis diagram and 4-PSK signal assignment.

5.6 / Trellis Coding with Asymmetric Modulations

channels are equal and the symbol transition times are aligned but the powers are unbalanced. The ratio of energies between the in-phase and quadrature channels, denoted E_I and E_Q, respectively, can be related to the angle ϕ that defines the asymmetry. In particular, letting $\alpha = E_Q/E_I$, we have

$$\alpha = \tan^2 \frac{\phi}{2}. \tag{5.31}$$

The trellis of Fig. 5.35 can be implemented by a constraint length 2, rate 1/2 linear convolutional code. The superstate transition diagram for this code is illustrated in Fig. 5.36 and has the transfer function bound

$$T(D, I) = \frac{ID^{4(1+2\alpha)/(1+\alpha)}}{1 - ID^{4/(1+\alpha)}} \tag{5.32}$$

where D is defined by (5.25). Using (5.32) in (5.28) gives the upper bound on the average bit error probability,

$$P_b \leq \frac{0.5 \operatorname{erfc}\left\{\sqrt{-\left[4(1+2\alpha)/(1+\alpha)\right] \ln D}\right\}}{\left[1 - D^{4/(1+\alpha)}\right]^2} \tag{5.33}$$

where our assumption of unit-energy signals implies that $E_I + E_Q = 1$.

The optimum value of α (or equivalently ϕ), that is, the value that minimizes the bound on P_b of (5.33), is

$$\alpha_{\text{opt}} \approx -4\left(\frac{\ln D}{\ln 3}\right) - 1. \tag{5.34}$$

Equation (5.34) is the exact optimum value for (5.27), but only an asymptotic optimum for (5.28). Substituting (5.34) in (5.33) gives the optimum (in the sense

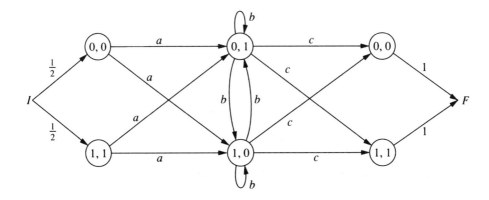

$T(D,I) = \dfrac{4ac}{1-2b}$ $a = \dfrac{I}{2} D^4$ $b = \dfrac{I}{2} D^{4/(1+\alpha)}$ $c = \dfrac{1}{2} D^{4\alpha/(1+\alpha)}$

FIGURE 5.36 Pair-state transition diagram for the trellis diagram of Fig. 5.35.

of the best asymmetric 4-PSK signal design) upper bound on the average bit error probability,

$$P_b \leq \frac{9}{8} \operatorname{erfc}\left(\sqrt{\frac{2E_b}{N_0} - \ln 3}\right). \tag{5.35}$$

For the symmetric signal design ($\phi = \pi/2, \alpha = 1$), the upper bound in (5.33) becomes

$$P_b \leq \frac{0.5 \operatorname{erfc}\left(\sqrt{3E_b/2N_0}\right)}{\left[1 - \exp(E_b/2N_0)\right]^2}. \tag{5.36}$$

Finally, for uncoded PSK, the corresponding P_b would be

$$P_b \leq 0.5 \operatorname{erfc}\left(\sqrt{\frac{E_b}{N_0}}\right). \tag{5.37}$$

Figure 5.37 illustrates the three upper bounds of (5.35), (5.36), and (5.37) versus E_b/N_0. For sufficiently large values of E_b/N_0, the denominator of (5.36) can be approximated by unity. Thus, asymptotically, the gain in E_b/N_0 of the coded symmetric 4-PSK system over the uncoded PSK system is $10 \log_{10} \frac{3}{2} = 1.76$ dB. To determine how much additional gain due to asymmetry is achievable in the same asymptotic limit, we turn to a discussion of the free-distance behavior of the coded system.

Let δ_j^2 denote the squared distance from signal point 0 to signal point $j = 1, 2, 3$. Then for the asymmetric constellation of Fig. 5.34,

$$\delta_1^2 = 4 \sin^2 \frac{\phi}{2}, \quad \delta_2^2 = 4, \quad \delta_3^2 = 4 \cos^2 \frac{\phi}{2}. \tag{5.38}$$

For the minimum-distance path of length 2, we have

$$d_{\text{free}}^2 = \delta_2^2 + \delta_1^2 = 4\left(1 + \sin^2 \frac{\phi}{2}\right) \tag{5.39}$$

which for the symmetric signal design ($\phi = \pi/2$) becomes

$$d_{\text{free}}^2 = 4(1 + \tfrac{1}{2}) = 6. \tag{5.40}$$

In the more general asymmetric case, substituting (5.38) into (5.36) gives

$$d_{\text{free}}^2 = 4\left(1 + \frac{\alpha}{1 + \alpha}\right). \tag{5.41}$$

Thus the improvement in d_{free}^2 due to the asymmetry is, from (5.40) and (5.41),

$$\eta = 10 \log_{10} \frac{d_{\text{free}}^2(\text{asymm})}{d_{\text{free}}^2(\text{symm})} = 10 \log_{10} \frac{2(1 + 2\alpha)}{3(1 + \alpha)}. \tag{5.42}$$

If instead of minimizing the bit error probability, we select the asymmetry angle that maximizes d_{free}^2 of (5.39), the value of this angle will be independent of

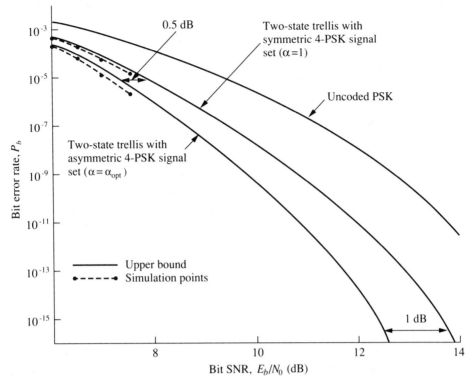

FIGURE 5.37 Ideal performance of rate 1/2, two-state, trellis-coded symmetric and optimum asymmetric 4-PSK.

the SNR. From (5.39) we see that d_{free}^2 is maximized when $\phi = \pi$; that is, signal points 1 and 2 merge together, and similarly for signal points 0 and 3. In this limiting case, $d_{\text{free}}^2 = 8$ and the gain (in squared free distance) relative to the symmetric constellation is $10 \log_{10}(8/6) = 1.25$ dB. It also represents the asymptotic improvement in the E_b/N_0 performance due to asymmetry, as would be obtained by letting the symmetric and asymmetric coded curves in Fig. 5.37 approach infinite E_b/N_0. With respect to the symmetric case, the gain is 0.5 and 1.00 dB for $E_b/N_0 = 7.5$ and 12.5 dB, respectively.

As mentioned previously, the type of limiting case described above results in a catastrophic trellis code. In practice, one would not use this limiting case any more than one would use an infinite-bandwidth expansion code to perform at the Shannon limit.

Since for uncoded 2-PSK (or simple PSK), the square of the minimum distance is 4 (two signal points diametrically opposed on a circle of diameter 2), the limiting gain of the two-state trellis-coded asymmetric 4-PSK relative to this equivalent bandwidth uncoded system is $10 \log_{10}(8/4) = 3.0$ dB. The relative gain of trellis-coded symmetric 4-PSK to uncoded 2-PSK would, from the discussion above, be 1.25 dB less, or 1.76 dB, which agrees with the statement above.

In Fig. 5.37 the results of a computer simulation of rate 1/2 trellis-coded symmetric and asymmetric 4-PSK signaling scheme have been shown for comparison to those obtained from the analysis. Also shown is a curve corresponding to the exact bit error probability performance of uncoded PSK according to the well-known result $0.5 \, \text{erfc} \left(\sqrt{E_b/N_0} \right)$. We observe that the simulations, which reflect the actual system behavior, preserve the same relative gains of the coded over the uncoded system as those predicted by the upper bounds. In particular, asymmetry modulation buys about 0.5 dB at a bit error probability of 10^{-5}. The buffer size of the decoder used in the simulation is 16 bits.

Four-state trellis

For a rate 1/2, four-state trellis code combined with 4-PSK, the assignment of signals to the branches according to steps 2 and 3 of Section 5.6.1 leads to the trellis illustrated in Fig. 5.38. Depending on the value of ϕ, there are two possibilities for the shortest path with the minimum free distance. For small values of ϕ, the length-4 path corresponding to M-PSK signals 2, 3, 3, 2 is the dominant one, whereas for values of ϕ near π, the length-3 path corresponding to M-PSK signals 2, 1, 2 is dominant. The squared Euclidean distances for these paths are

$$d^2(2, 3, 3, 2) = 4 + 8\cos^2 \frac{\phi}{2} + 4$$
$$d^2(2, 1, 2) = 4 + 4\sin^2 \frac{\phi}{2} + 4. \tag{5.43}$$

To find the optimum value of ϕ, we note that the two squared distances are functions, one increasing and one decreasing, of ϕ over the interval $(0, \pi)$. Their crossover point results in the maximum value of the smaller of the two evaluated at each ϕ. Equating the two squared distances in (5.43) results in

$$\tan^2 \frac{\phi}{2} = 2 \quad \text{or} \quad \phi = 1.9 \text{ rad} \tag{5.44}$$

with a corresponding value of d^2_{free}:

$$d^2_{\text{free}} = 4 + 8\left(\frac{1}{1+2}\right) + 4 = 10.67. \tag{5.45}$$

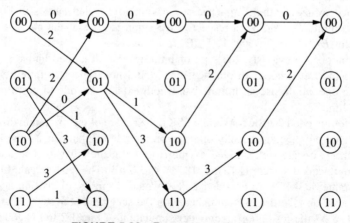

FIGURE 5.38 Four-state trellis diagram.

For the symmetric case ($\phi = \pi/2$), the length-3 path gives the smaller minimum distance, which from (5.43) is

$$d_{\text{free}}^2 = 4 + 4(\tfrac{1}{2}) + 4 = 10. \tag{5.46}$$

Thus, from (5.45) and (5.46), the gain in d_{free}^2 due to asymmetry is

$$\eta = 10 \log_{10} \frac{10.67}{10} = 0.28 \text{ dB}. \tag{5.47}$$

Again, relative to an uncoded PSK, the gains are as follows:

$$\eta(\text{asymm}) = 10 \log_{10} \frac{d_{\text{free}}^2(\text{asymm})}{d_{\min}^2} = 10 \log_{10} \frac{10.67}{4} = 4.26 \text{ dB}$$

$$\eta(\text{symm}) = 10 \log_{10} \frac{d_{\text{free}}^2(\text{symm})}{d_{\min}^2} = 10 \log_{10} \frac{10}{4} = 3.98 \text{ dB}. \tag{5.48}$$

Eight-state trellis

Following the steps discussed previously for the design of good codes, one arrives at the eight-state trellis diagram in Fig. 5.39 with state transition matrix, **T**, given by

$$\mathbf{T} = \begin{matrix} & \begin{matrix} 1 & 2 & 3 & 4 & 5 & 6 & 7 & 8 \end{matrix} \\ \begin{matrix} 1 \\ 2 \\ 3 \\ 4 \\ 5 \\ 6 \\ 7 \\ 8 \end{matrix} & \begin{bmatrix} 0 & 2 & & & & & & \\ & & 3 & 1 & & & & \\ & & & & 2 & 0 & & \\ & & & & & & 1 & 3 \\ 2 & 0 & & & & & & \\ & & 1 & 3 & & & & \\ & & & & 0 & 2 & & \\ & & & & & & 3 & 1 \end{bmatrix} \end{matrix} \tag{5.49}$$

As for the four-state trellis, there are two shortest-length paths (solid lines) that, depending on the value of ϕ, yield the minimum free distance. The squared distance of these paths is given by

$$d^2(2, 3, 2, 2) = 4 + 4\cos^2\frac{\phi}{2} + 4 + 4 = 12 + 4\cos^2\frac{\phi}{2}$$

$$d^2(2, 1, 1, 0, 2) = 4 + 4\sin^2\frac{\phi}{2} + 4\sin^2\frac{\phi}{2} + 0 + 4 \tag{5.50}$$

$$= 8 + 8\sin^2\frac{\phi}{2}.$$

Again, the two distances in (5.50) are monotonic with ϕ and thus equating them results in the maximum value of the smaller of the two over all ϕ over the interval $(0, \pi)$. When these distances are equated, the optimum value of ϕ is found to be

$$\sin^2\frac{\phi}{2} = \frac{2}{3} \quad \Longrightarrow \quad \phi = 1.23 \text{ rad} \tag{5.51}$$

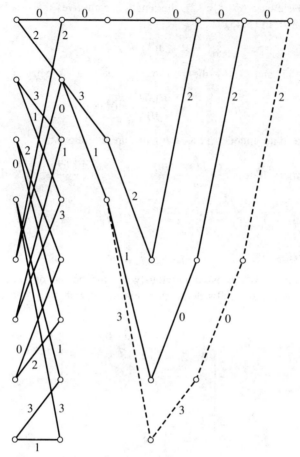

FIGURE 5.39 Eight-state trellis diagram for asymmetric 4-PSK.

and the corresponding squared minimum free distance is

$$d^2_{\text{free}} = 8 + 8\left(\frac{2}{3}\right) = \frac{40}{3} = 13.33. \tag{5.52}$$

For the symmetric signal design with $\phi = \pi/2$, the length-5 path provides the smaller distance with the value

$$d^2_{\text{free}} = 8 + 8\left(\frac{1}{2}\right) = 12. \tag{5.53}$$

Thus the gain due to the asymmetry is

$$\eta = 10 \log_{10} \frac{40/3}{12} = 0.46 \text{ dB} \tag{5.54}$$

and the gains of the asymmetric and symmetric eight-state trellis-coded 4-PSK system over the uncoded PSK system are

$$\eta(\text{asymm}) = 10 \log_{10} \frac{40/3}{4} = 5.23 \text{ dB}$$

$$\eta(\text{symm}) = 10 \log_{10} \frac{12}{4} = 4.77 \text{ dB}.$$
(5.55)

There is another path, illustrated by dashed lines in Fig. 5.39, which corresponds to the length-6 error event 2, 1, 3, 3, 0, 2. The squared distance of this path from the all-zeros path is identical to that of the length-4 path found above, and thus does not change the relative gains given in (5.54) and (5.55).

5.6.3 Best Rate 2/3 Codes Combined with Asymmetric 8-PSK (A8-PSK)

The signal partitioning for rate 2/3 trellis-coded A8-PSK is as shown in Fig. 5.33. Here there are four paths that diverge from each state. Thus one now has more flexibility as to how many parallel paths (e.g., one or two) should be assigned per transition between states. For the two-state trellis the choice is very simple. Thus we again begin our discussion with this case.

Two-state trellis

The two-state trellis used here is exactly of the form given in Fig. 5.35 except that now each branch represents two parallel paths (see Fig. 5.40). The minimum free-distance path is once again of length 2 and corresponds to error event 2, 1. Since from Fig. 5.33 the set of squared distances from signal point 0 to signal point $j = 1, 2, 3, \ldots, 7$ is now

$$\delta_1^2 = 4 \sin^2 \frac{\phi}{2} = 2(1 - \cos \phi)$$

$$\delta_3^2 = 4 \sin^2 \left(\frac{\pi}{4} + \frac{\phi}{2} \right) = 2(1 + \sin \phi)$$

$$\delta_5^2 = 4 \sin^2 \left(\frac{\pi}{2} - \frac{\phi}{2} \right) = 2(1 + \cos \phi) \quad (5.56)$$

$$\delta_7^2 = 4 \sin^2 \left(\frac{\pi}{4} - \frac{\phi}{2} \right) = 2(1 - \sin \phi)$$

$$\delta_2^2 = 2 \quad \delta_4^2 = 4 \quad \delta_6^2 = 2.$$

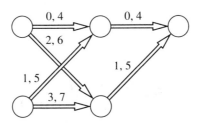

FIGURE 5.40 Two-state trellis diagram and signal assignment for 8-PSK.

the squared minimum free distance is given by

$$d_{\text{free}}^2 = \delta_2^2 + \delta_1^2 = 4 - 2\cos\phi \tag{5.57}$$

which is maximized when $\phi = \pi/2$ (i.e., the signal points 1, 3, 5, and 7 merge, respectively, with points 2, 4, 6, and 0). In this limiting case, the maximum value of (5.57) becomes

$$d_{\text{free}}^2 = 4. \tag{5.58}$$

For the symmetric 8-PSK constellation ($\phi = \pi/4$), (5.57) becomes

$$d_{\text{free}}^2 = 4 - \sqrt{2}. \tag{5.59}$$

Thus the gain due to asymmetry is

$$\eta = 10 \log_{10} \frac{4}{4 - \sqrt{2}} = 1.895. \tag{5.60}$$

Since rate 2/3 trellis-coded A8-PSK is equivalent in bandwidth to uncoded 4-PSK, and since the latter has $d_{\min}^2 = 2$, the relative gains for the asymmetric and symmetric coded signal designs are, respectively,

$$\eta(\text{asymm}) = 10 \log_{10} \frac{4}{2} = 3.01 \text{ dB}$$

$$\eta(\text{symm}) = 10 \log_{10} \frac{4 - \sqrt{2}}{2} = 1.116 \text{ dB}. \tag{5.61}$$

As was true for the two-state rate 1/2 trellis-coded A4-PSK case, the optimum asymmetric signal design corresponds to a merger of alternate signal points in the original symmetric set. This implies that the gain due to asymmetry translates into an equivalent E_b/N_0 gain only in the limit of infinite E_b/N_0. Thus we should obtain the practical gain achievable with asymmetry, once again by finding the superstate transition diagram for the trellis, evaluating its transfer function $T(D, I)$, and differentiating this result to find an upper bound on the average bit error probability. Minimization of this bit error rate bound with respect to the asymmetry angle ϕ then results in an optimum asymmetric signal point design as a function of E_b/N_0. The details of this procedure are as follows.

Figure 5.41 illustrates the superstate transition diagram for the rate 2/3 trellis code. The transfer function of this diagram is given by

$$T(D, I) = 2d + \frac{2(a_1 + a_2)c}{1 - 2b}. \tag{5.62}$$

Applying (5.27), after some algebra we get

$$P_b \leq \frac{1}{2} \frac{\partial}{\partial I} T(D, I)\Big|_{I=1}$$

$$= \frac{1}{2} D^{\delta_4^2} + \frac{D^{\delta_6^2}(D^{\delta_1^2} + D^{\delta_5^2})(2 - D^{\delta_7^2})}{(1 - D^{\delta_7^2} - D^{\delta_3^2})^2}. \tag{5.63}$$

The upper bound in (5.63) is implicitly a function of the asymmetry angle ϕ through the distances between signal points defined in (5.56). Minimizing (5.63)

$$a_1 = 1/2(D^{\delta_1^2}I^2 + D^{\delta_5^2}I)$$
$$a_2 = 1/2(D^{\delta_1^2}I + D^{\delta_5^2}I^2)$$
$$b = 1/2(D^{\delta_7^2}I + D^{\delta_3^2}I^2)$$
$$c = 1/2[D^{\delta_6^2}(I + 1)]$$
$$d = 1/2(D^{\delta_4^2}I)$$

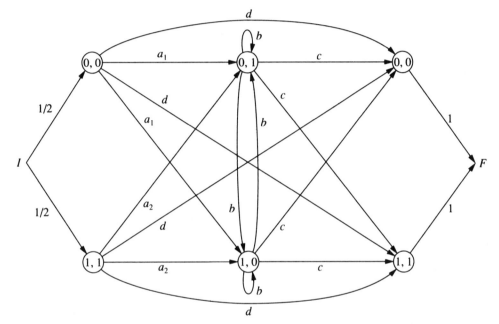

FIGURE 5.41 Pair-state transition diagram for rate 2/3 trellis code.

with respect to ϕ does not lead to an exact closed-form expression for the optimum asymmetry angle. Thus we have to perform the minimization by numerical analysis. The resulting values are tabulated in Table 5.9.

Substituting the values of ϕ from Table 5.9 into (5.63) results in the optimum upper bound on the average bit error rate and is illustrated in Fig. 5.42. Also

TABLE 5.9 Optimum Values of Asymmetry Angles Versus E_b/N_0.

E_b/N_0 (dB)	ϕ (rad)
5	0.7854
6	0.9189
7	1.037
8	1.139
9	1.217
10	1.280
11	1.327

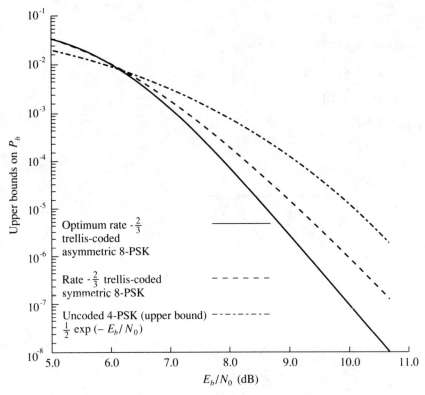

FIGURE 5.42 Upper bounds on average bit error probability performance for rate 2/3, trellis-coded symmetric and optimum asymmetric 8-PSK.

illustrated in that figure is the result for the symmetric case [i.e., (5.23) evaluated for $\phi = \pi/4$] and the corresponding upper bound for uncoded 4-PSK.

Four-state trellis

For four states we can either have a trellis with two parallel paths between states or one with no parallel paths. These two possibilities and their corresponding signal point assignments are illustrated in Fig. 5.43. The state transition matrix for the latter trellis is

$$\mathbf{T} = \begin{bmatrix} 0 & 4 & 2 & 6 \\ 1 & 5 & 3 & 7 \\ 4 & 0 & 6 & 2 \\ 5 & 1 & 7 & 3 \end{bmatrix}. \tag{5.64}$$

The minimum-distance path for the trellis in Fig. 5.43(b) is of length 3, corresponding to the M-PSK output symbols 2, 0, 1. The squared distance of this path from the all-zeros path is

$$d^2(2, 0, 1) = 4 - 2\cos\phi \tag{5.65}$$

which for every value of ϕ between 0 and $\pi/2$ is smaller than that corresponding to any other path of any length. In the limit, (5.65) achieves its maximum value (i.e., $d^2_{\text{free}} = 4$) when $\phi = \pi/2$. For the symmetric case where $\phi = \pi/4$, (5.65)

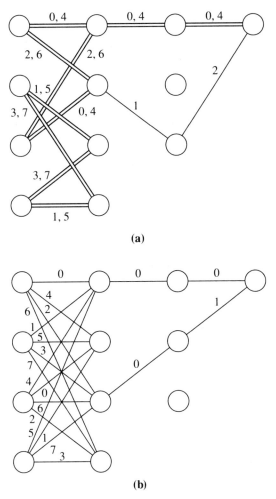

FIGURE 5.43 Four-state trellis diagram: (a) two parallel paths per transition; (b) no parallel paths per transition.

evaluates to $d_{\text{free}}^2 = 4 - \sqrt{2}$, which is the same result as for the two-state trellis, thus implying no gain by going to the additional complexity.

One might wonder at this point whether the selection of another signal point assignment for the trellis of Fig. 5.43(b) would lead to improved results. Many other combinations have exhaustively been tried, with the result that with the fully connected trellis structure of Fig. 5.43(b), no further improvement is possible.

To show that Fig. 5.43(a) is the preferred approach, we observe, as did Ungerboeck [1], that all paths of length greater than 1 have a squared distance larger than 4. In fact, the closest to this value would be achieved by the error event path 2, 1, 2, with squared distance $6 - 2\cos\phi$, which is greater than 4 for all values of ϕ (other than $\pi/2$). In conclusion, the maximum d_{free}^2 is achieved by the four-state trellis of Fig. 5.43(a) and has the value of 4, independent of the asymmetry angle. Stated another way, for rate 2/3, four-state trellis-coded 8-PSK, there exists no gain due to asymmetry, and the gain relative to the uncoded 4-PSK case is 3.01 dB.

Eight-state trellis

For eight states, we again have several options of signal assignment, according to whether or not the parallel transitions should exist. If parallel transitions are present, d_{free}^2 is limited to have a value of 4, regardless of asymmetry. Thus we should first investigate a trellis with no parallel transitions and see if, indeed, one can achieve a larger free distance. In that regard, consider the eight-state trellis of Ungerboeck [1] reproduced here in Fig. 5.44, with a state transition matrix given by

$$\mathbf{T} = \begin{array}{c} \\ 1 \\ 2 \\ 3 \\ 4 \\ 5 \\ 6 \\ 7 \\ 8 \end{array} \begin{array}{c} \begin{array}{cccccccc} 1 & 2 & 3 & 4 & 5 & 6 & 7 & 8 \end{array} \\ \left[\begin{array}{cccccccc} 0 & 4 & 2 & 6 & & & & \\ & & & & 1 & 5 & 3 & 7 \\ 4 & 0 & 6 & 2 & & & & \\ & & & & 5 & 1 & 7 & 3 \\ 2 & 6 & 0 & 4 & & & & \\ & & & & 3 & 7 & 1 & 5 \\ 6 & 2 & 4 & 0 & & & & \\ & & & & 7 & 3 & 5 & 1 \end{array} \right] \end{array}. \quad (5.66)$$

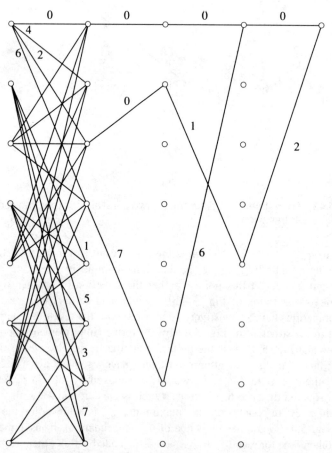

FIGURE 5.44 Eight-state trellis code for 8-PSK.

For this assignment the two shortest paths that, depending on the amount of asymmetry, yield the minimum distance from the all-zeros path are 6, 7, 6 and 2, 0, 1, 2. The squared distances for these paths are, respectively,

$$d^2(6,7,6) = 4 + 4\sin^2\left(\frac{\pi}{4} - \frac{\phi}{2}\right) = 6 - 2\sin\phi$$

$$d^2(2,0,1,2) = 4 + 4\sin^2\frac{\phi}{2} = 6 - 2\cos\phi.$$
(5.67)

Equating these distances and solving for ϕ, we again find that the optimum value corresponds to the symmetric constellation, $\phi = \pi/4$. Thus once again there is no gain due to asymmetry. Substituting $\phi = \pi/4$ into (5.67) gives

$$d_{\text{free}}^2 = 6 - \sqrt{2} = 4.586 \tag{5.68}$$

and a gain relative to an uncoded 4-PSK of

$$\eta = 10\log_{10}\frac{6-\sqrt{2}}{2} = 3.60 \text{ dB}. \tag{5.69}$$

Since d_{free}^2 of (5.68) is indeed larger than 4, the trellis of Fig. 5.44 is preferred over any configuration with parallel transitions.

Sixteen-state trellis

Since we have already demonstrated that an 8-state trellis with no parallel transitions has a d_{free}^2 that exceeds the maximum distance due to parallel transitions, it is not necessary to consider a 16-state trellis with parallel transitions. Instead, we go directly to the fully connected trellis of Fig. 5.45 as considered by Ungerboeck [1], with a state transition matrix given by

$$\mathbf{T} = \begin{bmatrix} 0 & 4 & 2 & 6 & & & & & & & & & & & & \\ & & & & 1 & 5 & 3 & 7 & & & & & & & & \\ & & & & & & & & 4 & 0 & 6 & 2 & & & & \\ & & & & & & & & & & & & 5 & 1 & 7 & 3 \\ 2 & 6 & 0 & 4 & & & & & & & & & & & & \\ & & & & 3 & 7 & 1 & 5 & & & & & & & & \\ & & & & & & & & 6 & 2 & 4 & 0 & & & & \\ & & & & & & & & & & & & 7 & 3 & 5 & 1 \\ 4 & 0 & 6 & 2 & & & & & & & & & & & & \\ & & & & 5 & 1 & 7 & 3 & & & & & & & & \\ & & & & & & & & 0 & 4 & 2 & 6 & & & & \\ & & & & & & & & & & & & 1 & 5 & 3 & 7 \\ 6 & 2 & 4 & 0 & & & & & & & & & & & & \\ & & & & 7 & 3 & 5 & 1 & & & & & & & & \\ & & & & & & & & 2 & 6 & 0 & 4 & & & & \\ & & & & & & & & & & & & 3 & 7 & 1 & 5 \end{bmatrix}. \tag{5.70}$$

194 Ch. 5 / One- and Two-Dimensional Modulations for TCM

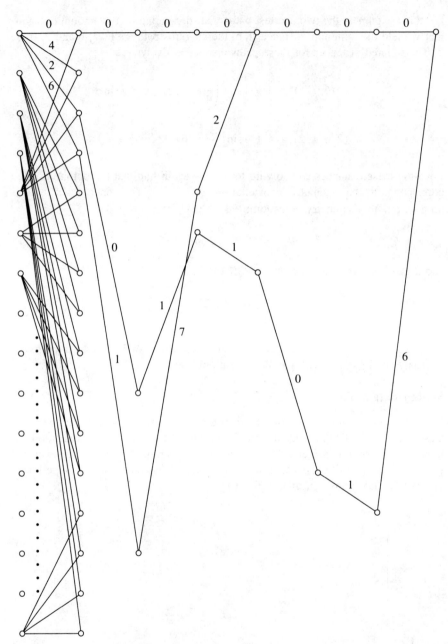

FIGURE 5.45 Sixteen-state code for 8-PSK.

For this assignment, the two shortest paths that, depending on the amount of asymmetry, yield the minimum distance from the all-zeros path are 6, 1, 7, 2 and 2, 0, 1, 1, 0, 1, 6. The first of these paths (the one of length 4), discovered by Ungerboeck [1], is concerned only with symmetric M-PSK constellations. The second one, which indeed allows a slight gain to be achieved with asymmetry, does not show up until one investigates paths of length 7. This once again emphasizes the point that paths of all lengths (up to some reasonable limit) must be looked at before deciding whether or not there can exist a gain due to asymmetry.

The squared distances for the two paths above are, respectively,

$$d^2(6, 1, 7, 2) = 8 - 2(\sin\phi + \cos\phi)$$
$$d^2(2, 0, 1, 1, 0, 1, 6) = 10 - 6\cos\phi. \tag{5.71}$$

Equating these two distances gives the optimum asymmetric 8-PSK design corresponding to

$$\cos\phi = \frac{4}{5} \Longrightarrow \phi = 0.6435 \text{ rad}$$
$$d^2_{\text{free}} = \frac{26}{5} = 5.20. \tag{5.72}$$

It should also be pointed out that the length-8 path 6, 7, 0, 0, 0, 7, 7, 6, which has the squared distance

$$d^2(6, 7, 0, 0, 0, 7, 7, 6) = 10 - 6\sin\phi \tag{5.73}$$

can be used to determine an alternative optimum asymmetric 16-PSK constellation with $\phi = (\pi/2) - 0.6435$ rad and the same value of d^2_{free}.

The gain due to asymmetry is

$$\eta = 10\log_{10}\frac{26/5}{8 - 2\sqrt{2}} = 0.024 \text{ dB} \tag{5.74}$$

and the gains relative to uncoded 4-PSK are

$$\eta(\text{asymm}) = 10\log_{10}\frac{26/5}{2} = 4.15 \text{ dB}$$
$$\eta(\text{symm}) = 10\log_{10}\frac{8 - 2\sqrt{2}}{2} = 4.126 \text{ dB}. \tag{5.75}$$

Although the gain due to asymmetry is so small as to be only of academic interest, it points out the curious fact that whereas asymmetry provided no advantage with 4- and 8-state trellises, a theoretical gain once again becomes achievable when the complexity is increased to 16 states.

5.6.4 Best Rate 3/4 Codes Combined with Asymmetric 16-PSK (A16-PSK)

The signal partitioning for trellis-coded A16-PSK follows the same steps as those leading to the partitionings in Figs. 5.33 and 5.34. For a rate 3/4 code, there will be eight transitions leaving each state. As before, we begin with the simple two-state case.

Two-state trellis

The two-state trellis for A16-PSK is identical in form to that shown in Fig. 5.35 except that now each branch represents four parallel paths. In particular, the transitions between like states correspond to signals 0, 4, 8, 12 and 3, 7, 11, 15, respectively, while the cross transitions correspond to 2, 6, 10, 14 and 1, 5, 9, 13. The minimum-distance path is of length 2 and corresponds to the error event 2, 1. The set of squared distances from signal point 0 to signal point $j = 1, 2, 3, \ldots, 15$ is now

$$\begin{aligned}
\delta_1^2 &= 2(1 - \cos\phi) & \delta_7^2 &= 2\left[1 - \cos\left(\frac{3\pi}{4} + \phi\right)\right] \\
\delta_2^2 &= 2 - \sqrt{2} = \delta_{14}^2 & \delta_8^2 &= 4 \\
\delta_3^2 &= 2\left[1 - \cos\left(\frac{\pi}{4} + \phi\right)\right] & \delta_9^2 &= 2(1 + \cos\phi) \\
\delta_4^2 &= 2 = \delta_{12}^2 & \delta_{11}^2 &= 2\left[1 - \cos\left(\frac{3\pi}{4} - \phi\right)\right] \\
\delta_5^2 &= 2(1 + \sin\phi) & \delta_{13}^2 &= 2(1 - \sin\phi) \\
\delta_6^2 &= 2 + \sqrt{2} = \delta_{10}^2 & \delta_3^2 &= 2\left[1 - \cos\left(\frac{\pi}{4} - \phi\right)\right].
\end{aligned} \quad (5.76)$$

Thus the squared free distance is given by

$$d_{\text{free}}^2 = \delta_2^2 + \delta_1^2 = 2 - \sqrt{2} + 2(1 - \cos\phi) \quad (5.77)$$

which is maximized when $\phi = \pi/4$ (i.e., signal points 1, 3, 5, 7, 9, 11, 13, 15 merge with points 0, 2, 4, 6, 8, 10, 12, 14, respectively). In this limiting case the maximum value of (5.77) becomes

$$d_{\text{free}}^2 = 4 - 2\sqrt{2} = 1.172 \quad (5.78)$$

while for the symmetric case ($\phi = \pi/8$), (5.77) evaluates to

$$d_{\text{free}}^2 = 4 - \sqrt{2} - 2\cos\frac{\pi}{8} = 0.738. \quad (5.79)$$

Thus the gain due to asymmetry is

$$\eta = 10\log_{10}\frac{1.172}{0.738} = 2.01 \text{ dB} \quad (5.80)$$

and the gains relative to the equivalent-bandwidth, uncoded 8-PSK system are

$$\eta(\text{asymm}) = 10 \log_{10} \frac{1.172}{2 - \sqrt{2}} = 3.01 \text{ dB}$$

$$\eta(\text{symm}) = 10 \log_{10} \frac{0.738}{2 - \sqrt{2}} = 1.00 \text{ dB} \quad (5.81)$$

where we have made use of the fact that the latter has $d_{\min}^2 = 2 - \sqrt{2}$. We note that for all two-state cases considered, the total gain of the trellis-coded asymmetric M-PSK constellation over the uncoded $M/2$-point constellation is 3.01 dB. Indeed, this can be shown to be true always, independent of M.

Four-state trellis

The four-state trellis for A16-PSK has the structure of Fig. 5.43(a) and is illustrated in Fig. 5.46. Unlike the A8-PSK case, the minimum distance is not determined by the length-1 path between like states; that is, there exist paths with length greater than 1 whose distance from the all-zeros path is less than the minimum distance among the parallel transitions. In particular, the squared minimum distance due to parallel transitions is determined by signal points 4 or 12 and has a value of 2. Depending on the value of ϕ, paths 2, 1, 2 and 2, 15, 15, 2 yield the optimum asymmetric design, which, as we shall see shortly, has a value of d_{free}^2 that is less than 2 but is still larger than that corresponding to a symmetric constellation.

From (5.76) we can determine the squared distances of the two paths above as

$$d^2(2, 1, 2) = 6 - 2\sqrt{2} - 2\cos\phi$$

$$d^2(2, 15, 15, 2) = 8 - 2\sqrt{2} - 4\cos\left(\frac{\pi}{4} - \phi\right) \quad (5.82)$$

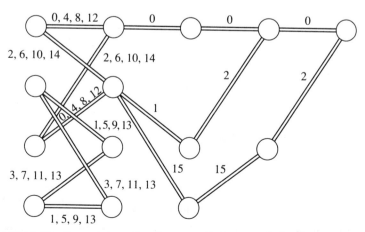

FIGURE 5.46 Four-state trellis diagram with four parallel paths per transition.

which, when equated, give the optimum A16-PSK design with

$$\tan\frac{\phi}{2} = 1 - \sqrt{2 - \sqrt{2}} \implies \phi = 0.46 \text{ rad} \tag{5.83}$$

$$d_{\text{free}}^2 = 1.38.$$

For the symmetric case, the path 2, 1, 2 has the shorter distance, which from (5.82) becomes

$$d_{\text{free}}^2 = 1.324. \tag{5.84}$$

Thus the gain due to asymmetry is

$$\eta = 10 \log_{10} \frac{1.38}{1.324} = 0.18 \text{ dB} \tag{5.85}$$

and the gains relative to the uncoded 8-PSK system are

$$\eta(\text{asymm}) = 10 \log_{10} \frac{1.38}{2 - \sqrt{2}} = 3.72 \text{ dB}$$

$$\eta(\text{symm}) = 10 \log_{10} \frac{1.324}{2 - \sqrt{2}} = 3.54 \text{ dB}. \tag{5.86}$$

Eight-state trellis

The eight-state trellis for A16-PSK is as illustrated in Fig. 5.39, except that the signal assignments are now defined by the state transition matrix given as

$$\mathbf{T} = \begin{matrix} & \begin{matrix} 1 & 2 & 3 & 4 & 5 & 6 & 7 & 8 \end{matrix} \\ \begin{matrix} 1 \\ 2 \\ 3 \\ 4 \\ 5 \\ 6 \\ 7 \\ 8 \end{matrix} & \begin{bmatrix} C_0 & C_2 & & & & & & \\ & & C_3 & C_1 & & & & \\ & & & & C_2 & C_0 & & \\ & & & & & & C_1 & C_3 \\ C_2 & C_0 & & & & & & \\ & & C_1 & C_3 & & & & \\ & & & & C_0 & C_2 & & \\ & & & & & & C_3 & C_1 \end{bmatrix} \end{matrix}. \tag{5.87}$$

$$C_0 = 0, 4, 8, 12 \qquad C_2 = 2, 6, 10, 14$$

$$C_3 = 1, 5, 9, 13 \qquad C_1 = 3, 7, 11, 15$$

and only the subscript of C was used as an index for the signal assignment in the trellis in Fig. 5.39. Since we are only interested in determining the minimum-distance paths through the trellis, we can simplify the state transition matrix (5.87) by considering only the signal points that are the minimum distance from signal point 0. As such, the "reduced" state transition matrix is

$$\mathbf{T} = \begin{array}{c} \\ 1 \\ 2 \\ 3 \\ 4 \\ 5 \\ 6 \\ 7 \\ 8 \end{array} \begin{array}{cccccccc} 1 & 2 & 3 & 4 & 5 & 6 & 7 & 8 \\ \left[\begin{array}{cccccccc} 0 & 2 & & & & & & \\ & & 1 & 15 & & & & \\ & & & & 2 & 0 & & \\ & & & & & & 15 & 1 \\ 2 & 0 & & & & & & \\ & & 15 & 1 & & & & \\ & & & & 0 & 2 & & \\ & & & & & & 1 & 15 \end{array}\right] \end{array}. \tag{5.88}$$

The minimum-distance paths are still the paths illustrated in Fig. 5.39, which, using (5.76), now have the distances

$$d^2(2, 1, 2, 2) = 8 - 3\sqrt{2} - 2\cos\phi$$

$$d^2(2, 15, 15, 0, 2) = 8 - 2\sqrt{2} - 4\cos\left(\frac{\pi}{4} - \phi\right) \tag{5.89}$$

$$d^2(2, 1, 0, 1, 15, 0, 2) = 10 - 2\sqrt{2} - 4\cos\phi - 2\cos\left(\frac{\pi}{4} - \phi\right).$$

We note that, unlike the 8-PSK case, the length-4 (solid) and the length-7 (dashed) paths do not have the same distance. In fact, the length-7 path is, for all values of ϕ, closer in distance to the all-zeros path. Thus, to find the optimum asymmetric design, we equate the distance of the length-5 and length-7 paths, which results in

$$\tan\frac{\phi}{2} = 0.1637 \Longrightarrow \phi = 0.3244 \text{ rad}$$

$$d^2_{\text{free}} = 1.589 \tag{5.90}$$

and a gain due to asymmetry of

$$\eta = 10\log_{10}\frac{1.589}{1.476} = 0.319 \text{ dB}. \tag{5.91}$$

Finally, the gains relative to an uncoded 8-PSK are

$$\eta(\text{asymm}) = 10\log_{10}\frac{1.589}{2 - \sqrt{2}} = 4.333 \text{ dB}$$

$$\eta(\text{symm}) = 10\log_{10}\frac{1.476}{2 - \sqrt{2}} = 4.014 \text{ dB}. \tag{5.92}$$

The results for the 2-, 4-, 8-, and 16-state cases of trellis-coded symmetric and asymmetric 4-, 8-, and 16-PSK modulations are summarized in Table 5.10.

TABLE 5.10 Performance of Rate $n/(n+1)$ Trellis-Coded M-PSK Versus Uncoded $M/2$-PSK.

Number of Modulation Levels, $M = 2^{n+1}$	Number of Parallel Transitions Between States	Number of States in Trellis	d_{min}^2 for Uncoded Modulation with $M/2$ Levels	d_{free}^2 for Coded Symmetric	d_{free}^2 for Coded Asymmetric	ϕ_{opt} (rad)	Gain (dB) of Symmetric Coded M-PSK over Uncoded $M/2$-PSK	Gain (dB) of Asymmetric Coded M-PSK over Symmetric Coded M-PSK
4	1	2	4	6	8	π	1.76	1.25
4	1	4	4	10	$32/3 = 10.67$	1.91	3.98	0.28
4	1	8	4	12	$40/3 = 13.33$	1.23	4.77	0.46
8	2	2	2	$4 - \sqrt{2} = 2.586$	4	$\pi/2$	1.116	1.895
8	2	4	2	4	4	$\pi/4$	3.01	0.00
8	1	8	2	$6 - \sqrt{2} = 4.586$	$6 - \sqrt{2} = 4.586$	$\pi/4$	3.60	0.00
8	1	16	2	$8 - 2\sqrt{2} = 5.172$	$26/5 = 5.20$	0.6435	4.13	0.024
16	4	2	$2 - \sqrt{2}$	0.738	1.172	$\pi/4$	1.00	2.01
16	4	4	$2 - \sqrt{2}$	1.324	1.380	0.46	3.54	0.18
16	4	8	$2 - \sqrt{2}$	1.476	1.589	0.3244	4.014	0.319
16	4	16	$2 - \sqrt{2}$	1.628	1.628	$\pi/8$	4.440	0.00

FIGURE 5.47 Asymmetric 4-AM signal set.

5.6.5 Best Rate 1/2 Codes Combined with Asymmetric 4-AM

Suppose that as in Section 5.6.2, we want to transmit one bit per symbol time. Another way of doing this is to use a rate 1/2 trellis code with 4-AM. In particular, we select the asymmetric 4-AM signal set illustrated in Fig. 5.47, where each of the four signal points has been moved by $\pm\Delta$ from their nominal positions; that is, setting $\Delta = 0$ produces the conventional symmetric constellation. The average power of the signal set in Fig. 5.47 is

$$P_{av} = 5 - 2\Delta + \Delta^2. \tag{5.93}$$

Consider the two-state trellis diagram illustrated in Fig. 5.48 with the assignment of 4-AM signals to different branches according to the set-partitioning method. The superstate transition diagram corresponding to Fig. 5.48 is derived from (5.29) and illustrated in Fig. 5.49. Calculating the transfer function $T(D, I)$, we get

$$T(D, I) = \frac{2a(c + d)}{1 - 2b}$$

$$a = \frac{I}{2} D^{16/P_{av}},$$

$$b = \frac{I}{2} D^{4(1-\Delta)^2/P_{av}} \tag{5.94}$$

$$c = \frac{1}{2} D^{4(3-\Delta)^2/P_{av}},$$

$$d = \frac{1}{2} D^{4(1+\Delta)^2/P_{av}}$$

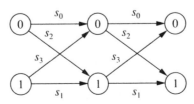

FIGURE 5.48 Two-state trellis diagram for asymmetric 4-AM.

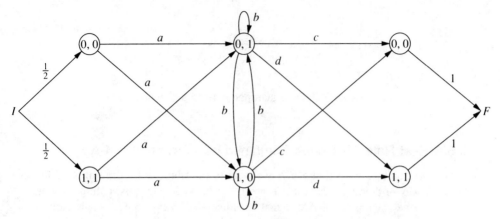

FIGURE 5.49 Pair-state transition diagram for the trellis diagram of Fig. 5.48.

Differentiating $T(D, I)$ with respect to I and evaluating at $I = 1$ gives the upper bound on average bit error probability:

$$P_b \leq \frac{1}{4} \operatorname{erfc}\left(\sqrt{\frac{E_b \, d_{\text{free}}^2}{N_0 \, 4}}\right) \frac{1 + y^8}{(1 - y)^2} \tag{5.95}$$

where

$$y = Z^{4(1-\Delta)^2/P_{\text{av}}}. \tag{5.96}$$

To find the optimum asymmetric 4-AM signal set, we may choose Δ to minimize P_b. Here the optimum Δ will depend on the Bhattacharyya parameter Z, defined in (5.25).

As mentioned previously, for large enough SNR, we may equivalently optimize the signal set by maximizing the normalized (by the average energy) free Euclidean distance. From the superstate transition diagram of Fig. 5.49, we get

$$d_{\text{free}}^2(\text{asymm}) = \frac{16 + 4(1 + \Delta)^2}{P_{\text{av}}} \tag{5.97}$$

which approaches the maximum value

$$d_{\text{free}}^2(\text{asymm}) = 8 \tag{5.98}$$

asymptotically as Δ approaches unity. Note that for $\Delta = 1$, the two pairs of signal points merge together. Thus, to avoid obtaining a catastrophic code, we can approach the limiting value of (5.98), but never reach it.

For the symmetric 4-AM, setting $\Delta = 0$ in (5.97) gives $d_{\text{free}}^2(\text{symm}) = 4$. Thus, for this simple case, the coding gain using an asymmetric signal set rather than a symmetric set asymptotically approaches 3 dB. In practice, Δ would be

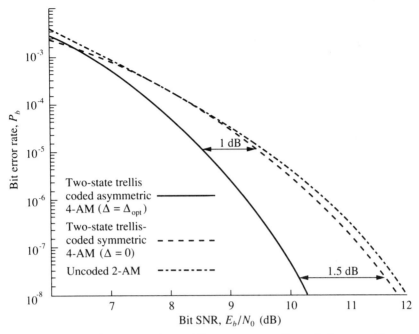

FIGURE 5.50 Performance of rate 1/2, two-state, trellis-coded symmetric and asymmetric 4-AM.

optimized by minimizing P_b of (5.95) as discussed previously. The bit error rate as a function of bit SNR is illustrated in Fig. 5.50.

Table 5.11 presents the performance of trellis-coded M-AM for symmetric and asymmetric signal sets and several different code memories. Included in this table are the relative gain of trellis-coded symmetric M-AM over uncoded $M/2$-AM and the additional gain obtained by making the signal set asymmetric. Note that for a two-state trellis code, the total gain of the trellis-coded asymmetric M-AM relative to uncoded $M/2$-AM is 3 dB, independent of the value of M.

5.6.6 Trellis-Coded Asymmetric 16-QAM

Consider again a two-state trellis code combined now with the asymmetric 16-QAM signal set illustrated in Fig. 5.32. Using the set-partitioning method, we get the four sets

$$\begin{aligned}
S_0 &= \{s_0,\ s_2,\ s_8,\ s_{10}\} \\
S_1 &= \{s_5,\ s_7,\ s_{13},\ s_{15}\} \\
S_2 &= \{s_6,\ s_{14},\ s_4,\ s_{16}\} \\
S_3 &= \{s_3,\ s_{11},\ s_9,\ s_1\}.
\end{aligned} \quad (5.99)$$

TABLE 5.11 Performance of Rate $n/(n+1)$ Trellis-Coded M-AM Versus Uncoded $M/2$-AM.[a]

Number of Modulation Levels, $M = 2^n$	Number of Transitions Between States	Number of States in Trellis	d_{min}^2 for Uncoded Modulation with $M/2$ Levels	d_{free}^2 for Coded Symmetric	d_{free}^2 for Coded Asymmetric	Δ_{opt}[b]	Gain (dB) of Symmetric Coded M-AM over Uncoded $M/2$-AM	Gain (dB) of Asymmetric Coded M-AM over Symmetric Coded M-AM
4	1	2	4	4	8	1	0	3.01
4	1	4	4	36/5	8	$3 - \sqrt{8}$	2.552	0.458
4	1	8	4	8	9.17	$3 - \sqrt{8}$	3	0.593
8	2	2	4/5	20/21	8/5	1	0.76	2.24
8	2	4	4/5	12/7	$\dfrac{32 - 8\sqrt{8}}{8 - \sqrt{8}}$	$3 - \sqrt{8}$	3.31	0.24
8	1	8	4/5	40/21	$\dfrac{56 - 16\sqrt{8}}{8 - \sqrt{8}}$	$3 - \sqrt{3}$	3.77	0.38
16	4	2	4/21	20/85	8/21	1	0.918	2.092
16	4	4	4/21	36/85	$\dfrac{32 - 8\sqrt{8}}{24 - \sqrt{8}}$	$3 - \sqrt{8}$	3.47	0.192
16	2	8	4/21	40/85	$\dfrac{56 - \sqrt{8}}{24 - \sqrt{8}}$	$3 - \sqrt{8}$	3.93	0.328

[a] Symmetric M-AM points are $\pm 1, \pm 3, \ldots, \pm(M-1)/2$.
[b] Δ, Shift of all M points away from their nominal positions to produce a symmetric set of M points (center of gravity remains fixed at zero).

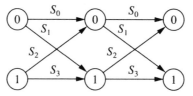

FIGURE 5.51 Two-state trellis diagram for asymmetric 16-QAM.

The assignment of these signal sets to the trellis diagram is as shown in Fig. 5.51. The normalized square distances between these sets are given by

$$d^2(S_0, S_1) = d^2(S_2, S_3) = \frac{8}{10}$$

$$d^2(S_0, S_3) = d^2(S_1, S_2) = \frac{8}{10} \sin^2 \frac{\phi}{2} \qquad (5.100)$$

$$d^2(S_0, S_2) = d^2(S_1, S_3) = \frac{8}{10}\left(1 + \sin\phi + \sin^2 \frac{\phi}{2}\right).$$

The free Euclidean distance is determined by an error event of length 2. Thus, using (5.100), we get

$$d^2_{\text{free}}(\text{asymm}) = d^2(S_0, S_1) + d^2(S_0, S_2). \qquad (5.101)$$

For the symmetric 16-QAM set ($\phi = -\pi/2$), (5.101) becomes $d^2_{\text{free}}(\text{symm}) = 12/10$, whereas the asymmetric case is optimized when $\phi = 0$, in which case (5.101) becomes $d^2_{\text{free}}(\text{asymm}) = 16/10$. Again, this limiting case cannot be achieved with a noncatastrophic code, so one must back away from $\phi = 0$. Nevertheless, the limiting value of coding gain is $10 \log_{10}(1.6/1.2) = 1.25$ dB.

In summary, in this section, some asymmetry into the constellation design of a combined modulation trellis-coding system has been introduced, which has resulted in improvement of system performance under certain circumstances. The performance gain has been analyzed for M-PSK, M-AM, and 16-QAM.

REFERENCES

1. G. Ungerboeck, "Channel coding with multi-level/phase signals," *IEEE Trans. Inf. Theory*, Vol. IT-28, pp. 55–67, Jan. 1982.
2. G. Ungerboeck, "Trellis-coded modulation with redundant signal sets. Part I: Introduction," *IEEE Commun. Mag.*, Vol. 25, pp. 5–12, Feb. 1978.
3. G. Ungerboeck, "Trellis-coded modulation with redundant signal sets. Part II: State of the art," *IEEE Commun. Mag.*, Vol. 25, pp. 12–22, Feb. 1987.
4. R. Calderbank and J. E. Mazo, "A new description of trellis codes," *IEEE Trans. Inf. Theory*, Vol. IT-30, pp. 784–791, Nov. 1984.
5. M. Turgeon, "Minimal complexity design of analytically described trellis codes," thesis for the degree of Master of Science, Queen's University,

Kingston, Ontario, Canada, Aug. 1987; see also, *Proc. International Conference on Communications*, Philadelphia, Pa., June 12–15, 1988; and *IEEE Trans. Comm.*, Sept. 1990.
6. M. K. SIMON and D. DIVSALAR, "A new description of combined trellis coding with asymmetric modulation," *JPL Publication*, Pasadena, Calif., pp. 45–85, July 15, 1985.
7. J.-E. PORATH, "Algorithms for converting convolutional codes from feedback to feedforward form and vice versa," *Electron. Lett.*, July 1989; see also *Technical Report 7*, Chalmers University of Technology, Division of Information Theory, Goteberg, Sweden, June 1987.
8. D. DIVSALAR and J. H. YUEN, "Performance of convolutionally coded unbalanced QPSK systems," *Proc. National Telecommunincations Conference*, Houston, Tex., pp. 14.5.1–14.5.7, Nov. 1980.
9. D. DIVSALAR and J. H. YUEN, "Asymmetric MPSK for trellis codes," *Proc. Globecom '84*, Atlanta, Ga., pp. 20.6.1–20.6.8, Nov. 1984.
10. M. K. SIMON and D. DIVSALAR, "Combined trellis coding with asymmetric MPSK modulation," *JPL Publication 85-24*, Pasadena, Calif., May 1, 1985.
11. D. DIVSALAR and M. K. SIMON, "Combined trellis coding with asymmetric modulations," *Proc. Globecom '85*, New Orleans, La., Dec. 2–5, 1985.
12. D. DIVSALAR, M. K. SIMON, and J. H. YUEN, "Trellis coding with asymmetric modulations," *IEEE Trans. Commun.*, Vol. COM-35, No. 2, Feb. 1987.
13. R. BOZOVIC, D. L. SCHILLING, and A. ISHAK, "Noncoherent detection of trellis coded continuous phase multilevel-FM," *Proc. ICC '87*, Seattle, Wash., June 7–10, 1987.

CHAPTER 6

Multidimensional Modulations

We have learned so far that to improve the performance of a TCM scheme based on a given signal constellation, the number of states should be increased. However, as this number exceeds a certain value, the coding gain increases more slowly. Thus, if the signal constellation is not changed, returns are progressively diminishing. Now, how should we change the constellation in order to increase the coding gain of a TCM scheme? In this chapter we consider using *multidimensional signals*; Chapter 7 is devoted to a different solution, *multiple TCM*.

Motivation for the use of multidimensional signals for digital transmission dates back to the work of Shannon. In his celebrated analysis of the limit performance achievable in digital communication over a given channel, he recognized [1] that the performance of a signal constellation used to transmit digital information over the additive white Gaussian noise channel can be improved by increasing N, the dimensionality of the signal set used for transmission. In particular, as the dimension number grows to infinity, the performance tends to an upper limit that defines the capacity of the channel (see also [2]). Heuristically, as the number of dimensions grows we have more space to accommodate the signals, and hence the distance between signal points increases. In turn, a greater distance between signal points means (at least for high-enough signal-to-noise ratios) a smaller error probability. The price to pay for an improvement in performance when the dimensionality is increased is essentially the increase in complexity of the modulator and the demodulator. If we are willing to pay this price, multidimensional signaling is certainly an attractive proposition.

Uncoded four-dimensional signal sets were considered by Welti and Lee [3], Zetterberg and Brändström [4], Wilson et al. [5], and Biglieri and Elia [6, 7]. Four-dimensional signals have a special appeal, because besides being the closest thing to the widely used two-dimensional sets, they can be implemented in radio communications without any increase in bandwidth, by utilizing two spatially orthogonal electric field polarizations for transmitting on the same carrier frequency. Welti–Lee codes are essentially subsets or translations of the lattice consisting of the points in the four-dimensional space whose integer coordinates have an even sum. Zetterberg–Brändström designs are based on quaternion groups, and their signals are constrained to have an equal energy. Wilson et al. consider four-dimensional signal sets based on subsets of lattice packings, which are known to afford simplification of encoding and decoding procedures. Gersho and Lawrence [8] consider four- and eight-dimensional signal sets with two information bits per dimension.

Their designs show a 1.2- to 2.4-dB gain in noise margin over conventional (two-dimensional) quadrature amplitude modulation.

If multidimensional alphabets are used in conjunction with TCM, a number of positive features can be reaped, as summarized in the following [5, 6, 9–12]:

- Spaces with larger dimensionality have more room for the signals, which are consequently spaced at larger Euclidean distances $d_{\min}$. Thus an increased noise margin may be derived from the multidimensional constellation itself.
- An inherent cost with one- and two-dimensional TCM schemes is that when the size of the constellation is doubled over that of uncoded schemes, it may occur that $E > E'$, that is, the average energy needed for signaling increases. For example, with two-dimensional rectangular constellations (QAM) doubling the constellation size makes us lose roughly 3 dB. Without this increase in energy expenditure, the coding gain of TCM would be greater. Now it happens that if we use a *multidimensional* rectangular constellation, this cost falls from 3 dB for two dimensions to 1.5 dB (four dimensions) or to 0.75 dB (eight dimensions).
- A simple way of generating a multidimensional constellation is by time division. For example, if N two-dimensional signals are sent in a time interval of duration T, and each has duration T/N, we obtain a $2N$-dimensional constellation. With respect to the case $N = 1$, it is possible to reduce the size of their constituent two-dimensional constellations. This may prove advantageous, since smaller two-dimensional constellations imply better performance when the received signal contains signal-dependent noise (linear or nonlinear distortions).
- Multidimensional constellations provide a great degree of flexibility in achieving higher effective information rates.
- As we shall see in Chapter 8, for some applications it is necessary to design TCM schemes that are transparent to phase rotations. Multidimensional constellations may simplify the design of these "rotationally invariant" schemes.
- Due to their byte-oriented nature, multidimensional schemes are suitable for use as inner codes in a concatenated coding system.

The requirements for a multidimensional signal set to be used in conjunction with TCM are the following.

- Inherent symmetries, possibly derived from an algebraic structure, which make it simple to describe the set.
- Easy description of the set-partitioning procedure.
- Flexibility of the signal set to the introduction of special constraints, such as transparency to phase rotations (see Chapter 8).
- Structural properties that simplify the decoding procedure. It should be kept in mind that TCM schemes with multidimensional signals are usually based on a large subset of signals associated with a single transition between two states. The first step in the application of the Viterbi algorithm involves the computation of the branch metrics, which in turn is based on selection of the subset element closest to the observed signal. This multihypothesis detection problem may have an exceedingly large complexity if not simplified by the inherent structure of the multidimensional set.

Partitioning multidimensional signal sets

In the following we consider the construction of TCM schemes based on multidimensional signal sets. Now, an integral part of the construction of TCM schemes is the geometric partition of the signal set, described in Chapter 3. Consequently, together with a description of the signal sets, we need to provide a set-partitioning procedure.

Let us first formalize the concept of a partition. This formal treatment is necessary here because with multidimensional signal sets it will not be possible to rely on simple pictures to describe the partition process. Given a constellation $\mathcal{X}$, we obtain its partition by dividing it into a family Γ of nonoverlapping subsets (subconstellations) such that the union of all subsets in Γ is equal to $\mathcal{X}$. An L-level partition of $\mathcal{X}$ is a sequence of partitions $\Gamma_1, \ldots, \Gamma_L$, where Γ_i is a refinement of Γ_{i-1}.

Given a partition Γ_i and a distance $d(\cdot, \cdot)$, we define in it an *intradistance* δ_i as the minimum distance among signals belonging to the same subconstellation in Γ_i. Formally,

$$\delta_i = \min d(a, b) \qquad (6.1)$$

where the minimum is taken over all the signals a, b, $a \neq b$, with $a, b \in \mathcal{S}$ and $\mathcal{S} \in \Gamma_i$. Similarly, we define an *interdistance* as the minimum distance between elements in two different subconstellations of the same partition level,

$$d(\mathcal{S}_1, \mathcal{S}_2) = \min d(a, b) \qquad (6.2)$$

where the minimum is taken over all the signals a, b chosen from different subconstellations in Γ_i. We shall call a partition *fair* if its subsets include the same number of signals, and their intradistances are equal.

A particularly fruitful approach to partitioning of multidimensional constellations is through the definition of a suitable algebraic structure on which the original signal set is based. Sometimes the constellation itself forms a group (this is the case of *lattices*); sometimes it is generated from a group (this is the case of *group alphabets*). In both cases we shall see that the concept of the partition of a group into its cosets provides a suitable algebraic tool for partitioning.

6.1 Lattices

We start this section by providing an overview of lattice theory. A thorough treatment of this topic can be found in the encyclopedic book by Conway and Sloane [13], from which most of the material that follows is taken (see also [14] and [15]). In general, a lattice Λ is defined as a set closed under ordinary addition and multiplication by integers. Specifically, an N-dimensional lattice Λ is a set of points, or m-dimensional vectors, of the form

$$\mathbf{x} = u_1 \mathbf{i}_1 + \cdots + u_N \mathbf{i}_N \qquad (6.3)$$

where $\mathbf{x}$ is an m-dimensional vector, u_i are integers, and $\mathbf{i}_k$, $k = 1, \ldots, N$ are linearly independent vectors in the real Euclidean m-dimensional space $\mathbf{R}^m$. Note that $m \geq N$. The vectors $\mathbf{i}_k$ are a *basis* for the lattice in an integer-coefficient expansion. If the basis vectors are

$$\mathbf{i}_1 = (i_{11}, \ldots, i_{1m})$$
$$\vdots$$
$$\mathbf{i}_N = (i_{N1}, \ldots, i_{Nm})$$

then the matrix

$$\mathbf{M} = \begin{bmatrix} i_{11} & \cdots & i_{1m} \\ & \cdots & \\ i_{N1} & \cdots & i_{Nm} \end{bmatrix} \quad (6.4)$$

is called a *generator matrix* for the lattice, and the lattice points have coordinates **uM**, where **u** is any vector with integer components. The *dual lattice* Λ^* consists of all points **y** spanned by $\mathbf{i}_1, \ldots, \mathbf{i}_N$ such that the scalar product $\mathbf{x} \cdot \mathbf{y}$ is integer-valued.

Two lattices are *equivalent* if one of them can be obtained from the other by a rotation, reflection, and scaling. If Λ is equivalent to Λ', we write

$$\Lambda \simeq \Lambda'.$$

The generator matrices **M** and **M'** of two equivalent lattices are related by

$$\mathbf{M}' = c\mathbf{UMB}$$

where c is a nonzero constant, **U** a matrix with integer entries and determinant ± 1, and **B** a real orthogonal matrix with $\mathbf{BB}^T = 1$ (as usual, the superscript T denotes transpose).

If $d_{\min}$ is the minimum distance between any two points in the lattice, the *kissing number* τ is the number of adjacent lattice points located at distance $d_{\min}$ (i.e., the number of nearest neighbors of any lattice point).

Transformation of lattices

Given a lattice Λ with vectors **x**, new lattices can be generated by the following operations.

- *Scaling.* If r is any real number, then $r\Lambda$ is the lattice with vectors $r\mathbf{x}$.
- *Orthogonal transformation.* If T is a scaled orthogonal transformation of $\mathbf{R}^n$, then $T\Lambda$ is the lattice with vectors $T\mathbf{x}$.
- *Direct product.* The n-fold direct product of Λ with itself, that is, the set of all nm-tuples $(\mathbf{x}_1, \mathbf{x}_2, \ldots, \mathbf{x}_n)$, where each $\mathbf{x}_i$ is in Λ, is a lattice denoted by Λ^n.

6.1.1 Some Examples of Lattices

The lattice Z^n

The set Z^n of all n-tuples with integer coordinates is called the *cubic lattice*, or *integer lattice*. Its generator matrix is the $n \times n$ identity matrix. The minimum distance is $d_{\min} = 1$, and the kissing number is $\tau = 2n$.

The n-dimensional lattices A_n and A_n^*

A_n is the set of all vectors with $(n+1)$ integer coordinates whose sum is zero. This lattice may be viewed as the intersection of Z^{n+1} and a hyperplane cutting the origin. The minimum distance is $d_{\min} = \sqrt{2}$, and the kissing number is $\tau = n(n+1)$.

A generator matrix for the dual lattice A_n^* is

$$\mathbf{M} = \begin{bmatrix} 1 & -1 & 0 & \cdots & 0 & 0 \\ 1 & 0 & -1 & \cdots & 0 & 0 \\ & & & \cdots & & \\ 1 & 0 & 0 & \cdots & -1 & 0 \\ -\frac{n}{n+1} & \frac{1}{n+1} & \frac{1}{n+1} & \cdots & \frac{1}{n+1} & \frac{1}{n+1} \end{bmatrix}.$$

We have $\tau = 2$ for $n = 1$, and $\tau = 2n + 2$ for $n \geq 2$. The minimum distance is $d_{\min} = \sqrt{n/(n+1)}$.

The n-dimensional lattices D_n and D_n^*

D_n is the set of all n-dimensional points whose integer coordinates have an even sum. It may be viewed as a punctured version of Z^n, in which the points are colored alternately red and white with a checkerboard coloring, and the white points (those with odd sums) are removed. The generator matrix is

$$\mathbf{M} = \begin{bmatrix} -1 & -1 & 0 & \cdots & 0 & 0 \\ 1 & -1 & 0 & \cdots & 0 & 0 \\ 0 & 1 & -1 & \cdots & 0 & 0 \\ & & & \cdots & & \\ 0 & 0 & 0 & \cdots & 1 & -1 \end{bmatrix}.$$

We have $d_{\min} = \sqrt{2}$, and $\tau = 2n(n-1)$.

The dual lattice D_n^* has generator matrix

$$\mathbf{M} = \begin{bmatrix} 1 & 0 & 0 & \cdots & 0 & 0 \\ 0 & 1 & 0 & \cdots & 0 & 0 \\ 0 & 0 & 1 & \cdots & 0 & 0 \\ & & & \cdots & & \\ 0 & 0 & 0 & \cdots & 1 & 0 \\ \frac{1}{2} & \frac{1}{2} & \frac{1}{2} & \cdots & \frac{1}{2} & \frac{1}{2} \end{bmatrix}.$$

It has kissing number $\tau = 8$ for $n = 3$, $\tau = 24$ for $n = 4$, and $\tau = 2n$ for $n \geq 5$. The minimum distance is $d_{\min} = \sqrt{3}/2$ ($n = 3$) or $d_{\min} = 1$ ($n \geq 4$).

D_4 represents the densest lattice packing in $\mathbf{R}^4$. This means that if unit-radius, four-dimensional spheres with centers in the lattice points are used to pack $\mathbf{R}^4$, D_4 is the lattice with the largest number of spheres per unit volume. Its dual D_4^* can also be described as

$$D_4^* = Z^4 \cup \{Z^4 + (\tfrac{1}{2}, \tfrac{1}{2}, \tfrac{1}{2}, \tfrac{1}{2})\}$$

that is, as the union of Z^4 and a translate of Z^4.

The Gosset lattice E_8

In the *even-coordinate system*, E_8 consists of the points

$$\left\{ (x_1, \ldots, x_8) : \forall x_i \in Z \text{ or } \forall x_i \in Z + \tfrac{1}{2}, \sum_{i=1}^{8} x_i \equiv 0 \text{ mod } 2 \right\}.$$

In words, E_8 consists of the 8-vectors whose components are all integers, or all halves of odd integers, and whose sum is even.

The *odd-coordinate system* is obtained by changing the sign of any coordinate; the points are

$$\left\{ (x_1, \ldots, x_8) : \forall x_i \in Z \text{ or } \forall x_i \in Z + \tfrac{1}{2}, \sum_{i=1}^{8} x_i \equiv 2x_8 \text{ mod } 2 \right\}.$$

This lattice has $d_{\min} = \sqrt{2}$ and $\tau = 240$.

Other lattices

The description and the properties of other important lattices, such as the 16-dimensional Barnes–Wall lattice Λ_{16} and the 24-dimensional Leech lattice Λ_{24}, can be found in Chapter 4 of [13].

6.1.2 Structural Characteristics of Lattices

Algebraically, a lattice is a group, that is, the sum or difference of any two vectors in Λ is in Λ. Thus Λ necessarily includes the null vector $\mathbf{0}$, and if $\mathbf{x} \in \Lambda$, then also $-\mathbf{x} \in \Lambda$.

A *sublattice* Λ' of a lattice Λ is a subset of Λ that is itself a lattice. Since Λ is a group, Λ' is a subgroup of Λ. Thus Λ' induces a partition of Λ into subsets, which are Λ' and its cosets.

Equivalences between lattices

We have the following results:

$$Z \simeq A_1 \simeq A_1^*$$
$$A_3 \simeq D_3$$
$$A_2 \simeq A_2^*$$
$$D_4 \simeq D_4^*.$$

6.1.3 Examples of Lattice Constellations for TCM

We shall now consider some simple examples of signal constellations based on lattices, and TCM schemes based on them.

Our first examples are taken from [15] and are based on the use of four-dimensional signals. As mentioned before, four-dimensional signals appear to be well suited for *frequency reuse*, a technique that utilizes two spatially orthogonal electric field polarizations for transmission over the same frequency band. If the two fields can be kept orthogonal over the channel, the bandwidth efficiency is doubled.

Modulation schemes for the four-dimensional channel can be generated by transmitting two-dimensional signals independently over the two pairs of dimensions, or by designing more general four-dimensional constellations. We shall refer to these schemes as *product schemes*. Figure 6.1 shows two-dimensional lattice constellations with 2^m signals. They are subsets of Z^2. Figure 6.2 plots the energy versus spectrum efficiency of four-dimensional schemes obtained as products of these two-dimensional constellations. The energy efficiency of a constellation with M points is defined as the ratio between $d_{min}^2 \log_2 M$ and the average energy E of the constellation, and is plotted here relative to that of a 4-PSK constellation used on every pair of dimensions (which has $d_{min}^2/E = 1$, and $\log_2 M = 4$, for an energy efficiency of 4). The bandwidth efficiency is the ratio between $\log_2 M$ and the number of dimensions N ($N = 4$ in these examples).

For example, the two-dimensional 32-ary constellation of Fig. 6.1 has $d_{min}^2/E = 0.2$. The product of two such constellations has the same minimum distance, while the energy is doubled. Moreover, $\log_2 M = \log_2 32^2 = 10$, which gives an energy efficiency of 1, that is, 6 dB below the reference, and a bandwidth efficiency of 2.5.

If the four-dimensional constellation is not restricted to be a product of two-dimensional sets, subsets of 2^m points of suitable lattices may be found with a

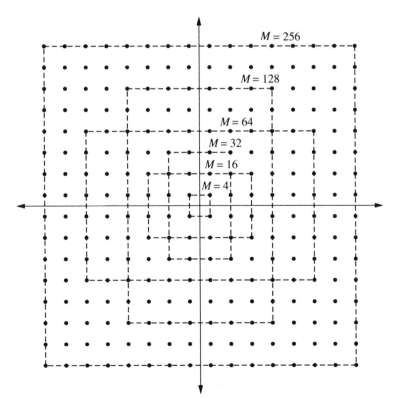

FIGURE 6.1 Two-dimensional constellations as subsets of the lattice Z^2.

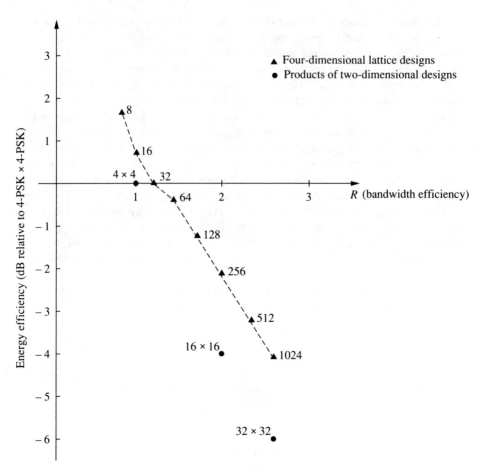

FIGURE 6.2 Energy efficiency versus bandwidth efficiency of four-dimensional constellations. From [15].

better energy efficiency. Figure 6.2 shows the efficiency of some of these constellations extracted from the lattice D_4. The subset selection was performed by a computer search whose details can be found in [15]. It is seen, for example, that the constellation of 1024 points from D_4 has virtually the same energy efficiency as the 16 × 16 product constellation, but a spectral efficiency 25% greater. Constellations with equal spectral efficiency show a gain of 1.5 to 2.5 dB for M in the range 64 to 1024.

Consider next a few simple examples of designs of TCM schemes based on four-dimensional lattices. (A more systematic approach to the design of multidimensional TCM schemes based on lattices will be provided later.)

Two states, 2 bits per symbol

Consider the scheme of Fig. 6.3. We assign symbols from an octonary set to the eight branches of the trellis as shown. It is seen that parallel transitions have $d^2 = 4$ because of antipodality, the same distance of error events with length 2. The free distance is 4 and the average energy per symbol is 1, so that $d_{\text{free}}^2/E = 4$. To evaluate this design, we must compare it with an uncoded scheme transmitting

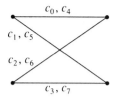

$$c_0 = 1\ 0\ 0\ 0 \quad c_4 = -1\ 0\ 0\ 0$$
$$c_1 = 0\ 1\ 0\ 0 \quad c_5 = 0\ -1\ 0\ 0$$
$$c_2 = 0\ 0\ 1\ 0 \quad c_6 = 0\ 0\ -1\ 0$$
$$c_3 = 0\ 0\ 0\ 1 \quad c_7 = 0\ 0\ 0\ -1$$

FIGURE 6.3 Two-state TCM scheme with 2 bits per symbol and four-dimensional signals.

2 bits per four-dimensional signal. If we use binary PSK in each pair of dimensions (admittedly, this is not the best scheme), we obtain $d_{\min}^2/E' = 4/2 = 2$, which shows an asymptotic coding gain of 3 dB.

Four states, 3 bits per symbol

Consider the scheme of Fig. 6.4. The squared interdistance is 16, while the squared intradistance is 8. Error events with length 2 give $d^2 = 12$. We get $d_{\text{free}}^2/E = 12/4 = 3$, corresponding to an energy occupancy of 9. The bandwidth efficiency is 0.75. If we compare this figure with uncoded 4×4 signals, we see that we have an improvement in energy efficiency of $9/4 = 3.6$ dB with a bandwidth occupancy that is 33% greater.

A similar design can be found with eight states and $d_{\text{free}}^2/E = 4$. It yields a 4.8-dB gain in energy efficiency over uncoded 16-ary, again with a 33% bandwidth expansion [15].

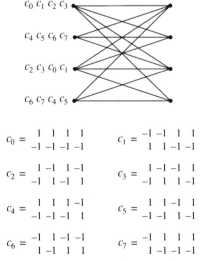

FIGURE 6.4 Four-state TCM scheme with 3 bits per symbol and four-dimensional signals.

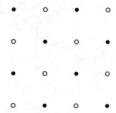

FIGURE 6.5 Lattice Z^2 and its sublattice RZ^2.

6.1.4 Partition of Lattices

As we have seen before, a sublattice Λ' induces a partition of a lattice Λ into subsets. This partition is denoted Λ/Λ'. One simple way of generating a sublattice is via an endomorphism R of Λ that magnifies the distances. An important example is provided by the endomorphism of Z^2 induced by the *rotation operator* R defined by the matrix

$$\mathbf{R} = \begin{bmatrix} 1 & 1 \\ 1 & -1 \end{bmatrix}. \tag{6.5}$$

The effect of $\mathbf{R}$ is to rotate Z^2 by 45° and scale it by $\sqrt{2}$ (see Fig. 6.5). The resulting partition Z^2/RZ^2 is two-way.

A one-dimensional example is provided by the four-way partition $Z/4Z$ of the integers into the four residue classes mod 4. A four-way partition of Z^2 is generated as $Z^2/2Z^2$, while $Z^2/2RZ^2$ [R as in (6.5)] generates an eight-way partition of the same lattice. The latter partition is shown in Fig. 6.6, where points belonging to the same coset are identified by the same shape.

The concept of a lattice partition can be fruitfully applied to the generation of constellations based on lattices to be used in conjunction with TCM schemes. The original lattice is first partitioned into sublattices with larger minimum distance, then finite subconstellations with the desired number of points are carved from

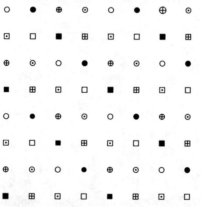

FIGURE 6.6 Eight-way partition $Z^2/2RZ^2$.

these sublattices. We first illustrate this statement with an example, then describe briefly a general design technique.

An eight-dimensional TCM scheme based on the Gosset lattice

An eight-dimensional design based on the Gosset lattice E_8 has been developed by Calderbank and Sloane [10]. Consider the matrix

$$\mathbf{R} = \begin{bmatrix} 1 & 1 & 0 & 0 & 0 & 0 & 0 & 0 \\ 1 & -1 & 0 & 0 & 0 & 0 & 0 & 0 \\ 0 & 0 & 1 & 1 & 0 & 0 & 0 & 0 \\ 0 & 0 & 1 & -1 & 0 & 0 & 0 & 0 \\ 0 & 0 & 0 & 0 & 1 & 1 & 0 & 0 \\ 0 & 0 & 0 & 0 & 1 & -1 & 0 & 0 \\ 0 & 0 & 0 & 0 & 0 & 0 & 1 & 1 \\ 0 & 0 & 0 & 0 & 0 & 0 & 1 & -1 \end{bmatrix}.$$

The norm of the vector $\mathbf{xR}$, $\mathbf{x} \in E_8$, is twice as large as the norm of $\mathbf{x}$, and the set of vectors $\mathbf{xR}$ forms a sublattice of E_8, called $R(E_8)$. $R(E_8)$ itself and its 15 cosets form a 16-way partition of E_8. Every vector in E_8 has 240 nearest neighbors at a squared distance 2 from it. They are divided equally among the cosets, resulting into a squared intradistance 2 (and 16 nearest neighbors in each coset).

The average distance of the constellation is minimized by moving the origin of E_8 to $(1, 0^7) = (1, 0, 0, 0, 0, 0, 0, 0)$. Then the vectors of the shifted E_8 are divided into shells, the nth shell consisting of the vectors with norm n. For example, the first shell contains 16 points $(\pm 1, 0^7)$ of norm 1 and the second shell contains 128 points with norm 2, expressed by $(\pm \frac{1}{2}^8)$, where the number of minus signs is odd.

The signal constellation used for the TCM scheme consists of $16 \cdot 2^k$ vectors from the shifted E_8, with 2^k vectors in each coset. It is formed by taking a union of successive shells, while omitting enough vectors from the outermost shell to obtain exactly $16 \cdot 2^k$ vectors. Different choices of the parameter k provide different designs. In [10] a linear convolutional code is exhibited, giving a family of non-catastrophic TCM schemes with eight states and $d_{\text{free}} = 4$. They have $k = 13, 21,$ and 29 (corresponding to rates 2, 3, and 4 bits per dimension), and exhibit asymptotic coding gains of about 6 dB with respect to uncoded transmission at the same rate based on a signal constellation in the shape of a hypercube. (Implementation considerations can also be found in [10].)

6.1.5 Calderbank–Sloane TCM Schemes Based on Lattices

These schemes [16] are based on the new concept of a generalized convolutional code whose output symbols are cosets of a lattice. The trellis code maps the q-ary source symbols into a signal constellation Ω that is a finite subset of a lattice Λ. This construction requires three ingredients in addition to the lattice Λ.

- A sublattice $\Lambda' \subseteq \Lambda$. Λ' is obtained from Λ by applying an endomorphism R of Λ that magnifies distances by a factor of $\sqrt{m}$. The number of cosets of Λ' in Λ, also called the *index* of Λ' in Λ, is $m^{N/2}$.

- A convolutional code that has k_1 input symbols and a single output symbol chosen from Λ/Λ'. Let the current output symbol depend on the current input block of size k_1 and on the previous v input blocks. Let $\mathcal{T}$ denote the set of possible cosets that occur as outputs of the code.
- The signal constellation, which consists of $M = |\mathcal{T}|q^{k_2}$ points of Λ, partitioned into q^{k_2} points in each coset in $\mathcal{T}$.

The complete trellis code is represented in Fig. 6.7. The q-ary source symbols are divided into blocks of $k = k_1 + k_2$ bits each. The first k_1 bits are fed to the convolutional encoder, producing a coset of Λ' in Λ. The remaining k_2 bits are used to select one of the q^{k_2} points in the coset. This scheme has a rate $k \log_2 q$ bits/symbol, or $(k \log_2 q)/N$ bits per dimension.

Denote by V the *memory* of this code, that is, the minimum number of previous inputs that are required to calculate the current output. The number of states in the encoder is q^V. Thus, in the trellis diagram describing the code, at each discrete time there are q^V possible states. Each state has q^{k_1} branches emanating from it, one branch for each possible input. The branches are labeled with the corresponding output symbols (i.e., by cosets of Λ').

A number of trellis codes based on this construction are described in [16]. Their rate ranges from 1.5 to 3 bits/dimension, with a number of states from 4 to 256 and asymptotic coding gains up to 6.4 dB. Parameters of a selected set of TCM schemes are summarized in tables in [14] and [16]. The coding gain is computed with reference to an uncoded system using constellations based on Z^N. It can be observed that schemes can be found having the same asymptotic coding gain but lower "error coefficients" $N(d_{\text{free}})$ for a larger number of states. For these schemes an increase in complexity results into a better error performance at *intermediate* values of signal-to-noise ratio.

An example of a Calderbank–Sloane TCM scheme is illustrated in the following. This scheme has 16 states. Here the linear convolutional code is described by the generator matrix

$$\mathbf{G} = [g_{vk_1}, \ldots, g_{v1}, |\cdots| g_{1k_1}, \ldots, g_{11}, | g_{0k_1}, \ldots, g_{01}]$$

whose entries are elements of Λ/Λ'. The source symbols fed to the convolutional encoder are labeled u_{ij}, with $u_{01}, u_{02}, \ldots, u_{0k_1}$ being the current input block,

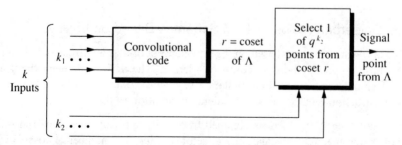

FIGURE 6.7 TCM encoder for the Calderbank–Sloane schemes.

TABLE 6.1 Coset Representatives and Labels for $Z^2/4Z^2$.

Coset Representative	Label of the Points	Coset Representative	Label of the Points
(0, 0)	3	(0, 2)	11
(1, 0)	0	(1, 2)	8
(2, 0)	1	(2, 2)	9
(−1, 0)	2	(−1, 2)	10
(0, −1)	7	(0, 1)	15
(1, −1)	4	(1, 1)	12
(2, −1)	5	(2, 1)	13
(−1, −1)	6	(−1, 1)	14

$u_{11}, u_{12}, \ldots, u_{1k_1}$ the previous block, and so on. The current output is the coset

$$r = \sum_{i=0}^{v} \sum_{j=1}^{k_1} u_{ij} g_{ij}.$$

Consider $N = 2$ and the partition $Z^2/4Z^2$. Here $m = 16$, so we get the 16 cosets described in Table 6.1.

Figure 6.8 shows a constellation with 64 points to be used in conjunction with this partition. Each coset has four points, and the average energy of the constellation

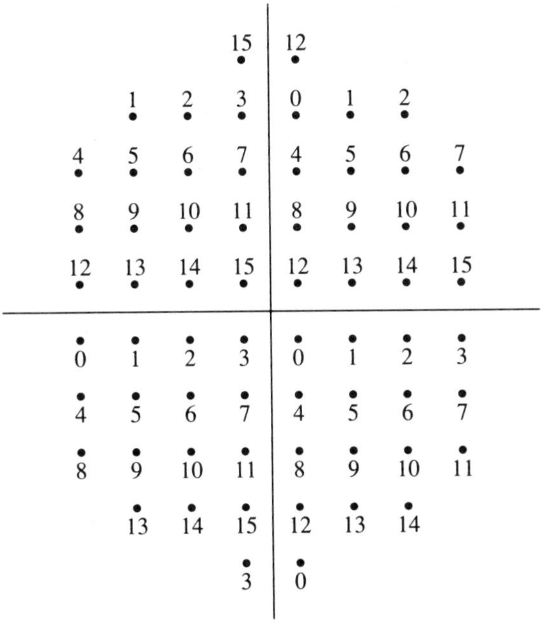

FIGURE 6.8 Constellation with 64 points for $Z^2/4Z^2$.

is 10.25. With a quaternary source (i.e., $q = 4$ and $k_1 = k_2 = 1$), we obtain a rate of 2 bits per dimension. The generator matrix of the convolutional code is

$$\mathbf{G} = \left[\begin{array}{c|c|c} 1 & 1 & 1 \\ 2 & 1 & 2 \end{array} \right]$$

with cosets as columns according to Table 6.1. The coset output at discrete time j is $\Lambda' + \mathbf{Gu}_j$, where

$$\mathbf{u}_j = \left(u_{(j-2)1}, u_{(j-1)1}, u_{j1} \right)^T.$$

Let $\mathbf{x}_j$ be the modulo 4 sum of $\mathbf{u}_j$ and $\mathbf{v}_j$. Then

$$(\Lambda' + \mathbf{Gu}_j) + (\Lambda' + \mathbf{Gv}_j) = \Lambda' + \mathbf{Gx}_j.$$

The squared intradistance here is 16, while the minimum squared distance of an error event with length greater than 1 is 12. Thus $d_{\text{free}}^2 = 12$. Comparison with an uncoded 16-QAM constellation, which has the same rate, gives the asymptotic coding gain

$$10 \log_{10} \frac{12/10.25}{4/10} = 4.664 \text{ dB}.$$

6.2 Group Alphabets

The basic problem when dealing with multidimensional signals is that set partitioning, a procedure required in the design of TCM schemes, may be difficult to perform if the signal set is not endowed with a particular structure. Here we consider a special class of signal sets that have good distance properties as well as a special structure that makes it possible to do set partitioning by using algebraic techniques. This class is based on the concept of "group codes for the Gaussian channel" introduced by Slepian [17], but also includes signals that, unlike Slepian's group codes, have more than one energy level. Although it offers a large degree of symmetry and of algebraic structure, this class of alphabets is exceedingly general and includes most of the signal sets proposed for application. For instance, as we shall soon see in two examples, two-dimensional group alphabets include the 16-QAM constellation, and asymmetric as well as symmetric PSK.

Generalized group alphabets are generated as follows. Consider a set of K N-dimensional vectors $\mathbf{X} = \{\mathbf{x}_1, \ldots, \mathbf{x}_K\}$, called the *initial set*, and L orthogonal $N \times N$ matrices $\mathbf{S}_1, \ldots, \mathbf{S}_L$ that form a finite group $\mathcal{G}$ under multiplication. The set of vectors $\{\mathcal{G}\mathbf{x}_1, \ldots, \mathcal{G}\mathbf{x}_K\}$ obtained from the action of $\mathcal{G}$ on the initial set is called a *group alphabet* (GA). A GA is called *separable* if the vectors of the initial set are transformed by $\mathcal{G}$ into either disjoint or coincident vector sets; that is, the intersection of the sets $\mathcal{G}\mathbf{x}_j$ and $\mathcal{G}\mathbf{x}_k$ is either empty (for $j \neq k$) or equal to $\mathcal{G}\mathbf{x}_j$ itself (for $j = k$).

A GA is called *regular* if the number of vectors in each subalphabet $\mathcal{G}\mathbf{x}_j$, $j = 1, \ldots, K$, does not depend on j; that is, each vector of the initial set is transformed into the same number of distinct vectors. A regular GA is called *strongly regular* if each set $\mathcal{G}\mathbf{x}_j$ contains exactly L distinct vectors.

It follows directly from the definitions that the number M of vectors in a regular GA is a multiple of K. If the GA is strongly regular, then $M = KL$.

EXAMPLE 6.1 ASYMMETRIC M-PSK

In this example we show that asymmetric M-PSK, a two-dimensional signal set with one energy level, can be generated as a GA. Choose an initial vector $\mathbf{x} = (\cos\theta, \sin\theta)$, θ a given constant, an integer $M = 2^\mu$, and consider the group of 2×2 orthogonal matrices of the form $\mathbf{R}^i \mathbf{T}^j$, $i = 0, 1, \ldots, M - 1$, $j = 1, 2$, where

$$\mathbf{R} = \begin{bmatrix} \cos\dfrac{2\pi}{M} & \sin\dfrac{2\pi}{M} \\ -\sin\dfrac{2\pi}{M} & \cos\dfrac{2\pi}{M} \end{bmatrix}$$

and

$$\mathbf{T} = \begin{bmatrix} 0 & 1 \\ 1 & 0 \end{bmatrix}.$$

It is seen that the effect of $\mathbf{R}$ on a two-dimensional vector is to rotate it by an angle $2\pi/M$, and the effect of $\mathbf{T}$ is to exchange its components. This group has $2M$ elements and gives rise to a separable alphabet of M or $2M$ vectors, according to the choice of the initial vector. This alphabet is strongly regular only if it has $2M$ elements. □

EXAMPLE 6.2 AN ALPHABET WITH FOUR DIMENSIONS AND ONE ENERGY LEVEL

Consider the group of matrices that act on a four-dimensional initial vector by permuting its components and replacing one or more of them with their negatives. This group has $4!\, 2^4$ elements. If the initial vector is $\mathbf{x}_1 = (a, a, a, 0)$, $a = 1/\sqrt{3}$, the resulting (separable) alphabet has $M = 32$ distinct unit-energy vectors (see Table 6.2). □

EXAMPLE 6.3 A GA WITH TWO DIMENSIONS AND THREE ENERGY LEVELS

Figure 6.9 shows 16-QAM as a GA. Points 1, 2, 3, and 4 denote the four vectors in the initial set. The matrices generating the GA are those associated with plane rotations by multiples of $\pi/2$. The result is a strongly regular, separable GA. □

EXAMPLE 6.4 A GA WITH FOUR DIMENSIONS AND TWO ENERGY LEVELS

A GA can be obtained from the initial set of vectors

$$\begin{matrix} c & c & c & 0 \\ -b & c & c & 0 \\ c & -b & c & 0 \\ c & c & -b & 0 \end{matrix}$$

TABLE 6.2 GA of Example 6.2 and Its Fair Partition in Four Subsets.

0	a	a	0	a	a	a	0	a	a	0	a	a	a	a	0
a	$-a$	a	a	$-a$	a	a	a	a	a	a	0	a	$-a$	0	a
a	$-a$	$-a$	$-a$	$-a$	0	$-a$	a	a	a	a	a	a	a	a	0
$-a$	$-a$	0	$-a$	0	$-a$	0	$-a$	$-a$	$-a$	0	a	0	$-a$	a	a
a	$-a$	$-a$	$-a$	$-a$	a	a	a	$-a$	$-a$	0	a	a	$-a$	a	a
$-a$	0	$-a$	0	$-a$	0	a	0	$-a$	0	a	$-a$	$-a$	$-a$	a	$-a$
$-a$	a	a	$-a$	a	a	a	$-a$	0	a	a	a	0	a	$-a$	$-a$
$-a$	a	0	a	$-a$	$-a$	$-a$	$-a$	a	$-a$	a	$-a$	$-a$	a	a	0

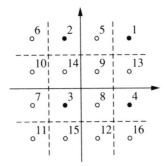

FIGURE 6.9 16-QAM as a GA and its fair partition.

with $c = 0.389$ and $b = 0.939$. If we apply to this initial set the same matrix group that generates the GA of Example 6.2, we get a separable alphabet with 128 vectors. Among them, 32 have energy $3c^2$, and 96 have energy $b^2 + c^2$. The average energy is 1. □

Some general classes of GA, as generated by special groups, are described in Appendix 6A at the end of this chapter.

6.2.1 Set Partitioning of a GA

To design a TCM scheme based on GAs, after choosing a constellation we must partition it into subsets with the same cardinality. We also want them to exhibit similar distance properties, in order to provide a maximum of symmetry. (Although no proof is available, it is widely believed that a good TCM scheme should exhibit a large amount of symmetries.)

To do this, the concept of *fair partition* is defined. Choose a partition of a GA into m subsets $\mathcal{X}_1, \mathcal{X}_2, \ldots, \mathcal{X}_m$. For each subset $\mathcal{X}_i$, we can define the *intradistance set* as the set of all the Euclidean distances among pairs of vectors in $\mathcal{X}_i$. For any pair of distinct subsets $\mathcal{X}_i, \mathcal{X}_j$, we define their *interdistance set* as the set of all Euclidean distances between a vector in $\mathcal{X}_i$ and a vector in $\mathcal{X}_j$. The partition of a separable GA is called *fair* if all the subsets are distinct, include the same number of vectors, and their intradistance sets are equal.

To generate a fair set partitioning of a group alphabet, consider the group $\mathcal{G}$ generating a GA, and assume that each set $\mathcal{G}X_j$ contains exactly L distinct vectors. Take one of its subgroups, say, $\mathcal{H}$, and the partition of $\mathcal{G}$ into left cosets of $\mathcal{H}$. If the left cosets of $\mathcal{H}$ are applied to the initial set of a GA, this procedure results into a fair partition of the GA. It is easily seen that this procedure can be repeated by using a subgroup of $\mathcal{H}$ to generate a further fair partition.

EXAMPLE 6.2 (CONTINUED) ─────────────────────────────

The GA of this example (a "permutation alphabet") is based on the group $\mathcal{G}$ of matrices that permute the components of a 4-vector and/or replace them by their

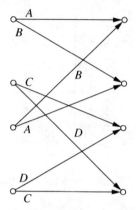

FIGURE 6.10 Four-state TCM scheme for the GA of Example 6.2.

negatives. Table 6.2 shows a fair partition of this alphabet in four subsets of eight vectors each. The partition is obtained as follows. Denote by **D** the orthogonal matrix whose effect on a vector is to cyclically shift its components to the right by one position, and to change sign to the second component. Then the set

$$\mathcal{H} = \{\mathbf{D}^0, \mathbf{D}^1, \mathbf{D}^2, \mathbf{D}^3, \mathbf{D}^4, \mathbf{D}^5, \mathbf{D}^6, \mathbf{D}^7\}$$

is a cyclic normal subgroup of $\mathcal{G}$, and its cosets generate the fair partition.

The minimum distance between vectors in this GA is $2a^2$, while the minimum intradistance in this partition is $6a^2$. By choosing a four-state trellis code with $k = 4$ and two branches stemming from each node (see Fig. 6.10), we obtain $d_{\text{free}}^2/E = 6a^2$, with a 3-dB gain with respect to the minimum distance achieved by using two independent, uncoded 4-PSK alphabets (which transmit the same amount of information and use the same number of dimensions). □

EXAMPLE 6.4 (CONTINUED)

As shown in [6], a four-state TCM scheme can be designed that provides $d_{\text{free}}^2 = 1.2$, a coding gain of about 4.3 dB with respect to the use of uncoded 8/4-PSK signals. □

6.3 Ginzburg Construction

Ginzburg [18] has described a multidimensional signal construction that makes it possible to design signals that have an arbitrary minimum distance and a regular structure by combining elementary signals and by employing algebraic properties of block codes (see also [19], [20], and [21]). Ginzburg construction makes it possible to convert the complicated mapping between source bits and N-dimensional signals into simple mappings involving lower-dimensional constellations. Moreover, it lends itself to a suboptimum decoding method that is amenable to a parallel/pipelined structure [22, 23].

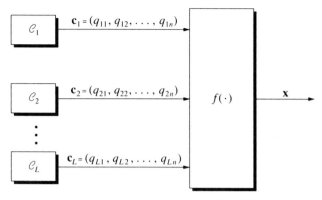

FIGURE 6.11 Ginzburg construction.

Figure 6.11 shows an L-stage construction. L block encoders $\mathscr{C}_1, \mathscr{C}_2, \ldots, \mathscr{C}_L$ accept source symbols and output L blocks

$$\mathbf{c}_i = (q_{i1}, q_{i2}, \ldots, q_{in}), \qquad i = 1, \ldots, L$$

of n symbols each. The symbols of $\mathscr{C}_i$ are chosen from an alphabet with M_i elements. The output signal $\mathbf{x}$ depends on these L n-tuples, that is, on the matrix

$$\mathbf{Q} = \begin{bmatrix} q_{11} & q_{12} & \cdots & q_{1n} \\ q_{21} & q_{22} & \cdots & q_{2n} \\ & & \cdots & \\ q_{L1} & q_{L2} & \cdots & q_{Ln} \end{bmatrix}$$

so that we can write $\mathbf{x}$ in one of the equivalent forms:

$$\mathbf{x} = f(\mathbf{Q})$$

$$\mathbf{x} = f(\mathbf{c}_1, \mathbf{c}_2, \ldots, \mathbf{c}_L)$$

$$\mathbf{x} = (F(\mathbf{q}_1), F(\mathbf{q}_2), \ldots, F(\mathbf{q}_n)) \tag{6.6}$$

where the code words $\mathbf{c}_i$, $i = 1, \ldots, L$, are the rows of $\mathbf{Q}$, and $\mathbf{q}_j$, $j = 1, \ldots, n$, are its columns. Each component of $\mathbf{x}$, called an *elementary signal*, is a function [as described by (6.6)] of a vector $\mathbf{q}_j$, referred to as the *label* of the elementary signal. The elementary signals $F(\mathbf{q}_j)$ are chosen from a set $\mathscr{A}$ of $M = M_1 M_2 \cdots M_L$ elements.

The mapping F is generated with the aim of getting a minimum Euclidean distance between two elementary signals, which depends only on the difference between their labels.[1] To this purpose, in the set $\mathscr{A}$ we define a system of L partitions. Each class of the ℓth partition includes M_ℓ classes of the $(\ell-1)$th partition, so that it will consist of $M_1 M_2 \cdots M_\ell$ elementary signals. By numbering the classes of the $(\ell-1)$th level occurring in a class of the ℓth level, we obtain a one-to-one mapping of the set of classes of the $(\ell-1)$th partition onto the set of

---

[1] Ginzburg contruction is not the only way of achieving this goal. Another construction is described by Tanner [23].

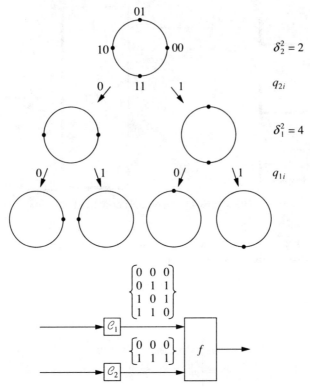

FIGURE 6.12 Example of Ginzburg construction with $L = 2$ and $M_1 = M_2 = 2$.

integers $\{0, \ldots, M_\ell - 1\}$. Therefore, if q_{ij} are chosen in the set $\{0, \ldots, M_\ell - 1\}$, $\ell = 1, \ldots, L$, then any L-tuple $(q_{1j}, \ldots, q_{Lj})$ defines a unique value of the jth elementary signal $F(q_{1j}, \ldots, q_{Lj})$ (see Fig. 6.12).

We associate with the ℓth level the minimum Euclidean distance among signals in the class. This is denoted by δ_ℓ. Since a class of a subsequent level can incorporate a class of a preceding level only in its entirety, we have

$$\delta_{\ell-1} \geq \delta_\ell.$$

If $\delta_{\ell-1} = \delta_\ell$, then two levels can be combined by omitting the $(\ell - 1)$th partition. Hence we can assume that

$$\delta_1 > \delta_2 > \cdots > \delta_L.$$

The signal set obtained with this construction has a number of dimensions

$$N = n\nu$$

where ν is the dimensionality of the elementary signals. The number of ν-dimensional elementary signals is M, while the total number of N-dimensional signals is equal to the product of the number of code words in $\mathscr{C}_1, \ldots, \mathscr{C}_L$. The minimum squared Euclidean distance of this alphabet, $d^2_{\min}$, satisfies

$$d^2_{\min} \geq \min(\delta_1^2 d_1^H, \delta_2^2 d_2^H, \ldots, \delta_L^2 d_L^H) \tag{6.7}$$

where $d_1^H, \ldots, d_L^H$ are the minimum Hamming distances of the L block codes $\mathscr{C}_1, \ldots, \mathscr{C}_L$.[2]

To prove (6.7), consider the distance between the two signals $\mathbf{x}' = (x_1', \ldots, x_n') = f(\mathbf{Q}')$ and $\mathbf{x}'' = (x_1'', \ldots, x_n'') = f(\mathbf{Q}'')$, where $x_i' = F(\mathbf{q}_i')$ and $x_i'' = F(\mathbf{q}_i'')$. For $\mathbf{x}' \neq \mathbf{x}''$, and hence $\mathbf{Q}' \neq \mathbf{Q}''$, there exists at least one ℓ and one i such that $q_{\ell i}' \neq q_{\ell i}''$. Let us denote by λ the largest such value of ℓ,

$$\lambda = \max\{\ell : q_{\ell i}' \neq q_{\ell i}'', 1 \leq \ell \leq L, 1 \leq i \leq n\}.$$

Then δ_λ is the smallest nonzero distance between the elementary signals that appear in $\mathbf{x}'$ and $\mathbf{x}''$, and any of these distances is greater than or equal to $\delta_\lambda \delta(q_{\lambda i}', q_{\lambda i}'')$, where $\delta(q_{\lambda i}', q_{\lambda i}'')$ is the Kronecker delta. From this we have

$$d^2(\mathbf{x}', \mathbf{x}'') \geq \delta_\lambda^2 \sum_{i=1}^n \delta(q_{\lambda i}', q_{\lambda i}'') \geq \delta_\lambda^2 d_\lambda^H$$

since signals from $\mathscr{A}$ for $q_{\lambda i}' \neq q_{\lambda i}''$ correspond to different words of λth level code. Since such a λ exists for any pair of different signals from $\mathscr{A}$, (6.7) follows from the last inequality.

As observed in [18], if a minimum Euclidean distance $d_{\min}^2$ is specified, then it is advisable to take the minimum Hamming distance of the codes to be

$$d_\ell^H = \lceil d_{\min}^2 / \delta_\ell^2 \rceil, \quad \ell = 1, \ldots, L \tag{6.8}$$

where $\lceil x \rceil$ is the smallest integer not less than x.

In the following we describe two examples of Ginzburg constructions. Other examples can be found in [18] and [22].

EXAMPLE 6.5

As a simple example of Ginzburg construction, consider $L = 2$ with a 4-PSK elementary constellation. This constellation has $\delta_2^2 = 2$ and $\delta_1^2 = 4$, if we assume unit-energy 4-PSK (Fig. 6.12). Choose as $\mathscr{C}_2$ the (3, 1) binary repetition code (which has two code words, and $d_2^H = 3$) and as $\mathscr{C}_1$ the (3, 2) binary parity check code (which has four code words, and $d_1^H = 2$). The components of $\mathbf{x}_n$ are complex numbers chosen from the set $\{\pm 1, \pm j\}$, representative of 4-PSK signals. The signal set resulting from this construction is six-dimensional. It has eight elements (Fig. 6.13) and its minimum squared Euclidean distance is bounded above by $\min\{4 \cdot 2, 2 \cdot 3\} = 6$. The correspondence between the mapper input and the signals is the following:

00 00 00 ⟶ $\mathbf{x}_1 = (1, 1, 1)$ 01 01 01 ⟶ $\mathbf{x}_5 = (j, j, j)$

00 10 10 ⟶ $\mathbf{x}_2 = (1, -1, -1)$ 01 11 11 ⟶ $\mathbf{x}_6 = (j, -j, -j)$

10 00 10 ⟶ $\mathbf{x}_3 = (-1, 1, -1)$ 11 01 11 ⟶ $\mathbf{x}_7 = (-j, j, -j)$

10 10 00 ⟶ $\mathbf{x}_4 = (-1, -1, 1)$ 11 11 01 ⟶ $\mathbf{x}_8 = (-j, -j, j)$.

---

[2] Equality is achieved in (6.7) if a pair of code words can be found for which equality holds, which is true in most cases.

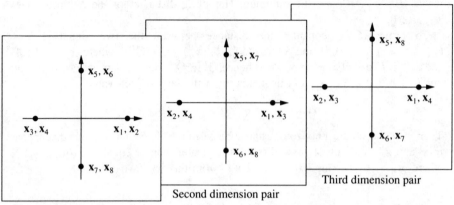

FIGURE 6.13 Simple example of a six-dimensional constellation with eight signals generated by Ginzburg construction. Here $L = 2$, and the elementary constellation is 4-PSK.

As can be proved by direct computation, the *exact* value of the minimum squared distance is actually 6. □

An example of Ginzburg construction based on Reed–Muller codes [24] is provided in the following.

EXAMPLE 6.6

This example [18] is based on the one-dimensional signals

$$x_i = -1.5 + 2q_{1i} + q_{2i}, \qquad q_{1i}, q_{2i} \in \{0.1\}.$$

Here we have $M = 4$, $L = 2$, $M_1 = M_2 = 2$. In each of the two classes of the first level, obtained for fixed q_{2i}, say, $\{-1.5, 0.5\}$ (for $q_{2i} = 0$) and $\{-0.5, 1.5\}$ (for $q_{2i} = 1$), the minimum squared distances are equal to $\delta_1^2 = 4$. In the second-level class $\delta_2^2 = 1$. If two binary codes (n, k_1, d_1^H) and $(n, k_2, 4d_1^H)$ can be found, then the minimum squared distance $d_{\min}^2$ of a signal set with $2^{k_1+k_2}$ signals is at least $4d_1^H = 0.8E_1 d_1^H$, where E_1 denotes the average energy of each component of x_i.

For $n = 2^\nu$, $d_{\min}^2 = 0.8E_1 2^{\nu-r}$, the codes $\mathscr{C}_1$ and $\mathscr{C}_2$ can be provided by Reed–Muller codes of orders $r + 2$ and r. In this case we get 2^κ signals, where

$$\kappa = 2 + 2\binom{m}{1} + \cdots + 2\binom{m}{r} + \binom{m}{r+1} + \binom{m}{r+2}.$$

Thus for $n = 32$, $r = 1$ codes with 2^{26} and 2^6 code words, respectively, provide a signal set with 2^{32} signals and minimum squared distance $d_{\min}^2 = 12.8$ with unit average energy per dimension. This is about 5 dB greater than in the case of uncoded ± 1 signals. This signal set may be suitable for use in conjunction with 16-QAM.

Notice also that to obtain the same result with a single block code we would need a quaternary code with $n = 32$, $k = 16$, and Hamming distance 16, which does not exist because it would exceed the Hamming bound. □

6.3.1 Set Partitioning

To perform set partitioning of an alphabet generated by the Ginzburg construction, a possible technique is to map the alphabet one-to-one onto a group $\mathcal{G}$. Then if

$$\mathcal{G}_1 \subset \mathcal{G}_2 \subset \cdots \subset \mathcal{G}_L = \mathcal{G}$$

denotes a chain of subgroups, then the coset of $\mathcal{G}$ with respect to subgroup $\mathcal{G}_\ell$ can be mapped onto a subconstellation at ℓth partition level. In the following we shall see examples of this partitioning technique.

This technique is based on the following property of linear block codes. Given a linear block code $\mathcal{C}$, let $\mathcal{C}'$ be one of its linear subcodes. Then $\mathcal{C}$ can be partitioned into the union of $\mathcal{C}'$ and its cosets, which are of the form $\mathcal{C}' \oplus \mathbf{c}$ for certain *coset representatives* $\mathbf{c} \in \mathcal{C}$. The set of Hamming distances in each one of the cosets is the same as in $\mathcal{C}'$.

The proof of the first part of this property follows immediately from the group property of linear block codes and the standard decomposition of a group into a subgroup and its cosets. To prove the second part, observe that the Hamming distance between two elements of the coset $\mathcal{C}' \oplus \mathbf{c}$ is given by

$$d^H(\mathbf{c}' \oplus \mathbf{c}, \mathbf{c}'' \oplus \mathbf{c}) = d^H(\mathbf{c}', \mathbf{c}'')$$

where $\mathbf{c}'$, $\mathbf{c}''$ are two code words in $\mathcal{C}'$.

As an example, consider the (7, 4, 3) Hamming code. If we take the subcode obtained as the set of linear combinations of the code words

$$\begin{array}{ccccccc} 0 & 0 & 0 & 1 & 1 & 1 & 1 \\ 0 & 1 & 1 & 0 & 1 & 1 & 0 \\ 1 & 0 & 1 & 1 & 0 & 1 & 0 \end{array}$$

then we get a (7, 3, 4) code $\mathcal{C}'$, and the partition

$$\mathcal{C} = \mathcal{C}' \cup \{\mathcal{C}' \oplus (1111111)\}.$$

The partition is shown in Table 6.3.

TABLE 6.3 Partition of the (7, 4, 3) Hamming Code into a (7, 3, 4) Subcode $\mathcal{C}'$ and Its Coset $\mathcal{C}' \oplus (1111111)$.

Subcode							Coset						
0	0	0	0	0	0	0	1	1	1	1	1	1	1
1	0	1	1	0	1	0	0	1	0	0	1	0	1
0	1	1	0	1	1	0	1	0	0	1	0	0	1
0	0	0	1	1	1	1	1	1	1	0	0	0	0
1	0	1	0	1	0	1	0	1	0	1	0	1	0
0	1	1	1	0	0	1	1	0	0	0	1	1	0
1	1	0	1	1	0	0	0	0	1	0	0	1	1
1	1	0	0	0	1	1	0	0	1	1	1	0	0

Reed–Muller codes

Reed–Muller codes [24] are a class of binary codes for which finding a subcode is relatively easy. In fact, if $\mathcal{R}(r, m)$ denotes the Reed–Muller code of order r, we have the property [24, p. 377]

- $\mathcal{R}(r, m)$ is a subcode *of* $\mathcal{R}(r + 1, m)$.

By using this property, we get a partition of the $\mathcal{R}(r + 1, r)$ Reed–Muller code into $2^{\binom{m}{r+1}}$ subsets, each of them with a minimum Hamming distance that is twice as large as in the original code. For example, we can partition the (16, 11, 4) code into the (16, 5, 8) code and its cosets, which provides a 2^6-way partition.

The partition of the (8, 4, 4) Reed–Muller code into an (8, 3, 4) subcode $\mathcal{C}'$ and its coset $\{\mathcal{C}' \oplus (1111111)\}$ is shown in Table 6.4.

Cyclic codes

For a cyclic code with generator polynomial $g(D)$, a subcode can easily be found by adding a factor to $g(D)$ and using this new polynomial as the generator of the subcode. However, although k decreases, in general one cannot say a priori that d_{min}^H will increase.

An important special case is provided by a BCH code with designed distance d^H. By designing a code with the same n and a larger designed distance, the construction itself shows that a subcode is obtained.

A special case: Ungerboeck's set partitioning

It may be interesting to observe here that Ungerboeck's set-partitioning technique, described in Chapter 3, can be described as an application of the concept of partitions generated by subcodes. Consider, in fact, L trivial $(N, N, 1)$ binary codes (their code words are all the 2^N binary N-tuples) and the signal set generated by using them in an L-level Ginzburg construction. The minimum squared distance is obviously δ_L^2. Now, a two-way partition of this signal set can be obtained by generating a partition of $\mathcal{C}_L$ based on its subcode $(N, N - 1, 2)$ (the single-parity-check code). We get two subconstellations whose signals have labels that differ by 1 bit and whose minimum distance has increased to

$$\min(\delta_{L-1}^2, 2\delta_L^2).$$

TABLE 6.4 Partition of the (8, 4, 4) Reed–Muller Code into an (8, 3, 4) Subcode and Its Coset.

Subcode								Coset							
0	0	0	0	0	0	0	0	1	1	1	1	1	1	1	1
0	0	0	0	1	1	1	1	1	1	1	1	0	0	0	0
0	0	1	1	0	0	1	1	1	1	0	0	1	1	0	0
0	1	0	1	0	1	0	1	1	0	1	0	1	0	1	0
0	0	1	1	1	1	0	0	1	1	0	0	0	0	1	1
0	1	0	1	1	0	1	0	1	0	1	0	0	1	0	1
0	1	1	0	0	1	1	0	1	0	0	1	1	0	0	1
0	1	1	0	1	0	0	1	1	0	0	1	0	1	1	0

6.3.2 Designing a TCM Scheme

A general guideline for the design of the TCM scheme to be used in conjunction with multidimensional signals is the following. Suppose that we have m' subconstellations of m'' signals each, for a total of $M = m'm''$ signals. Suppose that the minimum squared *intradistance* (i.e., the distance among signals in the same subconstellation) is D^2, and the minimum squared *interdistance* (i.e., the distance among signals in different subconstellations) is d^2. Then a TCM scheme can be found that has rate $r/r+1$, $m'/2$ states, and its squared free distance is bounded below by

$$d_{\text{free}}^2 \geq \min(2d^2, D^2). \tag{6.9}$$

In fact, if $m'm''/2$ signals are associated with each node of the trellis, we achieve the rate sought for the TCM scheme. This is obtained by associating with each node of a fully connected trellis $m'/2$ subconstellations of m'' signals each.[3] If parallel transitions determine the free distance, then $d_{\text{free}}^2 = D^2$. Otherwise, an error event with length 2 will generate a squared free distance of at least $2d^2$.

If a larger squared free distance is desired, we need to reduce the connectivity of the trellis by increasing the number of states. By doubling the number of states from $m'/2$ to m', we can increase by one the minimum length of an error event. If a suitable assignment of subconstellations to the trellis branches can be obtained, a squared free distance $3d^2$ will be achieved.

EXAMPLE 6.7

Consider an 8-PSK constellation with energy E. From (6.7) we have

$$\frac{d_{\min}^2}{0.586E} \geq \min(6.8 d_1^H, 3.4 d_2^H, d_3^H). \tag{6.10}$$

With $\mathscr{C}_1$ the $(8, 8, 1)$ trivial code, $\mathscr{C}_2$ the $(8, 7, 2)$ single-parity-check code, and $\mathscr{C}_3$ the $(8, 1, 8)$ repetition code, we have a constellation with $2^{8+7+1} = 2^{16}$ signals, $d_{\min}^2 = 4E$, and 1 bit per dimension. It provides an energy gain of 3 dB with respect to 4-PSK, which has the same number of bits per dimension but a minimum squared Euclidean distance $2E$.

Consider now the partition of this alphabet. Choose the $(8, 7, 2)$ single-parity-check code as a subcode of $\mathscr{C}_1$, the $(8, 4, 4)$ Reed–Muller code as a subcode of $\mathscr{C}_2$, and the $(8, 0, \infty)$ "code" with a single code word as a subcode of $\mathscr{C}_3$. The partition based on these subcodes gives $2^{16}/2^{11} = 2^5 = 32$ subsets with 2^{11} symbols each, a squared intradistance $8E$, and a squared interdistance $4E$. A 16-state TCM scheme, with 16 subsets associated with each state, will give $15/16 \simeq 0.94$ bits per dimension and squared free distance $8E$, with a gain of 6 dB with respect to 4-PSK. □

Construction of rotationally invariant multidimensional schemes

A systematic code search algorithm to find the best TCM schemes for a given constellation is described in [25]. The criteria used for optimization are d_{free}^2, the

---

[3] In this scheme, every state transition is associated with a subconstellation in such a way that transitions from or to a node use different subconstellations.

"error coefficient" $N(d_{\text{free}})$, and the rotational-invariance properties of the scheme (see Chapter 8). Tables describing the best designs based on M-PSK signals can also be found in [25].

6.4 Wei Construction

This construction, due to L.-F. Wei [12], generates a $2N$-dimensional alphabet by using a two-dimensional constellation in each one of N adjacent baud intervals, and adding one redundant bit every N intervals. With a two-dimensional constellation, 2^{Q+1} signals must be used to transmit Q bits per baud with a TCM scheme of rate $m/m+1$. If we choose to use $2N$-dimensional signals instead, a constellation of 2^{NQ+1} signals is needed. In fact, instead of adding one redundant bit every pair of dimensions, the redundant bit is added every N pairs. For large values of $NQ+1$, and in the presence of transmission channels where linear or nonlinear distortion is present, it is important to construct the constellation so that

1. The mapping between the $NQ+1$ bits and the signals may be converted to N simple constituent two-dimensional mappings.
2. The size of the constituent two-dimensional mappings is kept as small as possible.
3. The peak-to-average power ratio is kept as small as possible.

Since we need to transmit a noninteger number of bits per signaling interval (specifically, $NQ+1$ bits in N intervals, i.e., $Q+1/N$ per each interval), one may use the technique proposed in [26]. To understand this concept, a simple example should suffice. Suppose that we must send $n + \frac{1}{2}$ bits per interval. We use a signal constellation with $(1 + \frac{1}{2})2^n$ signals. 2^n "inner" signals are drawn from a regular two-dimensional lattice, and an additional $\frac{1}{2}2^n$ signals (the "outer" signals) are drawn from the same lattice and have as little average energy as possible. Incoming bits are then grouped into blocks of $2(n + \frac{1}{2}) = 2n + 1$ bits, and sent in two successive symbol intervals as follows. One bit in the block determines whether any outer signal is to be used. If not, the remaining $2n$ bits are used, n at a time, to select two inner signals. If so, then, one additional bit selects which of the two signals is to be an outer signal, $n - 1$ bits select which outer signal, and the remaining n bits select which inner signal for the second symbol interval [26, p. 637].

To implement the technique described above in a $2N$-dimensional constellation of 2^{NQ+1} signals, we can proceed as follows.

1. Start from a constituent two-dimensional rectangular constellation with $(1 + \frac{1}{N})2^Q$ signals.
2. Divide this constellation into two groups, an inner group and an outer group. The number of signals in the inner group is 2^Q, the number of signals in the outer group is $1/N$ of that in the inner group. The inner group is selected from the original constellation so that its average power is kept as small as possible. The outer group is selected from the rest of the original constellation so that its average power is kept as small as possible.

3. The constellation of 2^{NQ+1} signals is then constructed by concatenating N such two-dimensional constellations, and excluding those $2N$-dimensional signals corresponding to more that one two-dimensional outer signal. For each constituent two-dimensional constellation the inner group is used $2N - 1$ times as often as the outer group.

6.4.1 A Design Example

Based on the general principles described before, we now describe the design of a TCM scheme to transmit 19,200 bits/s at a rate of 2743 baud [12]. Here 7 bits/baud are necessary. With coding, if we assume that $N = 4$ (i.e., an eight-dimensional signal constellation), we must send $7 \times 4 + 1 = 29$ bits in four baud intervals. Hence the signal constellation must include $2^{NQ+1} = 2^{29}$ signals.

With these parameters, the constituent two-dimensional constellation will include

$$\left(1 + \tfrac{1}{4}\right)2^7 = 160$$

signals, instead of the $2^{7+1} = 256$ that would be used if one redundant bit were added in each interval. Thus, from our choice of using multidimensional signals we may expect an additional gain and a reduction of peak-to-average power ratio. Consider in particular the 160-signal two-dimensional constellation shown in Fig. 6.14. The inner group has 128 signals. The outer group has a quarter as many signals (i.e., 32). Inner and outer groups are also shown in Fig. 6.14. The eight-dimensional constellation is formed by concatenating four such 160-signal two-dimensional constellations, and excluding those eight-dimensional signals corresponding to more than one two-dimensional outer signal. The peak-to-average power ratio of the eight-dimensional constellation turns out to be 2.14.

Partitioning the eight-dimensional constellation

Now consider the partition of the resulting eight-dimensional constellation. This partition should have a hierarchical structure, in a way that makes maximum-likelihood decoding feasible. If we choose a rate 3/4 convolutional code for the TCM scheme, our goal is to partition the eight-dimensional constellation into 16 eight-dimensional constellations whose minimum squared distance (MSD) is $4d^2$, where d^2 denotes the MSD of the original two-dimensional constellation (see Fig. 6.14). The partition is done, according to the guidelines of [12], through the following steps.

1. Partition the original two-dimensional constellation into four two-dimensional subconstellations A, B, C, and D as shown in Fig. 6.14 (A includes the signals designated a, etc.). Each one of the subconstellations includes $160/4 = 40$ signals, 32 of which are inner and 8 outer, and has MSD $4d^2$.

2. Define 16 four-dimensional "types" by concatenating pairs of these two-dimensional constellations, say, (A, A), (A, B), ..., (D, D). The MSD of each type is $4d^2$, and each type includes $40 \times 40 = 1600$ signals.

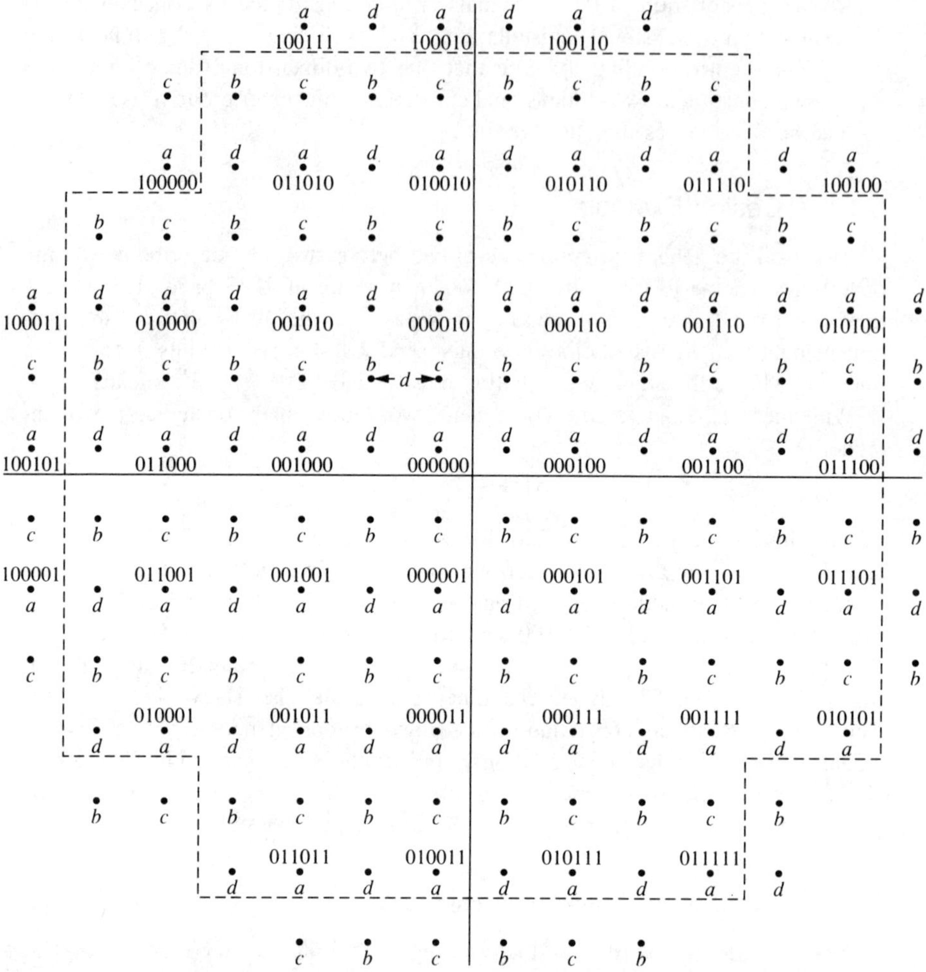

FIGURE 6.14 160-signal two-dimensional constellation. The number beneath each signal is $Z2_{n+i}Z3_{n+i}Z4_{n+i}Z5_{n+i}Z6_{n+i}Z7_{n+i}$, $i = 0, 1, 2, 3$. The same 6-bit value is associated with each of the four points that can be obtained from each other through 90° rotations.

3. Group these 16 types into eight four-dimensional subconstellations, which we denote 0, 1, 2, ..., 7. The grouping is done so as to maintain the MSD of each subconstellation at $4d^2$. Each such subconstellation has 3200 signals. We have

$$0: (A, A) \cup (B, B)$$

$$1: (C, C) \cup (D, D)$$

$$2: (A, B) \cup (B, A)$$

$$3: (C, D) \cup (D, C)$$

4: $(A, C) \cup (B, D)$

5: $(C, B) \cup (D, A)$

6: $(A, D) \cup (B, C)$

7: $(C, A) \cup (D, B)$.

4. Form 64 eight-dimensional types by concatenating pairs of these subconstellations: $(0, 0)$, $(0, 1)$, ..., $(7, 7)$. Each type has $3200^2 = 10.24 \times 10^6$ signals.
5. Group these 64 eight-dimensional types into 16 eight-dimensional subconstellations with MSD $4d^2$ as in Table 6.5. Each subconstellation has $10.24 \times 10^6 \times 4 = 40.96 \times 10^6$ signals.

Encoder design

A rate 3/4, 64-state code based on the eight-dimensional constellation of 2^{29} signals is shown in Fig. 6.15. The eight-dimensional constellation is partitioned into 16 subsets as in Table 6.5. The table provides the association of the four convolutionally encoded bits $Y0_n$, $I1_n$, $I2_n$, and $I3_n$ with the eight-dimensional subconstellations. The MSD between two allowed sequences of eight-dimensional signals is $4d^2$, which gives a coding gain over uncoded signals of 5.4 dB. Actually, 6 dB come from the distance increase, while 0.6 dB is the energy penalty due to the expansion of the constellation.

To map the four convolutionally encoded bits and the 25 remaining (nonconvolutionally encoded) information bits into the eight-dimensional constellation, a "bit converter" and a "block encoder" are used to convert those 29 bits into four

TABLE 6.5 16-Subconstellation Partitioning of an Eight-Dimensional Constellation.

Eight-Dimensional Subconstellation	$Y0_n$	$I1_n$	$I2_n$	$I3_n$	Eight-Dimensional Types
0	0	0	0	0	(0, 0), (1, 1), (2, 2), (3, 3)
1	0	0	0	1	(0, 1), (1, 0), (2, 3), (3, 2)
2	0	0	1	0	(0, 2), (1, 3), (2, 0), (3, 1)
3	0	0	1	1	(0, 3), (1, 2), (2, 1), (3, 0)
4	0	1	0	0	(4, 4), (5, 5), (6, 6), (7, 7)
5	0	1	0	1	(4, 5), (5, 4), (6, 7), (7, 6)
6	0	1	1	0	(4, 6), (5, 7), (6, 4), (7, 5)
7	0	1	1	1	(4, 7), (5, 6), (6, 5), (7, 4)
8	1	0	0	0	(0, 4), (1, 5), (2, 6), (3, 7)
9	1	0	0	1	(0, 5), (1, 4), (2, 7), (3, 6)
10	1	0	1	0	(0, 6), (1, 7), (2, 4), (3, 5)
11	1	0	1	1	(0, 7), (1, 6), (2, 5), (3, 4)
12	1	1	0	0	(4, 0), (5, 1), (6, 2), (7, 3)
13	1	1	0	1	(4, 1), (5, 0), (6, 3), (7, 2)
14	1	1	1	0	(4, 2), (5, 3), (6, 0), (7, 1)
15	1	1	1	1	(4, 3), (5, 2), (6, 1), (7, 0)

236 Ch. 6 / Multidimensional Modulations

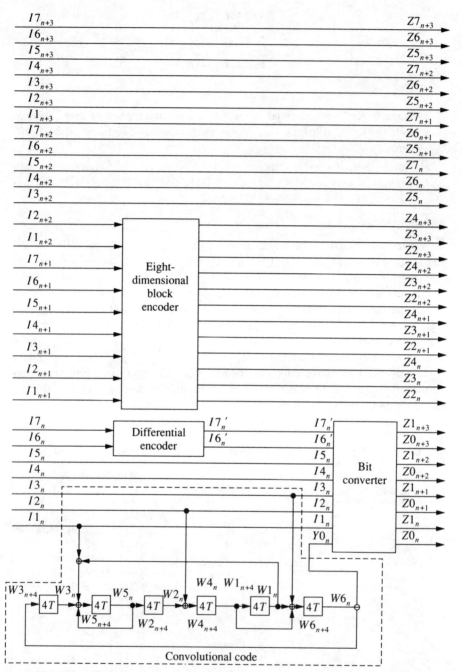

FIGURE 6.15 64-state TCM scheme with eight-dimensional signal constellation.

groups of eight "selection bits" each, say,

$$Z2_p, Z3_p, Z4_p, Z5_p, Z6_p, Z7_p, Z0_p, Z1_p$$

with $p = n, n + 1, n + 2, n + 3$. Each group is then used to address the same two-dimensional mapping table to obtain a two-dimensional signal. The eight-dimensional signal corresponding to those four two-dimensional signals is the one selected for transmission. The two-dimensional mapping is based on Fig. 6.14 and Table 6.6. In Fig. 6.14, *the same 6-bit value* is associated with each of the four signals, which can be obtained from each other through 90° rotations.

The table for the bit converter can be obtained from Tables 6.6 and 6.7. The operation of the block encoder is shown in Table 6.8. It takes nine uncoded information bits and generates four groups of three "selection bits" each. Each bit group is used to select the inner or outer group of signals in a two-dimensional subset corresponding to the previously selected eight-dimensional subtype. If the value of the bit group is 000, 001, 010, or 011, one-fourth of the *inner* group of signals is selected. If it is 100, the *outer* group of signals is selected. The differential encoder has the form

$$I6'_n I7'_n = (I6'_{n-4} I7'_{n-4} + I6_n I7_n) \bmod 100_{\text{base2}} .$$

Decoder design

As described previously, the preliminary step in the application of Viterbi algorithm is determining which signal x_i, in each of the multidimensional constellations corresponding to one branch in the code trellis, is closest to the received signal z_i. (The subscript i denotes the discrete time at which the signal is received.) Then the Euclidean distance between x_i and z_i is used as the metric for that branch.

Because of the way in which a multidimensional constellation is partitioned, the signal x_i may be obtained as follows [12]. Given the four two-dimensional signals associated with z_i, the decoder determines the closest two-dimensional signal in each of the four 160-signal, two-dimensional constellations, and calculates its associated metric. These metrics are called *two-dimensional subconstellation metrics*. Note that there are only 40 signals in each of the four two-dimensional subconstellations, so this step is no more complex than that required for a two-dimensional constellation when no coding is used.

TABLE 6.6 Correspondence Between $Z0_p Z1_p$ and Four Two-Dimensional Subconstellations.

Two-Dimensional Subconstellation	$Z0_p$	$Z1_p$
A	0	0
B	0	1
C	1	0
D	1	1

TABLE 6.7 Eight-Dimensional "Subtype" Selection Procedure for the TCM Scheme with 64 States [12].

Step 1. Use $Y0_n\, I1_n\, I2_n\, I3_n$ to obtain an eight-dimensional subtype.

$Y0_n$	$I1_n$	$I2_n$	$I3_n$	Eight-Dimensionsal Subtype
0	0	0	0	(A A A A)
0	0	0	1	(A A C C)
0	0	1	0	(A A A B)
0	0	1	1	(A A C D)
0	1	0	0	(A C A C)
0	1	0	1	(A C C B)
0	1	1	0	(A C A D)
0	1	1	1	(A C C A)
1	0	0	0	(A A A C)
1	0	0	1	(A A C B)
1	0	1	0	(A A A D)
1	0	1	1	(A A C A)
1	1	0	0	(A C A A)
1	1	0	1	(A C C C)
1	1	1	0	(A C A B)
1	1	1	1	(A C C D)

Step 2. Rotate the $\begin{Bmatrix} \text{3RD \& 4TH} \\ \text{2ND \& 4TH} \\ \text{2ND \& 3RD} \end{Bmatrix}$ two-dimensional subsets of the eight-dimensional subtype by 180° if $I4_n I5_n = \begin{Bmatrix} 01 \\ 10 \\ 11 \end{Bmatrix}$.

Step 3. Rotate all four two-dimensional subsets of the eight-dimensional subtype obtained in Step 2 by $\begin{Bmatrix} 90° \\ 180° \\ 270° \end{Bmatrix}$ clockwise if $I6'_n I7'_n = \begin{Bmatrix} 01 \\ 10 \\ 11 \end{Bmatrix}$.

Next the decoder determines the four-dimensional signal in each of the 16 four-dimensional "types" that is closest to the first received four-dimensional signal (the four-dimensional signal corresponding to the first and second two-dimensional signal of z_i), and calculates its associate metric. These metrics are called *four-dimensional type metrics*. (The four-dimensional type metrics are obtained merely by adding the two two-dimensional corresponding subconstellation metrics.) The decoder then compares the two four-dimensional metrics for the pair of four-dimensional types within each four-dimensional subconstellation. The smaller four-dimensional metric becomes the four-dimensional metric associated with that four-dimensional subconstellation, and the four-dimensional signal associated with the smaller four-

TABLE 6.8 Coding Table for the Eight-Dimensional Block Encoder.

$I1_{n+1}$	$I2_{n+2}$	$I3_{n+1}$	$Z2_n$	$Z3_n$	$Z4_n$	$Z2_{n+1}$	$Z3_{n+1}$	$Z4_{n+1}$
0	X	X	0	$I2_{n+1}$	$I3_{n+1}$	0	$I4_{n+1}$	$I5_{n+1}$
1	0	0	1	0	0	0	$I4_{n+1}$	$I5_{n+1}$
1	0	1	0	$I4_{n+1}$	$I5_{n+1}$	1	0	0
1	1	0	0	$I4_{n+1}$	$I5_{n+1}$	0	$I6_{n+1}$	$I7_{n+1}$
1	1	1	0	$I4_{n+1}$	$I5_{n+1}$	0	$I6_{n+1}$	$I7_{n+1}$

$I1_{n+1}$	$I2_{n+2}$	$I3_{n+1}$	$Z2_{n+2}$	$Z3_{n+2}$	$Z4_{n+2}$	$Z2_{n+3}$	$Z3_{n+3}$	$Z4_{n+3}$
0	X	X	0	$I6_{n+1}$	$I7_{n+1}$	0	$I1_{n+2}$	$I2_{n+2}$
1	0	0	0	$I6_{n+1}$	$I7_{n+1}$	0	$I1_{n+2}$	$I2_{n+2}$
1	0	1	0	$I6_{n+1}$	$I7_{n+1}$	0	$I1_{n+2}$	$I2_{n+2}$
1	1	0	1	0	0	0	$I1_{n+2}$	$I2_{n+2}$
1	1	1	0	$I1_{n+2}$	$I2_{n+2}$	1	0	0

dimensional metric is the four-dimensional signal in the four-dimensional subconstellation that is closest to the first received four-dimensional signal. The same process is repeated for the second received four-dimensional signal.

The decoder then determines the closest eight-dimensional signal in each of the 64 eight-dimensional types (Table 6.5), and calculates its associated metric. These metrics are called *eight-dimensional type metrics*. The eight-dimensional type metric for an eight-dimensional type is obtained by adding the two four-dimensional subset metrics for the pair of four-dimensional subconstellation metrics corresponding to that eight-dimensional type.

Finally, the decoder compares the four eight-dimensional type metrics corresponding to the four eight-dimensional types within each eight-dimensional subconstellation (see Table 6.5). The smallest eight-dimensional type metric becomes the eight-dimensional subconstellation metric associated with that eight-dimensional subconstellation, and the eight-dimensional signal associated with the smallest eight-dimensional metric is the closest signal in that eight-dimensional subconstellation. These eight-dimensional subconstellation metrics are then used in the Viterbi algorithm in the usual way.

The final decision on a transmitted eight-dimensional signal based on the foregoing procedure may not be a valid signal of the eight-dimensional constellation, because more than one of the four two-dimensional signals in the decision may be outer signals in the two-dimensional constellation. When this happens, a modification must be made in the procedure to arrive at a valid eight-dimensional signal.

Once the Viterbi algorithm outputs a decision, one needs to transform it into a set of information bits. To map the decoded signal back to the information bits, we can proceed as follows. Each of the four two-dimensional signals corresponding to the final decision is first mapped back to eight Z bits (see Fig. 6.14 and Table 6.6). By performing the inverse conversions corresponding to the bit converter and the eight-dimensional block encoder, followed by a differential decoding operation, we obtain the desired 28 information bits.

6.5 Trellis-Encoded CPM

Consider digital transmission over a channel that is simultaneously band-width- and power-limited. Here a bandwidth- *and* power-efficient coding/modulation scheme must be used. For example, digital transmission over satellite-based mobile radio channels provides this kind of environment.

Trellis-coded modulation offers indeed a feasible solution to increase the reliability of the transmission system without increasing the transmitted power or the required bandwidth. Furthermore, in a bandlimited environment the signals to be used in conjunction with TCM must be chosen very carefully. An additional constraint may come from the requirement of constant-envelope signals, which sometimes are needed for use in conjunction with nonlinear amplifiers. A class of bandwidth-efficient signals that satisfies these constraints is offered by continuous-phase modulated (CPM) signals, based on phase modulation where phase continuity is introduced to reduce the bandwidth occupancy.

The synergy between TCM, which improves error probability, and CPM signals, which provide constant envelope and low spectral occupancy, may provide a satisfactory solution to the problem of transmitting over nonlinear channels that are both bandwidth- and power-limited. To implement this idea, the multilevel symbols at the output of the TCM encoder are used as the inputs to the continuous-phase modulator. The design of the coding scheme and of the modulator scheme should be performed jointly in order to maximize the Euclidean distance resulting from the combination of the two. As observed in [27], to maximize the Euclidean distance of the coding/modulation scheme the trellis encoder should reduce the connectivity of its trellis in such a way that

- The Euclidean distance between signals leaving the same state is maximized.
- The Euclidean distance between signals merging into the same state is maximized.

To implement TCM/CPM, the receiver combines the trellis structure of TCM with that of CPM. The number of states necessary for a trellis representation of these signals, and hence for their optimum demodulation, turns out to be the product of the number of states needed by TCM and the number of states needed by CPM. (See [28] for a suboptimum TCM/CPM scheme based on precoding and differential detection that avoids this expansion of the number of states. Other suboptimum schemes are studied in [29].)

6.5.1 Review of CPM

In the following we recall the basic features of continuous-phase modulated signals, mainly for reference sake. Additional details can be found in [30] and [31].

A continuous-phase modulated signal is defined by

$$s(t, \mathbf{a}) = \sqrt{\frac{2E_s}{T}} \cos(2\pi f_0 t + \theta(t, \mathbf{a}) + \theta_0) \qquad t \geq 0 \qquad (6.11)$$

where E_s is the symbol energy, T is the symbol interval, f_0 the carrier frequency, and θ_0 the initial carrier phase. The transmitted information is contained in the

6.5 / Trellis-Encoded CPM

phase

$$\theta(t, \mathbf{a}) = 2\pi h \sum_{n=0}^{\infty} a_n q(t - nT), \qquad t \geq 0 \qquad (6.12)$$

with $q(t)$ the phase-shaping pulse given by

$$q(t) = \int_{-\infty}^{t} g(\tau) \, d\tau \qquad (6.13)$$

where $g(t)$ is the frequency pulse with finite duration [i.e., $g(t)$ is nonzero only for $0 \leq t \leq LT$, L the pulse length]. The value of L contributes to the taxonomy of CPM in the following way:

- $L = 1$: full-response CPM.
- $L > 1$: partial-response CPM.

The frequency pulse $g(t)$ is usually normalized in amplitude so as to have

$$\int_0^\infty g(\tau) \, d\tau = \tfrac{1}{2}.$$

In (6.12),

$$\mathbf{a} = \ldots, a_{-2}, a_{-1}, a_0, a_1, \ldots$$

denotes the symbol sequence sent to the CPM modulator. The symbols a_n take values $\pm 1, \pm 3, \ldots, \pm(M - 1)$, where M is an even positive integer. The parameter h is called the *modulation index*; when this parameter is a rational number, the number of states in CPM is finite. Thus it is customary to assume that h is the ratio of two integer numbers, and, in particular, that it has the form

$$h = \frac{p}{q} \qquad (6.14)$$

with p, q relatively prime integers.

The phase function $\theta(t, \mathbf{a})$ during a symbol interval may be written in the form

$$\theta(t, \mathbf{a}) = 2\pi h a_n q(t-nT) + 2\pi h \sum_{i=n-L+1}^{n-1} a_i q(t-iT) + \theta_n, \qquad nT \leq t \leq (n+1)T$$

$$(6.15)$$

where

$$\theta_n = \pi h \sum_{j=-\infty}^{n-L} a_j \bmod 2\pi. \qquad (6.16)$$

With the choice (6.14), it can be seen from (6.16) that θ_n can take on q (if p is even) or $2q$ (if p is odd) different values:

$$\theta_n \in \left\{ 0, \pi \frac{p}{q}, 2\pi \frac{p}{q}, \ldots, (q-1)\pi \frac{p}{q} \right\} \qquad (6.17)$$

if p is even, and

$$\theta_n \in \left\{ 0, \pi\frac{p}{q}, 2\pi\frac{p}{q}, \ldots, (2q-1)\pi\frac{p}{q} \right\} \tag{6.18}$$

if p is odd. Thus we can say that the phase function during any given interval depends on the actual transmitted symbol a_n, on the $L-1$ previously transmitted symbols $a_{n-1}, \ldots, a_{n-L+1}$, and on the "phase state" θ_n of the modulator. Since the $(L-1)$-tuple $(a_{n-1}, \ldots, a_{n-L+1})$ can take on M^{L-1} values and θ_n can take on q (or $2q$) values, the phase trajectories during any T-second interval can take on qM^{L-1} (or $2qM^{L-1}$) different shapes.

A model often used to describe the CPM modulator follows directly from (6.11) and is shown in Fig. 6.16. A different model, based on the decomposition of the CPM modulator into an encoder and a memoryless modulator, is described in Appendix 6B at the end of this chapter.

6.5.2 Parameter Selection

In this section we follow the notations and the numerical results contained in [27] and [32]. The selection of modulation and coding formats is based on the following parameters, whose definition was given in Section 3.2 and is repeated here for convenience:

- $\mathcal{R}$, the information rate, that is, the number of bits per second that we want to transmit on the channel.
- B, the transmission bandwidth available (in hertz).
- The error probability to be achieved at a given signal-to-noise ratio.

The ratio $r = \mathcal{R}/B$, in bits/s/Hz, is the throughput (i.e., the number of bits per second we can transmit in 1 Hz). Hence the number of bits per channel use is given by

$$\mathcal{R} = TBr.$$

EXAMPLE 6.8

Consider the following parameters:

- $B = 5$ kHz.
- $\mathcal{R} = 4.8$ kbits/s.
- Bandwidth definition = 99.9% of power.
- $P(e) = 10^{-3}$ @ $E_b/N_0 = 9$ dB (unfaded).

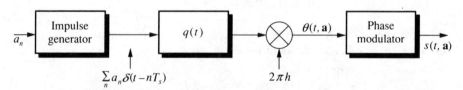

FIGURE 6.16 Model for the CPM modulator.

In this case we get

$$r = \frac{\mathcal{R}}{B} = \frac{4.8}{5} \approx 1 \text{ bit/s/Hz}.$$

If we consider using M-ary CPFSK (i.e., a full-response CPM scheme with rectangular frequency pulse, with $h = 1/M$), spectrum computations show that we have $B = 2.7/T$. Hence we have $R = 2.7$ bits/channel use. This shows that *octonary* CPFSK may be adequate since it gives $R = 3$. For octonary CPFSK, error probability over the additive white Gaussian noise channel with power spectral density $N_0/2$ is given approximately by

$$P(e) \approx \frac{1}{2} \text{erfc}\left(\sqrt{\frac{D_{\text{free}}^2}{4N_0}}\right)$$

where D_{free}^2 is the free distance of the CPFSK signals. Hence for $P(e) = 10^{-3}$ we must have

$$\frac{D_{\text{free}}^2}{4N_0} = 4.7.$$

For octonary CPFSK we have $D_{\min}^2/2E_s = 0.2$, where E_s is the energy per symbol. Hence

$$\frac{E_s}{N_0} = \frac{9.4}{0.2} = 47$$

and

$$\frac{E_b}{N_0} = \frac{E_s}{RN_0} = \frac{47}{3} = 15.7$$

which corresponds to 11.9 dB. Thus coding is necessary, and we need a gain of about 3 dB. □

6.5.3 Designing the TCM Scheme

We consider now the choice of a convolutional encoder to be put in front of the CPM encoder [27, 32–35]. Two possible perspectives are possible to understand the interaction of CPM with the convolutional code. On one side, we observe that the code introduces a correlation between the symbols at the input of the CPM modulator. As a result, some transitions in the CPM trellis are pruned out, thus decreasing the connectivity of the trellis. As a result, the minimum merge length is increased. Since generally larger merge lengths involve larger Euclidean free distances, the error performance of the CPM system is improved. On the other side, the system can be viewed as a trellis-coded scheme using CPM signals in order to take advantage of their high bandwidth efficiency.

The block diagram of the overall system is represented in Fig. 6.17. The source generates a sequence of binary symbols. During the ith time interval, the binary μ-tuple $b_i^{(1)}, \ldots, b_i^{(\mu)}$ enters the rate μ/η convolutional encoder. The output of the encoder is the binary η-tuple $c_i^{(1)}, \ldots, c_i^{(\eta)}$, with

$$c_i^{(m)} = \sum_{\ell=0}^{\mu} \sum_{j=0}^{\nu_\ell} b_{i-j}^{(\ell)} g_j^{(\ell,m)} \quad \text{mod } 2$$

where $m = 1, \ldots, \eta$, and the numbers $g_j^{(\ell,m)}$ describe the connections between the inputs and the outputs of the convolutional encoder.

With reference to Fig. 6.17, the selection of a rate μ/η convolutional code is equivalent to the selection of the following quantities:

- The constraint length ν, or, equivalently, the number of states.
- A partition of ν in the form

$$\nu = \nu_1 + \cdots + \nu_\mu.$$

- The set of binary numbers $g_j^{(\ell,m)}$.

Since the main purpose of pairing TCM with CPM is to provide good spectral characteristics for the signal sent through the channel, the performance of the TCM/CPM combination should be evaluated by considering jointly the power gain and the bandwidth occupancy, as well as the receiver complexity, as expressed in terms of the number of decoder states for a coherent Viterbi detector. This number is given by

$$N_{\text{TCM/CPM}} = N_{\text{TCM}} \cdot N_{\text{CPM}}.$$

Lindell and Sundberg [33] and Pizzi and Wilson [35] have considered coded quaternary CPM with partial response signaling, with code searches in large ensembles. Ho and McLane [36] have introduced a set of code design rules to limit the size of the code ensembles, and hence reducing the amount of searches. Since for good codes all output symbols should occur with equal frequency, regularity, and symmetry, the code search in [36] was limited to within the ensemble of codes that possess these properties. These design rules are

1. The M channel symbols are divided into two subsets, $\mathcal{S}_1$ and $\mathcal{S}_2$. Each subset contains $M/2$ channel symbols and $\mathcal{S}_1, \mathcal{S}_2$ are disjoint.
2. The transitions originating from the same state receive symbols from either $\mathcal{S}_1$ or $\mathcal{S}_2$.

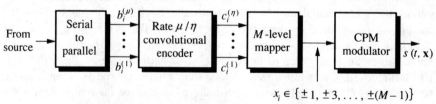

FIGURE 6.17 Convolutionally encoded CPM.

3. The transitions joining in the same state receive symbols from either $\mathcal{S}_1$ or $\mathcal{S}_2$.
4. Let the output of the encoder/mapper at the convolutional encoder state σ_n be x_n. Then the output at state $\bar{\sigma}_n$ should be $-x_n$, where $\bar{\sigma}_n$ denotes a binary vector whose elements are the complements of the elements of σ_n.

Rules 1 to 3 are based on set partitioning, while rule 4 is based on a symmetry requirement. These rules guarantee that all the states are on an equal footing in spanning the free distance error events.

EXAMPLE 6.9

A rate 1/2 code that obeys rules 1 to 4 can easily be identified. Let $\mathbf{g}_1 = (g_0^{(1,1)}, \ldots, g_v^{(1,1)})$ and $\mathbf{g}_2 = (g_0^{(1,2)}, \ldots, g_v^{(1,2)})$ be the vectors containing the coefficients of the code generator polynomials. Then it was shown in [36] that

- The number of nonzero coefficients in $\mathbf{g}_1$ and $\mathbf{g}_2$ must be odd.
- $g_0^{(1,1)} = g_v^{(1,1)}$ and $g_0^{(1,2)} = g_v^{(1,2)}$.

For instance, consider the encoder–mapper combination of Fig. 6.18. The eight

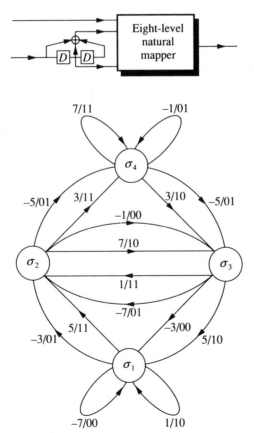

FIGURE 6.18 Rate 2/3, (7, 2) convolutional encoder.

symbols, ± 1, ± 3, ± 5, ± 7, are divided into the two subsets

$$\mathcal{S}_1 = \{-7, -3, 1, 5\} \qquad \mathcal{S}_2 = \{7, 3, -1, -5\}.$$

This code obeys rules 1 to 4. □

6.5.4 Performance Examples

In [36] various TCM/CPM schemes are analyzed and compared in terms of free distance, bandwidth efficiency, and receiver complexity. The performance of selected TCM/CPM schemes is presented in the form of curves on the power–bandwidth plane. The abscissa of the power–bandwidth plane is the 99% *bandwidth-efficiency* η_{99}, defined as the bit rate divided by the bandwidth B_{99} which contains 99% of the total signal power. Explicitly, if f_0 denotes the carrier frequency, T_b^{-1} the bit rate, and $\mathcal{G}(f)$ the power spectral density of the encoded signal, we have

$$\eta_{99} = \frac{1}{2B_{99}T_b}$$

where B_{99} is the solution of the equation

$$0.99 = \frac{\int_{f_c - B_{99}}^{f_c + B_{99}} \mathcal{G}(f)\, df}{\int_0^\infty \mathcal{G}(f)\, df}.$$

The power spectral density of a trellis-encoded CPM signal can be evaluated by using the techniques described in [37] to [39].

Among the results described in [36], it was observed that rate 2/3 convolutional codes in conjunction with eight-level CPM schemes have better joint power–bandwidth efficiency, and that partial response is better than full-response signaling. When 99% bandwidth is considered, the coded eight-level 2RC class (CPM with $L = 2$ and raised-cosine frequency pulse [30]) has the best joint power–bandwidth efficiency followed by coded eight-level CPFSK, coded four-level 2REC (CPM with $L = 2$ and rectangular frequency pulse [30]), coded four-level 2RC, and coded four-level CPFSK. However, the receiver complexity, given in terms of the number of matched filters needed and the number of decoder states, also increases in that order. Note that if the 99.9% bandwidth is considered instead, then four-level 2RC is better than four-level 2REC.

Usually, partial-response schemes result into a larger number of matched filters and decoding states. The scheme with the best compromise between joint power–bandwidth efficiency and receiver complexity seems to be the $\nu = 3$ coded eight-level CPFSK scheme with modulation index $h = \frac{1}{6}$. This scheme is 4 dB more energy efficient, and 15% more bandwidth efficient (based on the 99% bandwidth) than the well-known MSK scheme [30, 31], and requires 16 matched filters and 48 decoder states for its receiver implementation.

APPENDIX 6A EXAMPLES OF GROUP ALPHABETS

In this appendix we describe two examples of group alphabets: permutation alphabets and cyclic-group alphabets.

6A.1 Permutation Alphabets

Permutation alphabets (PA) [40, 41] are an example of group alphabets with one energy level. The set of signal vectors is chosen as follows. In the first type of PA, called variant I, a prescribed sequence of N real numbers, not necessarily all different, is taken as the first vector. The remaining vectors consist of all distinct sequences that can be formed by permuting the order of the N numbers that form the first vector. In the second type of alphabet, called variant II, a prescribed sequence of N nonnegative numbers, not necessarily all different, is taken as the first vector. The remaining vectors consist of all the distinct sequences that can be formed by permuting the order and/or changing the sign of the N numbers that form the first vector.

Let the first vector of a variant I PA be

$$\mathbf{x}_1 = [\underbrace{\mu_1, \ldots, \mu_1}_{m_1}, \underbrace{\mu_2, \ldots, \mu_2}_{m_2}, \ldots, \underbrace{\mu_k, \ldots, \mu_k}_{m_k}] \quad (6A.1)$$

that is, its successive components consist of μ_1 repeated m_1 times, μ_2 repeated m_2 times, ..., μ_k repeated m_k times. Here we assume that

$$\mu_1 < \mu_2 < \cdots < \mu_k$$

and have

$$N = m_1 + m_2 + \cdots + m_k$$

with the m's and k all positive integers. The other vectors are obtained by permuting the order of the μ's in all possible ways. We obtain

$$M = \frac{N!}{m_1! m_2! \cdots m_k!}$$

vectors. All signals have the same energy, and

$$E = \| \mathbf{x}_i \|^2 .$$

For variant II PA, the following notation is adopted. The first vector is still given by (6A.1), where now we take

$$0 \leq \mu_1 < \mu_2 < \cdots < \mu_k .$$

If $\mu_1 = 0$, k must be at least 2. The other words of the variant II PA are obtained by permuting the order of the μ's in (6A.1) in all possible ways and by making all the possible assignments of sign to the components of the resulting vectors. The code has

$$M = \frac{n! 2^k}{m_1! m_2! \cdots m_k!}$$

vectors, where $h = n - m_1$ if $\mu_1 = 0$, and $h = n$ otherwise. In this case all the signals again have the same power.

We now consider the following problem [42]. Given the set $\{m_1, \ldots, m_k\}$ in a variant I PA, choose the set $\{\mu_1, \ldots, \mu_k\}$ so that the minimum distance between any two vectors in a PA is a maximum. In other words, we wish to find an N-vector

x, satisfying the constraint $\| \mathbf{x} \|^2 = 1$, such that

$$g(\mathbf{x}) = \min_{S \neq I_N} \| \mathbf{x} - S\mathbf{x} \|^2 \qquad (6A.2)$$

is a maximum. Here I_N denotes the identity permutation and S is a permutation on N objects. Since **x** is completely defined, once $m_1, \ldots, m_k$ are given, by the numbers $\mu_1, \ldots, \mu_k$ and by a correspondence between the m's and the μ's, we can write

$$\max_{\mathbf{x}} = \max_{\tau} \max_{\{\mu_1, \ldots, \mu_k\}} g(\mathbf{x})$$

where τ is a one-to-one mapping of the set of integers $\{1, 2, \ldots, k\}$ onto itself. Thus we may take **x** in this form:

$$\mathbf{x} = [\underbrace{\mu_1, \ldots, \mu_1}_{m_{\tau(1)}}, \underbrace{\mu_2, \ldots, \mu_2}_{m_{\tau(2)}}, \ldots, \underbrace{\mu_k, \ldots, \mu_k}_{m_{\tau(k)}}]. \qquad (6A.3)$$

We shall approach the problem in two steps: (1) find the optimum set $\{\mu_1, \ldots, \mu_k\}$ for a given τ, and (2) find the optimum τ.

Before proceeding further, we observe from (6A.2) that if $\mathbf{x} + \mathbf{h}$ denotes the vector obtained by adding an equal quantity $\mathbf{h}$ to all the components of **x** (i.e., a translation of **x**), then

$$g(\mathbf{x} + \mathbf{h}) = g(\mathbf{x}).$$

Moreover, the minimum value of $\| \mathbf{x} + \mathbf{h} \|$ is obtained when

$$\mathbf{h} = -\sum_{i=1}^{k} m_{\tau(i)} \mu_i$$

and hence we shall impose the condition

$$\sum_{i=1}^{k} m_{\tau(i)} \mu_i = 0. \qquad (6A.4)$$

We are now ready to solve our main problem of finding the maximum value of $g(\mathbf{x})$ under the constraints (6A.4) and

$$\sum_{i=1}^{k} m_{\tau(i)} \mu_i^2 = 1. \qquad (6A.5)$$

Suppose that masses $m_{\tau(1)}, \ldots, m_{\tau(k)}$ are located at points $\mu_1, \ldots, \mu_k$ along the μ-axis. We seek to slide the points along the line (without passing each other) to maximize the nearest neighbor distance keeping a fixed central moment of inertia. Assume that the maximization problem were solved by an arrangement such that $\mu_{i+1} - \mu_i$ is not a constant, and let Δ denote the smallest separation between two adjacent masses in this configuration. Now slide all the masses along so that a configuration results with every adjacent pair of masses separated by a distance Δ. The central moment of inertia, η^2, of this new configuration is smaller than 1, since we have packed the masses closer together. Now multiply the coordinates of all the masses in this new configuration by $1/\eta$; an equally spaced configuration of masses is obtained with central moment of inertia 1 and minimum distance between points $\Delta/\eta > \Delta$. Thus the original configuration with unequal separation was not the best possible configuration. We have proved that the optimum $\mathbf{x}$ must have components satisfying

$$\mu_\ell = \mu_1 + (\ell - 1)\omega. \tag{6A.6}$$

By defining

$$A_1 = \frac{1}{N}\sum_{j=1}^{k} m_{\tau(j)}(j - 1)$$

$$A_2 = \frac{1}{N}\sum_{j=1}^{k} m_{\tau(j)}(j - 1)^2$$

we have

$$\mu_1 = -\frac{A_1}{\sqrt{N(A_2 - A_1^2)}}$$

$$\omega = -\frac{\mu_1}{A_1}$$

$$d_{\min}^2 = \frac{2/N}{A_2 - A_1^2}$$

$$= \frac{4N}{\sum_{i=1}^{N}\sum_{j=1}^{N} m_{\tau(i)} m_{\tau(j)}(i - j)^2}.$$

Our second step is finding the optimum mapping τ. Since every one-to-one mapping of a finite set onto itself is equivalent to a permutation, we want to find the permutation τ such that

$$Q(\tau) = \sum_{i=1}^{k}\sum_{j=1}^{k} m_{\tau(i)} m_{\tau(j)}(i - j)^2$$

is a minimum. This problem was solved by Slepian [40], who showed that $Q(\tau)$ attains its minimum value when τ is such that

$$m_{\tau(1)} \leq m_{\tau(k)} \leq m_{\tau(2)} \leq m_{\tau(k-1)} \leq \cdots.$$

In words, we must choose the pairing of m and μ in the initial vector $\mathbf{x}$ in such a way that the least m is paired with the least μ, the second least m with the largest μ, the third least m with the second least μ, and so on.

We can also observe that the denominator of the last expression of $d_{\min}^2$ can be interpreted as the variance of a random variable that takes values $(j - 1)$ with probability $m_{\tau(j)}/N$, $j = 1, \ldots, k$. Thus we can expect to get higher values of $d_{\min}^2$ when the probability distribution of this random variable is concentrated around its mean value.

The same arguments leading to the optimum initial vector for variant I permutation alphabets can be used to solve the same problem for variant II.

Some examples of PA

Consider the special case of PA in which we assume that the first vector has $N - h$ zero components and that its h remaining components are equal. In our notations, we have $\mu_1 = 0$, $m_1 = N - h$, and $m_2 = h$. This case is analyzed in some detail in [43]. We shall refer to these signal sets as (N, h) PA.

- $(N, 1)$ *alphabet*. We have $M = N$ for variant I PA, which gives rise to an *orthogonal* signal set. For variant II we have $M = 2N$ and a *biorthogonal* set.
- (N, N) *alphabet*. With variant II alphabets, we have $M = 2^N$, and the signal vectors lie at the vertices of an N-dimensional hypercube.

6A.2 Cyclic-Group Alphabets

The process of generating a variant I permutation alphabet can also be interpreted as follows. Consider the collection $\mathscr{S}$ of all $N \times N$ permutation matrices, that is, the matrices having a single unit entry in each row and column, all other entries being zero. These matrices are orthogonal, and they form a finite group with $N!$ elements. The action of this group on the initial vector is to generate a set of vectors, the *orbit* of the group, which form the PA. Similar considerations hold for variant II alphabets, where $\mathscr{S}$ is now replaced by the finite group of $2^N n!$ $N \times N$ orthogonal matrices having a single nonzero entry in each row and column, that entry being $+1$ or -1.

This argument can be applied to give rise to a family of alphabets generated by cyclic groups. Consider a cyclic group of orthogonal matrices, that is, a set of orthogonal matrices

$$\mathbf{G} = \{\mathbf{C}^i\}_{i=0}^{M-1}$$

with $\mathbf{C}^M = \mathbf{I}$, $\mathbf{I}$ the identity matrix. It has been shown in [44] that a vector $\mathbf{x} \in \mathbf{R}^N$ can always be found such that the set of vectors

$$\{\mathbf{C}^i \mathbf{x}\}_{i=0}^{M-1}$$

spans a space whose dimensionality equals the number of distinct eigenvalues of the generator matrix $\mathbf{C}$.

In [45] a computational technique to optimize cyclic-group alphabets is given, along with a table of the best designs found.

APPENDIX 6B DECOMPOSITION OF THE CPM MODULATOR

In this appendix we describe a different approach to CPM modeling, suggested by Rimoldi [27, 46], based on the decomposition of the CPM modulator into a linear encoder and a memoryless modulator.

6B.1 Phase Trellis and Tilted Phase Trellis

Consider first the *physical phase* $\tilde{\theta}(t, \mathbf{a})$, defined as the reduction mod 2π of the actual phase (6.15). If we draw all the possible trajectories of the physical phase, we get the *physical phase trellis* of the CPM scheme. For example, for $M = 2$, $h = \frac{1}{2}$, and $g(t) = 1/2T$ (this is known as MSK), we have the physical phase trellis depicted in Fig. 6B.1. It is seen that this trellis is not time-invariant: The trajectories in even-numbered symbol intervals are not time translates of those in the odd-numbered symbol intervals. This is true in general for all CPM schemes.

To achieve time invariance, we measure the phase by taking the lowest phase trajectory as a reference. Define the *tilted phase*

$$\psi(t, \mathbf{a}) = \theta(t, \mathbf{a}) + \frac{\pi h(M - 1)t}{T}. \tag{6B.1}$$

The physical tilted phase $\tilde{\psi}(t, \mathbf{a})$ has a time-invariant trellis. To prove this, observe that by defining the modified data sequence $\mathbf{u} = (u_i)_{i=0}^{\infty}$ with

$$u_i = \frac{a_i + (M - 1)}{2}$$

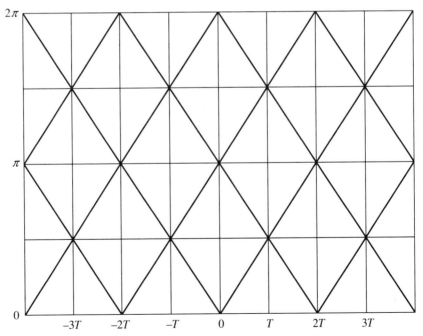

FIGURE 6B.1 Physical phase trellis for MSK.

so that $u_i \in \{0, 1, \ldots, M-1\}$, we obtain, for $t = \tau + nT$ and $0 \le \tau < T$,

$$\psi(\tau + nT, \mathbf{u}) = 2\pi h \sum_{j=0}^{n-L} u_j + 4\pi h \sum_{j=0}^{L-1} u_{n-j} q(\tau + jT) + \pi h(M-1)\frac{\tau}{T}$$

$$- 2\pi h(M-1) \sum_{j=0}^{L-1} q(\tau + jT) + (L-1)(M-1)\pi h.$$

We see that the time-dependent part of the right-hand side of the last equation depends only on τ and not on n. Thus after an initial transient, the possible physical phases $\psi(\tau + nT, \mathbf{u})$ in any two symbol intervals will be such that one is the translate of the other.

6B.2 Decomposing the CPM Modulator

We shall now show, following again [27] and [46], how the CPM modulator can be decomposed as the cascade of a discrete CPM encoder and of a memoryless modulator. Rewrite the transmitted signal in the form

$$s(t, \mathbf{u}) = \sqrt{\frac{2E_s}{T}} \cos(2\pi f_1 t + \tilde{\psi}(t, \mathbf{u}) + \theta_0)$$

where the new frequency variable

$$f_1 = f_0 - \frac{h(M-1)}{2T}$$

was introduced to compensate for the phase tilting, and the physical tilted phase is, for $0 \le \tau < T$,

$$\tilde{\psi}(\tau + nT, \mathbf{u}) = R_{2\pi} \left[2\pi h R_q \left(\sum_{j=0}^{n-L} u_j \right) + 4\pi h \sum_{j=0}^{L-1} u_{n-j} q(\tau + jT) + w(\tau) \right]$$

(6B.2)

where

$$w(\tau) = \frac{\pi h(M-1)\tau}{T} - 2\pi h(M-1) \sum_{j=0}^{M-1} q(\tau + jT) + (L-1)(M-1)\pi h$$

(6B.3)

is data-independent.

Equation (6B.2) shows that the modulated signal is specified completely by the pair

$$\sigma_n = (u_n, u_{n-1}, \ldots, u_{n-L+1}, v_n)$$

where

$$v_n = R_q\left(\sum_{j=0}^{n-1} u_j\right).$$

In conclusion, we obtain, for $nT \leq t < (n+1)T$,

$$s(\tau + nT, \sigma_n) = \sqrt{\frac{2E_s}{T}} \cos\left[2\pi f_1(\tau + nT) + \tilde{\psi}(\tau, \sigma_n) + \theta_0\right], 0 \leq \tau < T.$$

If the right-hand side of the last equation is decomposed into in-phase and quadrature components, we obtain the set of equations that describe the operation of the memoryless modulator in a symbol interval (choose $0 \leq \tau < T$):

$s(\tau + nT, \sigma_n)$

$$= \frac{1}{\sqrt{2}} I(t, \sigma_n) \cos\left[2\pi f_1(\tau + nT) + \theta_0\right]$$

$$- \frac{1}{\sqrt{2}} Q(t, \sigma_n) \sin\left[2\pi f_1(\tau + nT) + \theta_0\right]$$

$$I(\tau, \sigma_n) = \sqrt{\frac{E_s}{T}} \cos \tilde{\psi}(\tau + nT, \sigma_n)$$

$$Q(\tau, \sigma_n) = \sqrt{\frac{E_s}{T}} \sin \tilde{\psi}(\tau + nT, \sigma_n)$$

$$\sigma_n = (u_n, u_{n-1}, \ldots, u_{n-L+1}, v_n)$$

$$u_n = \frac{a_n + (M-1)}{2}$$

$$v_n = R_q\left(\sum_{j=0}^{n-1} u_j\right)$$

$\tilde{\psi}(\tau + nT, \sigma_n)$

$$= R_{2\pi}\left[2\pi h R_q\left(\sum_{j=0}^{n-L} u_{n-j}\right) + 4\pi h \sum_{j=0}^{L-1} u_{n-j} q(\tau + jT) + w(\tau)\right]$$

and

$$w(\tau) = \frac{\pi h(M-1)\tau}{T} - 2\pi h(M-1) \sum_{j=0}^{L-1} q(\tau + jT) + (L-1)(M-1)\pi h.$$

Figure 6B.2 shows the memoryless modulator.

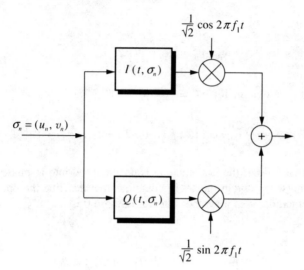

FIGURE 6B.2 Memoryless modulator in the decomposition of CPM transmitter.

The task of the CPM encoder is to update the memoryless modulator input σ_n to produce the next input, σ_{n+1}. We have

$$v_{n+1} = R_q \left(\sum_{j=0}^{n-L+1} u_j \right)$$

$$= R_q \left[R_q \left(\sum_{j=0}^{n-L} u_j \right) + u_{n-L+1} \right]$$

$$= R_q [v_n + u_{n-L+1}].$$

The scheme of the CPM encoder is represented in Fig. 6B.3. Figure 6B.4 shows the implementation of the CPM encoder when the input and output quantities are split into their binary components. In particular, it is assumed that u_n is represented by μ binary digits, and v_n by λ binary digits. Since v_n can take on q values, the

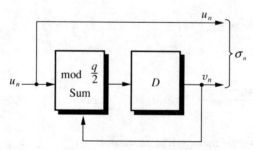

FIGURE 6B.3 CPM encoder.

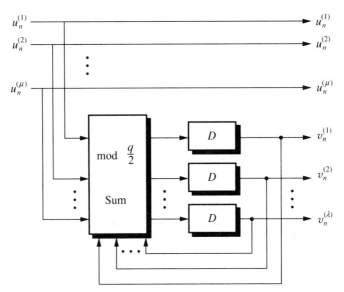

FIGURE 6B.4 Scheme of the CPM encoder in binary form.

actual value of q influences the structure of the CPM encoder in the form of Fig. 6B.4. For example, if $q = 4$, then v_n is represented by a single binary digit, and

$$v_{n+1} = R_2[u_n + v_n] = R_2[R_2[u_n] + v_n]$$

which shows that only the least significant bit of u_n is involved in the summation, and only one binary delay element is needed in the scheme of Fig. 6B.4.

REFERENCES

1. C. E. SHANNON, "Communication in the presence of noise," *Proc. IRE*, Vol. 37, pp. 10–21, Jan. 1949.
2. D. SLEPIAN, "Bounds on communications," *Bell Syst. Tech. J.*, Vol. 42, pp. 681–707, 1963.
3. G. R. WELTI and S. L. LEE, "Digital transmission with coherent four-dimensional modulation," *IEEE Trans. Inf. Theory*, Vol. IT-20, No. 4, pp. 397–402, July 1974.
4. L. ZETTERBERG and H. BRÄNDSTRÖM, "Codes for combined phase and amplitude modulated signals in four-dimensional space," *IEEE Trans. Commun.*, Vol. COM-25, pp. 943–950, Sept. 1977.
5. S. G. WILSON, H. A. SLEEPER, and N. K. SRINATH, "Four-dimensional modulation and coding: An alternate to frequency reuse," *Proc. ICC'84*, Amsterdam, The Netherlands, pp. 919–923, May 1984.
6. E. BIGLIERI and M. ELIA, "Multidimensional modulation and coding for band-limited digital channels," *IEEE Trans. Inf. Theory*, Vol. IT-34, pp. 803–809, July 1988.

7. E. BIGLIERI and M. ELIA, "Multidimensional modulation and coding for digital transmission," *Conference on Digital Processing of Signal in Communications*, Loughborough, England, Apr. 1985.
8. A. GERSHO and V.B. LAWRENCE, "Multidimensional signal constellations for voiceband data transmission," *IEEE Select. Areas Commun.*, Vol. SAC-2, No. 5, pp. 687–702, Sept. 1984.
9. A. R. CALDERBANK and N. J. A. SLOANE, "Four-dimensional modulation with an eight-state trellis code," *Bell Syst. Tech. J.*, Vol. 64, pp. 1005–1018, 1985.
10. A. R. CALDERBANK and N. J. A. SLOANE, "An eight-dimensional trellis code," *Proc. IEEE*, Vol. 74, pp. 757–759, 1986.
11. R. H. DENG and D. J. COSTELLO, JR., "High rate concatenated coding systems using multi-dimensional bandwidth efficient trellis inner codes," *IEEE Trans. Commun.*, Vol. 37, No. 5, pp. 420–427, May 1989.
12. L.-F. WEI, "Trellis-coded modulation with multidimensional constellations," *IEEE Trans. Inf. Theory*, Vol. IT-33, No. 4, pp. 483–501, July 1987.
13. J. H. CONWAY and N. J. A. SLOANE, *Sphere Packings, Lattices and Groups*. Springer-Verlag, New York, N.Y., 1987.
14. G. D. FORNEY, JR., "Coset codes. Part I: Introduction and geometrical classification," *IEEE Trans. Inf. Theory*, Vol. 34, No. 5, pp. 1123–1187, Sept. 1988.
15. S. G. WILSON and H. A. SLEEPER, "Four-dimensional modulation and coding: An alternate to frequency reuse," *Technical Report*, Communications System Laboratory, University of Virginia, Charlottesville, Va., Sept. 1983.
16. A. R. CALDERBANK and N. J. A. SLOANE, "New trellis codes based on lattices and cosets," *IEEE Trans. Inf. Theory*, Vol. IT-33, No. 2, pp. 177–195, Mar. 1987.
17. D. SLEPIAN, "Group codes for the Gaussian channel," *Bell Syst. Tech. J.*, pp. 575–602, Apr. 1968.
18. V. V. GINZBURG, "Mnogomerniye signaly dlya nepreryvnogo kanala" (in Russian), *Probl. Peredachi Inf.*, pp. 28–46, Jan.–Mar. 1984. English translation: "Multidimensional signals for a continuous channel," *Probl. Inf. Transmission*, Vol. 23, No. 4, pp. 20–34, Jan.–Mar. 1984.
19. E. L. CUSACK, "Error control codes for QAM signalling," *Electron. Lett.*, Vol. 20, No. 2, pp. 62–63, Jan. 19, 1984.
20. H. IMAI and S. HIRAKAWA, "A new multilevel coding method using error-correcting codes," *IEEE Trans. Inf. Theory*, Vol. IT-23, pp. 371–377, 1977.
21. S. L. SAYEGH, "A class of optimum block codes in signal space," *IEEE Trans. Commun.*, Vol. COM-34, pp. 1043–1045, Oct. 1986.
22. E. BIGLIERI, "Multidimensional signaling with parallel demodulation," submitted for publication, 1989.
23. R. M. TANNER, "Algebraic construction of large Euclidean distance combined coding/modulation systems," *IEEE Trans. Inf. Theory*, to be published.
24. F. J. MACWILLIAMS and N. J. A. SLOANE, *The Theory of Error-Correcting Codes*. North-Holland, Amsterdam, The Netherlands, 1977.
25. S. S. PIETROBON, R. H. DENG, A. LAFANECHÈRE, G. UNGERBOECK, and D. J. COSTELLO, JR., "Trellis-coded multidimensional phase modulation," *IEEE Trans. Inf. Theory*, Vol. 36, No. 1, pp. 63–89, Jan. 1990.

26. G. D. FORNEY, JR., et al., "Efficient modulation for band-limited channels," *IEEE Select. Areas Commun.*, Vol. SAC-2, No. 5, pp. 632–647, Sept. 1984.
27. B. RIMOLDI, "Continuous-phase modulation and coding for bandwidth and energy efficiency," Dr. Sc. dissertation, Swiss Federal Institute of Technology, Zürich, Switzerland, 1988.
28. T. M. THESKEN and E. BIGLIERI, "Cutoff rate for channel having precoding, continuous-phase modulation, and differential detection," *MILCOM'89*, Boston, Mass., pp. 46.3.1–46.3.5, Oct. 15–18, 1989.
29. F. ABRISHAMKAR and E. BIGLIERI, "Suboptimum detection of trellis-coded CPM for transmission on bandwidth- and power-limited channels," to appear, *IEEE Trans. Commun.*
30. J. B. ANDERSON, T. AULIN, and C.-E. SUNDBERG, *Digital Phase Modulation*. Plenum Press, New York, N.Y., 1986.
31. S. BENEDETTO, E. BIGLIERI, and V. CASTELLANI, *Digital Transmission Theory*. Prentice-Hall, Englewood Cliffs, N.J., 1987.
32. B. RIMOLDI, "Design of coded CPFSK modulation systems for bandwidth and energy efficiency," *IEEE Trans. Commun.*, Vol. COM-37, No. 9, pp. 897–905, Sept. 1989.
33. G. LINDELL and C.-E. W. SUNDBERG, "Power and bandwidth efficient coded modulation schemes with constant amplitude," *Arch. Elektron Übertragungstech*. Vol. 39, No. 1, pp. 45–56, Jan.–Feb. 1985.
34. G. LINDELL, C.-E. W. SUNDBERG, and T. AULIN, "Minimum Euclidean distance for the best combination of short rate 1/2 convolutional codes and CPFSK modulation," *IEEE Trans. Inf. Theory*, Vol. IT-30, pp. 509–520, May 1984.
35. S. V. PIZZI and S. G. WILSON, "Convolutional coding combined with continuous phase modulation," *IEEE Trans. Commun.*, Vol. COM-33, pp. 20–29, Jan. 1985.
36. P. HO and P. J. MCLANE, "Spectrum, distance and receiver complexity of encoded continuous phase modulation," *IEEE Trans. Inf. Theory*, Vol. IT-34, No. 5, pp. 1021–1032, Sept. 1988.
37. P. HO, "The power spectral density of digital continuous phase modulation with correlated data symbols," Ph.D thesis, Department of Electrical Engineering, Queen's University, Kingston, Ontario, Canada, Sept. 1985.
38. P. HO and P. J. MCLANE, "The power spectral density of digital continuous phase modulation with correlated data symbols. Part 1: The correlation function method," *IEE Proc.*, Vol. 133, Part F, No. 1, pp. 95–105, Feb. 1986.
39. P. HO and P. J. MCLANE, "The power spectral density of digital continuous phase modulation with correlated data symbols. Part 2: The Rowe–Prabhu method," *IEE Proc.*, Vol. 133, Part F, No. 1, pp. 106–114, Feb. 1986.
40. D. SLEPIAN, "Several new families of alphabets for signalling," Bell Telephone Laboratories, unpublished memorandum, 1951.
41. D. SLEPIAN, "Permutation modulation," *Proc. IEEE*, Vol. 53, pp. 228–236, Mar. 1965.
42. E. BIGLIERI and M. ELIA, "Optimum permutation modulation codes and their asymptotic performance," *IEEE Trans. Inf. Theory*, Vol. IT-22, pp. 751–753, Nov. 1976.

43. A. G. Zyuko, A. I. Fal'ko, I. P. Panfilov, B. L. Banket, and I. V. Ivastchenko, *Pomekhoustoychivost' i Effektivnost' Sistem Peredachi Informatsii*. Radio i Svyaz', Moscow, USSR, 1985 (in Russian).
44. I. Ingemarsson, "Signal sets generated by orthogonal transformations of the signal space," *Technical Report 21*, Royal Institute of Technology, Stockholm, Sweden, Feb. 1969.
45. E. Biglieri and M. Elia, "Cyclic-group codes for the Gaussian channel," *IEEE Trans. Inf. Theory*, Vol. IT-22, pp. 624–629, Sept. 1976.
46. B. Rimoldi, "A decomposition approach to CPM," *IEEE Trans. Inf. Theory*, Vol. 34, No. 2, pp. 260–270, Mar. 1988.

CHAPTER 7

Multiple TCM

In Chapter 4 we demonstrated that an asymptotic measure of performance gain of a TCM system is the comparison of the minimum free Euclidean distance d_{free} of the trellis code relative to the minimum distance $d_{\min}$ of the uncoded modulation. This performance measure is an indication of the maximum reduction in required bit energy-to-noise ratio E_b/N_0 that can be achieved for arbitrarily small system bit error rates. Although at practical bit error rate values, this measure can often be misleading since the "real" gain in E_b/N_0 reduction due to coding could be significantly less, techniques that increase d_{free} are still highly desirable.

In Chapter 5 we saw that introducing *asymmetry* into the modulation signal point constellation was one method for increasing d_{free}. Unfortunately, however, in certain cases of asymmetry, the asymptotic improvement in E_b/N_0 reduction as measured by d_{free} could only be achieved in the limit as the points in the signal constellation merged together; that is, the trellis code becomes *catastrophic*, namely, a finite number of channel symbol errors produces an infinite number of decoded bit errors.

An example of such is the two-state code, which for all values of m (the number of input bits) asymptotically results in a 3-dB gain over the same bandwidth uncoded system. In reality, however, this 3-dB performance gain can never be achieved. Another problem with having signal points too close together in the constellation is that the system becomes less robust in the presence of carrier phase synchronization errors (i.e., the TCM is much less tolerant of carrier phase jitter).

The question then arises: Is it possible to achieve the same type of asymptotic performance gains without having to resort to modulation asymmetry? One positive answer to this question was provided by the use of multidimensional constellations studied in Chapter 6. Another is provided by the technique referred to as *multiple trellis-coded modulation* (MTCM) (the significance of this acronym will be apparent shortly) [1–3].

In its most general form, MTCM is implemented by an encoder with b binary input bits and s binary output symbols that are mapped into k M-ary symbols in each transmission interval (see Fig. 7.1).[1] The parameter k is referred to as the

---

[1] Analogous to the results in Chapter 6, MTCM may also be thought of as a coding onto a multidimensional signal set constructed from successive channel symbols.

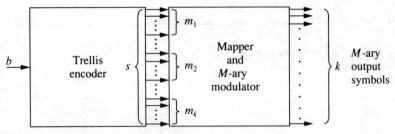

FIGURE 7.1 Generalized MTCM transmitter. Special case: $m_1 = m_2 = \cdots = m_k = m$. (The k multiple symbols per trellis branch are chosen from the same-size alphabet.)

multiplicity of the code since it represents the number of M-ary symbols allocated to each branch in the trellis diagram ($k = 1$ corresponds to the original Ungerboeck scheme which we shall refer to herein as *conventional* TCM). To produce such a result, we partition the s binary encoder output symbols into k groups containing $m_1, m_2, \ldots, m_k$ symbols. Each of these groups, through a suitable mapping function, results in an M_i-ary output symbol where $M_i = 2^{m_i}$; $i = 1, 2, \ldots, k$. Thus the transmitter parameters are constrained such that

$$s = \sum_{i=1}^{k} m_i = \sum_{i=1}^{k} \log_2 M_i = \log_2 \prod_{i=1}^{k} M_i. \qquad (7.1)$$

For the special and more common case where all the M-ary alphabet sizes are equal (later on we shall discuss reasons why they might not want to be), that is, $M_1 = M_2 = \cdots = M_k = M$, (7.1) simplifies to

$$s = k \log_2 M. \qquad (7.2)$$

For the transmitter of Fig. 7.1, the throughput (the ratio of bit rate to transmission bandwidth) is $r = b/k$ bits/s/Hz, which, depending on the choice of b and k, may or may not be integer-valued. Also, since b is not required to be an integer multiple of the multiplicity k, the ratio b/s is not constrained to be the ratio of adjacent integers as suggested in the original Ungerboeck scheme. It is interesting to note that the noninteger throughput MTCMs do not have equivalent uncoded counterparts. Thus, in these cases, the notion of unity-bandwidth expansion of the trellis-coded scheme relative to the uncoded scheme has no meaning. Later in the chapter we shall use the computational cutoff rate R_0 of the channel, defined in Chapter 1, as a performance measure that *directly* characterizes the MTCM technique independent of its relation to an uncoded modulation.

The relation in (7.1) can be rewritten as a constraint among the multiplicity, throughput, and alphabet sizes as follows. Since the ratio b/s must be less than unity (or, equivalently, $b + 1 \leq s$), then using (7.1) and the fact that $b = rk$, we have

$$2^{rk+1} \leq \prod_{i=1}^{k} M_i. \qquad (7.3)$$

It is interesting to note that (7.3) also characterizes the MTCM transmitter in cases where M_i is not necessarily an integer power of 2.

A further specialization of the general MTCM technique, which is in fact the one originally proposed by Divsalar and Simon [1, 2], corresponds to the case where $b = mk$ and $s = (m + 1)k$ with k groups of $(m + 1)$ symbols each being mapped into an M-ary symbol chosen from an alphabet of size $M = 2^{m+1}$. Here, in each transmission interval, mk bits enter the encoder and k M-ary symbols leave the modulator; thus the throughput is n bits/s/Hz (i.e., integer) and results in a unity-bandwidth expansion relative to an uncoded modulation with 2^m signal points. Since this special case allows for a direct comparison with uncoded modulation and is perhaps simpler to conceptualize, we begin this chapter with a detailed discussion of its behavior starting with its application to M-PSK and M-AM modulations and a two-state trellis diagram.

7.1 Two-State MTCM

To set the stage properly for the general two-state case, we begin with two simple examples.

EXAMPLE 7.1

Figure 7.2(a) is the two-state trellis diagram for conventional rate 1/2 trellis-coded QPSK and Fig. 7.2(b) is the corresponding multiple trellis diagram for the same

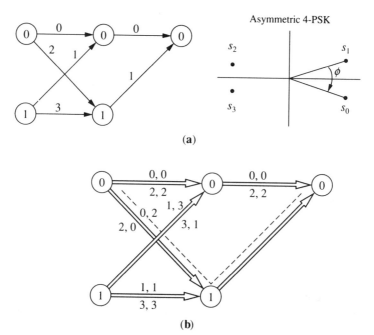

FIGURE 7.2. (a) Trellis diagram for conventional rate 1/2 trellis-coded QPSK; (b) trellis diagram for rate 2/4 multiple trellis-coded QPSK; $k = 2$.

coded modulation. In Fig. 7.2(b), we have $m = 1$ and $k = 2$ ($b = 2$) and thus there are $2^{mk} = 4$ branches emanating from each state. Since there are only two states in the diagram, this implies that there must be two parallel branches between each pair of states. Also, since $k = 2$, we have two output QPSK symbols[2] assigned to each branch. The assignment of these symbols to each branch is made to maximize the minimum Euclidean distance between the path through the trellis corresponding to correct reception of the transmitted symbols and that corresponding to an error event path. Also, the assignment must be made in such a way as to prevent the code from becoming *catastrophic*.

Figure 7.2(b) illustrates the appropriate assignment of QPSK symbol pairs for each branch in the trellis diagram. Without loss of generality, we assume the all-zero sequence as the transmitted bit sequence with corresponding all-zero QPSK output symbols. Then the error event path of length 1 (i.e., the parallel path between successive zero states) produces a squared Euclidean distance

$$d^2 = 2d^2(0, 2) = 2(4) = 8 \tag{7.4}$$

where $d^2(i, j)$ denotes the squared Euclidean distance between QPSK symbols s_i and s_j. For the error event path of length 2 [illustrated by dashed lines in Fig. 7.2(b)], we see that the squared Euclidean distance is

$$d^2 = d^2(0, 0) + d^2(0, 2) + d^2(0, 1) + d^2(0, 3)$$
$$= 0 + 4 + 4\sin^2\frac{\phi}{2} + 4\cos^2\frac{\phi}{2} = 8 \tag{7.5}$$

independent of the asymmetry angle ϕ. Thus we may choose $\phi = \pi/2$ (symmetric QPSK) and obtain a rate 2/4 trellis-coded QPSK modulation that achieves a squared free distance equal to 8. We recall from Chapter 5 that for the conventional TCM of Fig. 7.2(a), we can achieve a squared free distance equal to 6 for the symmetric QPSK constellation and a squared free distance equal to 8 for the asymmetric QPSK constellation whose adjacent signal points merge together (i.e., a catastrophic code). □

EXAMPLE 7.2

Figure 7.3 is the two-state trellis diagram for conventional rate 1/2 trellis-coded 4-AM (note that the output symbols assigned to the transitions emanating from state "1" are reversed with respect to those in Fig. 7.2(a) so as to get maximum gain from the asymmetry of the modulation). The appropriate trellis diagram for rate 2/4 multiple trellis coding of *symmetric* 4-AM (distance 2 between adjacent signal points) is identical to Fig. 7.2(b) with the understanding that the branch assignments now correspond to two 4-AM symbols per branch. If we assume that the all-zero 4-AM sequence transmitted, the error event of length 1 (i.e., the parallel path between successive zero states) produces a normalized (by the average energy of the signal set, which, in this case, has value equal to 5) squared Euclidean

---

[2] For convenience, we denote the QPSK symbol s_j by its subscript j on the branch labels.

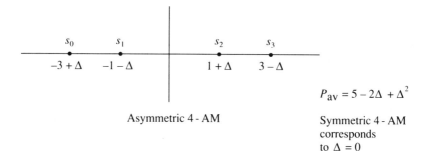

$P_{av} = 5 - 2\Delta + \Delta^2$

Symmetric 4-AM corresponds to $\Delta = 0$

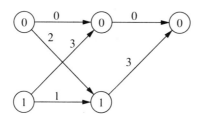

FIGURE 7.3 Trellis diagram for conventional rate 1/2 trellis-coded 4-AM.

distance

$$d^2 = \frac{d^2((0,0)(2,2))}{E_{av}} = \frac{2(4)^2}{5} = 6.4. \quad (7.6a)$$

For the error event path of length 2, the squared Euclidean distance is given by

$$d^2 = \frac{d^2((0,0),(0,2)) + d^2((2,2),(1,3))}{E_{av}} = \frac{(4)^2 + 2(2)^2}{5} = 4.8. \quad (7.6b)$$

Thus the minimum squared free distance is the smaller of (7.6a) and (7.6b),

$$d^2_{free} = 4.8. \quad (7.7)$$

Relative to the squared minimum distance of uncoded 2-AM (same bandwidth as rate 1/2 trellis-coded 4-AM) that has value 4, we achieve a gain of 0.792 dB. We recall from Chapter 5 that conventional rate 1/2 trellis-coded symmetric 4-AM produces no gain relative to uncoded 2-AM. Thus even for just a multiplicity of 2, MTCM has produced an advantage.

We further recall from Chapter 5 that when asymmetry was introduced into the 4-AM modulation, then asymptotically the squared Euclidean distance achieved by conventional TCM could approach a value equal to $32/4 = 8$ (i.e., a gain of 3 dB over the uncoded system). Once again to achieve that gain it was necessary to merge signal points together (i.e., s_0 with s_1 and s_2 with s_3), which results in a catastrophic code. With MTCM, we shall see shortly that with a multiplicity $k = 4$, we can achieve a squared Euclidean distance equal to $32/5 = 6.4$ or a gain of 2.041 dB over the uncoded system. For larger values of k, we are limited by the Euclidean distance of the error event path of length 1, and thus the squared free distance cannot be increased beyond the above value. ☐

The reason that we cannot achieve with multiple trellis coding of symmetric M-AM the same maximum gain "achieved" with conventional trellis coding of asymmetric M-AM is explained as follows. We see from the simple example above that, *ignoring the normalization by the average energy of the signal set*, we can in either case achieve a squared free distance equal to 32. With MTCM, the signal set remains symmetric and hence the normalization is constant (e.g., a value of 5). When asymmetry is introduced into M-AM modulation, as in Fig. 7.3 for example, the average energy is reduced, for example, to a value of 4 in the limit as adjacent 4-AM signal points merge together. Thus, *the difference in performance gain between the two schemes relative to an equivalent bandwidth uncoded system is attributed to the reduction in the average energy of the latter*. As the number of levels, M, gets larger, the reduction of the average energy of the signal set due to asymmetry becomes smaller; thus, in the limit of large M, the multiple trellis-coded symmetric M-AM will approach the 3-dB gain over the uncoded system. Numerical justification of this discussion will appear later in Table 7.2.

7.1.1 Mapping Procedure for Two-State MTCM

As in conventional TCM, we begin by partitioning the original signal point constellation into two subconstellations each with maximum distance among its signal points (see Fig. 7.4, for example). Signals in partition 1 are assigned to transitions

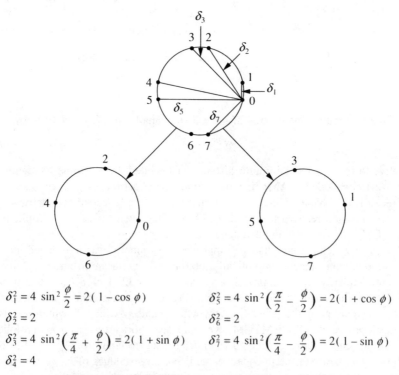

$\delta_1^2 = 4 \sin^2 \frac{\phi}{2} = 2(1 - \cos \phi)$ $\delta_5^2 = 4 \sin^2 \left(\frac{\pi}{2} - \frac{\phi}{2} \right) = 2(1 + \cos \phi)$

$\delta_2^2 = 2$ $\delta_6^2 = 2$

$\delta_3^2 = 4 \sin^2 \left(\frac{\pi}{4} + \frac{\phi}{2} \right) = 2(1 + \sin \phi)$ $\delta_7^2 = 4 \sin^2 \left(\frac{\pi}{4} - \frac{\phi}{2} \right) = 2(1 - \sin \phi)$

$\delta_4^2 = 4$

FIGURE 7.4 Set partitioning for asymmetric 8-PSK.

emanating from state 0, and signals in partition 2 are assigned to transitions emanating from state 1.

From the discussion above of MTCM with multiplicity k, we note that there will be 2^{mk-1} parallel paths between like states (e.g., "0" and "0" or "1" and "1"), and the same number of parallel paths between unlike states (e.g., "0" and "1" or "1" and "0"). For the transition between like states, we assign to each parallel path a *sequence* of k symbols (herein referred to as a *k-tuple*) all chosen from a fixed partition (of 2^m points) such that the minimum squared distance between any two of these parallel paths is equal to *twice* the minimum squared distance between points in the partition (i.e., $8\sin^2 \pi/2^m$ for M-PSK and $32/[(2^{2m+2}-1)/3]$ for M-AM).[3] The remaining 2^{mk-1} k-tuples formed from symbols in the same partition are assigned to the parallel paths corresponding to a transition to an unlike state. The minimum squared distance among all pairs of parallel paths between unlike states will also be *twice* the minimum squared distance between points in the partition. However, the minimum squared distance among all pairs of paths consisting of a path between like states and a path between unlike states *both originating from the same state* is only equal to the minimum squared distance between the points in the partition, that is, $4\sin^2 \pi/2^m$ for M-PSK and $16/[(2^{2m+2}-1)/3]$ for M-AM. Note that thus far the distances discussed have been *independent of the multiplicity k*.

The place where the trellis multiplicity k has its influence is in regard to the minimum squared distance among all pairs of paths consisting of a path between like states and one between unlike states *where the two paths originate from two different states*. With the k-tuple assignments above, this minimum squared distance is k times the minimum squared distance between points in one partition and points in the other (i.e., $4k\sin^2 \pi/2^{m+1}$ for M-PSK and $4k/[(2^{2m+2}-1)/3]$ for M-AM). As will be seen in the next section, it is the increase in these distances with k that increases the minimum distance associated with the error event path of length 2 and thus allows for an improvement in d_{free} performance. The examples provided by the trellis diagrams of Fig. 7.5 are further illustrations of the general mapping procedure.

7.1.2 Evaluation of Minimum Squared Free Distance

If one constructs a two-state trellis based on the mapping procedure above, then clearly the minimum squared distance for an error event path of length 1 is the minimum squared distance among parallel paths between like states (i.e., $d_1^2 = 8\sin^2 \pi/2^m$ for M-PSK and $32/[(2^{2m+2}-1)/3]$ for M-AM). For the error event path of length 2 [see, e.g., the dashed curve in Fig. 7.2(b)], the minimum squared distance is made up of two parts. The first part corresponds to the minimum squared distance between two paths originating from the same state, one of which terminates in a like state and the other in an unlike state. As discussed above, this is given by $4\sin^2 \pi/2^m$ for M-PSK and $32/[(2^{2m+2}-1)/3]$ for M-AM. The second part corresponds to the minimum squared distance between two paths that originate from two

---

[3] Herein in our discussion of M-AM distances, it will be assumed that the signal point constellation and its partitions have been normalized by $E_{\text{av}} = (2^{m+2}-1)/3$.

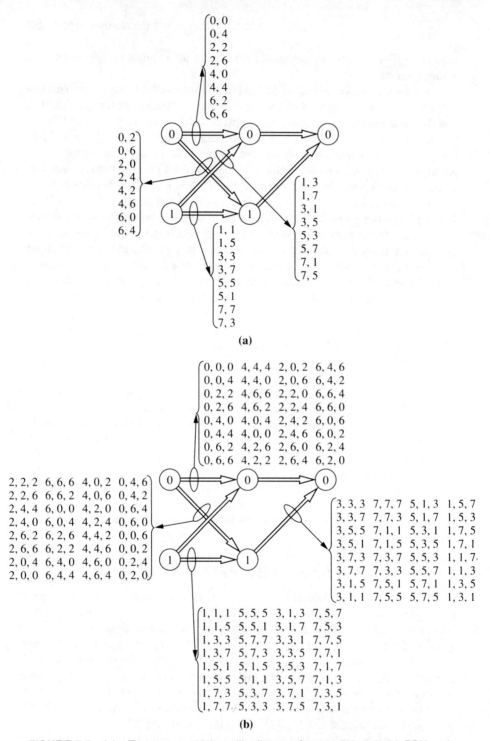

FIGURE 7.5 (a) Two-state multiple trellis diagram for rate 2/3 coded 8-PSK and 8-AM; $k = 2$; (b) two-state multiple trellis diagram for rate 2/3 coded 8-PSK and 8-AM; $k = 3$.

different states and terminate in the same state. Again from the discussion above, this is given by $4k \sin^2 \pi/2^{m+1}$ for M-PSK and $4k/[(2^{2m+2} - 1)/3]$ for M-AM. Thus the minimum squared distance for the error event path of length 2 is $d_1^2 = 4\sin^2 \pi/2^m + 4k \sin^2 \pi/2^{m+1}$ for M-PSK and $d_2^2 = (16 + 4k)/[(2^{2m+2} - 1)/3]$ for M-AM. Finally, then, the minimum squared free distance for the two-state multiple trellis-coding scheme is the smaller of d_1^2 and d_2^2,

$$d_{\text{free}}^2 = \begin{cases} \min\left\{ 8 \sin^2 \frac{\pi}{2^m},\ 4 \sin^2 \frac{\pi}{2^m} + 4k \sin^2 \frac{\pi}{2^{m+1}} \right\} & \text{for } M\text{-PSK} \\ \min\left\{ \frac{32}{(2^{2m+2} - 1)/3},\ \frac{16 + 4k}{(2^{2m+2} - 1)/3} \right\} & \text{for } M\text{-AM.} \end{cases} \quad (7.8)$$

It is interesting to investigate, as a function of m, the value of trellis multiplicity k, which if increased causes d_2^2 to become greater than d_1^2 (i.e., the largest value of k beyond which there is no performance improvement in d_{free}). In particular, we seek the largest integer k for which $d_1^2 > d_2^2$. From the above, using straightforward trigonometric manipulations, we arrive at the result

$$k_{\max} = \begin{cases} \left\lfloor 2\left(1 + \cos \frac{\pi}{2^m}\right) \right\rfloor \leq 4 & \text{for } M\text{-PSK} \\ 4 & \text{for } M\text{-AM} \end{cases} \quad (7.9)$$

where $\lfloor x \rfloor$ denotes the largest integer less than or equal to x. From (7.9) we see the following interesting results. For M-PSK, a value of $m = 1$ yields $k_{\max} = 2$, whereas for any $m > 1$, $k_{\max} = 4$. For M-AM, we have $k_{\max} = 4$ for all $m \geq 1$.

Tables 7.1 and 7.2 tabulate, for M-PSK and M-AM, respectively, the squared free distance, as computed from (7.8), versus m for values of k from 1 to $k_{\max}$.

TABLE 7.1 Minimum Squared Free Distance Performance of Multiple Trellis-Coded M-PSK: Two States.[a]

d_{free}^2	n	k	Performance Gain Relative to Conventional TCM ($k = 1$) (dB)	Performance Gain Relative to Uncoded 2^n−PSK (dB)
6.0	1	1	0.0	1.76
8.0	1	2	1.25	3.01
2.586	2	1	0.0	1.116
3.172	2	2	0.887	2.003
3.757	2	3	1.623	2.739
4.0	2	4	1.895	3.01
0.738	3	1	0.0	1.00
0.8903	3	2	0.814	1.814
1.0425	3	3	1.50	2.50
1.172	3	4	2.01	3.01

[a] Unit energy M-PSK signal constellation is assumed.

TABLE 7.2 Minimum Squared Free Distance Performance of Multiple Trellis-Coded M-AM: Two States.

d^2_{free}	n	k	Performance Gain Relative to Conventional TCM ($k = 1$) (dB)	Performance Gain Relative to Uncoded 2^n–AM (dB)
4.0	1	1	0.0	0.0
4.8	1	2	0.792	0.792
5.6	1	3	1.461	1.461
6.4	1	4	2.041	2.041
20/21	2	1	0.0	0.757
24/21	2	2	0.792	1.549
28/21	2	3	1.461	2.218
32/21	2	4	2.041	2.798
20/85	3	1	0.0	0.918
24/85	3	2	0.792	1.71
28/85	3	3	1.461	2.379
32/85	3	4	2.041	2.959

Also tabulated are (1) the performance gain (in dB) of multiple trellis-coded M-PSK (M-AM) relative to conventional TCM obtained by taking the ratio of squared free distance for the given value of k to squared free distance for $k = 1$, and (2) the performance gain (in dB) of multiple trellis-coded M-PSK (M-AM) relative to uncoded symmetric M-PSK (M-AM) with 2^m symbols.

In summary, then, for M-PSK with a symmetric signal point constellation, using $k = 2$ when $m = 1$ [i.e., rate 2/4(= 1/2) trellis-coded QPSK], and using $k = 4$ for $m > 1$ (i.e., rate $4m/4(m + 1)[= m/(m + 1)]$ trellis-coded 2^{m+1} – PSK), one can achieve a 3-dB asymptotic performance gain improvement. Also, in the latter case ($m > 1$), values of $k = 2, 3$ still give performance improvement relative to conventional TCM, but by an amount less than the maximum achievable 3 dB.

For M-AM, a value of $k = 4$ for all $m \geq 1$ is required to achieve the maximum performance gain from the multiple trellis-coding scheme and a symmetric modulation. In this case, however, we achieve a 3-dB asymptotic gain theoretically only as M approaches *infinity*. *From a practical standpoint, however, a value of $M = 16$ brings us arbitrarily close* (i.e., a gain of 2.96 dB). Also, as for M-PSK, values of $k = 2, 3$ give proportionate gains relative to the conventional TCM approach.

7.2 Generalized MTCM

Here we consider the general MTCM case illustrated in Fig. 7.1, where the throughput is not restricted to be integer-valued. We first discuss the general set-partitioning method for this technique and then illustrate the mapping of these sets and the free

Euclidean distance evaluation for a number of examples corresponding to trellises with more than two states.

7.2.1 Set-Partitioning Method for Generalized MTCM

In this section we describe a set-partitioning method for generating k-tuple signal sets of M-PSK symbols to be assigned to the branches of the trellis diagram. The method will be explained in detail for the case of $M = 8$ and $k = 2, 3, 4$. An examination of the arguments given will clearly indicate the details of the procedure for other values of M.

Set partitioning for $k = 2$, 8-PSK signal set

Start by defining the set[4]

$$A_0 = \begin{bmatrix} 0 & 0 \\ 4 & 4 \end{bmatrix} \qquad (7.10)$$

which represents a set of two 2-tuples with maximum squared Euclidean distance between them given by $d^2(A_0) = 8$. Next define the set

$$A_1 = A_0 + [0 \; 4] = \begin{bmatrix} 0 & 4 \\ 4 & 0 \end{bmatrix} \qquad (7.11)$$

where the addition in (7.11) is performed (component-wise) modulo 8. This set also has the maximum squared Euclidean distance between its two 2-tuples equal to 8.

The next-*higher*[5] partition level is started by forming the union of A_0 and A_1:

$$B_0 = A_0 \cup A_1 = \begin{bmatrix} 0 & 0 \\ 4 & 4 \\ 0 & 4 \\ 4 & 0 \end{bmatrix}. \qquad (7.12)$$

---

[4] For simplicity, we define the M-PSK symbols merely by their subscript. Also, for convenience of notation, we arrange the sets of 2-tuples as rows of a matrix, for example, the set of 2-tuples $\{(a, b), (c, d), (e, f)\}$ is denoted by

$$\begin{bmatrix} a & b \\ c & d \\ e & f \end{bmatrix}.$$

Furthermore, all operations (e.g., addition, union, etc.) performed on the matrices are understood to be performed on the sets of 2-tuples. Finally, since the groups of 2-tuples are in reality sets and not matrices, we shall not use boldface type for their representation.

[5] Note that the set-partitioning method described here proceeds in the opposite direction to that given by Ungerboeck (i.e., we built the tree from leaf to root).

The minimum squared distance between 2-tuples of B_0 is $d_{min}^2(B_0) = 4$. The next set appropriate to this partition level is

$$B_1 = B_0 + [2 \quad 2] = \begin{bmatrix} 2 & 2 \\ 6 & 6 \\ 2 & 6 \\ 6 & 2 \end{bmatrix}. \tag{7.13}$$

which also has a minimum squared distance between its 2-tuples equal to 4.

Taking the union of B_0 and B_1 starts us off on the next-higher partition level with

$$C_0 = B_0 \cup B_1 = \begin{bmatrix} 0 & 0 \\ 4 & 4 \\ 0 & 4 \\ 4 & 0 \\ 2 & 2 \\ 6 & 6 \\ 2 & 6 \\ 6 & 2 \end{bmatrix} \tag{7.14}$$

which maintains the minimum squared distance between its 2-tuples equal to 4. To get to the next-higher level, we first form $C_1 = C_0 + [0 \quad 2]$ where $d_{min}^2(C_1) = 4$ and then take the union of C_0 and C_1, resulting in

$$D_0 = C_0 \cup C_1 = \begin{bmatrix} 0 & 0 & 0 & 2 \\ 4 & 4 & 4 & 6 \\ 0 & 4 & 0 & 6 \\ 4 & 0 & 4 & 2 \\ 2 & 2 & 2 & 4 \\ 6 & 6 & 6 & 0 \\ 2 & 6 & 2 & 0 \\ 6 & 2 & 6 & 4 \end{bmatrix} \tag{7.15}$$

which has a cardinality equal to 16 and $d_{min}^2(D_0) = 2$. Note that the addition of $[2 \quad 0]$, $[0 \quad 6]$, or $[6 \quad 0]$ instead of $[0 \quad 2]$ to C_0 generates a set identical to C_1. Finally, we generate the set

$$D_1 = D_0 + [1 \quad 1] = \begin{bmatrix} 1 & 1 & 1 & 3 \\ 5 & 5 & 5 & 7 \\ 1 & 5 & 1 & 7 \\ 5 & 1 & 5 & 3 \\ 3 & 3 & 3 & 5 \\ 7 & 7 & 7 & 1 \\ 3 & 7 & 3 & 1 \\ 7 & 3 & 7 & 5 \end{bmatrix} \tag{7.16}$$

which also has a cardinality of 16 and $d_{min}^2(D_1) = 2$.

7.2 / Generalized MTCM

To design an MTCM with, for example, throughput $b/k = 4/2 = 2$ (i.e., a rate 4/6 trellis-coded 8-PSK), we need four sets each with cardinality equal to 8. We start on the highest partition level with C_0, C_1 and then generate the two additional sets

$$C_2 + C_0 + [1 \ 1] = \begin{bmatrix} 1 & 1 \\ 5 & 5 \\ 1 & 5 \\ 5 & 1 \\ 3 & 3 \\ 7 & 7 \\ 3 & 7 \\ 7 & 3 \end{bmatrix} \quad C_3 = C_1 + [1 \ 1] = \begin{bmatrix} 1 & 3 \\ 5 & 7 \\ 1 & 7 \\ 5 & 3 \\ 3 & 5 \\ 7 & 1 \\ 3 & 1 \\ 7 & 5 \end{bmatrix}.$$

(7.17)

Note that $d^2_{\min}(C_i) = 4$ for $i = 0, 1, 2, 3$, and the minimum distance between a 2-tuple in C_i and a 2-tuple in C_j, $i \neq j$, is $d^2_{\min}(C_i, C_j) = 2$. These sets will be used in the next section in the design of a two-state MTCM example.

Similarly, to design an MTCM with throughput $b/k = 5/2 = 2.5$ (i.e., a rate 5/6 trellis-coded 8-PSK), we need to start with four sets each having cardinality 16. The appropriate sets are D_0, D_1, and the two sets generated are

$$D_2 = D_0 + [0 \ 1] = \begin{bmatrix} 0 & 1 \\ 4 & 5 \\ 0 & 5 \\ 4 & 1 \\ 2 & 3 \\ 6 & 7 \\ 2 & 7 \\ 6 & 3 \end{bmatrix} \begin{matrix} 0 & 3 \\ 4 & 7 \\ 0 & 7 \\ 4 & 3 \\ 2 & 5 \\ 6 & 1 \\ 2 & 1 \\ 6 & 5 \end{matrix} \quad D_3 = D_1 + [0 \ 1] = \begin{bmatrix} 1 & 3 \\ 5 & 6 \\ 1 & 6 \\ 5 & 2 \\ 3 & 4 \\ 7 & 0 \\ 3 & 0 \\ 7 & 4 \end{bmatrix} \begin{matrix} 1 & 4 \\ 5 & 0 \\ 1 & 0 \\ 5 & 4 \\ 3 & 6 \\ 7 & 2 \\ 3 & 2 \\ 7 & 6 \end{matrix}.$$

(7.18)

Set partitioning for $k = 3$, 8-PSK signal set

Here we start with the equivalent to (7.10) using 3-tuples:

$$A_0 = \begin{bmatrix} 0 & 0 & 0 \\ 4 & 4 & 4 \end{bmatrix}. \tag{7.19}$$

The squared distance between the two 3-tuples is $d^2(A_0) = 12$. Next form the sets

$$A_1 = A_0 + [0 \ 0 \ 4]$$
$$A_2 = A_0 + [0 \ 4 \ 0] \tag{7.20}$$
$$A_3 = A_0 + [4 \ 0 \ 0].$$

These sets all have the same squared distance between their own 3-tuples equal to that of A_0.

The first set on the next-higher partition level is formed from the union of the sets in (7.19) and (7.20):

$$B_0 = A_0 \cup A_1 \cup A_2 \cup A_3 = \begin{bmatrix} 0 & 0 & 0 \\ 4 & 4 & 4 \\ 0 & 0 & 4 \\ 4 & 4 & 0 \\ 0 & 4 & 0 \\ 4 & 0 & 4 \\ 4 & 0 & 0 \\ 0 & 4 & 4 \end{bmatrix}. \qquad (7.21)$$

The minimum distance between 3-tuples of B_0 is $d^2_{min}(B_0) = 4$ and the set has a cardinality equal to 8. The remaining sets on this partition level containing *even*-numbered symbols are given by

$$B_1 = B_0 + [0 \ 2 \ 2]$$
$$B_2 = B_0 + [2 \ 0 \ 2] \qquad (7.22)$$
$$B_3 = B_0 + [2 \ 2 \ 0]$$

and have the same cardinality and minimum distance between their members as B_0.

Proceeding to the next-higher partition level, we start with the union of the sets in (7.21) and (7.22):

$$C_0 = B_0 \cup B_1 \cup B_2 \cup B_3 = \begin{bmatrix} 0 & 0 & 0 & 0 & 2 & 2 & 2 & 0 & 2 & 2 & 2 & 0 \\ 4 & 4 & 4 & 4 & 6 & 6 & 6 & 4 & 6 & 6 & 6 & 4 \\ 0 & 0 & 4 & 0 & 2 & 6 & 2 & 0 & 6 & 2 & 2 & 4 \\ 4 & 4 & 0 & 4 & 6 & 2 & 6 & 4 & 2 & 6 & 6 & 0 \\ 0 & 4 & 0 & 0 & 6 & 2 & 2 & 4 & 2 & 2 & 6 & 0 \\ 4 & 0 & 4 & 4 & 2 & 6 & 6 & 0 & 6 & 6 & 2 & 4 \\ 4 & 0 & 0 & 4 & 2 & 2 & 6 & 0 & 2 & 6 & 2 & 0 \\ 0 & 4 & 4 & 0 & 6 & 6 & 2 & 4 & 6 & 2 & 6 & 4 \end{bmatrix}.$$

$$(7.23)$$

This set has a cardinality equal to 32 and $d^2_{min}(C_0) = 4$. Similarly, we generate

$$C_1 = C_0 + [0 \ 0 \ 2] \qquad (7.24)$$

and then to start the next-higher partition level,

$$D_0 = C_0 \cup C_1. \qquad (7.25)$$

The set D_0 has a cardinality of 64 and a minimum squared distance between its 3-tuples given by $d^2_{min}(D_0) = 2$.

To generate the sets with *odd*-numbered symbols on this level, we start with

$$E_0 = D_0 + [1 \ 1 \ 1] \qquad (7.26)$$

and then for the next-higher partition level

$$F_0 = E_0 \cup D_0. \tag{7.27}$$

The set E_0 has the same minimum distance between its 3-tuples as D_0; however, $d_{\min}^2(F_0) = 1.75$. Also, F_0 has a cardinality equal to 128.

To design, for example, an MTCM with throughput $b/k = 6/3 = 2$ (i.e., a rate 6/9 trellis-coded 8-PSK), we need four sets each having cardinality equal to 32. For this purpose, we can use C_0, C_1, and the two additional sets containing *odd*-numbered symbols,

$$\begin{aligned} C_2 &= C_0 + [1 \ 1 \ 1] \\ C_3 &= C_1 + [1 \ 1 \ 1]. \end{aligned} \tag{7.28}$$

These sets all have minimum squared *intra*distances equal to 4 and minimum squared *inter*distances $d_{\min}^2(C_0, C_1) = 2$ and $d_{\min}^2(C_0, C_2) = 1.75$.

To construct an MTCM with throughput equal, say, to $b/k = 7/3$ (i.e., a rate 7/9 trellis-coded 8-PSK), we need to start with four sets each having cardinality of 64. For this purpose, we can use D_0 and the three additional sets

$$\begin{aligned} D_1 &= D_0 + [0 \ 1 \ 1] \\ D_2 &= D_0 + [1 \ 0 \ 1] \\ D_3 &= D_0 + [1 \ 1 \ 0]. \end{aligned} \tag{7.29}$$

All of these sets have a minimum squared *intra*distance equal to 2.

Finally, to design an MTCM with, say, a throughput equal to $b/k = 8/3$ (i.e., a rate 8/9 trellis-coded 8-PSK), we start with F_0 and generate the three additional sets

$$\begin{aligned} F_1 &= F_0 + [0 \ 0 \ 1] \\ F_2 &= F_0 + [0 \ 1 \ 0] \\ F_3 &= F_0 + [1 \ 0 \ 0] \end{aligned} \tag{7.30}$$

all having cardinality equal to 128 and $d_{\min}^2(F_i) = 1.75$ for $i = 0, 1, 2, 3$.

Set partitioning for $k = 4$, 8-PSK signal set

By now the procedure should be apparent. The details of the set-partitioning method for $k = 4$ are presented in Table 7.3.

7.2.2 Set Mapping and Evaluation of Squared Free Distance

For the generalized MTCM case and an arbitrary number of states, it is not easy to arrive at a simple expression for d_{free} analogous to (7.8). Thus we shall illustrate the evaluation of squared free distance by considering various different trellis structures derived from the set-partitioning method described above.

TABLE 7.3 Set-Partitioning Method for $k = 4$.

	Signal Sets	Number of Elements per Set	Intra- and Inter-Set Distances
1.	$A_0 = \begin{cases} 0000 \\ 4444 \end{cases}$	2	$d^2(A_i) = 16$
	$A_1 = A_0 + 0044$	2	$d^2(A_i, A_j) = 8$
	$A_2 = A_0 + 0404$	2	$i \neq j = 0, 1, 2, \ldots, 7$
	$A_3 = A_0 + 4004$	2	
	$A_{i+4} = A_i + 2222$	2	
	$i = 0, 1, 2, 3$		
2.	$B_0 = \bigcup_{i=0}^{7} A_i$	16	$d^2(B_{mi}) = 8$
	$B_1 = B_0 + 0022$	16	$d^2(B_{mi}, B_{mj}) = 4$
	$B_2 = B_0 + 0202$	16	$m = 0, 1, 2, 3$
	$B_3 = B_0 + 2002$	16	$i \neq j = 0, 1, 2, \ldots, 7$
	$B_{i+4} = B_i + 0004$	16	$d^2(B_{mi}, B_{(m+1)j}) = 4(2 - \sqrt{2})$
	$i = 0, 1, 2, 3$		$m = 0, 2$
			$i, j = 0, 1, 2, \ldots, 7$
(2a)	$B_{0i} = B_i$	16	$d^2(B_{mi}, B_{(n+2)j}) = 2$
	$B_{1i} = B_i + 1111$		$m, n = 0, 1$
	$i = 0, 1, 2, \ldots, 7$		$i, j = 0, 1, 2, \ldots, 7$
(2b)	$B_{2i} = B_{0i} + 0002$	16	
	$B_{3i} = B_{1i} + 0002$		
	$i = 0, 1, 2, \ldots, 7$		
3.	$C_0 = \bigcup_{i=0}^{7} B_i$	128	$d^2(C_i) = 4$
	$C_1 = C_0 + 1111$	128	$d^2(C_i, C_{i+1}) = 4(2 - \sqrt{2})$
			$i = 0, 2$
	$C_{i+2} = C_i + 0002$	128	$d^2(C_i, C_{j+2}) = 2$
	$i = 0, 1$		$i, j = 0, 1$
(3a)	$C_{i+4} = C_i + 0011$	128	$d^2(C_i, C_{j+4}) = 2(2 - \sqrt{2})$
	$i = 0, 1, 2, 3$		$i, j = 0, 1, 2, 3$
4.	$D_0 = C_0 \cup C_1$	256	$d^2(D_i) = 4(2 - \sqrt{2})$
	$D_1 = C_2 \cup C_3$	256	$d^2(D_i, D_{i+1}) = 2$
(4a)	$D_{i+2} = D_i + 0011$	256	$i = 0, 2, 4, 6$
	$D_{i+4} = D_i + 0101$	256	$d^2(D_i, D_{j+m}) = 2(2 - \sqrt{2})$
	$D_{i+6} = D_i + 1001$	256	$m = 2, 4, 6$
	$i = 0, 1$		$i, j = 0, 1, 2, \ldots, 7$
5.	$E_0 = D_0 \cup D_1$	512	$d^2(E_i) = 2$
			$i = 0, 1, 2, \ldots, 7$
	$E_1 = E_0 + 0011$	512	$d^2(E_{i+m}, E_{j+m}) = 2(2 - \sqrt{2})$
	$E_2 = E_0 + 0101$	512	$i, j = 0, 1, 2, 3$
	$E_3 = E_0 + 1001$	512	$m = 0, 4$
(5a)	$E_{i+4} = E_i + 0001$	512	$d^2(E_i, E_{j+4}) = 2 - \sqrt{2}$
	$i = 0, 1, 2, 3$		$i, j = 0, 1, 2, 3$

EXAMPLE 7.3

Consider a trellis encoder with $b = 3$, $s = 6$, whose binary output symbols are mapped into 8-PSK symbols with multiplicity $k = 2$ in accordance with (7.2). The throughput of this MTCM scheme is thus $r = b/k = 1.5$ bits/s/Hz. We again note that there is no equivalent (same throughput) uncoded modulation with 2^n signal points. In fact, the MTCM scheme above is, from the standpoint of throughput, midway between PSK (throughput $=1$ bit/s/Hz) and QPSK (throughput $= 2$ bits/s/Hz). For the example above, the number of transitions emanating from each state in the trellis diagram is $2^b = 8$. If we postulate that there are to be no parallel paths between transitions (later examples will relax this requirement), the minimum number of states for the trellis must be 8 (i.e., the trellis is fully connected). Thus we begin by considering this specific case.

Eight-State Trellis

In accordance with Section 7.2.1, we must assign a pair ($k = 2$) of 8-PSK symbols to each trellis branch in such a way as to maximize the free distance of the code. For the 8-PSK signal set of Fig. 7.4 (assuming symmetry), define the sets (pairs of 8-PSK symbols)

$$A_0 = 00 \quad A_4 = 22 \quad B_0 = 02 \quad B_4 = 20$$
$$A_1 = 44 \quad A_5 = 66 \quad B_1 = 46 \quad B_5 = 64 \qquad (7.31)$$
$$A_2 = 04 \quad A_6 = 26 \quad B_2 = 06 \quad B_6 = 24$$
$$A_3 = 40 \quad A_7 = 62 \quad B_3 = 42 \quad B_7 = 60.$$

These sets have the following minimum squared Euclidean distances:

$$d^2_{\min}(A_i, A_j) = 4$$
$$d^2_{\min}(B_i, B_j) = 4 \qquad (7.32)$$
$$d^2_{\min}(A_i, B_j) = 2.$$

We assign the A_i's to the paths leaving the *odd*-numbered states, each time permuting the assignment by one. Similarly, we assign the B_i's to the paths leaving the *even*-numbered states with the same permutation (Fig. 7.6). When this is done, the minimum distance path will be of length 2 (Fig. 7.6), and thus the squared free distance for the code is

$$d^2_{\text{free}} = d^2_{\min}(A_i, A_j) + d^2_{\min}(A_i, B_j) = 4 + 2 = 6. \qquad (7.33)$$

Note that, in effect, we require only a QPSK signaling set to achieve the above. We remind the reader that conventional rate 1/2 trellis-coded QPSK (throughput $=$ 1 bits/s/Hz) with an eight-state trellis resulted in $d^2_{\text{free}} = 2.0$, whereas conventional rate 2/3 trellis-coded 8-PSK (throughput $= 2$ bits/s/Hz) with an eight-state trellis produced $d^2_{\text{free}} = 6 - \sqrt{2} = 4.586$.

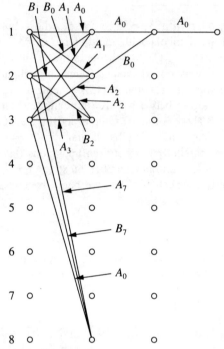

FIGURE 7.6 Eight-state trellis diagram for Example 7.3.

Sixteen-State Trellis

Here we still assume no parallel paths between states but use a half-connected trellis (each state transitions to only half the total number of states). The trellis and multiple 8-PSK symbol set assignment are illustrated in Fig. 7.7. In particular, we do as before; namely, we assign the A_i's to the paths leaving the odd-numbered states and the B_i's to the paths leaving the even-numbered states. With the assignment of Fig. 7.7, the minimum distance path is of length 2 and

$$d^2_{\text{free}} = d^2_{\min}(A_i, A_j) + d^2_{\min}(A_i, A_j) = 4 + 4 = 8. \quad (7.34)$$

This is to be compared with values of $d^2_{\text{free}} = 14.0$ and $d^2_{\text{free}} = 5.172$ for 16-state conventional rate 1/2 trellis-coded QPSK and rate 2/3 trellis-coded 8-PSK, respectively. Again, we note that, in effect, only QPSK signaling is used. □

EXAMPLE 7.4

The free distance of the MTCM schemes of Example 7.3 can be increased by defining the mapping sets such that transitions between states contain parallel paths. In particular, we define sets containing two elements per set as follows:

7.2 / Generalized MTCM

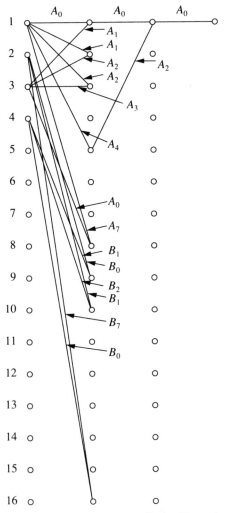

FIGURE 7.7 Sixteen-state trellis for Example 7.3.

$$C_0 = \begin{cases} 0 & 0 \\ 4 & 4 \end{cases} \quad D_0 = \begin{cases} 0 & 2 \\ 4 & 6 \end{cases} \quad E_0 = \begin{cases} 1 & 1 \\ 5 & 5 \end{cases} \quad F_0 = \begin{cases} 1 & 3 \\ 5 & 7 \end{cases}$$

$$C_1 = \begin{cases} 0 & 4 \\ 4 & 0 \end{cases} \quad D_1 = \begin{cases} 0 & 6 \\ 4 & 2 \end{cases} \quad E_1 = \begin{cases} 1 & 5 \\ 5 & 1 \end{cases} \quad F_1 = \begin{cases} 1 & 7 \\ 5 & 3 \end{cases}$$

$$C_2 = \begin{cases} 2 & 2 \\ 6 & 6 \end{cases} \quad D_2 = \begin{cases} 2 & 0 \\ 6 & 4 \end{cases} \quad E_2 = \begin{cases} 3 & 3 \\ 7 & 7 \end{cases} \quad F_2 = \begin{cases} 3 & 1 \\ 7 & 5 \end{cases}$$

$$C_3 = \begin{cases} 2 & 6 \\ 6 & 2 \end{cases} \quad D_3 = \begin{cases} 2 & 4 \\ 6 & 0 \end{cases} \quad E_3 = \begin{cases} 3 & 7 \\ 7 & 3 \end{cases} \quad F_3 = \begin{cases} 3 & 5 \\ 7 & 1. \end{cases}$$

(7.35)

These sets have the following squared Euclidean distances:

$$d^2(C_i) = 8 \qquad d^2(E_i) = 8$$

$$d^2(C_i, C_j)\big|_{i \neq j} = 4 \qquad d^2(E_i, E_j)\big|_{i \neq j} = 4$$

$$d^2_{\min}(C_i, D_j) = 2 \qquad d^2_{\min}(E_i, F_j) = 2$$

$$d^2(D_i) = 8 \qquad d^2(F_i) = 8$$

$$d^2(D_i, D_j)\big|_{i \neq j} = 4 \qquad d^2(F_i, F_j)\big|_{i \neq j} = 4$$

(7.36)

where again $d^2(X_i)$ denotes the squared Euclidean distance between the two elements in the set, $d^2(X_i, Y_j)$ is the squared Euclidean distance between either element in X_i and either element in Y_j, and $d^2_{\min}(X_i, Y_j)$ is the *minimum* of the squared Euclidean distances between either element in X_i and either element in Y_j.

Since $b = 3$ ($2^b = 8$ paths emanating from a given state) and there are two parallel paths per transition (i.e., each corresponding to one of the two elements in a given set), each state will now have a transition to only four other states. We begin by considering a trellis with four states, which implies a fully connected trellis.

Four-State Trellis

Consider the trellis of Fig. 7.8, where the C_i sets have been assigned to the paths leaving states 1 and 3 and the D_i sets have been assigned to the paths leaving states 2 and 4. Again we permute the assignment by one between paths leaving state 1 and paths leaving state 3, and similarly for states 2 and 4. By inspection of Fig. 7.8, we immediately find that the minimum distance path is of length 2 with squared Euclidean distance

$$d^2(C_i, C_j)\big|_{i \neq j} + d^2_{\min}(C_i, D_j) = 4 + 2 = 6. \qquad (7.37)$$

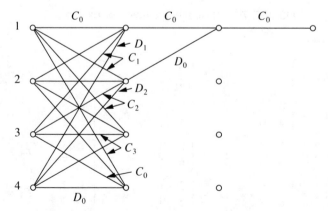

FIGURE 7.8 Four-state trellis for Example 7.4.

Since this squared distance is smaller than the squared distance between parallel paths (i.e., $d^2 = 8$), then $d^2_{\text{free}} = 6$. *We note that by using parallel paths between transitions, we are able to achieve a larger free distance with only four states than we achieved previously in Example 7.3 using eight states.* Also, the set assignment in Fig. 7.8 does not require the use of the E_i and F_i sets. Thus, in effect, we require only a QPSK signaling set to achieve the above.

Eight-State Trellis

Here we have a half-connected trellis, as illustrated in Fig. 7.9, with the C_i's assigned to the *odd*-numbered states and the D_i's to the *even*-numbered states. Again the minimum distance path is of length 2 and achieves

$$d^2(C_i, C_j)|_{i \neq j} + d^2_{\min}(C_i, C_j) = 4 + 4 = 8. \tag{7.38}$$

Since this is identical to the squared distance between parallel paths, we have $d^2_{\text{free}} = 8.0$, which is the same as that achieved with 16 states and no parallel paths in Example 7.3. Again, only a QPSK signaling set is needed, since sets E_i and F_i are not assigned to the trellis.

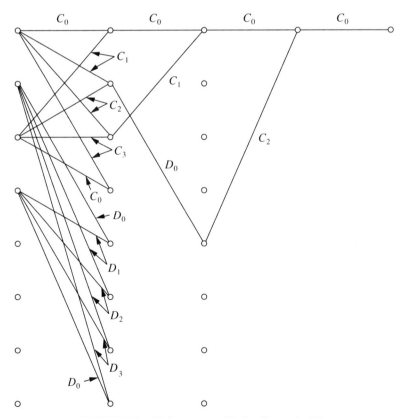

FIGURE 7.9 Eight-state trellis for Example 7.5.

Since the maximum free distance achievable is limited to the distance between parallel paths, we cannot achieve any further improvement by going to 16 states.

□

EXAMPLE 7.5

Consider next a trellis encoder with $b = 7$, $s = 12$, whose binary output symbols are mapped into 8-PSK with multiplicity $k = 4$ in accordance with (7.2). The throughput of this MTCM scheme is $b/k = 1.75$ with no equivalent uncoded system. There are now $2^b = 128$ transitions emanating from each state, so that for any number of trellis states less than 128, we must have parallel paths between states. The first case we consider is again that of a fully connected eight-state trellis, which implies that the number of parallel paths between states is $128/8 = 16$.

Eight-State Trellis

In accordance with the above, we must assign 16 4-tuples of 8-PSK symbols to each trellis branch in such a way as to maximize the free distance of the code. We can use the trellis diagram of Fig. 7.6, but first we must define and then assign the sets of 8-PSK 4-tuples. The construction of these sets is as follows:

Set	Number of elements per set	
$A_0 = \begin{cases} 0\ 0\ 0\ 0 \\ 4\ 4\ 4\ 4 \end{cases}$	2	
$A_1 = A_0 + 0\ 0\ 4\ 4$	2	
$A_2 = A_0 + 0\ 4\ 0\ 4$	2	
$A_3 = A_0 + 4\ 0\ 0\ 4$	2	
$B_0 = A_0 \cup A_1 \cup A_2 \cup A_3$	8	
$B_1 = B_0 + 2\ 2\ 2\ 2$	8	
$C_0 = B_0 \cup B_1$	16	(7.39)
$C_1 = C_0 + 0\ 0\ 0\ 4$	16	
$C_2 = C_0 + 0\ 0\ 2\ 2$	16	
$C_3 = C_0 + 0\ 0\ 2\ 6$	16	
$C_4 = C_0 + 2\ 0\ 0\ 2$	16	
$C_5 = C_0 + 2\ 0\ 0\ 6$	16	
$C_6 = C_0 + 0\ 2\ 0\ 2$	16	
$C_7 = C_0 + 0\ 2\ 0\ 6$	16	

Only the final sets C_i, D_i, E_i, and F_i, with 16 4-tuples each, are of interest insofar as assignment to the trellis. These sets have the following squared Euclidean distances:

$$d^2_{\min}(C_i) = 8 \qquad\qquad d^2_{\min}(E_i) = 8$$

$$d^2_{\min}(C_i, C_j)\big|_{i \neq j} = 4 \qquad d^2_{\min}(E_i, E_j)\big|_{i \neq j} = 4$$

$$d^2_{\min}(C_i, E_j) = 4(2 - \sqrt{2}) \qquad i, j = 0, 1, 2, \ldots, 7.$$

(7.40)

If we replace the trellis branch assignments A_i and B_i of Fig. 7.6 with C_i and E_i, respectively, then, since the distance between parallel branches is 8, the free distance is determined by the minimum distance path of length 2:

$$\begin{aligned} d^2_{\text{free}} &= d^2_{\min}(C_i, C_j)\big|_{i \neq j} + d^2_{\min}(E_i, E_j)\big|_{i \neq j} \\ &= 4 + 4(2 - \sqrt{2}) = 6.343. \end{aligned}$$

(7.41)

Sixteen-State Trellis

By using the half-connected trellis of Fig. 7.7 and again replacing the A_i and B_i branch assignments with C_i and E_i, respectively, the minimum distance path of length 2 has a squared Euclidean distance

$$\begin{aligned} d^2 &= d^2_{\min}(C_i, C_j)\big|_{i \neq j} + d^2_{\min}(C_i, C_j)\big|_{i \neq j} \\ &= 4 + 4 = 8. \end{aligned}$$

(7.42)

Since d^2 is equal to the minimum squared distance between parallel paths (i.e., 8), the squared free distance of the trellis is equal to 8. □

EXAMPLE 7.6

As a last example, we consider $b = 6$, $k = 4$ and a four-state trellis. This example has the same throughput as Examples 7.3 and 7.4, namely, $b/k = 1.5$, but with an increase in multiplicity from 2 to 4. The appropriate trellis diagram is illustrated in Fig. 7.10. Here the *even*-numbered C_i sets of Example 7.5 are

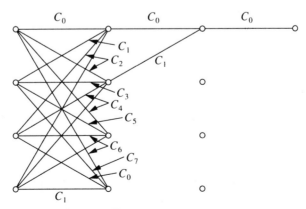

FIGURE 7.10 Four-state trellis for Example 7.6.

assigned to the paths leaving states 2 and 4. The minimum distance path is of length 2 and achieves a squared distance equal to 8, which is equal to the squared distance between parallel paths. Thus $d_{\text{free}}^2 = 8.0$. Hence we can achieve with four states and multiplicity 4 the equivalent performance to what required eight states when the multiplicity was only 2 (see Example 7.4). Once again going to a larger number of states will result in no gain, since a free distance equal to the distance between parallel paths has already been achieved. □

7.3 Analytical Representation of MTCM

In Chapter 3 we discussed an analytical representation of conventional TCM, introduced by Calderbank and Mazo [4], which expresses the single modulator output as a Volterra series expansion of products of the encoder input bits. This representation has the advantage that the traditional two-step design process [i.e., specifying an underlying trellis rate $m/(m + 1)$ code and then mapping the output code symbols into the fixed 2^{m+1}-point signal constellation] is now combined into a single operation. Furthermore, the analytical representation allows the implementation of the transmitter to be drawn by inspection and is also convenient for studying the behavior of TCM in the presence of intersymbol interference.

Here we demonstrate how to generalize the analytical representation so as to apply to MTCM. For convenience, we first present a brief review of this representation for conventional ($k = 1$) TCM and then show, primarily by example, how such a description is appropriate for each of the k elements in the k-tuples assigned to the trellis branches in MTCM.

Let $\{b_i\}$ be a sequence of ± 1-valued real variables with a mapping of the 0, 1-valued encoder input sequence $\{a_i\}$ according to the linear transformation $b_i = 1 - 2a_i$. From Chapter 3 we know that for a conventional trellis code, the modulator output $x(b_1, b_2, \ldots, b_n)$ may be written as a sum of products of the b_i's:

$$x(b_1, b_2, \ldots, b_n) = d_0 + \sum_{i=1}^{n} d_i b_i + \sum_{\substack{i,j=1 \\ j>1}}^{n} d_{ij} b_i b_j \\ + \sum_{\substack{i,j,k=1 \\ k>j>1}}^{n} d_{ijk} b_i b_j b_k + \cdots + d_{1\ldots n} b_1 b_2 \cdots b_n \quad (7.43)$$

where the sequence length n is equal to the sum of m, the number of input bits per output channel symbol, and v, the memory of the code (2^v is the number of states of the encoder). Also, the d's are a set of constraints that can be determined by a simple vector multiplication as follows.

Let $\mathbf{x} = (x_1, x_2, \ldots, x_{2^n})$ denote a column vector of length 2^n whose components represent the 2^n values that $x(b_1, b_2, \ldots, b_n)$ can take on. Next let $\mathbf{d} = (d_0, d_1, \ldots, d_n, d_{12}, d_{13}, \ldots, d_{1\ldots n})$ denote a 2^n-length column vector of the unknown constants. Finally, let $\mathbf{B}$ be a $2^n \times 2^n$ matrix where each row represents the 2^n values taken by all products of the b_i's called for in (7.43) for each sequence $b_1, b_2, \ldots, b_n$. In terms of these definitions, (7.10) can be written in the matrix form

$$\mathbf{x} = \mathbf{B}\mathbf{d}. \quad (7.44)$$

7.3 / Analytical Representation of MTCM

If $\boldsymbol{\beta}$ is a vector corresponding to a particular product of the b_i's (i.e., a column of **B**), then the corresponding coefficient of that product in the expansion of (7.43) is simply obtained from

$$d = \frac{1}{2^n}\boldsymbol{\beta}^T \mathbf{x} \tag{7.45}$$

that is, the Hadamard transform of the vector **x**. As elsewhere in the book, the superscript T in (7.45) denotes the transpose operation.

To apply the analytical description to MTCM, we observe that a representation such as (7.43) is appropriate for each of the k output symbols along any trellis branch. Thus, letting $\mathbf{x}^{(i)}$, $i = 1, 2, \ldots, k$, denote the ith modulator output corresponding to a particular input sequence of length b, the matrix representation of (7.44) is appropriate to the associated vector $\mathbf{x}^{(i)}$ and yields a vector $\mathbf{d}^{(i)}$ in accordance with (7.45) where now $n = b + v$. The set of vectors $\{\mathbf{d}^{(i)}\}$; $i = 1, 2, \ldots, k$, determined as above completely describes the multiple trellis code.

EXAMPLE 7.7

As a simple example of the above, consider the case of rate 2/4 multiple trellis-coded 4-AM with a two-state trellis as illustrated in Fig. 7.2(b). Here $b = 2$, $v = 1$, and hence $n = 3$. As such, (7.43) simplifies to[6]

$$\begin{aligned}x(b_1, b_2, b_3) = &\; d_1 b_1 + d_2 b_2 + d_3 b_3 + d_{12} b_1 b_2 \\ &+ d_{13} b_1 b_3 + d_{23} b_2 b_3 + d_{123} b_1 b_2 b_3.\end{aligned} \tag{7.46}$$

Here b_3 denotes the previous state and b_1 the present state. (Note that neither the previous nor the present state depends on b_2. Rather, b_2 is used to decide between the parallel paths between states.) From Fig. 7.2(b) (with state "0" and "1", respectively, replaced by "1" and "-1" in accordance with the relation between the a_i's and the b_i's and the signal constellation of Fig. 7.3 for symmetric modulation ($\Delta = 0$), we have

$$\begin{aligned}x^{(1)}(1, 1, 1) &= -3 & x^{(2)}(1, 1, 1) &= -3 \\ x^{(1)}(1, -1, 1) &= 1 & x^{(2)}(1, -1, 1) &= 1 \\ x^{(1)}(-1, 1, 1) &= -3 & x^{(2)}(-1, 1, 1) &= 1 \\ x^{(1)}(-1, -1, 1) &= 1 & x^{(2)}(-1, -1, 1) &= -3 \\ x^{(1)}(-1, 1, -1) &= -1 & x^{(2)}(-1, 1, -1) &= -1 \\ x^{(1)}(-1, -1, -1) &= 3 & x^{(2)}(-1, -1, -1) &= 3 \\ x^{(1)}(1, 1, -1) &= -1 & x^{(2)}(1, 1, -1) &= 3 \\ x^{(1)}(1, -1, -1) &= 3 & x^{(2)}(1, -1, -1) &= -1.\end{aligned} \tag{7.47}$$

---

[6] With no loss in generality, we choose the additive constant $d_0 = 0$. This assumption reduces the dimensionality of **d** and **B** to $2^n - 1$ and $2^n \times 2^n - 1$.

Then, using (7.47), the expansion of (7.43) can be put in the matrix form of (7.44), where

$$\mathbf{x}^{(1)} = \begin{bmatrix} -3 \\ 1 \\ -3 \\ 1 \\ -1 \\ 3 \\ -1 \\ 3 \end{bmatrix} \quad \mathbf{x}^{(2)} = \begin{bmatrix} -3 \\ 1 \\ 1 \\ -3 \\ -1 \\ 3 \\ 3 \\ -1 \end{bmatrix} \quad \mathbf{d} = \begin{bmatrix} d_1 \\ d_2 \\ d_3 \\ d_{12} \\ d_{13} \\ d_{23} \\ d_{123} \end{bmatrix} \quad (7.48)$$

and

$$\mathbf{B} = \begin{bmatrix} 1 & 1 & 1 & 1 & 1 & 1 & 1 \\ 1 & -1 & 1 & -1 & 1 & -1 & -1 \\ -1 & 1 & 1 & -1 & -1 & 1 & -1 \\ -1 & -1 & 1 & 1 & -1 & -1 & 1 \\ -1 & 1 & -1 & -1 & 1 & -1 & 1 \\ -1 & -1 & -1 & 1 & 1 & 1 & -1 \\ 1 & 1 & -1 & 1 & -1 & -1 & -1 \\ 1 & -1 & -1 & -1 & -1 & 1 & 1 \end{bmatrix}. \quad (7.49)$$

To solve for the elements of $\mathbf{d}^{(i)}$; $i = 1, 2$, we make use of (7.45). For example, using the second and third columns of $\mathbf{B}$ for $\mathbf{d}^{(1)}$ and the third and seventh columns of $\mathbf{B}$ for $\mathbf{d}^{(2)}$, we would have

$$d_2^{(1)} = \frac{1}{8}[1 \quad -1 \quad 1 \quad -1 \quad 1 \quad -1 \quad 1 \quad -1] \begin{bmatrix} -3 \\ 1 \\ -3 \\ 1 \\ -1 \\ 3 \\ -1 \\ 3 \end{bmatrix} = -2$$

(7.50a)

$$d_3^{(1)} = \frac{1}{8}[1 \quad 1 \quad 1 \quad 1 \quad -1 \quad -1 \quad -1 \quad -1] \begin{bmatrix} -3 \\ 1 \\ -3 \\ 1 \\ -1 \\ 3 \\ -1 \\ 3 \end{bmatrix} = -1$$

and

$$d^{(2)}_3 = \frac{1}{8} [\begin{array}{cccccccc} 1 & 1 & 1 & 1 & -1 & -1 & -1 & -1 \end{array}] \begin{bmatrix} -3 \\ 1 \\ 1 \\ -3 \\ -1 \\ 3 \\ 3 \\ -1 \end{bmatrix} = -1$$

(7.50b)

$$d^{(2)}_{123} = \frac{1}{8} [\begin{array}{cccccccc} 1 & -1 & -1 & 1 & 1 & -1 & -1 & 1 \end{array}] \begin{bmatrix} -3 \\ 1 \\ 1 \\ -3 \\ -1 \\ 3 \\ 3 \\ -1 \end{bmatrix} = -2.$$

Application of (7.45) to the remaining columns of **B** results in zero values for all other d's. Thus the trellis of Fig. 7.2(b) is represented by the relations

$$x^{(1)} = -2b_2 - b_3$$
$$x^{(2)} = -b_3 - 2b_1 b_2 b_3.$$
(7.51)

A simple implementation of (7.51) as a transmitter is illustrated in Fig. 7.11. Note that Fig. 7.11 represents the *combined* modulation/coding process without the necessity of separating it into its component parts (i.e., a trellis code followed by a rule for mapping and an AM modulator). □

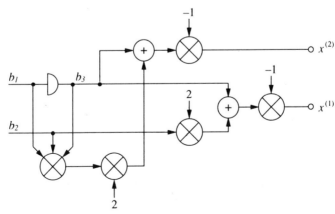

FIGURE 7.11 Transmitter implementation for rate 2/4 multiple trellis-coded 4-AM (two states); $k = 2$.

EXAMPLE 7.8

As a second example, consider the case of rate 2/4 multiple trellis-coded QPSK with a two-state trellis. Since once again $n = 3$, (7.46) holds for the input - output relationship of the trellis encoder. Here, however, the output symbol, x, represents phase instead of amplitude and the true encoder output, y, is given (in complex notation) by $y = e^{jx}$. Similarly, the trellis diagram of Fig. 7.2(b) is appropriate with the signal constellation of Fig. 7.2(a) (with $\phi = 0$). In particular,

$$\begin{aligned}
x^{(1)}(1, 1, 1) &= -\tfrac{\pi}{4} & x^{(2)}(1, 1, 1) &= -\tfrac{\pi}{4} \\
x^{(1)}(1, -1, 1) &= \tfrac{3\pi}{4} & x^{(2)}(1, -1, 1) &= \tfrac{3\pi}{4} \\
x^{(1)}(-1, 1, 1) &= -\tfrac{\pi}{4} & x^{(2)}(-1, 1, 1) &= \tfrac{3\pi}{4} \\
x^{(1)}(-1, -1, 1) &= \tfrac{3\pi}{4} & x^{(2)}(-1, -1, 1) &= -\tfrac{\pi}{4} \\
x^{(1)}(-1, 1, -1) &= \tfrac{\pi}{4} & x^{(2)}(-1, 1, -1) &= \tfrac{\pi}{4} & (7.52) \\
x^{(1)}(-1, -1, -1) &= -\tfrac{3\pi}{4} & x^{(2)}(-1, -1, -1) &= -\tfrac{3\pi}{4} \\
x^{(1)}(1, 1, -1) &= \tfrac{\pi}{4} & x^{(2)}(1, 1, -1) &= -\tfrac{3\pi}{4} \\
x^{(1)}(1, -1, -1) &= -\tfrac{3\pi}{4} & x^{(2)}(1, -1, -1) &= \tfrac{\pi}{4}.
\end{aligned}$$

Putting (7.49) in vector form as in (7.48), then using (7.49) and solving for $d^{(i)}$, $i = 1, 2$ from (7.45) gives the desired result:

$$\begin{aligned}
x^{(1)} &= \tfrac{\pi}{4}(b_3 - 2b_2 b_3) \\
x^{(2)} &= \tfrac{\pi}{4}(b_3 - 2b_1 b_2).
\end{aligned} \quad (7.53)$$

Figure 7.12 is an implementation of (7.53), where the phase modulator is used to convert x to y in accordance with the relation given above. □

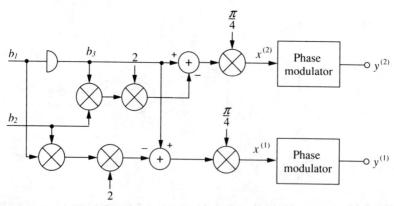

FIGURE 7.12 Transmitter implementation for rate 2/4 multiple trellis-coded QPSK (two states); $k = 2$.

7.4 Bit Error Probability Performance

Thus far our entire discussion has focused on performance gain as measured by improvement in minimum free distance of the trellis code. In the limit as the bit error probability becomes arbitrarily small, this measure is equivalent to the improvement in required bit energy-to-noise spectral density ratio. From a more practical standpoint, one is often interested in the reduction of bit energy-to-noise spectral density ratio for a given bit error probability. Using superstate diagrams and upper bounds on bit error probability computed from the transfer functions of these diagrams, we shall now determine the amount of the above-mentioned performance gains for MTCM.

Analogous to the analytical techniques discussed in Chapter 4 for upper bounding the bit error probability performance of conventional trellis codes, an upper bound on the bit error probability performance of MTCM is given by [1–3]

$$P_b \le \frac{1}{2b}\text{erfc}\left(\sqrt{\frac{bE_b}{kN_0}\frac{d^2_{\text{free}}}{4}}\right) Z^{-d^2_{\text{free}}} \frac{\partial}{\partial I} T(D, I)\Big|_{I=1, D=Z} \quad (7.54)$$

where Z is the Bhattacharyya parameter defined by

$$Z = \exp\left(-\frac{1}{4}\frac{E_s}{N_0}\right) = \exp\left(-\frac{1}{4}\frac{bE_b}{kN_0}\right) \quad (7.55)$$

with E_s the energy per trellis code symbol and $T(D, I)$ the transfer function of the superstate diagram associated with the multiple trellis code.

As an example consider the rate 2/4 multiple trellis-coded QPSK system with trellis diagram as in Fig. 7.2(b). The corresponding state diagram is illustrated in Fig. 7.13 and the equivalent superstate diagram for computing $T(D, I)$ is shown in Fig. 7.14. In Fig. 7.13 the branches are labeled with the input bit and output QPSK symbol pairs that cause that particular transition, whereas in Fig. 7.14 the branches are labeled with a gain of the form

$$G = \sum \frac{1}{b} I^\Omega D^{\delta^2}. \quad (7.56)$$

Here I is an index, Ω is the Hamming distance between input bit sequences, and δ^2 is the total squared Euclidean distance between the k M-PSK output sym-

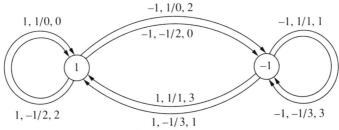

FIGURE 7.13 State diagram for rate 2/4 multiple ($k = 2$) trellis-coded QPSK

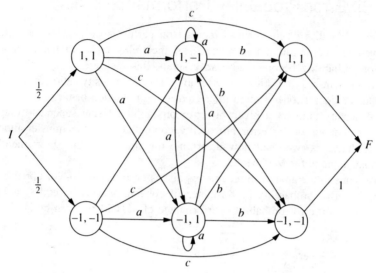

FIGURE 7.14 Superstate diagram corresponding to Fig. 7.13.

bols corresponding to the transition between the superstates. Also, the summation in (7.56) accounts for the possible existence of parallel paths in the trellis diagram.

The transfer function of Fig. 7.14 is easily computed as

$$T(D, I) = \frac{(2I + 2I^2 + I^3)D^8 - (I^2 + I^3)D^{12}}{1 - (I + I^2)D^4} \quad (7.57)$$

where, in accordance with (7.56),

$$a = \tfrac{1}{2}(I + I^2)D^4$$

$$b = \tfrac{1}{2}(1 + I)D^4 \quad (7.58)$$

$$c = \tfrac{1}{2}ID^8.$$

Substituting (7.57) into (7.54) and performing the differentiation required in (7.54) yields the desired upper bound on P_b:

$$P_b \leq \frac{1}{4}\,\text{erfc}\left(\sqrt{\frac{2E_b}{N_0}}\right) \frac{9 - 8Z^4 + 4Z^8}{9(1 - 2Z^4)^2}. \quad (7.59)$$

Figure 7.15 is an illustration of the upper bound of (7.59). Also shown in this figure, for purpose of comparison, are the upper bounds on P_b for uncoded PSK and conventional ($k = 1$) rate 1/2 trellis-coded symmetric and optimum asymmetric QPSK modulations. These results are obtained originally from [5] in accordance

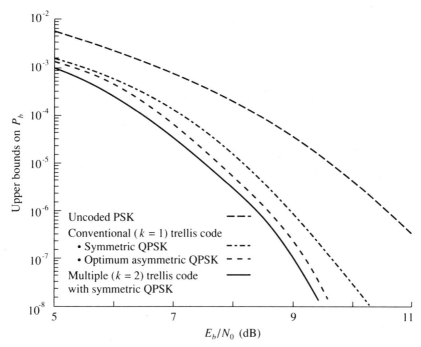

FIGURE 7.15 Comparison of the performance of several rate 1/2 trellis-coded QPSK modulations.

with the relations

$$P_b = \frac{1}{2} \operatorname{erfc} \sqrt{\frac{E_b}{N_0}} \quad \text{(uncoded PSK)}$$

$$P_b \leq \frac{\frac{1}{2} \operatorname{erfc} \sqrt{\frac{3E_b}{2N_0}}}{\left(1 - \exp(-E_b/2N_0)\right)^2} \quad \text{(coded symmetric)}$$

$$P_b \leq \frac{\frac{1}{2} \operatorname{erfc} \sqrt{\frac{(1+2\alpha)E_b}{(1+\alpha)N_0}}}{\left(1 - \exp\left\{-E_b/[(1+\alpha)N_0]\right\}\right)^2} \quad \text{(coded optimum asymmetric)}$$

(7.60)

with[7]

$$\alpha = \frac{E_b/N_0}{\ln 3} - 1. \tag{7.61}$$

[7] The value of asymmetry in (7.61) is exactly optimum for a slightly looser upper bound (see [5]) but only approximately optimum for (7.60).

We observe from these results that, over the range of E_b/N_0 illustrated, the multiple trellis scheme is slightly better in performance than the conventional trellis code with optimum asymmetry.

7.5 Computational Cutoff Rate and MTCM Performance

As discussed in Chapter 1, the computational cutoff rate, R_0, is dependent only on the coding channel and not on the coding scheme. In particular, for M-PSK modulation and discrete memoryless channels, R_0 is given by

$$R_0 = \log_2 M - \log_2 \left[1 + \sum_{i=1}^{M-1} Z^{4\sin^2 i\pi/M} \right] \tag{7.62}$$

where Z is again the Bhattacharyya parameter defined in (7.55) as a function of E_s/N_0. Here E_s represents the energy of an M-ary channel symbol. Figure 7.16 is a plot of R_0 versus E_s/N_0 (in dB) for various values of M.

The quantity R_0 represents the maximum throughput (in bits/s/Hz) achievable with a practical modulation/coding scheme. To compare the real performance of such a practical scheme with R_0, one must determine the value of required E_s/N_0 to achieve a given small error rate. This value of E_s/N_0 (abscissa) coupled with the actual throughput (ordinate) of the modulation/coding scheme (e.g., $r = b/k$ for MTCM) then gives a point that can be superimposed on the family of curves in Fig. 7.16. Clearly, the closer this point is to one of the R_0 versus E_s/N_0 curves, the more efficient is the combined modulation/coding scheme.

In (7.54) we presented an upper bound on the bit error probability of MTCM. A similar upper bound on first error event probability P_e [6] is given by

$$P_e \leq \frac{1}{2} \operatorname{erfc}\left(\sqrt{\frac{bE_b\, d_{\text{free}}^2}{kN_0\; 4}} \right) Z^{-d_{\text{free}}^2} T(Z) \tag{7.63}$$

where, for simplicity of notation, we denote $T(D, I)$ evaluated at $I = 1, D = Z$ by $T(Z)$. Both (7.54) and (7.63) require determining the transfer function of the superstate diagram associated with the trellis.

For trellises with a large number of states, the foregoing process can become analytically cumbersome. Instead, we consider an approximate (asymptotically approached at high SNR) lower bound, which for first error event probability is given by

$$P_e \geq \frac{1}{2} \operatorname{erfc}\left(\sqrt{\frac{bE_b\, d_{\text{free}}^2}{kN_0\; 4}} \right) N(d_{\text{free}}) = \frac{1}{2} \operatorname{erfc}\left(\sqrt{\frac{E_s\, d_{\text{free}}^2}{N_0\; 4}} \right) N(d_{\text{free}}) \tag{7.64}$$

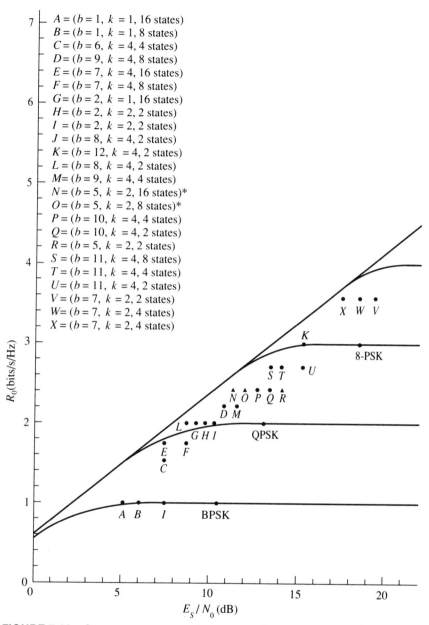

FIGURE 7.16 Comparison of computational cutoff rate of M-PSK with throughput performance of trellis-coded M-PSK. Points (dots and triangles) correspond to $P_e = 10^{-6}$.* See S. G. Wilson, "Rate 5/6 trellis-coded 8-PSK," *IEEE Trans. Commun.*, Vol. COM-34, No. 10, pp.104–1049, Oct. 1986.

where $N(d_{\text{free}})$ is the number of error event paths at distance d_{free} from the all-zeros path. In effect, (7.64) represents the result that would be obtained from (7.63) by keeping only the first term in the power series expansion of $T(D)$. Furthermore, (7.64) can be simplified (at a slight expense in tightness) by ignoring $N(d_{\text{free}})$. Thus, for a given P_e, we can readily compute the required E_s/N_0 for any particular MTCM scheme with given values of b, k, and M once we determine its d_{free}.

Superimposed on the curves of R_0 versus E_s/N_0 in Fig. 7.16 are points corresponding to the examples of Section 7.2.2 as well as many others. The abscissa of each point is the required E_s/N_0 to achieve the upper bound on P_e [ignoring the factor $N(d_{\text{free}})$] equal to 10^{-6}. The ordinates are obtained by evaluating $r = b/k$ for each case at hand. Also superimposed are points corresponding to uncoded BPSK, QPSK, and 8-PSK.

7.6 Complexity Considerations

In this section, we discuss the complexity of multiple trellis codes designed for the fading channel using a definition of complexity given by Ungerboeck [7] for multidimensional trellis codes.

From the diagram of the multiple trellis-encoded M-PSK transmitter (Fig. 7.1), the number of paths that emanate from a given node is 2^b. If 2^b exceeds the number of states, 2^v, then depending on the degree of connectivity, there may or may not be parallel paths in the trellis. Parallel paths in the trellis represent transitions from a given state to the *identical* state and thus insofar as the implementation of the encoder is concerned, represent input bits that are not encoded. Paths (transitions) from one state to another in the trellis represent input bits that are indeed encoded. In defining *normalized complexity* (i.e., complexity per two-dimensional signal), Ungerboeck [7, part II, Fig. 1] distinguishes between input bits that pass through the encoder and those that do not. In particular, only those input bits that are encoded enter into his trellis complexity definition. Thus the foregoing demarkation between parallel paths and distinct transitions between different states in the trellis is an important consideration when examining the complexity of a trellis-encoded scheme.

In our terminology, let $\tilde{b}$ represent the portion of the total number of input bits b that are indeed encoded; equivalently, $2^{\tilde{b}}$ is the number of paths leaving a given node in the trellis, *excluding parallel paths*. Then, from an implementation point of view, the trellis encoder (Fig. 7.1) with b input bits and s output symbols will have a structure wherein $b - \tilde{b}$ bits pass directly through (i.e., result in $b - \tilde{b}$ identical output symbols) and $\tilde{b}$ bits pass through an encoder resulting in $s - (b - \tilde{b})$ coded output symbols. Thus the *actual* encoder (contained within the box labeled "trellis encoder") would have a rate $\tilde{b}/[s - (b - \tilde{b})]$.

Following Ungerboeck [7, part II, p. 18], for a multiple trellis-coded M-PSK modulation of multiplicity k, the *normalized complexity (per M-PSK two-dimensional signal)* is given by $C = 2^{\tilde{b}+v}/k$. Table 7.4 evaluates this complexity for the various trellis-code designs given in Section 7.2.2.

TABLE 7.4 Complexity of Various Multiple Trellis Codes.

b	k	b/k	b	Number of Trellis States	Normalized Complexity, C
Example 7.3					
3	2	1.5	3	8	32
3	2	1.5	3	16	64
Example 7.4					
3	2	1.5	2	4	8
3	2	1.5	2	8	16
Example 7.5					
7	4	1.75	3	8	16
7	4	1.75	3	16	32
Example 7.6					
6	4	1.5	2	4	4

7.7 Concluding Remarks

In the introduction to this chapter, we alluded to the fact that in certain situations there might be reason to design a generalized MTCM with different M-ary alphabet sizes for the transmitted symbols. One practical case where such a design might be desirable corresponds to a coherent receiver with a noisy carrier demodulation reference signal, in which case the receiver performance is degraded by carrier phase jitter. This is explained as follows. It is well known that for uncoded M-PSK, the sensitivity of the receiver performance to phase jitter increases with the number of phases, M.[8] For example, an MTCM system with multiplicity 2, whose output symbols alternate between, say, QPSK and 8-PSK would allow the carrier synchronization information to be obtained from only the QPSK symbols (which are less sensitive to phase jitter). An example of such might be a rate 4/5 trellis code, where in each transmission interval two of the five coded bits are assigned to a QPSK symbol and the remaining three to an 8-PSK symbol. While the throughput of this system is 2, its performance in the absence of phase jitter will be inferior to, say, a rate 2/3 coded 8-PSK system with the same number of states, which has a single M-ary signal set for its output symbols and also achieves a throughput equal to 2. However, *in the presence of phase jitter*, the rate 4/5 coded hybrid QPSK/8-PSK system could conceivably outperform the rate 2/3 coded 8-PSK one. This is an area that requires further investigation although some work along these lines has been recently reported in [8, 9].

---

[8] The fact that this is true is obvious when one realizes that for a given power, the signal points in the constellation are closer together as M gets larger, which for a given phase perturbation is more likely to cause an error in detection.

REFERENCES

1. D. DIVSALAR and M. K. SIMON, "Multiple trellis-coded modulation (MTCM)," *JPL Publication 86-44*, Pasadena, Calif., Nov. 15, 1986.
2. D. DIVSALAR and M. K. SIMON, "Multiple trellis-coded modulation (MTCM)," *IEEE Trans. Commun.*, Vol. 36, No. 4, pp. 410–419, Apr. 1988. See also *Globecom'86 Conf. Rec.*, Houston, Tex., pp. 30.8.1–30.8.7, Dec. 1986.
3. D. DIVSALAR and M. K. SIMON, "Generalized multiple trellis-coded modulation," *ICC'87 Conf. Rec.*, Seattle, Wash., pp. 20.3.1–20.3.7, June 7–10, 1987.
4. R. CALDERBANK and J. E. MAZO, "A new description of trellis codes," *IEEE Trans. Inf. Theory*, Vol. IT-30, No. 6, pp. 784–791, Nov. 1984.
5. D. DIVSALAR, M. K. SIMON, and J. H. YUEN, "Trellis coding with asymmetric modulations," *IEEE Trans. Commun.*, Vol. COM-35, No. 2, pp. 130–141, Feb. 1987.
6. A. J. VITERBI and J. K. OMURA, *Principles of Digital Communication and Coding*. McGraw-Hill, New York, N.Y., 1979.
7. G. UNGERBOECK, "Trellis-coded modulation with redundant signal sets. Part I: Introduction; Part II: State of the art," *IEEE Commun. Mag.*, Vol. 25, No. 2, pp. 5–21, Feb. 1987.
8. J. HAGENAUER and C.-E. W. SUNDBERG, "On hybrid trellis-coded 8/4-PSK modulation systems," *ICC'89 Conf. Rec.*, Boston, Mass., pp. 18.4.1–18.4.5, June 11–14, 1989.
9. M. BERTELSMEIER and G. KOMP, "Trellis-coded 8PSK with embedded QPSK," *IEEE J. Select Areas Commun.*, Vol. 7, No. 9, pp. 1296–1306, Dec. 1989.

CHAPTER 8

Rotationally Invariant Trellis Codes

8.1 Introduction

The need for rotationally invariant trellis codes occurs in coherent detection using suppressed carrier modulation. It is sufficient to view the received signal as

$$r(t) = A \cos(\omega_c t + \theta_m(t) + \theta) \tag{8.1}$$

where $\theta_m(t)$ is the modulation phase due to trellis coding, ω_c is the carrier frequency in radians per second, and θ is the offset angle that must be estimated for coherent detection. A block diagram of a coherent demodulator is shown in Fig. 8.1. There is a requirement to estimate ω_c and θ given the signal in (8.1). These estimates are denoted as $\hat{\omega}_c$ and $\hat{\theta}$, respectively. Most techniques for estimating ω_c and θ require that the modulation phase be removed. Once the modulation is removed, the estimator for ω_c and θ can base its estimate on an unmodulated carrier. To consider a simple example, let the modulation be 4-PSK. The signal constellation is thus four phases on a circle with the spacing between adjacent phases set at 90°. Consider the carrier recovery system shown in Fig. 8.2. The modulation is passed through a fourth-power device, and the fourth harmonic of the output is passed to a phase-locked-loop circuit. The operation of such systems, including the voltage-controlled oscillator (VCO), is given in [1]. It is clear that $\hat{\theta}$ in Fig. 8.2 could be equal to $\theta + \theta_{mi}$, where θ_{mi} is any one of the four modulation phases for 4-PSK. We see from what has been stated above that even though the transmission started at, say, a modulation phase angle of 0°, the receiver could start in any of the other three modulation angles, $\pm\pi/2$ and π. This is for 4-PSK and we say that the carrier recovery system has a phase ambiguity of 90°. In general, for an M-PSK system the phase ambiguity is $360/M$ in degrees. It is clear that this ambiguity must be resolved if we are to use the trellis-coded carrier modulations presented to date.

To compensate for a phase ambiguity in the receiver, there are essentially two approaches. We can estimate the phase ambiguity by sending a fixed sequence of modulation phases to initialize data communication. On the other hand, we can design trellis codes that are transparent to phase offsets at multiples of the smallest difference between two modulation angles in the signal constellation. The latter approach is the subject of this chapter.

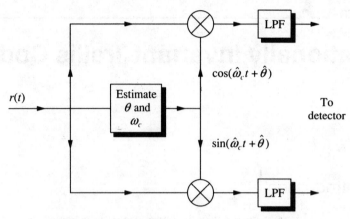

FIGURE 8.1 Structure of a coherent demodulator.

8.1.1 Rotational Invariance

Up to now, information has been encoded in an absolute phase modulation format. In rotationally invariant trellis codes, or indeed in any discrete phase modulation system that requires rotational invariance, the information is transmitted in terms of phase differences. To illustrate this concept, consider the encoding rule

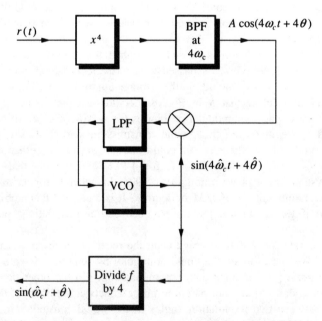

FIGURE 8.2 Structure of a carrier recovery technique for 4-PSK modulation.

for 2-PSK:

$$\beta_k = \beta_{k-1} \oplus \alpha_k \tag{8.2}$$

where $\alpha_k \in \{0, 1\}$ is the source bit, $k \geq 1$ is discrete time, and we set $\beta_0 = 1$. The bits β_k enter the phase modulator and determine the modulation phase for the kth symbol interval. A rule like (8.2) is known as differential encoding.

From equation (8.2) we see that $\alpha_k = 0$ implies that $\beta_k = \beta_{k-1}$ or no phase change from symbol interval $k - 1$ to symbol interval k. Conversely, $\alpha_k = 1$ implies that the phase does change over these two intervals. Thus we see that the rule in (8.2) results in the source bits being transmitted in phase changes, not in absolute phase values.

We have not yet explained, however, why such a differential modulation approach is needed. Recall that for 2-PSK modulation the phase ambiguity will be 180°. Thus upon detection, but with no differential encoding, our decisions will be either all correct or all wrong. This is because the receiver is uncertain of whether the initial phase of the transmitter was 0 or 180°. Thus, if (011010) was the transmitted word and a 180° ambiguity occurred, we decode (100101).

Let us invert equation (8.2) to derive the differential decoding rule:

$$\hat{\alpha}_k = \hat{\beta}_k \oplus \hat{\beta}_{k-1} \tag{8.3}$$

where $\hat{\beta}_k$ is the estimate of the transmitted bit. We further assume that the sequence of α's, (011010), is encoded via (8.2) to get the sequence of β's (1101100). Note that $\beta_0 = 1$ and $\beta_k = \beta_{k-1}$ if $\alpha_k = 0$ and $\beta_k \neq \beta_{k-1}$ if $\alpha_k = 1$. If we set $\hat{\beta}_k = \beta_k$ and apply this sequence of $\hat{\beta}$'s to (8.3), we get the correct α sequence. Now invert all the $\hat{\beta}$'s due to a 180° phase ambiguity to get (0010011). Use of (8.3) still gives the correct α sequence because a 180° phase ambiguity (namely, complementing the received bits) has no effect on the rule in (8.3) since information was sent in phase differences. If the information is sent in differential form, complementing it has no effect on the result. We say that such a scheme is rotationally invariant to 180°.

8.1.2 Rotational Invariant Code

Let us assume that we are considering the 8-PSK signal constellation in Fig. 8.3 and a rate 2/3 code. That is, the uncoded signal constellation is 4-PSK. We seek to design the trellis code to be invariant to 90°. We would prefer it to be invariant to 45° because this is the minimum tolerable phase of rotation. Later, using multidimensional trellis codes we will show how this can be done. For now, let us consider the 8-PSK case that we require to be invariant to a phase of 90°.

The general structure of a trellis encoder is shown in Fig. 8.4. We have inputs, state variables, and outputs—in other words the general structure of a dynamic system. *If the inputs, encoder states, and outputs are invariant under a 90° rotation, the code must be so-invariant, because every entity used to define the code is*

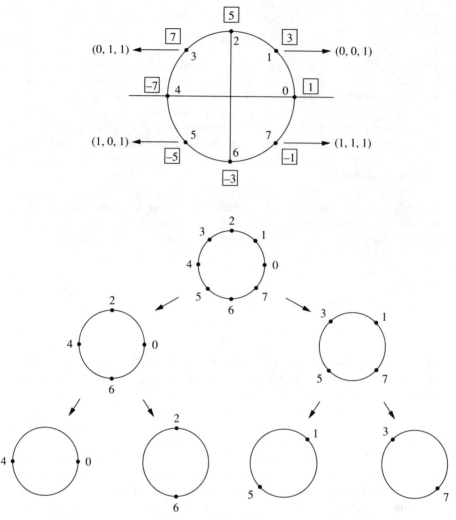

FIGURE 8.3 8-PSK signal set with natural mapper plus set partitioning.

so-invariant. This is the central theme of rotationally invariant trellis coding and forms the basis for this chapter.

Trellis-coded modulation was invented [2–4] to provide coding gain for voice-band modems. Redundant block coding was a competitor and was simply made rotationally invariant. Many thought that doing so would be difficult for trellis-

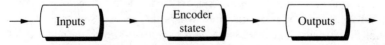

FIGURE 8.4 General structure of trellis encoder.

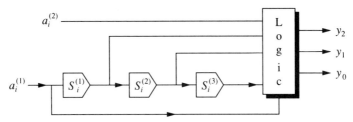

FIGURE 8.5 Eight-state, 8-PSK coder.

coded modulation. An elegant solution was provided by Wei [5, 6] and was put on a simpler basis in [7]. The approach given here closely follows [7].

In keeping with our central theme we must define a rotation rule for input bits that satisfy the 90°, say, phase rotation requirement. We are considering a rate 2/3 case with two input bits. The code structure is shown in Fig. 8.5. In this figure, a_i are the outputs of the differential encoder, whereas we denote the source bits as I_i. This structure is depicted in Fig. 8.6. The rotation sequence for two input bits is shown in Table 8.1.

In Table 8.2 we show the effect of $a_i^{(1)}$ alone because it is the only bit with any influence on the states $S_i^{(1)}$, $S_i^{(2)}$, and $S_i^{(3)}$ in Fig. 8.5; a_i will be the rotated version of the input bits $I_i^{(1)}$ and $I_i^{(2)}$.

The results in Table 8.2 can easily be represented in an expanded form of the differential encoding rule in (8.2). Let $I_i^{(2)} I_i^{(1)}$, $a_{i-1}^{(2)} a_{i-1}^{(1)}$, and $a_i^{(2)} a_i^{(1)}$ all be treated as symbols instead of bit pairs. Denoting these symbols as I_i, B_{i-1}, and B_i, respectively, each contained in the set $\{0, 1, 2, 3\}$, we have

$$B_i = (B_{i-1} + I_i) \bmod 4. \tag{8.4}$$

The logical equivalent is

$$a_i^{(1)} = I_i^{(1)} \oplus a_{i-1}^{(1)} \tag{8.5}$$

$$a_i^{(2)} = I_i^{(2)} \oplus a_{i-1}^{(2)} \oplus I_i^{(1)} \odot a_{i-1}^{(1)} \tag{8.6}$$

where now $a_i^{(\ )}$, $I_i^{(\ )} \in [0, 1]$. These equations define the differential precoder that is required for the general trellis-code structure shown in Fig. 8.6. The differential precoder is needed to supply the rotation to the inputs. For our example the inputs to the (two-dimensional) precoder satisfy (8.5) and (8.6); then the outputs from these equations are the inputs $\left(a_i^{(1)}, a_i^{(2)}\right)$ to the trellis-code structure in Fig. 8.5. This completes our investigation of the rotation of inputs in our general structure of Fig. 8.4.

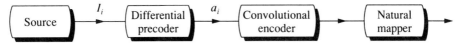

FIGURE 8.6 General structure of rotationally invariant trellis coder.

TABLE 8.1 Effects of Phase Rotation on Differentially Coded Vector at the Input of the Convolutional Encoder.

		\multicolumn{4}{c}{Phase Rotation (rad)}

$a_i^{(2)}$	$a_i^{(1)}$	0	$\frac{\pi}{2}$	π	$\frac{3\pi}{2}$
0	0	00	01	10	11
0	1	01	10	11	00
1	0	10	11	00	01
1	1	11	00	01	10

The next item that must be discussed is the rotation of code states. We have an eight-state code with the trellis structure given in Fig. 8.7. Actually, the results of our design are given there and we will proceed with the derivation of that structure. Two entities identify the states: the natural number value of $(S_i^{(1)}, S_i^{(2)}, S_i^{(3)})$ and the signals to be used in the state transitions. For now we use the former.

Recall from Fig. 8.5 that only $a_i^{(1)}$ has any influence on trellis state rotation. The effect on $a_i^{(1)}$ of a rotation is given in Table 8.2. Under $\pm\pi/2$ we have a complementation and otherwise no change in $a_i^{(1)}$. When $a_i^{(1)}$ is complemented, every state in Fig. 8.5 will eventually be complemented. These are the "would have been" states under rotation. The complete effects of these rotations are given in Table 8.3. For instance, $(S_i^{(3)}, S_i^{(2)}, S_i^{(1)}) = 000 \to 111$ under a $\pm\pi/2$ rotation due to the complementation of $a_i^{(1)}$. This table completes our study of the effects on trellis-code states under a 90° rotation.

The final consideration is the 90° rotational requirement for the output. The outputs in Fig. 8.5 are (y_0, y_1, y_2) and the mapping rule to 8-PSK symbols is given in Fig. 8.3. The Ungerboeck description of the code is via the trellis in Fig. 8.7. We will now derive the symbol assignments given there for each state. We have four transitions per state and we are using one parallel transition as $a_i^{(2)}$ in Fig. 8.5 is passed directly to the mapper. The assignment in $(a_i^{(2)}, a_i^{(1)})$ in Fig. 8.7 satisfies a 180° invariance on parallel transitions and a 90° invariance or split/merge transitions. This is because the transition $000 \to 001$ must be 90° away from $000 \to 000$. Thus the four input signals are $(0, 4)$ and then $(2, 6)$ because the latter set is 90° from the former

TABLE 8.2 Effects of Phase Rotation on $a_i^{(1)}$ Alone.

$a_i^{(1)}$	\multicolumn{4}{c}{Phase Rotation (rad)}

$a_i^{(1)}$	0	$\frac{\pi}{2}$	π	$\frac{3\pi}{2}$
0	0	1	0	1
1	1	0	1	0

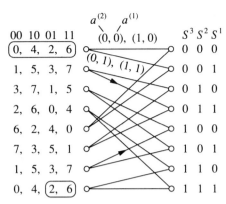

FIGURE 8.7 RIC trellis for eight-state, 8-PSK.

as follows from Fig. 8.3. This is essentially Wei's design rule [5, 6] for the outputs.

Now consider the level transition, 000 → 000, in the trellis diagram in Fig. 8.7. By Table 8.3, 111 is ±90° away from 000. Thus the last transition in Fig. 8.7 is ±90° away from the first transition in this figure. Thus on this last transition we assign the signals, (2, 6), because from Fig. 8.3 this set is ±90° away from the set (0, 4) used on the transition, 000 → 000. This is the rule to follow and the rest of the trellis description in Fig. 8.7 follows by using the same procedure.

To date, the encoded signals have been assigned. Let us assign two more just to make sure that the principle is clear. Consider the transition 001 → 011. From Table 8.3, a 90° rotation of 001 becomes 110 and under the same rotation 011 becomes 100. Thus the transition 110 → 100 is 90° away from 001 → 011. On 001 → 011 we use the signals (3, 7). This is because (0, 4, 2, 6) is used in state 1 and (1, 5, 3, 7) in state 2, in keeping with the Ungerboeck design rule U2 from Chapter 5. Thus on the transition 110 → 100 we use (1, 5) as it is −90°

TABLE 8.3 Effects of Phase Rotation on the State Vector.

			Phase Rotation (rad)			
$S_i^{(3)}$	$S_i^{(2)}$	$S_i^{(1)}$	0	$\frac{\pi}{2}$	π	$\frac{3\pi}{2}$
0	0	0	000	111	000	111
0	0	1	001	110	001	110
0	1	0	010	101	010	101
0	1	1	011	100	011	100
1	0	0	100	011	100	011
1	0	1	101	010	101	010
1	1	0	110	001	110	001
1	1	1	111	000	111	000

away from (3, 7) and $-90°$ is used because the input bits for (3, 7) are (01) and (11) and the input bits for the required transition are (00) and (01), which are $-90°$ away from (01) and (11).

One may wonder why (3, 7, 1, 5) defines the signals used on state 3 and not like (2, 6, 0, 4) in the optimum, or nonrotationally invariant 8-PSK code covered in earlier chapters. From Table 8.3, and the work just completed, we see a 90° rotational invariance has required that signals (1, 5) be used for the transition $110 \to 100$. Thus, by Ungerboeck's rule U2 of Chapter 5, signals (3, 7) should be used for the transition $010 \to 100$.

Repeated application of the ideas presented above gives the trellis in Fig. 8.7. If the transmitted sequence is rotated by ± 90 or ± 180 we still get a valid code sequence as long as we differentially decode the output decisions. This is just the inverse of the (two-dimensional) differential encoding in (8.4). The outputs are $\hat{a}_i$ and thus from (8.4),

$$\hat{I}_i = (\hat{B}_i + \hat{B}_{i-1}) \bmod 4. \tag{8.7}$$

If we compare (8.4) and (8.7), we see that (8.4) becomes (8.7) under the replacements, $a_i \to \hat{I}_i$, $I_i \to \hat{a}_i$, and $a_{i-1} \to \hat{a}_{i-1}$. Thus for the logical equivalent of (8.7), using these replacements in (8.5) and (8.6) we have

$$\hat{I}_i^{(1)} = \hat{a}_i^{(1)} \oplus \hat{a}_{i-1}^{(1)} \tag{8.8}$$

$$\hat{I}_i^{(2)} = \hat{a}_i^{(2)} \oplus \hat{a}_{i-1}^{(2)} \oplus \hat{a}_i^{(1)} \odot \hat{a}_{i-1}^{(1)} \tag{8.9}$$

which is our differential decoding rule.

We now summarize the example just considered by some general design rules. These are taken from [8].

8.1.3 Design Rules

There are two sets of rules to follow, according to the coder's functions. There is the convolutional encoder plus mapping rule, the design of which should follow Ungerboeck's approach [2–4] to obtain the optimum code. That is, wherever possible we should use rules U1 to U3 of Chapter 5. Herein the natural binary mapping rule is used. There are also some rotational invariance conditions due to Wei [5, 6] that must be satisfied. In some cases these two sets of rules are in conflict with one another. Therefore, the performance of the optimum codes is not always obtainable as a rotationally invariant code (RIC).

The design rules for optimum codes have been stated earlier and will not be repeated unless compromising modifications are made. The conditions of rotational invariance are as follows:

1. The magnitude of the smallest tolerable phase of rotation (STPR) depends on the number of input bits in Fig. 8.6 being differentially encoded. If f is the number of input bits to be differentially encoded, then the tolerable phase of rotation (TPR) is any multiple of $2\pi/(2^f)$ radians, provided that the RIC and the signal constellation described in rule 2 can be found.

2. For rate $m/(m + 1)$, q-state codes, let f be the number of input lines being differentially encoded, $f \leq m$; also the following assumptions are made:
   (a) $(I_n^{(1)}, I_n^{(2)}, \ldots, I_n^{(f)})$ is the current f-tuple being differentially encoded at the input of the convolutional encoder, $I_n^{(1)}$ being the least significant bit.
   (b) $S_n^{(1)}, S_n^{(2)}, \ldots, S_n^{(P)}$ is the current state vector determined by the previous input bits, P is a positive integer such that $2^P = q$.
   (c) $Q_{n,j}^{(1)}, Q_{n,j}^{(2)}, \ldots, Q_{n,j}^{(P)}$ is the state vector that results if $I_n^{(1)}, I_n^{(2)}, \ldots, I_n^{(f)}$ is rotated by j positions, $j = 1, 2, \ldots, 2^f - 1$.
   (d) U is the set of signal elements associated with the transitions from the current state $S_{n+1}^{(1)}, S_{n+1}^{(2)}, \ldots, S_{n+1}^{(P)}$.
   (e) V_j is the set of signal elements that results when U is rotated by $2j\pi/2^f$ radians, $j = 1, 2, \ldots, 2^f - 1$.
   Then the second rule requires that the set of signal elements associated with transitions from state $Q_{n,j}^{(i)}, i = 1, 2, \ldots, P$, to state $Q_{n+1,j}^{(i)}, i = 1, 2, \ldots, P$, be V_j, $j = 1, 2, \ldots, 2^f - 1$. This rule is sometimes known as the *rule of equal rotation*.
3. All the signal elements in each set obtained by set partitioning should be associated with the convolutional encoder output bits. The bits form the lower portion (the least significant bits) of the output symbol and are denoted as y_i, $i = 0, 1, \ldots, k - 1$, $k \leq m - 1$.
4. Any set of signal elements that have the same radius but are $2j\pi/2^f$ radians apart have to be assigned the same combinations of encoder output bits that are not used in rule 3. These output bits normally form the upper portion (the most significant bits) of the output symbol and are denoted as y_i, $i = k, k + 1, \ldots, m$.

8.1.4 Design Procedure

1. First, examine the signal constellation (SC) and determine whether the sets described in rule 2 can be found by set partitioning the SC. The minimum STPR is given by rule 1 with $f = m$. Such SCs do not always exist. Some might be rotationally invariant to a larger STPR. f is then determined by the smallest value of STPR: for example, 8-AMPM has two input bits but the STPR of its SC is 180°, which corresponds to $f = 1$.
2. Map the SC according to rules 3 and 4.
3. Set partition the SC into 2^{m+1-k} subsets according to the rules specified in rule 4.
4. Set up a table that contains every possible combination of the input bits and its images under different degrees of phase rotation [STPR(j) = $0, 1, \ldots, 2^f - 1$].
5. From the table in step 4, the effect of phase rotation on each of the input bits is known. The state that would have been reached if these input bits had really been rotated can then be determined. The "would-have-been" state is called the *image of the current state under rotation*. There are $(2^f - 1)$

different rotational images for each state. Again, the effect of phase rotation on the state of the vector can be shown in a table.
6. The table in step 5 can be used to assign the sets of signal elements obtained in step 2 to the state transitions according to both rule 2 and the Ungerboeck rules U1 → U3 of Chapter 5. Failure to do this will degrade the performance of the code. The number of the encoded states can be varied so as to satisfy both sets of rules. To simplify the trellis diagram, take as many output bits as possible from the input lines directly without breaking the assignment rules. This rule is operative on the AWGN channel. It may not be optimum for other channels (e.g., fading channels).
7. If the physical convolutional encoder is desired, design it with the procedure to follow in Section 8.2.
8. Design a differential encoder that differentially encodes the f least significant input bits and thus outputs y_i, $i = 0, 1, \ldots, k-1$. The encoder is a modulo-(2^f) adder that adds its input to its previous output, symbol by symbol, to obtain the next output. The bit-by-bit equivalent can also be expressed with a truth table, or in an equation, which employs the concept of symbol rotation. Once the f-input differential encoder is found, it can be reused in the design of a new code that requires a rotational invariance to $2\pi/(2^f)$ radians.

The following are some examples that illustrate this design procedure.

8.1.5 16-Point Examples

EXAMPLE 8.1 16-PSK, EIGHT-STATE RIC WITH PARALLEL TRANSITIONS

Since three input lines can be differentially encoded, the code is rotationally invariant to phase changes of $\pm\pi/4$, $\pm\pi/2$, and π. The convolutional encoder encodes the least significant bit only. The two most significant bits are Gray-coded. The structure of the encoder is shown in Fig. 8.8. Since there are four parallel

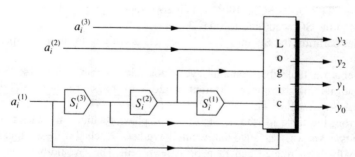

FIGURE 8.8 Code structure of 16-PSK RIC.

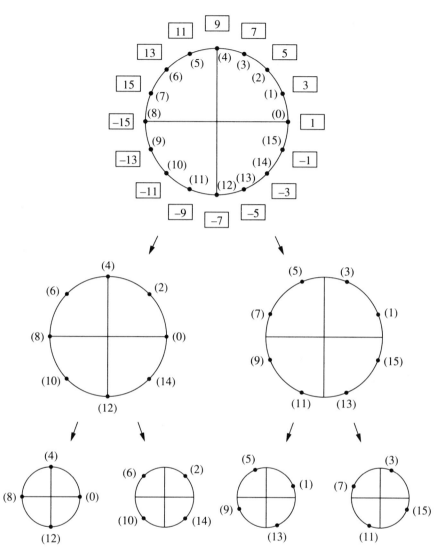

FIGURE 8.9 Set partitioning for 16-PSK.

transitions, the signal set is to be partitioned into four subsets and this is illustrated in Fig. 8.9. That means two stages of set partitioning. Inputs to the convolutional encoder in Fig. 8.8 are shown in Table 8.4.

Using the wraparound scheme, one can get any image of rotation for any a_i symbol. Consider $a_i = 011$. A rotation of $3\pi/2$ (six positions down) will get to 001 as follows from the entries in Table 8.4. The state vector in Fig. 8.8 is determined by $a_i^{(1)}$ only. From Table 8.4 we get Table 8.5.

The state vector is only affected when the phase rotation is an odd multiple of $\pi/4$ radians. This produces the results given in Table 8.6. From Table 8.6, the

TABLE 8.4 Effects of Phase Rotation on the Differentially Coded Vector.

Phase Rotation	$a_i^{(3)}$	$a_i^{(2)}$	$a_i^{(1)} = 0\ 0\ 0$
0	0	0	0
$\dfrac{\pi}{4}$	0	0	1
$\dfrac{\pi}{2}$	0	1	0
$\dfrac{3\pi}{4}$	0	1	1
π	1	0	0
$\dfrac{5\pi}{4}$	1	0	1
$\dfrac{3\pi}{2}$	1	1	0
$\dfrac{7\pi}{4}$	1	1	1

trellis diagram can now be derived and is shown in Fig. 8.10. Each solid line represents four parallel transitions. This completes the specification of the trellis for the present example.

The differential precoder has to encode all input lines differentially so that the code can be made rotationally invariant to phase changes of any multiple of $\pi/4$ radians. Denoting $I_i^{(3)} I_i^{(2)} I_i^{(1)}$, $a_{i-1}^{(3)} a_{i-1}^{(2)} a_{i-1}^{(1)}$, and $a_i^{(3)} a_i^{(2)} a_i^{(1)}$ as the symbols I_i, a_{i-1}, and a_i, respectively, we have

$$B_i = (B_{i-1} + I_i) \bmod 8.$$

The logical equivalent is

$$a_i^{(1)} = I_i^{(1)} \oplus a_{i-1}^{(1)}$$

$$a_i^{(2)} = I_i^{(2)} \oplus a_{i-1}^{(2)} \oplus I_i^{(1)} \oplus a_{i-1}^{(1)}$$

$$a_i^{(3)} = I_i^{(3)} \oplus a_{i-1}^{(3)} \oplus \left(I_i^{(2)} \odot a_{i-1}^{(2)} \oplus (I_i^{(2)} \oplus a_{i-1}^{(2)}) \odot I_i^{(1)} \odot a_{i-1}^{(1)} \right)$$

TABLE 8.5 Effects of Phase Rotation on the $a_i^{(1)}$ Alone.

	Phase Rotation (rad)							
$a_i^{(1)}$	0	$\dfrac{\pi}{4}$	$\dfrac{\pi}{2}$	$\dfrac{3\pi}{4}$	π	$\dfrac{5\pi}{4}$	$\dfrac{3\pi}{2}$	$\dfrac{7\pi}{4}$
0	0	1	0	1	0	1	0	1
1	1	0	1	0	1	0	1	0

TABLE 8.6 Effects of Phase Rotation on the State Vector.

$S_i^{(3)}$	$S_i^{(2)}$	$S_i^{(1)}$	Even Multiples of $\frac{\pi}{4}$ rad			Odd Multiples of $\frac{\pi}{4}$		
0	0	0	0	0	0	1	1	1
0	0	1	0	0	1	1	1	0
0	1	0	0	1	0	1	0	1
0	1	1	0	1	1	1	0	0
1	0	0	1	0	0	0	1	1
1	0	1	1	0	1	0	1	0
1	1	0	1	1	0	0	0	1
1	1	1	1	1	1	0	0	0

which is not difficult to derive if we note that the mathematical relationship is only a simple bit-by-bit add with carry.

For example,

$$a_i^{(3)} = I_i^{(3)} \oplus a_{i-1}^{(3)} \oplus \left(I_i^{(2)} \odot a_{i-1}^{(2)} \oplus (I_i^{(2)} \oplus a_{i-1}^{(2)}) \odot I_i^{(1)} \odot a_{i-1}^{(1)}\right).$$

Note that if $I_i^{(2)}$ and $a_{i-1}^{(2)}$ are 1 in their first occurrence in this equation, there is a carry that *must be included in the third bit addition*. For the next occurrence of these bits a carry is generated if either one is a 1 and both $I_i^{(1)}$ and $a_{i-1}^{(1)}$ are 1, which in turn generates a carry for the second bit addition.

The asymptotic coding gain is the same as that of the optimum code. Ungerboeck's rules can be fully satisfied. The sequence 2, 1, 15, 2 in the trellis in Fig. 8.10 gives the shortest path length possible relative to the all-zero path. Thus the coding gain with respect to uncoded 8-PSK is

$$\gamma = \frac{0.586 + 0.152 + 0.152 + 0.586}{0.586}$$

$$= 2.52 \Rightarrow 4.01 \text{ dB.} \qquad \square$$

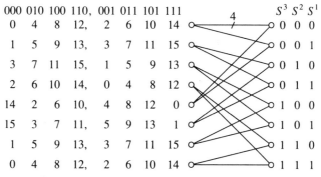

FIGURE 8.10 Trellis for eight-state, 16-PSK RIC.

EXAMPLE 8.2   16-QAM, EIGHT-STATE RIC

The signal constellation only allows a rotationally invariant code to tolerate phase changes of $\pm\pi/2$ radians and π radians. Set partitioning is shown in Fig. 8.11. The uncoded modulation is 8-AMPM whose signal points are also marked in Fig. 8.11. The convolutional encoder's front end is shown in Fig. 8.12.

The differential precoder is a two-input type. Hence the precoder given in Table 8.1 and equations (8.5) and (8.6) can be used in 16-QAM RIC. When the rotations in Table 8.1 are applied to the state variables shown in Fig. 8.12, the result is shown

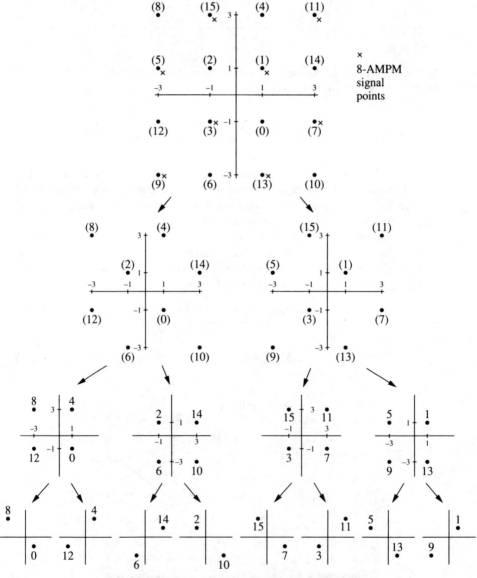

FIGURE 8.11 Set partitioning for 16-QAM.

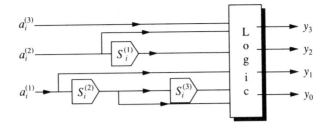

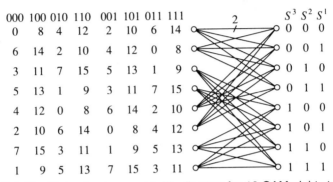

FIGURE 8.12 State structure and trellis diagram for 16-QAM eight-state RIC.

in tabular form in Table 8.7. There are two things to consider, set partitioning and the trellis diagram. Since there are two parallel transitions per state transition, the signal constellation has to be partitioned into eight subsets, each of two signal elements. The trellis diagram is shown in Fig. 8.12.

The coding gain is determined by the sequence of state transitions that is of the shortest path length. In this case, 272 is the "shortest sequence" found. The distance squared is 20, and thus the coding gain is as below, since the average

TABLE 8.7 Effects of Phase Rotation on the State Vector.

			Phase Rotation (rad)			
$S_i^{(3)}$	$S_i^{(2)}$	$S_i^{(1)}$	0	$\dfrac{\pi}{2}$	π	$\dfrac{3\pi}{2}$
0	0	0	000	110	001	111
0	0	1	001	111	000	110
0	1	0	010	101	011	100
0	1	1	011	100	010	101
1	0	0	100	010	101	011
1	0	1	101	011	100	010
1	1	0	110	001	111	000
1	1	1	111	000	110	001

energy is 10 for both 8-AMPM and 16-QAM signal constellations:

$$\gamma = \frac{20/10}{8/10} = 2.5.$$

The coding gain is 4.01 dB, the same as the optimum code. □

8.2 Generation: Rotationally Invariant Codes

In Chapter 5, we showed how to realize a trellis code with a convolutional encoder followed by a mapper. With rotationally invariant codes the convolutional encoder is usually nonlinear. Nonlinear convolutional encoders arise from the restrictions imposed by rotationally invariant codes. In [5] and [6] Wei described design rules to obtain rotationally invariant codes. Usually, one cannot obey all of Ungerboeck's rules (U1 to U3 of Section 5.1) or Turgeon's rules (T1 and T2 of Section 5.1), although one should try, when designing rotationally invariant codes. When the method outlined in this section is applied to codes that violate the design rules stated, the input–output bit relationship may be very difficult to discover. However, with common sense and a little trial and error, the input–output bit relationship can usually be determined. In this section we deal with some of the techniques employed to obtain this evasive relationship in the rotationally invariant code context. The method used for nonlinear codes proceeds the same as the method used for linear codes with the exception of a modification to the process of determining of the input–output bit relationships. This modification employs a general plan of attack that involves assigning the output bits to certain input bits so that each expression for the output bits are functions of only a few input bits. In addition, it is a necessary condition that each expression for the output bits yield either a +1 or −1. The derivation of a nonlinear convolutional encoder can best be shown through examples.

We consider first an 8-PSK coder with 90° rotational invariance. Then a 16-PSK code with a 45° invariance is treated. Both phase invariances should be lower; such an invariance is derived in a later section on multidimensional codes.

8.2.1 Eight-Point Example

EXAMPLE 8.3

In designing the code for this example, the rules given in Section 8.2 were used to achieve a 90° rotational invariant code. This is the example given in Fig. 8.7. Since this is an 8-PSK code, it can be treated as one-dimensional in phase. In this trellis modulation scheme there are three memory bits and their relation to the data bits is shown in Fig. 8.13. Also shown is the trellis structure in Fig. 8.7 in terms of the signal numbers in the squares in the constellation in Fig. 8.3. There is a parallel transition in this design. With $d_{\text{free}}^2 = 4$, this code has an asymptotic coding gain of 3.01 dB over uncoded QPSK modulation. Figure 8.13 uses the notation of our step-by-step design technique of Chapter 5. The sliding block is $(b_1, b_2, b_3, b_4, b_5)$. The d coefficients are determined through solving equation

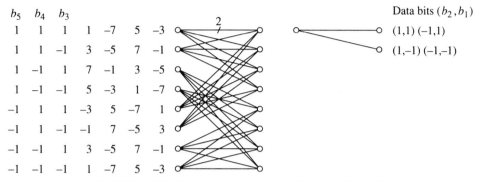

FIGURE 8.13 State structure and trellis for 8-PSK RIC.

(5.3) using the computer program in Appendix C for the trellis state variables in Fig. 8.13. By taking account of results just given for the b_i, the result is

$$d_2 = 2 \qquad d_{34} = -1$$
$$d_{24} = -2 \qquad d_{145} = -2$$
$$d_{25} = 2 \qquad d_{245} = 2.$$

Thus the resulting analytic description is found to be

$$x = 2b_2 - b_3 b_4 - 2b_2 b_4 + 2b_2 b_5 - 2b_1 b_4 b_5 + 2b_2 b_4 b_5. \qquad (8.10)$$

It should be noted that the channel signal is equal to $e^{j(x-1)\pi/8}$. As in Chapter 5, a table will be constructed in order to determine signal differences. From Table 8.8

TABLE 8.8

		z_2	1	1	1	1	−1	−1	−1	−1
Output		z_1	1	1	−1	−1	1	1	−1	−1
bits		z_0	1	−1	1	−1	1	−1	1	−1
Output signal		x	1	3	5	7	−7	−5	−3	−1

Effect on x due to z_i:

z_2: $|8|$

z_1: $|4|$

z_0: $|2|$

the following equation is obtained for the signal x:

$$x = 4z_2 - 2z_1 - z_0. \qquad (8.11)$$

When equation (8.11) is compared to equation (8.10), there is not a one-to-one relationship. One obvious equivalence is that $z_0 = b_3 b_4$, for these are the only two expressions that have coefficients of -1. However, it is not so obvious what z_1 and z_2 equal. If it is chosen that $z_1 = b_1 b_4 b_5$, it remains that $z_2 = (2b_2 - 2b_2 b_4 + 2b_2 b_5 + 2b_2 b_4 b_5)/4$. The choice of z_1 can be justified by the fact that this assignment leaves z_2 as an expression in only three variables and more important, that the expression for z_2 can only result in a ± 1. It is obvious from Chapter 5 what z_0 and z_1 are in the 0/1 convention: for instance, $z_0 = b_3 b_4 \Longrightarrow y_0 = a_3 \oplus a_4$, where a_i is 0 or 1. However, the expression for z_2 requires further investigation. Consider the expression for z_2:

$$z_2 = \frac{b_2 - b_2 b_4 + b_2 b_5 + b_2 b_4 b_5}{2}$$

$$= b_2 \left(\frac{1 - b_4 + b_5 + b_2 b_4 b_5}{2} \right)$$

$$= b_2 z_2' \Longrightarrow y_2 = a_2 \oplus y_2'.$$

To determine what z_2' represents in the 0 and 1 convention, one can evaluate z_2' for all combinations for b_4 and b_5. By substituting 1 for 0 and -1 for 1, Table 8.9 leads to the conclusion that

$$\frac{1 - b_4 + b_5 + b_4 b_5}{2} \Longrightarrow \overline{a_4} \odot a_5$$

where $\bar{a}$ is the complement of a. Therefore, in summary,

$$z_0 = b_3 b_4 \Longrightarrow y_0 = a_3 \oplus a_4$$

$$z_1 = b_1 b_4 b_5 \Longrightarrow y_1 = a_1 \oplus a_4 \oplus a_5$$

$$z_2 = b_2 \left(\frac{1 - b_4 + b_5 + b_4 b_5}{2} \right) \Longrightarrow y_2 = a_2 \oplus (\overline{a_4} \odot a_5)$$

TABLE 8.9 Bit Assignment for Eight-Point Signal Set.

b_4	b_5	$\dfrac{1 - b_4 + b_5 + b_4 b_5}{2}$
1	1	1
1	-1	-1
-1	1	1
-1	-1	1

8.2 / Generation: Rotationally Invariant Codes

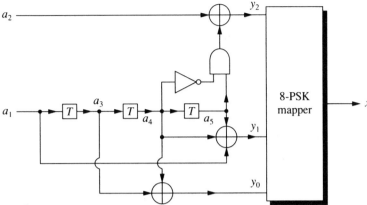

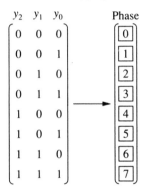

FIGURE 8.14 Nonlinear 8-PSK RIC.

The encoder in Fig. 8.14 was constructed from the relationships given just above. □

8.2.2 Sixteen-Point Examples

EXAMPLE 8.4 16-PSK, EIGHT-STATE, 45° ROTATIONAL INVARIANCE

Again in the design of this code the rules outlined earlier were followed to achieve a 45° rotational invariance and this was derived in Example 8.1. Since the signal constellation contains signals that vary only in phase, this code can be treated as a one-dimensional code. In this trellis modulation scheme there are three memory bits and their relation to the data bits is shown in Fig. 8.15. There are four parallel transitions and the signal constellation and mapping is the same as the one used in Example 8.1 and can be seen in Fig. 8.9. The signal values in the "squares" in Fig. 8.9 are used in this design. The trellis from Fig. 8.10 in terms of the sliding block (b_1, b_2, b_3, b_4, b_5, b_6) is also shown in Fig. 8.15. The minimum squared Euclidean

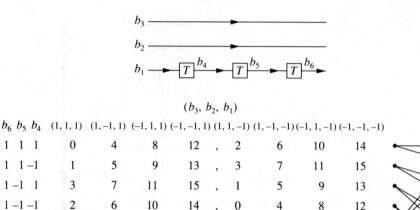

FIGURE 8.15 State structure and trellis of 16-PSK RIC encoder.

distance can be shown to be $d_{free}^2 = 1.48$. Compared to uncoded 8-PSK, this coding scheme exhibits a gain of 4.01 dB. The d coefficients were determined to be

$$d_2 = -2 \qquad d_{36} = 2 \qquad d_{356} = 2$$
$$d_3 = 6 \qquad d_{45} = -1 \qquad d_{1235} = -2$$
$$d_{25} = 2 \qquad d_{123} = -2 \qquad d_{1236} = 2$$
$$d_{26} = -2 \qquad d_{156} = -2 \qquad d_{12356} = 2$$
$$d_{35} = -2 \qquad d_{256} = -2.$$

With all these coefficients one can see that the analytic transmitter implementation would be quite awkward. The coefficients above reveal that

$$x = 6b_3 - b_4 b_5 + 2(-b_2 + b_2 b_5 - b_2 b_6 - b_3 b_5 + b_3 b_6 \\
- b_1 b_2 b_3 - b_1 b_5 b_6 - b_2 b_5 b_6 + b_3 b_5 b_6 \\
- b_1 b_2 b_3 b_5 + b_1 b_2 b_3 b_6 + b_1 b_2 b_3 b_5 b_6) \qquad (8.12)$$

where the signal sent $= e^{j(x-1)\pi/16}$. A table is now constructed to determine the relationship between x and the output bits. The generation of the 16-PSK signal constellation is shown in Table 8.10. From the signal differences given in Table 8.10,

$$x = 8z_3 - 4z_2 - 2z_1 - z_0. \qquad (8.13)$$

It appears to be quite an arduous task to determine the input–output bit relationship by comparing (8.12) to (8.13), although some assignments can be made quickly. It is obvious that $z_0 = b_4 b_5$, since both coefficients are equal to -1.

TABLE 8.10 16-Point Signal Assignment.

Output bits																	
z_3		1	1	1	1	1	1	1	1	-1	-1	-1	-1	-1	-1	-1	-1
z_2		1	1	1	1	-1	-1	-1	-1	1	1	1	1	-1	-1	-1	-1
z_1		1	1	-1	-1	1	1	-1	-1	1	1	-1	-1	1	1	-1	-1
z_0		1	-1	1	-1	1	-1	1	-1	1	-1	1	-1	1	-1	1	-1
Output signal x		1	3	5	7	9	11	13	15	-15	-13	-11	-9	-7	-5	-3	-1

Effect on x due to z_i:

z_3: $|16|$

z_2: $|8|$

z_1: $|4|$

z_0: $|2|$

TABLE 8.11
Relationship for z_2'.

b_5	b_6	z_2'
1	1	1
1	-1	-1
-1	1	1
-1	-1	1

The assignment for z_1 can contain only one term with a coefficient of 2. It was decided to let $z_1 = b_1 b_5 b_6$ because with some examination this term is the most "awkward" when compared with the rest of the terms. It was then decided to let z_2 contain only b_2, b_5, and b_6 terms. This assignment results in

$$z_3 = \frac{[6b_3 - 2b_3 b_5 + 2b_3 b_6 + 2b_3 b_5 b_6 - 2b_1 b_2 b_3 - 2b_1 b_2 b_3 b_6 + 2b_1 b_2 b_3 b_5 + 2b_1 b_2 b_3 b_5 b_6)]}{8}$$

The expressions for z_0 and z_1 can readily be transformed into expression for y_0 and y_1 in terms of a_i. However, the expressions for z_2 and z_3 must be analyzed with truth tables to determine the proper transformation. Consider

$$z_2 = \frac{b_2 + b_2 b_6 - b_2 b_5 + b_2 b_5 b_6}{2}$$

$$= b_2\{(1/2)[1 - b_5 + b_6(1 + b_5)]\}$$

$$= b_2 z_2' \Longrightarrow y_2 = a_2 \oplus y_2'.$$

From Table 8.11 it can be seen that $z_2' \Longrightarrow y_2' = \overline{a_5} \odot a_6$.

Consider the factored expression for z_3:

$$z_3 = b_3(\tfrac{1}{4})[3 - b_5 + b_6 + b_5 b_6 + b_1 b_2(-1 - b_5 + b_6 + b_5 b_6)]$$

$$= b_3(\tfrac{1}{4})[4 + (-1 - b_5 + b_6 + b_5 b_6) + b_1 b_2(-1 - b_5 + b_6 + b_5 b_6)]$$

$$= b_3(\tfrac{1}{4})[4 + (1 + b_1 b_2)(-1 - b_5 + b_6 + b_5 b_6)]$$

$$= b_3 z_3' \Longrightarrow y_3 = a_3 \oplus y_3'.$$

A few observations about z_3' can be made. When b_1 and b_2 are different or when $b_5 = -1$, then $z_3' = 1$. With these observations in mind a truth table can easily be constructed. From Table 8.12 one can deduce that $z_3' \Longrightarrow y_3' = (a_1 \oplus a_2) \odot \overline{a_5} \odot a_6$.

In summary,

$y_0 = a_4 \oplus a_5$

$y_1 = a_1 \oplus a_5 \oplus a_6$

$y_2 = a_2 \oplus (\overline{a_5} \odot a_6)$

$y_3 = a_3 \oplus [(a_1 \oplus a_2) \odot \overline{a_5} \odot a_6]$.

TABLE 8.12 Relationship for z_3'.

b_1	b_2	b_5	b_6	z_3'
1	1	1	1	1
1	1	1	-1	-1
1	1	-1	1	1
1	1	-1	-1	1
1	-1	1	1	1
1	-1	1	-1	1
1	-1	-1	1	1
1	-1	-1	-1	1
-1	1	1	1	1
-1	1	1	-1	1
-1	1	-1	1	1
-1	1	-1	-1	1
-1	-1	1	1	1
-1	-1	1	-1	-1
-1	-1	-1	1	1
-1	-1	-1	-1	1

The structure of the convolutional encoder with its mapper is illustrated in Fig. 8.16. □

EXAMPLE 8.5 16-QAM, EIGHT-STATE, 90° ROTATIONAL INVARIANCE

Although in this example there are three data bits, only a 90° rotational invariance can be achieved because of the shape of the signal constellation. In this particular

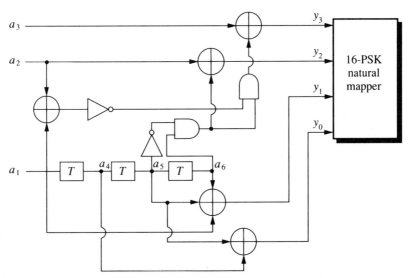

FIGURE 8.16 Nonlinear convolutional code generation of 16-PSK RIC.

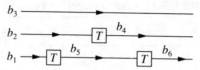

FIGURE 8.17 Input bit relationship for 16-QAM, eight-state, 90° rotationally invariant code.

code there are three memory bits (eight-state) and a parallel transition. The relationship between the input bits is shown in Fig. 8.17. Figure 8.18 shows the signal constellation and mapping, while Fig. 8.19 shows the trellis structure and signal assignment. The set partitioning for this case can be deduced from Fig. 8.11; only the signal numbers have been changed. The minimum squared Euclidean distance, normalized by average signal power, is $d_{\text{free}}^2 = 2$. Compared to uncoded 8-AMPM modulation, this results in a coding gain of 3.98 dB. The d coefficients were determined to be

For x_1:
$d_{134} = 1$
$d_{145} = -1$
$d_{236} = -1$
$d_{1345} = 1$
$d_{2356} = 1$

For x_2:
$d_{14} = 1$
$d_{134} = -1$
$d_{1236} = -1$
$d_{1345} = 1$
$d_{12356} = -1$

where x_1 is the real part and x_2 the imaginary part of the QAM phasor. Therefore,

$$x_1 = b_1 b_3 b_4 - b_1 b_4 b_5 - b_2 b_3 b_6 + b_1 b_3 b_4 b_5 + b_2 b_3 b_5 b_6$$
$$x_2 = b_1 b_4 - b_1 b_3 b_4 - b_1 b_2 b_3 b_6 + b_1 b_3 b_4 b_5 - b_1 b_2 b_3 b_5 b_6.$$

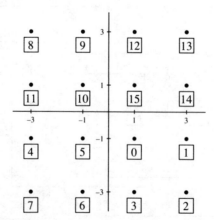

FIGURE 8.18 Signal constellation for 16-QAM.

8.2 / Generation: Rotationally Invariant Codes

(b_3, b_2, b_1)

b_6 b_5 b_4	(1, 1, 1)	(−1, 1, 1)	(1, −1, 1)	(−1, −1, 1)	(1, 1, −1)	(−1, 1, −1)	(1, −1, −1)	(−1, −1, −1)	
1 1 1	0	8,	12	4,	10	2,	6	14	
1 1 −1	6	14,	10	2,	12	4,	0	8	
1 −1 1	5	13,	1	9,	11	3,	15	7	
1 −1 −1	11	3,	15	7,	5	13,	1	9	
−1 1 1	12	4,	0	8,	6	14,	10	2	
−1 1 −1	10	2,	6	14,	0	8,	12	4	
−1 −1 1	1	9,	5	13,	15	7,	11	3	
−1 −1 −1	15	7,	11	3,	1	9,	5	13	

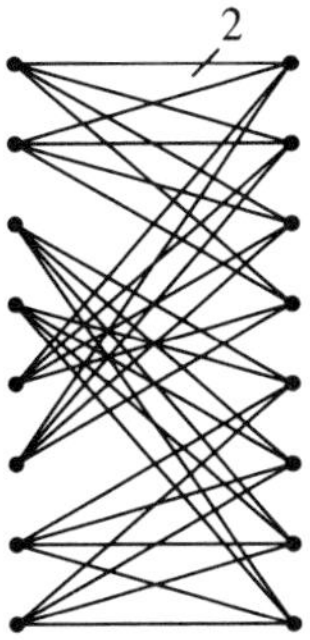

FIGURE 8.19 Trellis diagram for 16-QAM, eight-state, 90° rotationally invariant code.

The next step will determine how x is related to the output bits. From Table 8.13 the following relations can be derived:

$$x_1 = -z_0 z_1 + 2z_2 z_3$$

$$x_2 = -2z_3 + z_1.$$

It would be wiser to deal with the expression for x_2 first because of its relative simplicity. Let $z_1 = b_1 b_4$, resulting in

$$z_3 = \frac{-b_1 b_3 b_4 - b_1 b_2 b_3 b_6 + b_1 b_3 b_4 b_5 - b_1 b_2 b_3 b_5 b_6}{(-2)}$$

$$= b_1 b_3 \{(1/2)[b_2 b_6 (b_5 + 1) + b_4 (1 - b_5)]\}$$

$$= b_1 b_3 z_3'.$$

The following observations about z_3' are useful in determining the equivalent expression for y_3'. If $b_5 = -1$, then $z_3' = b_4$, and if $b_5 = 1$, then $z_3' = b_2 b_6$. From this information it is not too difficult to see that

$$y_3' = (a_4 \odot a_5) \oplus [\overline{a_5} \odot (a_2 \oplus a_6)].$$

Now x_1 must be considered. Let $z_0 z_1 = b_1 b_4 b_5$, resulting in $z_0 = b_5$ and

$$2z_2 z_3 = b_1 b_3 b_4 - b_2 b_3 b_6 + b_1 b_3 b_4 b_5 + b_2 b_3 b_5 b_6.$$

Since z_3 is known, the resulting expression for z_2 is

$$z_2 = \frac{b_1 b_3 b_4 - b_2 b_3 b_6 + b_1 b_3 b_4 b_5 + b_2 b_3 b_5 b_6}{2z_3}$$

$$= \frac{b_1 b_4 - b_2 b_6 + b_1 b_4 b_5 + b_2 b_5 b_6}{b_1 [b_2 b_6 (b_5 + 1) - b_4 (b_5 - 1)]}.$$

It should be noted that the expression $b_1[b_2 b_6 (b_5 + 1) - b_4 (b_5 - 1)]$ equals ± 2 for all values. Thus $1/2\{b_1[b_2 b_6 (b_5 + 1) - b_4 (b_5 - 1)]\}$ will only equal ± 1. Dividing

TABLE 8.13 Formula for 16-QAM Constellation.

Output bits	z_3	1	1	1	1	1	1	1	1	-1	-1	-1	-1	-1	-1	-1	-1
	z_2	1	1	1	1	-1	-1	-1	-1	1	1	1	1	-1	-1	-1	-1
	z_1	1	1	-1	-1	1	1	-1	-1	1	1	-1	-1	1	1	-1	-1
	z_0	1	-1	1	-1	1	-1	1	-1	1	-1	1	-1	1	-1	1	-1
Output signals	x_1	1	3	3	-1	1	3	3	-1	-1	-3	-3	1	-1	-3	-3	1
	x_2	-1	-1	-3	-3	1	1	3	3	-1	-1	-3	-3	1	1	3	3

Effect on x_i due to z_j:

	x_1	x_2
z_3:	$|4|$	$|4|$
z_2:	$|4|$	No effect
z_1:	$|2|$	$|2|$
z_0:	$|2|$	No effect

TABLE 8.14
Truth Table for z_2' in Example 8.5.

b_1	b_5	z_2'
1	1	1
1	−1	−1
−1	1	1
−1	−1	1

by ± 1 is the same as multiplying by ± 1. Also note that $b_1^2 = 1$. Therefore,

$$\begin{aligned} z_2 &= \frac{b_1 b_4 - b_2 b_6 + b_1 b_4 b_5 + b_2 b_5 b_6}{2(1/2) b_1 [b_2 b_6 (b_5 + 1) - b_4 (b_5 - 1)]} \\ &= \frac{(b_1 b_4 - b_2 b_6 + b_1 b_4 b_5 + b_2 b_5 b_6)\{(1/2) b_1 [b_2 b_6 (b_5 + 1) - b_4 (b_5 - 1)]\}}{2} \\ &= b_2 b_4 b_6 \left[1/2 (1 + b_5 - b_1 + b_1 b_5) \right] \\ &= b_2 b_4 b_6 z_2' . \end{aligned}$$

To determine the equivalent expression for z_2', a short truth table is constructed. From Table 8.14 $y_2' = \overline{a_1} \odot a_5$. In summary,

$$y_0 = a_5 \qquad y_2 = a_2 \oplus a_4 \oplus a_6 \oplus (\overline{a_1} \odot a_5)$$
$$y_1 = a_1 \oplus a_4 \qquad y_3 = a_1 \oplus a_3 \odot \{(a_4 \odot a_5) \oplus [\overline{a_5} \odot (a_2 \oplus a_6)]\}.$$

The convolutional encoder and natural mapper are shown in Fig. 8.20. □

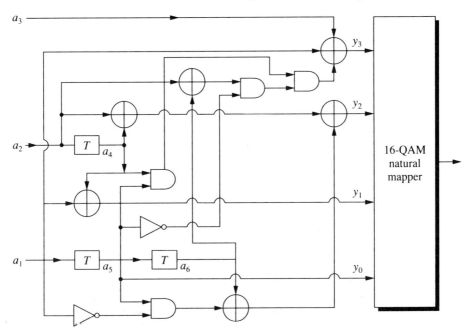

FIGURE 8.20 Nonlinear convolutional code generation of 16-QAM RIC.

8.3 Multidimensional RIC

We examined multidimensional trellis codes in Chapters 6 and 7. In this chapter we regard a trellis interval to be made up from a number of symbol intervals. The notation M/N-PSK means that M-PSK is used on the first segment of a trellis interval and N-PSK is used on the second interval. Thus we have a four-dimensional signal. The concept is easily extended to any number of signal intervals per trellis interval. Finally, this section will also serve as our multidimensional examples of our step-by-step design procedure given in Chapter 5.

So far in two-dimensional signaling, if a rate $m/m+1$ encoder is used, the degree of rotational invariance is restricted to $360°/2^m$ when all m inputs are differentially encoded. Thus we observed in Section 8.1.2 that a rate 2/3, 8-PSK code was restricted to a 90° rotational invariance, as only $m = 2$ bits were available for differential encoding. However, the smallest angle of rotation was 45°, and thus extra effort in the decoder is required to resolve this phase ambiguity [7].

There are more input bits per trellis interval in multidimensional trellis codes. We illustrate this situation in Fig. 8.21. For instance, a 4/8-PSK trellis code has 4 input bits per trellis interval. Thus a $22.5° = 360°/2^4$ rotational invariance could be achieved. However, only a 90° invariance is required if we decide to phase track only over the 4-PSK symbol, or the first symbol, in each trellis interval. We will see that the required rotational invariance can be achieved with a linear convolutional encoder, not a nonlinear one as we have had to use up to now. This will occur because the codes designed with optimum Ungerboeck rules will inherently have the desired rotational invariance. In this case a 90° invariance suffices, as the carrier tracking takes place only over the 4-PSK signal in the 4/8-PSK signal set. An additional bonus is that performance analysis will be possible as the convolutional code that generates the trellis code is linear. These concepts are taken from [9] and [10] and our examples are taken from [10].

We will close this section with an example involving multidimensional codes where carrier recovery can operate continuously. This will require a nonlinear code, but the required amount of rotational invariance will be achieved.

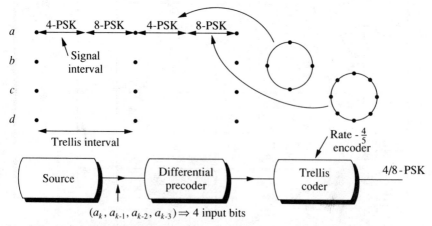

FIGURE 8.21 4/8-PSK, four-dimensional signals.

8.3.1 Linear Examples

For multidimensional constellations, rotations are considered in terms of the constituent symbols. We assume that if the receiver locks on a phase that is a rotated version of the actual phase, all consecutive channel symbols in the signal transmission will be rotated by this amount. With multidimensional signals rotational invariance can be inherent in the signal constellation. An example is the 4/8-PSK signal set in Fig. 8.22. We begin by considering the 8/8-PSK signal

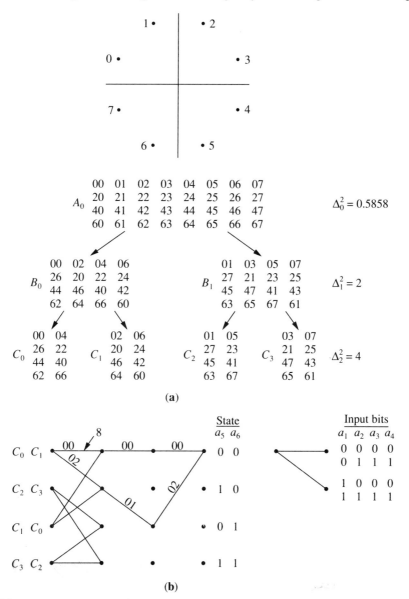

FIGURE 8.22 Partitioning of 4/8-PSK and trellis for four-state rotational invariant code: (a) Partitioning of 4/8-PSK signal set; (b) trellis diagram for rotationally invariant, four-state, 4/8-PSK code.

set in Fig. 8.23. By using the signal 00 as a reference, a 180° rotation yields 44, a −90° rotation yields 22, and a −45° rotation yields 11. If the signal set is comprised of different constellations, the symbols with the larger phase ambiguity are rotated as close as possible without exceeding the rotation of the symbols with the smaller phase ambiguity. Take the 4/8-PSK signal set in Fig. 8.22 as an example. The 4-PSK signal set is the phases 0, 2, 4, 6 in Fig. 8.22. Taking 00 as the reference, a rotation of 135° results in signal 23, since the 8-PSK symbol can be rotated through this angle, but the QPSK symbol can only be rotated through 90°.

In PSK-based signal sets, the greatest distance occurs between a signal and a 180° rotation of itself. This means that if parallel transitions occur, the bit respon-

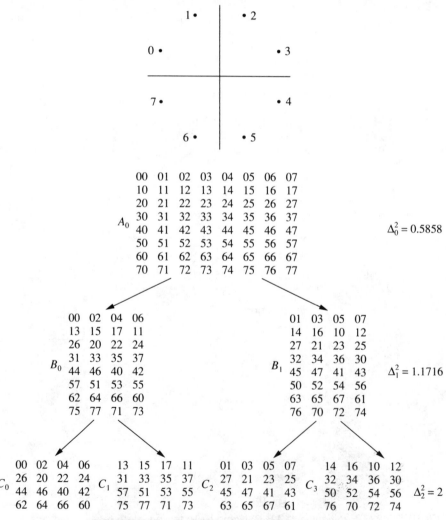

FIGURE 8.23 Partitioning of 8/8-PSK signal set.

sible for rotation will be one of the inputs that goes directly into the signal mapper. After partitioning is complete, the subset of signals that occur between like states should be checked to see if any rotated versions of signals are also in the subset. If this occurs, the bits responsible for these rotations are shown from those input directly to the mapper. The design procedure for multidimensional codes will start with the assignment of signals to the trellis according to Ungerboeck's design rules. As the signals are assigned, the input bits responsible for the various rotations can be determined. A table can then be drawn to determine the rotated states, and the rest of the signal assignment can be made, giving Wei's rule priority over the optimum design rules. The input bits $I_1, I_2, \ldots, I_f$ can then be differentially encoded with a binary modulo-2^f adder, where I_j is the bit responsible for a $180°/(f - j + 1)$ rotation. It is evident that with the large number of input bits and the smaller phase ambiguities present in the constituent constellations of the multidimensional codes, the potential to remove all phase ambiguities exists. Let us first consider a simple example.

EXAMPLE 8.6 2/4-PSK CODE, TWO-STATE

The signal set used for this code consists of two constituent symbols, the first having a 180° phase ambiguity and the second having a 90° phase ambiguity. Since the number of input bits is $m = 2$, the smallest angle the code can be made rotationally invariant to is 90°. Since this is a two-state code, there is only $v = 1$ memory bit. The signal set is partitioned in Fig. 8.24. The number of coded input bits is constrained by v to be $\tilde{m} = 1$, causing the trellis to have two parallel transitions. Examining the signal subsets that are assigned to parallel transitions

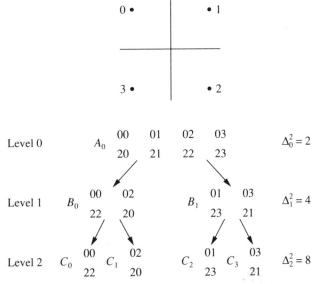

FIGURE 8.24 Partitioning of 2/4-PSK signal set.

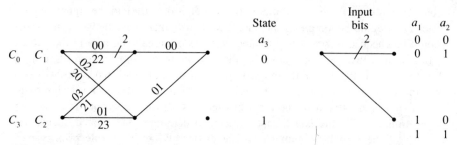

FIGURE 8.25 Trellis diagram for rotationally invariant, two-state, 2/4-PSK code.

in the trellis diagram in Fig. 8.25, it can be seen that each subset contains two signals that are 180° rotations of each other. This will cause the data bit input to the mapper to be responsible for 180° rotations. This leaves the bit input to the memory element to be responsible for the 90° rotations. The signals are assigned to the trellis in Fig. 8.25. Since subset C_0 is assigned to the transition from state 0 to state 0, the 90° rotated signals must be assigned to the rotated state. In this case the rotated set, C_2, is assigned to the transition from state 1 to state 1. This code obeys Ungerboeck design rules (U1 to U3 of Chapter 5), so it is termed an *optimum code*.

Although the code has an optimum coding gain, the signal assignment in the trellis in Fig. 8.25 is different from that for the optimum code based on Ungerboeck's rules. This is because the rules (T1 and T2 of Chapter 5) for minimal complexity of the analytic description are obeyed. The sliding block of input bits will be (b_1, b_2, b_3). The corresponding 0/1 sliding block is (a_1, a_2, a_3) and is used in the state definition in Fig. 8.25. From this figure and the definition of the sets C_i, $i = 0, \ldots, 3$, in Fig. 8.24, we can solve equation (5.3) for the analytic description. This gives $x_1 = b_2$ and $x_2 = b_3 + 2b_1b_2b_3$, where the first symbol transmitted is $\exp(jx_1\pi/2)$ and the second is $\exp(jx_2\pi/4)$. The encoder/modulator based on this description is shown in Fig. 8.26(a). To convert the analytic description to the Ungerboeck form, three output bits are introduced (z_0, z_1, z_2), where z_2 is mapped to the BPSK symbol and (z_0, z_1) is mapped to the QPSK symbol. Since the natural mapping is being used, Table 8.15 can be used to show the output symbols resulting from all possible combinations of output bits. Table 8.15 shows the signal differences with respect to these output bits. Using the half-signal difference rule from Chapter 5 yields $x_1 = z_1$ and $x_2 = z_0 + 2z_1$. Comparing these equations with those of the analytic description yields the relationships $z_0 = b_3$, $z_1 = b_1b_2b_3$, and $z_2 = b_2$, which when converted to the 0, 1 convention for data bits results in $y_0 = a_3$, $y_1 = a_1 \oplus a_2 \oplus a_3$, and $y_2 = a_2$. The encoder/modulator based on these equations is shown in Fig. 8.26(b). □

We now consider the case discussed in the introduction to this section.

EXAMPLE 8.7 4/8-PSK, FOUR-STATE, 90° RIC

This code uses a four-dimensional signal set comprised of an interval of QPSK followed by an interval of 8-PSK. The number of input bits to the trellis encoder is $m = 4$, and the number of memory elements is $v = 2$. The 4/8-PSK signal

TABLE 8.15 Mapping Rule and Signal Differences for 2/4-PSK.

Output bits	z_2	1	1	1	1	-1	-1	-1	-1
	z_1	1	1	-1	-1	1	1	-1	-1
	z_0	1	-1	1	-1	1	-1	1	-1
Output signals	x_1	1	1	1	1	-1	-1	-1	-1
	x_2	3	1	-1	-3	3	1	-1	-3

Effect on x_i due to z_j:

	x_1	x_2
z_0:	No effect	$|2|$
z_1:	No effect	$|4|$
z_2:	$|2|$	No effect

set is partitioned in Fig. 8.22(a). The best optimum code is obtained by choosing $\tilde{m} = 1$, which results in a half-connected trellis as shown in Fig. 8.22(b). The optimum signal assignment is also shown in Fig. 8.22(b), from which it can be seen that parallel transitions consist of the signals in the subsets at the second level of set partitioning. Examining these subsets, it is observed that each subset contains the 180° and 90° rotations of each signal in the subset. Due to this occurrence, this code is inherently rotationally invariant to 90° rotations. Thus by differentially

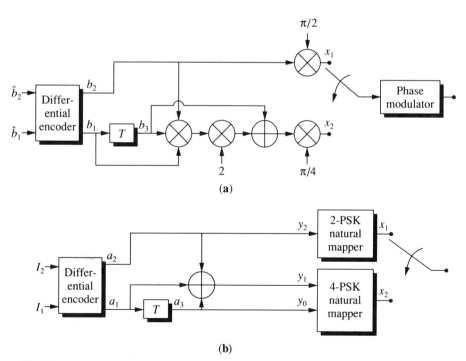

FIGURE 8.26 Encoders for rotationally invariant, two-state, 2/4-PSK TCM: (a) Analytic description; (b) Ungerboeck description.

encoding two of the input bits responsible for the rotations, an optimum code that is rotationally invariant to 90° rotations is obtained. As discussed in the introduction to this section, phase recovery can then take place over the first symbol interval, which has all phase ambiguity removed. From the signal assignment to the trellis, it can be seen that the bit a_4 is responsible for 180° rotations and that a_2 is responsible for 90° rotations. Once again a modulo-4 binary adder is required to differentially encode these input bits. This operation is defined by (8.5) and (8.6) with the replacements $I_1 \rightarrow I_2$, $a_1 \rightarrow a_2$, $I_2 \rightarrow I_4$, and $a_2 \rightarrow a_4$. For this trellis, $d_{\text{free}}^2 = 4$. This code transmits 2 bits per symbol and exhibits an asymptotic gain over QPSK of $\gamma = 10 \log((4)(4)/(2)(2)(2)) = 3.01$ dB. This code has the same trellis complexity, throughout, and asymptotic performance as the 8-PSK, four-state rotationally invariant code used in Section 8.1.2. However, 8-PSK has a 45° phase ambiguity requirement, and this code can only be made invariant to 90° rotations. Also, both data bits must be differentially encoded in this code, whereas half of the data bits must be encoded for the 4/8-PSK code, resulting in a smaller increase in bit error rate due to differential decoding. Herein, we see an advantage for the four-dimensional code. It should also be noted that when the trellis is increased to eight states, the free distance of the 4/8-PSK code does not increase.

Solving for the analytic description for this code yields

$$x_1 = b_2 + 2b_4$$

and

$$x_2 = b_5 + 4b_3b_4 + 2b_1b_2b_6$$

where the first symbol sent is $\exp(jx_1\pi/4)$ and the second symbol sent is $\exp(jx_2\pi/8)$. The encoder based on this description is shown in Fig. 8.27(a). To obtain the underlying convolutional encoder, five output bits $(z_0, \ldots, z_4)$ are introduced. The binary number z_4z_3 will be mapped to the QPSK constellation, and the number $z_2z_1z_0$ will be mapped to the 8-PSK constellation. Table 8.16 shows the resultant outputs for every possible combination of output bits; also shown are the signal differences as a function of the output bits. From this the output symbols as a function of output bits are described as

$$x_1 = z_3 + 2z_4$$

and

$$x_2 = z_0 + 2z_1 + 4z_2.$$

Comparing these equations with those of the analytic description yields $z_0 = b_5$, $z_1 = b_1b_2b_6$, $z_2 = b_3b_4$, $z_3 = b_2$, and $z_4 = b_4$. Converting these relationships back to the 0, 1 convention for data bits results in

$$y_0 = a_5, \quad y_1 = a_1 \oplus a_2 \oplus a_6, \quad y_2 = a_3 \oplus a_4, \quad y_3 = a_2$$

and

$$y_4 = a_4.$$

From these relationships, the Ungerboeck encoder/modulation can be created, as shown in Fig. 8.27(b). □

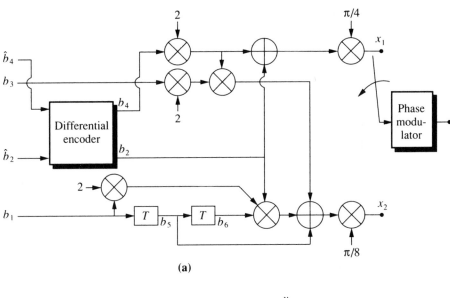

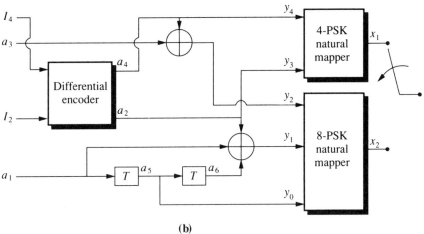

FIGURE 8.27 Encoders for rotationally invariant, four-state, 4/8-PSK TCM: (a) analytic description; (b) Ungerboeck description.

8.3.2 Nonlinear Example

We now present a four-dimensional 8-PSK code that has the required 45° invariance. The resulting code will be nonlinear. The procedure that is followed to derive the block diagram to generate this code closely follows Examples 5.2 and 5.3.

EXAMPLE 8.8 2 × 8-PSK, FOUR-STATE, 45° ROTATIONAL INVARIANCE

In this example the desired 45° rotational invariance is achieved because with this multidimensional code there are four data bits (this is a rate 4/6 code). However, only three of the data bits need to be differentially encoded to achieve 45° rotational invariance. In this modulation scheme there are two memory bits (four-state) and

TABLE 8.16 Effect of Output Bits on 4/8-PSK Channel Signals: (a) All Possible Combinations; (b) Signal Differences.

Output bits	
z_4	...
z_3	...
z_2	...
z_1	...
z_0	...

Output signals	
x_1	3, 3, 3, 3, 3, 3, 3, 3, 3, 3, 3, 3, 3, 3, 3, 3, -1, -1, -1, -1, -1, -1, -1, -1, -1, -1, -1, -1, -1, -1, -1, -3
x_2	7, 5, 3, 1, -1, -3, -5, -7, 7, 5, 3, 1, -1, -3, -5, -7, 7, 5, 3, 1, -1, -3, -5, -7, 7, 5, 3, 1, -1, -3, -5, -7

Effect on x_i due to z_j:

	x_1	x_2
z_0:	No effect	$\|2\|$
z_1:	No effect	$\|4\|$
z_2:	No effect	$\|8\|$
z_3:	$\|2\|$	No effect
z_4:	$\|4\|$	No effect

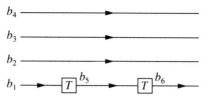

FIGURE 8.28 Input bit relationship for 2 × 8 PSK, four-state, 45° rotational invariant code.

the relation between input bits is shown in Fig. 8.28. The signal constellation and natural mapping appear in Fig. 8.3. The trellis structure and signal assignment appear in Fig. 8.29. This code transmits two bits per symbol, and with $d_{\text{free}}^2 = 4$, asymptotically gains 3.01 dB over uncoded QPSK. The d coefficients were determined by solving equation (5.3) for the signals $0, 1, \ldots, 7$ and sliding block $(b_1, b_2, \ldots, b_6)$ in the trellis in Fig. 8.29. The following expressions for x_1 and x_2 results:

$$x_1 = 2b_2 - b_5 - b_2b_4 + b_2b_6 - b_4b_6 - b_1b_2b_4 + b_1b_2b_6 - b_1b_4b_6 + b_2b_4b_5$$
$$+ b_2b_5b_6 - b_4b_5b_6 - b_1b_2b_4b_5 + 2b_1b_2b_4b_6 - b_1b_2b_5b_6 + b_1b_4b_5b_6$$

(8.14)

$$x_2 = -b_4 - b_5 - b_1b_4 + 3b_2b_3 + b_4b_5 + b_1b_2b_3 - b_1b_4b_5 + b_2b_3b_4$$
$$- b_2b_3b_5 - b_1b_2b_3b_4 + b_1b_2b_3b_5 + b_2b_3b_4b_5 - b_1b_2b_3b_4b_5.$$

(8.15)

Again as in our M-PSK examples in Chapter 5, the first signal sent $= e^{j(x_1-1)/8}$ and the second signal sent $= e^{j(x_2-1)/8}$. The table for the signal assignment in terms of the z_i, $i = 0, 1, 2, 3$ is given in Table 8.17. From this table the relationship

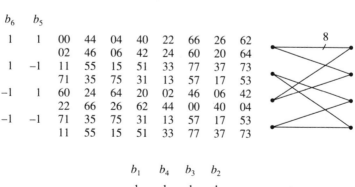

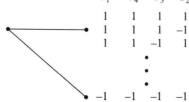

FIGURE 8.29 Trellis diagram for 2 × 8 PSK, four-state, 45° rotational invariant code.

TABLE 8.17 Signal Assignment and Set Differences for 2 × 8 PSK.

Output bits																
z_5	1	1	1	1	1	1	1	1	...	1	1	1	1	...	1	1
z_4	1	1	1	1	1	1	1	1	...	1	1	1	1	...	1	1
z_3	1	1	1	1	1	1	1	1	...	-1	1	1	1	...	1	1
z_2	1	1	1	1	-1	1	1	1	...	1	-1	1	1	...	1	1
z_1	1	1	-1	1	1	-1	1	1	...	1	-1	1	1	...	1	1
z_0	1	-1	1	1	1	1	-1	1	...	1	-1	1	1	...	1	1
Output signals																
x_1	1	3	5	7	-7	-5	-3	-1	...	3		5		...	7	
x_2	1		7				-3	-1	...	1		1		...	1	

Effect on x_i due to z_j:

	x_1	x_2
z_5:	$|8|$	No effect
z_4:	$|4|$	No effect
z_3:	$|2|$	No effect
z_2:	No effect	$|8|$
z_1:	No effect	$|4|$
z_0:	No effect	$|2|$

TABLE 8.18
Generation of z_4' in Example 8.8.

b_1	b_5	z_4'
1	1	1
1	-1	1
-1	1	1
-1	-1	-1

between the z_i and (x_1, x_2) above is

$$x_1 = 4z_5 - 2z_4 - z_3$$
$$x_2 = 4z_2 - 2z_1 - z_0.$$

The expressions for x_1 are considered first. Let $z_3 = b_5$ and let

$$z_4 = \frac{-b_4 b_6 - b_4 b_5 b_6 - b_1 b_4 b_6 + b_1 b_4 b_5 b_6}{(-2)}$$

$$= b_4 b_6 [(1/2)(1 + b_5 + b_1 - b_1 b_5)]$$

$$= b_4 b_6 z_4'.$$

Consider the equivalent expression for z_4' in Table 8.18. Therefore, $z_4' \Rightarrow y_4' = a_1 \odot a_5$. These two assignments above leave

$$z_5 = [2b_2 - b_2 b_4 + b_2 b_6 - b_1 b_2 b_4 + b_1 b_2 b_6 + b_2 b_4 b_5$$
$$+ b_2 b_5 b_6 - b_1 b_2 b_4 b_5 + 2b_1 b_2 b_4 b_6 - b_1 b_2 b_3 b_6]/4$$

$$= b_2 [(1/4)(2 - b_4 + b_6 - b_1 b_4 + b_1 b_6 + b_4 b_5 + b_5 b_6$$
$$- b_1 b_4 b_5 + 2b_1 b_4 b_6 - b_1 b_5 b_6)]$$

$$= b_2 z_5'.$$

With the information above a truth table can be constructed for z_5' in Table 8.19. From this table one can deduce that

$$z_5' \Longrightarrow y_5' = \overline{a_1} \odot \overline{a_4} \odot a_6 \oplus a_1 \odot (a_4 \odot \overline{a_5} \odot a_6 \oplus \overline{a_4} \odot a_5 \odot \overline{a_6}).$$

Now consider the second signal represented by x_2. Let $z_0 = b_5$ and choose

$$z_1 = b_1 b_4 + b_1 b_4 b_5 - b_4 b_5 + \frac{b_4}{2}$$

$$= b_4 [1/2(1 + b_1 - b_5 + b_1 b_5)]$$

$$= b_4 z_1'.$$

TABLE 8.19 Generation of z_5' in Example 8.8.

b_1	b_4	b_5	b_6	z_5'
1	1	1	1	1
1	1	1	-1	-1
1	1	-1	1	1
1	1	-1	-1	-1
1	-1	1	1	1
1	-1	1	-1	1
1	-1	-1	1	1
1	-1	-1	-1	1
-1	1	1	1	1
-1	1	1	-1	1
-1	1	-1	1	-1
-1	1	-1	-1	1
-1	-1	1	1	1
-1	-1	1	-1	-1
-1	-1	-1	1	1
-1	-1	-1	-1	1

From Table 8.20,

$$z_1' \Longrightarrow y_1' = a_1 \odot \overline{a_5}.$$

These first two assignments just leave:

$$z_2 = b_2 b_3 \left[\frac{b_1(1 - b_4 + b_5 - b_4 b_5)}{4} + \frac{(3 + b_4 - b_5 + b_4 b_5)}{4} \right]$$

$$= b_2 b_3 [b_1 z_2' + z_2''].$$

Consider z_2, z_2', and z_2''. From Table 8.21 it can be deduced that

$$z_2 \Longrightarrow y_2 = a_2 \oplus a_3 \oplus (a_1 \odot a_4 \odot \overline{a_5}).$$

TABLE 8.20 Generation of z_1 in Example 8.8.

b_1	b_5	z_1'
1	1	1
1	-1	1
-1	1	-1
-1	-1	1

TABLE 8.21 Generation of z_2 in Example 8.8.

b_4	b_5	z_2'	z_2''	z_2
1	1	0	1	$b_2 b_3$
1	-1	0	1	$b_2 b_3$
-1	1	1	0	$b_1 b_2 b_3$
-1	-1	0	1	$b_2 b_3$

In summary,

$$y_5 = a_2 \oplus [\overline{a_1} \odot \overline{a_4} \odot a_6 \oplus a_1 \odot (a_4 \odot \overline{a_5} \odot a_6 \oplus \overline{a_4} \odot a_5 \odot \overline{a_6})]$$

$$y_4 = a_4 \oplus a_6 \oplus (a_1 \odot a_5)$$

$$y_3 = a_5$$

$$y_2 = a_2 \oplus a_3 \oplus (a_1 \odot a_4 \odot \overline{a_5})$$

$$y_1 = a_4 \oplus (a_1 \odot \overline{a_5})$$

$$y_0 = a_5.$$

This particular convolutional encoder was not drawn because of the difficulty, especially in the y_5 term. However, the defining equations are given above. ☐

8.4 Bit Error Rate Performance

The performance of nonrotationally invariant trellis was covered in Chapter 4. A basic assumption leading to a simplified procedure for performance evaluation was that the underlying convolutional code is *linear*, and we comment on this further below. In the case of rotationally invariant codes the underlying convolutional code may not be linear. If this code is nonlinear, the theory on simplified performance analysis developed in Chapter 4 does not hold. To get a feel for the performance of such cases, we present some digital computer simulation results. When the underlying convolutional encoder is linear, we can analyze it with the techniques presented in Chapter 4. We begin with the nonlinear case.

8.4.1 Nonlinear Codes

We consider the eight-state, 8-PSK code studied in Section 8.1.2. The trellis diagram is given in Fig. 8.7. The code is rotationally invariant to a 90° rotation. The realization of this code was presented in Example 8.3. The encoder structure is given in Fig. 8.14 and the encoder is nonlinear.

As stated above, for nonlinear structure like that in Fig. 8.14, the theory for bit error rate analysis in Chapter 4 does not hold. Instead, we use the digital computer simulation program that was used to verify the goodness of the bounds in that chapter to determine the bit error rate performance of nonlinear trellis codes. We

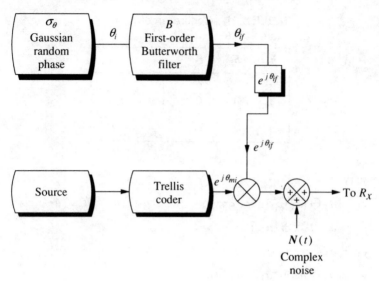

FIGURE 8.30 Model for first-order phase jitter and noise channel.

simulate the AWGN channel with phase jitter shown in Fig. 8.30. As such, we demonstrate the phase jitter sensitivity of both optimum codes and RIC. The results are taken from [8]. Note that the phase jitter is slowly varying. The phase jitter is a Gaussian random variable [11] with zero mean and standard deviation σ_θ. The time variation of the θ is governed by a first-order Butterworth filter with bandwidth B.

The bit error rate of the optimum eight-state, 8-PSK code in the presence of phase jitter is shown in Fig. 8.31. The trellis for this code is given in Example 5.2. The bandwidth of the phase jitter, normalized to the data rate, is 0.05. Also shown is the phase jitter sensitivity of 4-PSK, the uncoded modulation. These results for 4-PSK agree with Prabhu's [11] analytical results. One notes that an rms phase jitter of 7° causes a 1.7-dB loss in SNR at $P_b = 10^{-5}$. This loss is about the same as that for the uncoded case, 4-PSK.

The sensitivity of the eight-state, 8-PSK RIC is given in Fig. 8.32. The trellis for this code is given in Fig. 8.7. One notes that the RIC is no more sensitive to phase jitter than the optimum code shown in Fig. 8.31. The phase jitter sensitivities of other codes are given in [8].

8.4.2 Linear Codes

In Section 8.3.1 we found that multidimensional codes could yield rotational invariance and still retain a linear convolutional encoder. This makes them the same as optimum codes and they are also "uniform" in the sense defined in Chapter 4. Thus they can be analyzed using the transfer function bound and an error-state diagram having the same number of states as the encoder. The results to be given are taken from [10] and are based on the error weight profile technique of Zehavi and Wolf [12].

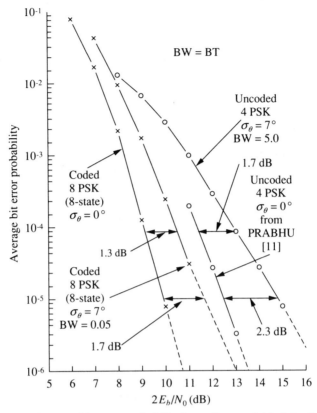

FIGURE 8.31 Average bit error probability for optimum eight-state, 8-PSK code.

As an example consider the four-dimensional RIC considered in Example 8.7. This is a 4/8-PSK code and the trellis diagram is shown in Fig. 8.22. From the trellis diagram in Fig. 8.22 we derive the error-state diagram in Fig. 8.33. The error weight profile for 4/8-PSK is given in Table 8.22. This table is used to derive the branch gains in the error-state diagram in Fig. 8.33. Note that [10] the error weight profile of the Zehavi–Wolf technique [12] can be specified quite simply in a single table.

The solution for transfer function $T(D, z)$ in the error-state diagram in Fig. 8.33 is

$$T(D, z) = g_1 + \frac{g_3 g_5 [g_6 g_7 + g_4(1 - g_8)]}{(1 - g_2 g_4)(1 - g_8) - g_2 g_8 g_6 g_7}.$$

The theory for deriving $T(D, z)$ was given in Chapter 4. Thus if we neglect the precoder, we obtain

$$P_b \leq \frac{\lambda}{8} \operatorname{erfc}\left(\sqrt{\frac{2E_b}{N_o}}\right) D^{-4} \frac{d}{dz} T(D, z)\bigg|_{z=1} \qquad (8.16)$$

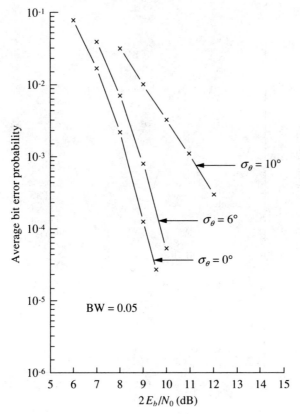

FIGURE 8.32 Average bit error probability for rotationally invariant, eight-state, 8-PSK code.

where λ is the effect of the precoder. Now [13] shows that for $\epsilon \ll 1$

$$\frac{d}{dz}T(D, z)\Big|_{z=1} < \frac{T(D, 1+\epsilon) - T(D, 1)}{\epsilon}.$$

The derivative of the transfer function $T(D, z)$ can be difficult to derive and the finite difference above actually preserves the upper bound. We usually use $\epsilon = 10^{-6}$ or 10^{-7}. This allows the bound in (8.16) to be evaluated numerically. A lower bound to P_b is

$$P_b \geq \frac{1}{8}\,\mathrm{erfc}\left(\sqrt{\frac{2E_b}{N_0}}\right). \tag{8.17}$$

Our final result will give both an upper and lower bound to P_b.

We now determine the effect of the required precoder on the result in (8.16); that is, we find λ. The lower bound in (8.17) remains so under precoding as this operation only increases the error rate. The required rotational invariance is 90°, as phase tracking occurs only over the 4-PSK symbol in the 4/8-PSK signal constellation. Thus the required precoder is the two-dimensional one given in (8.5)

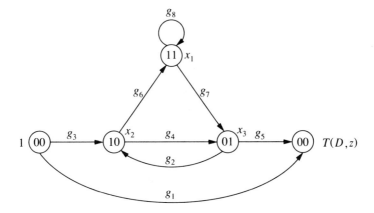

$$g_1 = \frac{g_2}{z} - 1 = (2z + 3z^2 + z^3)D^4 + zD^8$$

$$g_3 = zg_5 = (z + 2z^2 + z^4)D^2 + (z^2 + 3z^3)D^6$$

$$g_4 = \frac{g_8}{z} = D^{0.58} + \frac{1}{2}(z + 2z^2 + z^3)(D^{2.58} + D^{5.41}) + z(D^{3.41} + D^{7.41}) + z^2 D^{4.58}$$

$$g_6 = zg_7 = \frac{1}{2}(z + z^2)(D^{0.58} + D^{3.41}) + \frac{1}{2}(z^2 + z^3)(D^{4.58} + D^{7.41})$$
$$+ (z^2 + z^4)D^{2.58} + 2z^3 D^{5.41}$$

FIGURE 8.33 Error-state diagram for rotationally invariant, four-state, 4/8-PSK TCM.

TABLE 8.22 Error Weight Profiles for 4/8-PSK.

Error Vector, E	Squared Euclidean Error Weight, $d^2(C;E)$	Weight Profile of Subsets B_0 and B_1
00000	0	16
00001	0.5858	$16D^{0.5858}$
00010	2	$16D^2$
00011	0.5858, 3.4142	$8D^{0.5858} + 8D^{3.4142}$
00100	4	$16D^4$
00101	3.4142	$16D^{3.4142}$
00110	2	$16D^2$
00111	0.5858, 3.4142	$8D^{0.5858} + 8D^{3.4142}$

To derive the remaining error weights,

$$d^2(01e_1e_2e_3) = d^2(11e_1e_2e_3) = d^2(00e_1e_2e_3) + 2$$
$$d^2(10e_1e_2e_3) = d^2(00e_1e_2e_3) + 4$$

where the e_i can be 0 or 1, and $d^2(E)$ is the squared Euclidean error weight due to vector E.

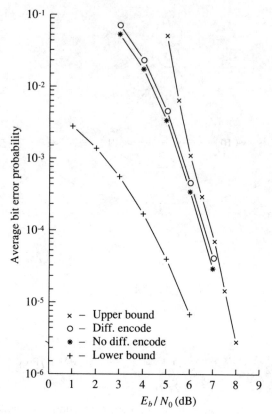

FIGURE 8.34 Performance of rotationally invariant, four-state, 4/8-PSK TCM in AWGN.

and (8.6). The decoder is given in (8.8) and (8.9). This decoder has a memory of 2, and at worst the error rate is doubled for the precoded bits. There is no effect for the 2 bits in Fig. 8.27 that are not precoded. Let $\tilde{P}_b$ be the bound in (8.16) with no precoding taken into account; that is, $\lambda = 1$. Then as 2 bits of 4 are precoded,

$$P_b \leq \frac{2\tilde{P}_b + 2\tilde{P}_b + \tilde{P}_b + \tilde{P}_b}{4} = \frac{3}{2}\tilde{P}_b$$

and thus $\lambda = 3/2$. The upper bound in (8.16) with $\lambda = 3/2$ and the lower bound in (8.17) are displayed in Fig. 8.34. Also given is a simulation result to see the quality of the error bounds. Other examples can be found in [9, 10].

REFERENCES

1. W. C. LINDSEY and M. K. SIMON, *Telecommunication Systems Engineering*. Prentice-Hall, Englewood Cliffs, N.J., 1973.

2. G. UNGERBOECK, "Channel coding with multi-level/phase signals," *IEEE Trans. Inf. Theory*, Vol. IT-28, pp. 55–67, Jan. 1982.
3. G. UNGERBOECK, "Trellis-coded modulation with redundant signal sets. Part I: Introduction," *IEEE Commun. Mag.*, Vol. 25, pp. 5–12, Feb. 1987.
4. G. UNGERBOECK, "Trellis-coded modulation with redundant signal sets. Part II: State of the art," *IEEE Commun. Mag.*, Vol. 25, pp. 12–22, Feb. 1987.
5. L.-F. WEI, "Rotationally invariant convolutional channel coding with expanded signal space. Part I: 180°," *IEEE J. Select. Areas Commun.*, Vol. SAC-2, pp. 659–671, Sept. 1984.
6. L.-F. WEI, "Rotationally invariant convolutional channel coding with expanded signal space. Part II: Nonlinear codes," *IEEE J. Select. Areas Commun.*, Vol. SAC-2, pp. 672–686, Sept. 1984.
7. Z. C. ZHU and A. P. CLARK, "Rotationally invariant coded PSK signals," *IEEE Proc.*, Part F, Vol. 134, pp. 43–52, Feb. 1987.
8. A. FUNG and P. MCLANE, "Phase jitter sensitivity of rotationally invariant 8- and 16-point trellis codes," *Globecom'88*, Hollywood, Fla., Nov. 28–Dec. 1, 1988.
9. R. BUZ and P. J. MCLANE, "Error bounds for multi-dimensional TCM," *Proc. IEEE International Conference on Communications*, Boston, Mass., June 11–14, 1989.
10. R. BUZ, "Design and performance analysis of multi-dimensional trellis coded modulation," M.Sc. thesis, Department of Electrical Engineering, Queen's University, Kingston, Ontario, Canada, Feb. 1989.
11. V. K. PRABHU, "PSK performance with imperfect carrier recovery," *IEEE Trans. Aero. Electron. Syst.*, Vol. AES-12, pp. 275–286, Mar. 1976.
12. E. ZEHAVI and J. K. WOLF, "On the performance evaluation of trellis codes," *IEEE Trans. Inf. Theory*, Vol. IT-33, pp. 196–202, Mar. 1987.
13. A. J. VITERBI and J. K. OMURA, *"Principles of Digital Communications and Coding.* McGraw-Hill, New York, N.Y., 1979.

Analysis and Performance of TCM for Fading Channels

Thus far in the book we have discussed the design and performance of TCM for the idealized additive white Gaussian noise (AWGN) channel. We have seen that the primary advantage of TCM over modulation schemes employing traditional error correction coding is its ability to achieve increased power efficiency without the customary expansion of bandwidth introduced by the coding process. Thus any channel that is both power-limited and bandwidth-limited would be ideally suited to TCM.

One such application is the mobile satellite channel, where bandwidth is constrained by the desire to accommodate a large number of users in a given transmission bandwidth (e.g., FDMA), and power is constrained by the flux density limitation of the satellite's radiated transmission and the physical size of the mobile's antenna. In addition to the usual additive thermal noise background, the mobile satellite channel is also impaired by Doppler frequency shift due to vehicle motion, the potential of a nonlinear channel due to the HPA in the transmitter, voice delay, and multipath fading and shadowing. Perhaps the most serious of these impairments is the latter, where for reliable performance the system must be able to combat short fades and recover quickly from long fades.

For most mobile satellite channels, fading can be assumed to be modeled by a Rician amplitude probability distribution [1], which not only introduces an error floor into the system (the level of which depends on the dynamics of the fading process) but also makes the problem of carrier recovery more difficult. Depending on the ratio of direct and specular (coherent component) to diffuse (noncoherent component) signal power, one might even be required to employ differentially coherent or noncoherent detection techniques, thus sacrificing the power savings associated with coherent detection. In heavy terrain, where severe shadowing is a problem, the channel might become log normally distributed [2] or even become bad enough to resemble a terrestrial channel characterized by Rayleigh fading. Thus, if TCM is to be employed on terrestrial or satellite mobile channels, it is essential that one assess its performance in such an environment.

In this chapter we present the performance of TCM when transmitted over a fading channel. In view of the discussion above, we consider both Rician and Rayleigh fading channels, coherent and differentially coherent detection, the presence and absence of channel state information (CSI) (i.e., information derived from the channel that can be used to design the decoding metric to give improved performance), and slow- and fast-fading dynamics. The results will be obtained from a combination of analysis and simulation. We consider only the case where the combination of interleaving and deinterleaving is employed to further combat the fading. This allows for considerable simplification of the analysis (as will be explained shortly) and is of great practical interest.

9.1 Coherent Detection of Trellis-Coded *M*-PSK on a Slow-Fading Rician Channel

9.1.1 Channel Model

The slow-fading Rician channel model is discussed in detail in Appendix A. For our purpose here it is sufficient to assume that the effect of the fading on the phase of the received signal is fully compensated for either by tracking it with some form of phase-locked loop or with pilot tone calibration techniques [3, 4]. Thus our results will reflect only the degradation due to the effect of the fading on the *amplitude* of the received signal. This amplitude is modeled by Rician statistics with parameter K representing the ratio of the power in the direct or line-of-sight (LOS) plus specular components to that in the diffuse component. If shadowing is severe, or if we are dealing with a terrestrial channel, a Rayleigh statistical model becomes appropriate, which can be looked upon as the limiting case of a Rician channel when K approaches zero. Of course, the case of no fading corresponds to a Rician channel with K approaching infinity.

Mathematically speaking, the statements correspond to a pdf for the normalized (unit mean-squared value) amplitude fading random variable, ρ, given by

$$p(\rho) = \begin{cases} 2\rho(1 + K)\exp\{-K - \rho^2(1 + K)\}I_0(2\rho\sqrt{K(1 + K)}); & \rho \geq 0 \\ 0; & \rho < 0 \end{cases}$$

(9.1)

where $I_0(x)$ is the zero-order modified Bessel function of the first kind.

9.1.2 System Model

Figure 9.1 is a block diagram of the end-to-end system under investigation. Input bits representing data or digitally encoded speech are passed through a rate b/s trellis encoder.[1] The appropriately mapped encoder output symbols are then block interleaved[2] to break up burst errors caused by amplitude fades of duration greater than one symbol time. Although in practice the depth of interleaving is finite and chosen in relation to the maximum fade duration anticipated, for the purpose of analysis we shall make the assumption of infinite interleaving depth. This assumption provides a memoryless channel for which well-known bit error probability bounding techniques can be applied. The simulation results will, however, reflect a finite interleaving depth. Thus these results will be slightly pessimistic when

---

[1] We shall consider the general case of multiple trellis encoding as discussed in Chapter 7. For the specific case of conventional trellis coding, we would set $b = m$ and $s = m + 1$.

[2] For simplicity of analysis, we assume *block* interleaving/deinterleaving although other forms of interleaving/deinterleaving would yield similar results [5].

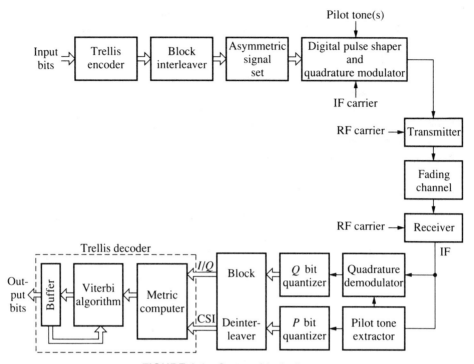

FIGURE 9.1 System block diagram.

compared with those derived from theory. The primary purpose of the analysis is to indicate the trend of the performance behavior as a function of the various system parameters leaving the actual numerical performance to be predicted by the software simulations. A further purpose of the analysis is to allow the derivation of a simple asymptotic (high-SNR) performance bound that clearly elucidates the meaningful system parameters that (1) affect performance and (2) influence the design of trellis codes for such channels. The latter is discussed in Chapter 10.

Continuing with the description of the system block diagram, we find that groups of s interleaving symbols are mapped into the M-PSK signal set according to the set-partitioning method discussed in earlier chapters for conventional and multiple trellis-coded modulations. The in-phase and quadrature components of the mapped signal point are digitally pulse shaped (to limit intersymbol interference) and modulated onto quadrature carriers for transmission over the channel. If pilot tone calibration techniques [3, 4] are used to recover the faded carrier at the receiver, the pilot tone (or tones) must be added to the data-modulated signal above before transmission.

At the receiver, the faded, noise-corrupted in-phase and quadrature signal components are demodulated with the extracted pilot tone(s), q-bit quantized for soft decision decoding, and then block deinterleaved. The metric chosen for the Viterbi decoding algorithm in the decoder depends on whether or not channel state information is provided [6]. As indicated in Fig. 9.1, a measure of CSI can be obtained from the power in the recovered pilot tone(s). Furthermore, the number of bits of quantization, p, for this operation can be much smaller than q since the accuracy of the CSI has only a secondary effect compared with that of the soft decisions themselves. Finally, the tentative soft decisions from the Viterbi decoder are then stored in a buffer whose size is a design parameter. For example, for the case of speech transmission, the total coding/decoding delay must be kept below about 60 ms so as not to be objectionable to the listener. Thus, for a given input bit rate, the decoder buffer and interleaving frame sizes must be limited so as to produce at most a 60-ms delay. Again for simplicity, we shall assume an infinite buffer in the analysis, whereas the simulations will reflect a finite buffer size in accordance with the delay limitation. Much of what follows in the remainder of Section 9.1 was originally described in [7].

9.1.3 Upper Bound on Pairwise Error Probability

The basic analysis model for the system of Fig. 9.1 is illustrated in Fig. 9.2. The input sequence, **u**, consists of binary data at a rate R_b bits/s, which are trellis encoded and mapped into a sequence, **x**, of M-ary symbols at a rate $R_s = (b/k)R_b$ symbols/s. These symbols serve as the input to the coding channel symbols, which, due to the assumption of adequate (theoretically infinite) interleaving and deinterleaving, is memoryless. Corresponding to **x**, the channel outputs the sequence **y**, whose kth element, y_k, is related to the kth element of **x**, namely, x_k, by

$$y_k = \rho_k x_k + n_k. \tag{9.2}$$

Here x_k represents the transmitted M-PSK symbol at time k, y_k represents the corresponding output, ρ_k is a normalized random amplitude with pdf given by (9.1), and n_k is a sample of a zero-mean Gaussian noise process with variance σ^2. If side information is available, the corresponding side information sequence will be denoted by **z**.

In view of the assumption of a memoryless channel, the ρ_k's are independent random variables and hence the channel probabilities satisfy

$$p_N(\mathbf{y} \mid \mathbf{x}, \mathbf{z}) = \prod_{n=1}^{N} p(y_n \mid x_n, z_n) \tag{9.3}$$

and

$$q_N(\mathbf{z}) = \prod_{n=1}^{N} q(z_n) \tag{9.4}$$

where N denotes the length of the sequences **x**, **y**, **n**, and **z**.

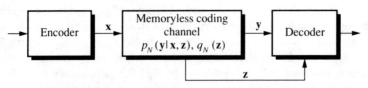

FIGURE 9.2 General memoryless channel.

For any coded communication system, the decoding process uses a metric of the form $m(\mathbf{y}, \mathbf{x}; \mathbf{z})$ if side information is available and $m(\mathbf{y}, \mathbf{x})$ if it is not. Whatever metric is selected, it is desirable from the standpoint of simplifying the decoding process that it have an additive property, namely, that the total metric for a sequence of symbols is the sum of the metrics for each channel input and output pair. In terms of the above, this takes the form

$$m(\mathbf{y}, \mathbf{x}; \mathbf{z}) = \sum_{n=1}^{N} m(y_n, x_n; z_n). \tag{9.5}$$

The maximum-likelihood metric

$$m(\mathbf{y}, \mathbf{x}; \mathbf{z}) = \ln p_N(\mathbf{y}, \mathbf{x}; \mathbf{z}) \tag{9.6a}$$

when side information is available or

$$m(\mathbf{y}, \mathbf{x}) = \ln p_N(\mathbf{y}, \mathbf{x}) \tag{9.6b}$$

when it is not satisfies the requirements in (9.5). Thus we shall use this metric in what follows. Also, for simplicity, we shall use only the notation of (9.6a), keeping in mind that the case where no side information is available is a special case where the metric $m(\mathbf{y}, \mathbf{x}; \mathbf{z})$ does not depend on $\mathbf{z}$.

The pairwise error probability $P(\mathbf{x} \rightarrow \hat{\mathbf{x}})$, which represents the probability of choosing the coded sequence $\hat{\mathbf{x}} = (\hat{x}_1, \hat{x}_2, \ldots, \hat{x}_N)$ when indeed $\mathbf{x} = (x_1, x_2, \ldots, x_N)$ was transmitted, with $\mathbf{x}$ and $\hat{\mathbf{x}}$ the only choices, is clearly given by

$$P(\mathbf{x} \rightarrow \hat{\mathbf{x}}) = \Pr\{m(\hat{\mathbf{y}}, \hat{\mathbf{x}}; \hat{\mathbf{z}}) \geq m(\mathbf{y}, \mathbf{x}; \mathbf{z}) \mid \mathbf{x}\}. \tag{9.7}$$

Using (9.5) in (9.7) and applying a Chernoff bound[3] results in

$$P(\mathbf{x} \rightarrow \hat{\mathbf{x}}) \leq \prod_{n \in \eta} E\{\exp(\lambda[m(y_n, \hat{x}_n; z_n) - m(y_n, x_n; z_n)]) \mid \mathbf{x}\} \tag{9.8}$$

where η is the set of all n such that $x_n \neq \hat{x}_n$. To simplify (9.8) any further, we must specify whether or not side information is available.

Ideal channel state information

The assumption of ideal channel state information is tantamount to assuming that the side information random variable z_n is equal to the fading amplitude ρ_n.

---

[3] In its simplest form, the Chernoff bound is given as follows. If x is a continuous random variable, then

$$\Pr\{x \geq 0\} \leq E\{\exp(\lambda x)\}$$

where $\lambda \geq 0$ is a parameter to be optimized. If x is itself the sum if N independent continuous random variables $x_1, x_2, \ldots, x_N$, then the above becomes

$$\Pr\left\{\sum_{n=1}^{N} x_i \geq 0\right\} \leq \prod_{n=1}^{N} E\{\exp(\lambda x_i)\}.$$

Thus, since n_n is Gaussian distributed and in general x_n and y_n are complex quantities, using (9.2) and (9.6a) gives

$$m(y_n, x_n; z_n) = -|y_n - \rho_n x_n|^2 \qquad (9.9)$$

where for simplicity we ignore the $1/2\sigma^2$ factor since later it would be absorbed in the Chernoff bound parameter anyway. Substituting (9.9) into (9.8) and conditioning on $\rho = (\rho_1, \rho_2, \ldots, \rho_N)$, we get

$$P(\mathbf{x} \to \hat{\mathbf{x}} | \rho) \leq \prod_{n \in \eta} \exp\{-\lambda \rho_n^2 |x_n - \hat{x}_n|^2\} E\{\exp[-2\lambda \rho_n \mathrm{Re}\{n_n(x_n - \hat{x}_n)*\}]\}. \qquad (9.10)$$

If we represent the complex noise n_n in terms of its real and imaginary parts where $\mathrm{Re}\{n_n\}$ and $\mathrm{Im}\{n_n\}$ are uncorrelated zero-mean Gaussian random variables, each with variance $\sigma_N^2 = \sigma^2/2$, it can be shown that

$$E\{\exp[-2\lambda \rho_n \mathrm{Re}\{n_n(x_n - \hat{x}_n)*\}]\} = \exp[2\lambda^2 \rho_n^2 \sigma_N^2 |x_n - \hat{x}_n|^2]. \qquad (9.11)$$

Substituting (9.11) into (9.10) gives after some simplification

$$P(\mathbf{x} \to \hat{\mathbf{x}} | \rho) \leq \prod_{n \in \eta} \exp\{-\lambda \rho_n^2 |x_n - \hat{x}_n|^2 (1 - 2\lambda \sigma_N^2)\}. \qquad (9.12)$$

Optimizing (9.12) over the Chernoff parameter yields

$$\lambda_{\mathrm{opt}} = \frac{1}{4\sigma_N^2} \qquad (9.13)$$

which when substituted in (9.12) produces the desired result,

$$P(\mathbf{x} \to \hat{\mathbf{x}} | \rho) \leq \exp\left\{-\frac{1}{8\sigma_N^2} d^2(\mathbf{x}, \hat{\mathbf{x}})\right\} \qquad (9.14)$$

where

$$d^2(\mathbf{x}, \hat{\mathbf{x}}) = \sum_{n \in \eta} \rho_n^2 |x_n - \hat{x}_n|^2 \qquad (9.15)$$

represents the square of the *weighted* Euclidean distance between the two symbol sequences $\mathbf{x}$ and $\hat{\mathbf{x}}$. Recalling that the elements of $\mathbf{x}$ are normalized to unit power, we find that the noise variance σ_N^2 can be related to the system symbol energy-to-noise spectral density ratio, E_s/N_0, by

$$\frac{1}{2\sigma_N^2} = \frac{E_s}{N_0} = \left(\frac{b}{k}\right) \frac{E_b}{N_0} \qquad (9.16)$$

where E_b/N_0 is the system bit energy-to-noise spectral density ratio.

Finally, using (9.16) in (9.14), the unconditional pairwise error probability upper bound of (9.8) is obtained by averaging (9.14) over the pdf of ρ [an N-fold product of (9.1)] resulting in

9.1 / Coherent Detection of Trellis-Coded M-PSK on a Slow-Fading Rician Channel

$$P(\mathbf{x} \to \hat{\mathbf{x}}) \leq \prod_{n \in \eta} \frac{1+K}{1+K+\dfrac{\overline{E_s}}{4N_0}|x_n - \hat{x}_n|^2} \exp\left\{-\frac{K\dfrac{\overline{E_s}}{4N_0}|x_n - \hat{x}_n|^2}{1+K+\dfrac{\overline{E_s}}{4N_0}|x_n - \hat{x}_n|^2}\right\} \tag{9.17}$$

which can be put in the form

$$P(\mathbf{x} \to \hat{\mathbf{x}}) \leq \exp\left(-\frac{\overline{E_s}}{4N_0}d^2\right) \tag{9.18}$$

with[4]

$$d^2 = \sum_{n \in \eta} \left\{ \underbrace{\frac{|x_n - \hat{x}_n|^2 K}{1+K+\dfrac{\overline{E_s}}{4N_0}|x_n - \hat{x}_n|^2}}_{d_{1n}^2} + \underbrace{\left(\frac{\overline{E_s}}{4N_0}\right)^{-1} \ln\left(\frac{1+K+\dfrac{\overline{E_s}}{4N_0}|x_n - \hat{x}_n|^2}{1+K}\right)}_{d_{2n}^2} \right\}. \tag{9.19}$$

Here the overbar on E_s/N_0 denotes the *average* symbol energy-to-noise spectral density ratio as a result of averaging over the normalized fading random amplitudes.

Note that for $K = \infty$ (no fading),

$$d_{1n}^2 = |x_n - \hat{x}_n|^2 \qquad d_{2n}^2 = 0 \tag{9.20}$$

and thus d^2 is merely the sum of the squared Euclidean distances along the error event path.

For $K = 0$ (Rayleigh fading),

$$d_{1n}^2 = 0$$

$$d_{2n}^2 = \left(\frac{\overline{E_s}}{4N_0}\right)^{-1} \ln\left(\frac{1+K+\dfrac{\overline{E_s}}{4N_0}|x_n - \hat{x}_n|^2}{1+K}\right) \tag{9.21}$$

and thus for reasonably large $\overline{E_s}/N_0$ values, d^2 is the sum of the *logarithms* of the squared Euclidean distances (each weighted by $\overline{E_s}/4N_0$). Equivalently, the upper bound on pairwise error probability for this special case (large $\overline{E_s}/N_0$) becomes

$$P(\mathbf{x} \to \hat{\mathbf{x}}) \leq \left(\prod_{n \in \eta} \frac{\overline{E_s}}{4N_0}|x_n - \hat{x}_n|^2\right)^{-1} \tag{9.22}$$

that is, it is inversely proportional to the *product* of the squared Euclidean distances along the error event path.

---

[4] Note that d satisfies the conditions for a distance metric (see Appendix 9A).

For values of K between 0 and ∞, the equivalent squared Euclidean distance of (9.19) will be a mixture of the two special cases above.

No channel state information

When no channel state information is available, then the metric of (9.6b) becomes

$$m(y_n, x_n) = -|y_n - x_n|^2. \tag{9.23}$$

Substituting (9.2) into (9.23), then, analogous to (9.10), we now get

$$P(\mathbf{x} \to \hat{\mathbf{x}}|\mathbf{\rho}) \leq \prod_{n \in \eta} \exp\{-\lambda |x_n - \hat{x}_n|^2 - 2\lambda(\rho_n - 1)\text{Re}\{x_n(x_n - \hat{x}_n)*\}\}$$

$$\times E\{\exp[-2\lambda\rho_n \text{Re}\{n_n(x_n - \hat{x}_n)*\}]\}. \tag{9.24}$$

Again using (9.11), (9.24) simplifies to

$$P(\mathbf{x} \to \hat{\mathbf{x}}|\mathbf{\rho}) \leq \prod_{n \in \eta} \exp[-\lambda|x_n - \hat{x}_n|^2(1 - 2\lambda\sigma_N^2) - 2\lambda(\rho_n - 1)\text{Re}\{x_n(x_n - \hat{x}_n)*\}]. \tag{9.25}$$

For constant envelope signal sets such as M-PSK where $|x|^2 = |\hat{x}|^2$, (9.25) can be further simplified by noting that

$$|x - \hat{x}|^2 = 2\,\text{Re}\{x(x - \hat{x})*\}. \tag{9.26}$$

Thus substituting (9.26) into (9.25) and renormalizing the Chernoff parameter (i.e., replacing λ by $2\lambda\sigma_N^2$), we get the desired result analogous to (9.12),

$$P(\mathbf{x} \to \hat{\mathbf{x}}|\mathbf{\rho}) \leq \exp\left\{-\frac{1}{8\sigma_N^2}c^2(\mathbf{x}, \hat{\mathbf{x}}|\mathbf{\rho}, \lambda)\right\} \tag{9.27}$$

where

$$c^2(\mathbf{x}, \hat{\mathbf{x}}|\mathbf{\rho}, \lambda) = \sum_{n \in \eta} 4\lambda(\rho_n - \lambda)|x_n - \hat{x}_n|^2$$

$$= 4\lambda \sum_{n \in \eta} \rho_n |x_n - \hat{x}_n|^2 - 4\lambda^2 d^2(\mathbf{x}, \hat{\mathbf{x}}). \tag{9.28}$$

Note that unlike (9.12), (9.27) cannot be optimized over λ to yield a value for this parameter that is independent of the summation index n. Thus, in this case, we must first average over the fading pdf.

9.1.4 Upper Bound on Bit Error Probability

To derive the upper bound on bit error probability from the pairwise error probability upper bound, we follow the transfer function approach discussed in Chapter 8 for the AWGN channel. In particular, we first find the unconditional pairwise error probability by averaging (9.14) or (9.27) over the pdf of ρ. For the ideal CSI case, this average can be obtained in closed form [i.e., (9.17) or (9.18)], whereas for the no CSI case, it cannot. In either case, however, the unconditional pairwise error probability upper bound can be put in the form

$$P(\mathbf{x} \to \hat{\mathbf{x}}) \leq \prod_{n \in \eta} \exp\left\{-\frac{E_s}{4N_0}\delta_n^2\right\} = \prod_{n \in \eta} \overline{Z^{\delta_n^2}} \qquad (9.29)$$

where, as in previous chapters, $Z = \exp\{-E_s/4N_0\}$ is the Bhattacharyya parameter,

$$\begin{aligned} \delta_n^2 &= \rho_n^2 |x_n - \hat{x}_n|^2 & \text{(ideal CSI)} \\ \delta_n^2 &= 4\lambda(\rho_n - \lambda)|x_n - \hat{x}_n|^2 & \text{(no CSI)} \end{aligned} \qquad (9.30)$$

and the overbar denotes averaging over ρ_n. Since, for no fading, (9.29) also applies with no overbar and $\delta_n^2 = |x_n - \hat{x}_n|^2$, then by analogy with previous results, the average bit error probability P_b for the slow-fading channel is upper bounded by[5]

$$P_b \leq \frac{1}{2b}\frac{\partial}{\partial I}\overline{T}(D, I)\bigg|_{I=1, D=Z} \qquad (9.31)$$

where $\overline{T}(D, I)$ is the transfer function of the superstate transition diagram [8] whose branch label gains are modified from those for the no-fading case as follows.

In the absence of fading, each branch is labeled with a gain of the form [see (7.56)][6]

$$G = \sum \frac{1}{b} I^\Omega \prod_{j=1}^{k} D^{\delta_j^2} \qquad (9.32)$$

where Ω is the Hamming distance between input bit sequences, and as mentioned above, $\delta_n^2 = |x_n - \hat{x}_n|^2$. Also, the summation in (9.32) accounts for the possible existence of parallel paths in the trellis diagram.

For the fading case, we simply replace G by $\overline{G}$, where[7]

$$\overline{G} = \sum \frac{1}{b} I^\Omega \prod_{j=1}^{k} \overline{D^{\delta_n^2}} \qquad (9.33)$$

with δ_n^2 given by either of the two equations in (9.30) as appropriate. Also, for the case of no CSI, we must further minimize the upper bound of (9.31) over the Chernoff parameter to obtain the tightest bound.

---

[5] This is a somewhat looser bound than that in (7.54) but easier to evaluate.

[6] For reasons that will soon become apparent, we write the factor D^{δ^2} (δ^2 being the total squared Euclidean distance between the k M-PSK symbols corresponding to the transition between the pair states) as a k-fold product of factors $D^{\delta_j^2}$, where the $\{\delta_j^2; j = 1, 2, \ldots, k\}$ are the squared Euclidean distances between *each* of the k symbols.

[7] It is important to emphasize that in the MTCM case, each $D^{\delta_j^2}$ in the product of these factors must be replaced by $\overline{D^{\delta_j^2}}$. The reason for this is that the interleaving is done *per M-PSK symbol*. Thus each symbol in the multiplicity is affected independently by the fading.

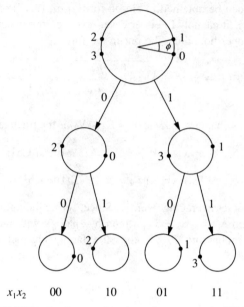

FIGURE 9.3 Set partitioning of asymmetric QPSK.

EXAMPLE 9.1

Consider the case of conventional rate 1/2 trellis-coded QPSK using a two-state trellis. The appropriate set partitioning is illustrated in Fig. 9.3, the trellis diagram in Fig. 9.4, and the superstate transition diagram in Fig. 9.5. The performance of this system in the absence of fading is treated in [9] with the following results:

$$T(D, I) = \frac{4ac}{1 - 2b} \qquad a = \frac{I}{2}D^4 \qquad b = \frac{I}{2}D^{4/(1+\alpha)} \qquad c = \frac{1}{2}D^{4\alpha/(1+\alpha)}$$

(9.34a)

or

$$T(D, I) = \frac{ID^{\frac{4(1+2\alpha)}{1+\alpha}}}{1 - ID^{\frac{4}{1+\alpha}}}$$

(9.34b)

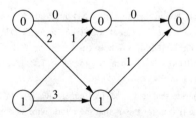

FIGURE 9.4 Trellis diagram and M-PSK signal assignment for QPSK.

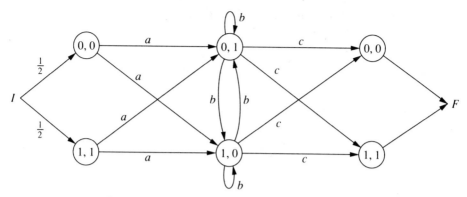

FIGURE 9.5 Pair-state transition diagram for Fig. 9.4.

where α is the ratio of powers between the I and Q channels, which is related to the angle ϕ that defines the asymmetry (see Fig. 9.3) by

$$\alpha = \tan^2 \frac{\phi}{2}. \tag{9.35}$$

Taking the derivative of (9.34b) with respect to I and substituting the result into (9.31) (letting $b = 1$ and ignoring the overbar) gives

$$P_b \leq \frac{Z^{\frac{4(1+2\alpha)}{1+\alpha}}}{\left(1 - Z^{\frac{4}{1+\alpha}}\right)^2}. \tag{9.36}$$

Optimizing (9.36) over the asymmetry produces

$$\alpha = \frac{E_b/N_0}{\ln 3} - 1 \tag{9.37}$$

which when substituted into (9.36) gives the desired upper bound

$$P_b \leq \frac{27}{4} \exp\left(-\frac{2E_b}{N_0}\right). \tag{9.38}$$

Note that here the bit energy E_b is equal to the symbol energy E_s.

Ideal Channel State Information

Recalling (9.15), we see that the transfer function $T(D, I)$ of the superstate transition diagram for the case of ideal CSI is obtained merely by replacing D^β by $\overline{D^{\rho^2\beta}}$ in the branch label gains of (9.34a) where $\beta = 4, 4/(1+\alpha)$, or $4\alpha/(1+\alpha)$ as appropriate. For the Rician pdf of (9.1), $\overline{D^{\rho^2\beta}}|_{D=Z}$ evaluates to

$$\overline{D^{\rho^2\beta}}|_{D=Z} = \left(\frac{1+K}{1+K+\beta\gamma}\right) Z^{\beta K/(1+K+\beta\gamma)} \qquad \gamma = \frac{E_b}{4N_0} \tag{9.39a}$$

where $\overline{E_b}$ denotes average bit energy. For the special case of Rayleigh fading ($K = 0$), (9.39a) becomes

$$\overline{D^{\rho^2 \beta}}\Big|_{D=Z} = \frac{1}{1+\beta\gamma}. \qquad (9.39b)$$

Evaluating $\overline{T}(D, I)\big|_{D=Z}$ of (9.34a) using (9.39) and performing the differentiation required in (9.31) gives the upper bound on P_b as[8]

$$P_b \leq \frac{\xi_1 \xi_3 Z^{\zeta_1 + \zeta_3}}{(1 - \xi_2 Z^{\zeta_2})^2} \qquad (9.40)$$

where

$$\zeta_i = \frac{\beta_i K}{1 + K + \beta_i \gamma} \qquad \xi_i = \frac{1+K}{1+K+\beta_i \gamma} \qquad i = 1, 2, 3$$

$$\beta_1 = 4 \qquad \beta_2 = \frac{4}{1+\alpha} \qquad \beta_3 = \frac{4\alpha}{1+\alpha}. \qquad (9.41)$$

To obtain the best performance in the presence of fading, one should optimize (9.40) over the asymmetry parameter α. Before doing this, however, we shall first examine the behavior of (9.40) for the symmetric case, that is, $\alpha = 1$, and the optimum asymmetry in the absence of fading as given by (9.37). Substituting $\alpha = 1$ in (9.41), the parameters ξ_i and ζ_i simplify to

$$\zeta_1 = \frac{4K}{1+K+4\gamma} \qquad \xi_1 = \frac{1+K}{1+K+4\gamma}$$

$$\zeta_2 = \zeta_3 = \frac{2K}{1+K+2\gamma} \qquad \xi_2 = \xi_3 = \frac{1+K}{1+K+2\gamma}. \qquad (9.42)$$

The curve labeled "symmetric" on Fig. 9.6 is a plot of the upper bound of (9.40) combined with (9.42) as a function of the average bit energy-to-noise ratio $\overline{E_b}/N_0$ with a Rician parameter $K = 10$ (typical of the mobile satellite channel).

When the value of α in (9.37) is substituted in (9.41), after some simplification (9.40) and (9.41) can be written as

$$P_b \leq \frac{\xi_1' \xi_3' \exp\left[-(\zeta_1' + \zeta_2')\right]}{(1 - \xi_2' \exp(-\zeta_2'))^2} \qquad (9.43)$$

with

$$\zeta_i' = \frac{\beta_i' K}{1+K+\beta_i'} \qquad \xi_i' = \frac{1+K}{1+K+\beta_i'} \qquad i = 1, 2, 3$$

$$\beta_1' = \frac{\overline{E_b}}{N_0} \qquad \beta_2' = \ln 3 \qquad \beta_3' = \frac{\overline{E_b}}{N_0} - \ln 3. \qquad (9.44)$$

---

[8] Here Z is the Bhattacharyya parameter expressed in terms of the *average* symbol energy-to-noise ratio $\overline{E_s}/N_0$. From this point on, wherever bit error probability in the presence of fading is discussed and an expression for P_b is given in terms of Z, we shall assume the above.

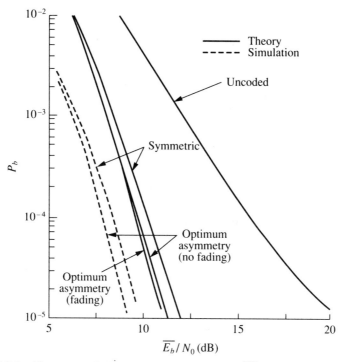

FIGURE 9.6 Bit error probability performance versus $\overline{E_b}/N_0$ for rate 1/2 trellis-coded QPSK in the presence of Rician fading; two states, $K = 10$; ideal channel state information.

The behavior of (9.43) combined with (9.44) is also illustrated in Fig. 9.6 by the curve labeled "optimum asymmetry (no fading)."

For the Rayleigh case, the results simplify even further. In particular, for the symmetric signal QPSK constellation, we get

$$P_b \leq \left(\frac{1 + \frac{\overline{E_b}}{2N_0}}{\frac{\overline{E_b}}{2N_0}}\right)^2 \left(\frac{1}{1 + \frac{\overline{E_b}}{N_0}}\right)\left(\frac{1}{1 + \frac{\overline{E_b}}{2N_0}}\right) \qquad (9.45)$$

whereas for the optimum asymmetry in the absence of fading,

$$P_b \leq \left(\frac{1 + \ln 3}{\ln 3}\right)^2 \left(\frac{1}{1 + \frac{\overline{E_b}}{N_0}}\right)\left(\frac{1}{1 + \frac{\overline{E_b}}{N_0} - \ln 3}\right). \qquad (9.46)$$

These results are illustrated in Fig. 9.7. Note that here the curve labeled "optimum asymmetry (no fading)" gives worse performance than that of the symmetric case. Thus, at least here, we clearly see the need for performing the asymmetry optimization in the presence of the fading.

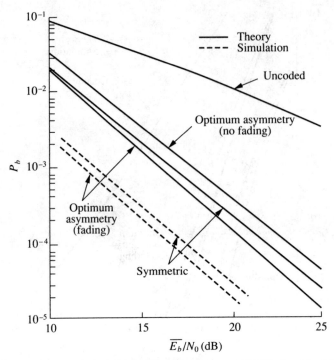

FIGURE 9.7 Bit error probability performance versus $\overline{E_b}/N_0$ for rate 1/2 trellis-coded QPSK in the presence of Rayleigh fading; two states; ideal channel state information.

To determine the optimum value of α for the Rician case, we need to differentiate (9.40) with respect to α and equate the result to zero. This leads to a transcendental equation that must be solved numerically. Rather than do that, it is more expedient to directly minimize (9.40) with respect to α using numerical techniques. When this is done, we obtain the optimum bit error probability bound labeled "optimum asymmetry (fading)" in Fig. 9.6. For $K = 10$, this curve lies quite close to the "optimum asymmetry (no fading)" curve. However, as we can already deduce from Fig. 9.7, this statement is not true for small values of K, in particular, the Rayleigh channel. To exhibit the sensitivity of the optimum asymmetry condition to the choice of K, Fig. 9.8 illustrates the optimum value of α as a function of $\overline{E_b}/N_0$ with K as a parameter.

For the Rayleigh case, we can indeed determine the optimum asymmetry condition in closed form. In particular, differentiating (9.40) with $K = 0$ in (9.41) and equating the result to zero has the solution

$$\alpha = \frac{-4 + \overline{E_b}/N_0 \left(\sqrt{17 + 8\,\overline{E_b}/N_0} - 1 \right)}{4(1 + \overline{E_b}/N_0)} \tag{9.47}$$

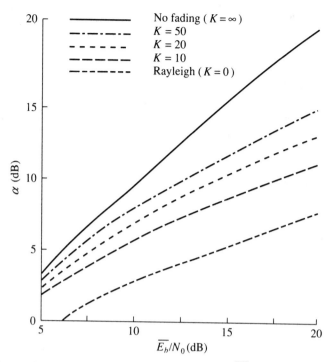

FIGURE 9.8 Optimum asymmetry parameter versus $\overline{E_b}/N_0$ for rate 1/2 trellis-coded QPSK in the presence of Rician fading.

which when substituted back in (9.40) gives

$$P_b \leq \frac{\left(3 + \sqrt{17 + 8\overline{E_b}/N_0}\right)\left[4\left(\overline{E_b}/N_0\right)^2 + \left(\overline{E_b}/N_0\right)\left(7 + \sqrt{17 + 8\overline{E_b}/N_0}\right)\right]^2}{16\left(\sqrt{17 + 8\overline{E_b}/N_0} - 1\right)\left(\overline{E_b}/N_0\right)^2 \left(1 + \overline{E_b}/N_0\right)^4}.$$

(9.48)

This result is illustrated by the curve labeled "optimum asymmetry (fading)" in Fig. 9.7 and is clearly superior to that corresponding to the symmetric signal constellation.

Finally, for purpose of comparison, the corresponding upper bound on the performance of uncoded BPSK (same bandwidth as rate 1/2 trellis-coded QPSK) in the presence of Rician and Rayleigh fading is also illustrated in Figs. 9.6 and 9.7. The analytical results for these curves are well known and are given by

$$P_b \leq \frac{1+K}{1+K+\overline{E_b}/N_0} \exp\left(-\frac{K\overline{E_b}/N_0}{1+K+\overline{E_b}/N_0}\right) \quad (9.49)$$

for the Rician channel and

$$P_b \leq \frac{1}{1 + \overline{E_b}/N_0} \qquad (9.50)$$

for the Rayleigh channel.

Depending on the shape of the bit error probability versus $\overline{E_b}/N_0$ curve, one can often deduce some important practical conclusions by examining the asymptotic behavior of the curve. Since for the Rayleigh case (Fig. 9.7), the error probability performance curves are essentially linear over a wide range of practical SNRs, one can approximately apply the asymptotic (large $\overline{E_b}/N_0$) result[9] over this domain. In particular, for large $\overline{E_b}/N_0$, (9.45), (9.46), (9.48), and (9.50), respectively, become

$$P_b \leq \frac{1}{\left(\overline{E_b}/\sqrt{2N_0}\right)^2} \quad \text{(symmetric)} \qquad (9.45')$$

$$P_b \leq \frac{1}{\left[\frac{\ln 3}{1 + \ln 3}\left(\frac{\overline{E_b}}{N_0}\right)\right]^2} \quad \text{(optimum asymmetry—no fading)} \qquad (9.46')$$

$$P_b \leq \frac{1}{(\overline{E_b}/N_0)^2} \quad \text{(optimum asymmetry—fading)} \qquad (9.48')$$

$$P_b \leq \frac{1}{\overline{E_b}/N_0} \quad \text{(uncoded)}. \qquad (9.50')$$

For example, comparing (9.45') with (9.50'), we see that the effect of coding is to change the rate of descent (later we relate this to the notion of *diversity*) of the error probability versus $\overline{E_b}/N_0$ performance from an inverse linear to an inverse square law behavior. If the QPSK constellation is now designed according to the optimum asymmetry for no fading, the performance is *worse* than that of the symmetric constellation by a factor of $(1 + \ln 3)/(\sqrt{2}\ln 3)$, or 1.3 dB. On the other hand, if the constellation is designed with the optimum asymmetry determined in the presence of fading, then, relative to the symmetric design, the performance is improved by a factor of $\sqrt{2} = 1.5$ dB. From Fig. 9.7, we see that these asymptotic results are almost achieved at an error rate of 10^{-5}.

No Channel State Information

Recalling (9.30), we see that the transfer function $\overline{T}(D, I)$ of the superstate transition diagram for the case of no channel state information is obtained by replacing D^β, this time by $\overline{D^{4\lambda(\rho - \lambda)\beta}} = D^{-4\lambda^2\beta}\overline{D^{4\lambda\beta\rho}}$ in the branch label gains of (9.34a), where again $\beta = 4, 4/(1 + \alpha)$, or $4\alpha/(1 + \alpha)$ as appropriate. Unfortunately,

---

[9] More general results for asymptotic behavior are given later in the chapter.

for the Rician distribution, the factor $\overline{D^{4\lambda\beta\rho}}$ cannot be evaluated in closed form. It can, however, be expressed as a single integral with finite limits as follows:

$$\overline{D^{4\lambda\beta\rho}}\Big|_{D=Z} = e^{-K}\left[1 - \frac{1}{\sqrt{\pi}}\int_0^\pi \eta(\theta)\exp\left[\eta^2(\theta)\right]\operatorname{erfc}\eta(\theta)\,d\theta\right]$$

$$\eta(\theta) = \frac{\lambda\beta(\overline{E_b}/2N_0)}{\sqrt{1+K}} - \sqrt{K}\cos\theta.$$

(9.51)

This integral is easily evaluated using Gauss–Chebyshev techniques; that is,

$$\int_0^\pi \eta(\theta)\exp[\eta^2(\theta)]\operatorname{erfc}\eta(\theta)\,d\theta \cong \frac{\pi}{N}\sum_{k=1}^N \eta(\theta_k)\exp\left[\eta^2(\theta_k)\right]\operatorname{erfc}\eta(\theta_k)$$

(9.52)

where $\theta_k = (2k-1)\pi/2N$.

For the Rayleigh case, we can obtain a closed-form result for this factor, since for $K = 0$, $\eta(\theta)$ becomes independent of θ. Thus

$$\overline{D^{4\lambda\beta\rho}}\Big|_{D=Z} = 1 - \sqrt{\pi}\lambda\beta\frac{\overline{E_b}}{2N_0}\exp\left[\left(\lambda\beta\frac{\overline{E_b}}{2N_0}\right)^2\right]\operatorname{erfc}\left(\lambda\beta\frac{\overline{E_b}}{2N_0}\right).$$

(9.53)

Evaluating $\overline{T}(D,I)|_{D=Z}$ of (9.34a) using (9.51) and performing the differentiation required in (9.31) gives the upper bound

$$P_b \leq \min_{\lambda \geq 0}\min_\alpha \frac{\xi_1\xi_3 Z^{-16\lambda^2(1+2\alpha)/(1+\alpha)}}{\left(1 - \xi_2 Z^{-16\lambda^2/(1+\alpha)}\right)^2}$$

(9.54)

where

$$\xi_i = e^{-K}\left[1 - \frac{1}{\sqrt{\pi}}\int_0^\pi \eta_i(\theta)\exp\left[\eta_i^2(\theta)\right]\operatorname{erfc}\eta_i(\theta)\,d\theta\right]$$

(9.55)

$$\eta_i(\theta) = \frac{\lambda\beta_i\left(\overline{E_b}/2N_0\right)}{\sqrt{1+K}} - \sqrt{K}\cos\theta$$

and β_i, $i = 1, 2, 3$, are defined in (9.41). For the Rayleigh case, one merely replaces ξ_i and $\eta_i(\theta)$ of (9.55) by

$$\xi_i = 1 - \sqrt{\pi}\eta_i\exp(\eta_i^2)\operatorname{erfc}\eta_i$$

$$\eta_i = \lambda\beta_i(\overline{E_b}/2N_0)$$

(9.56)

and performs the same minimizations required in (9.54).

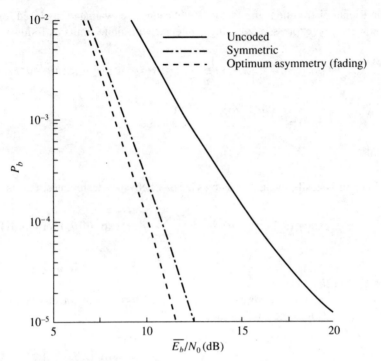

FIGURE 9.9 Bit error probability performance versus $\overline{E_b}/N_0$ for rate 1/2 trellis-coded QPSK in the presence of Rician fading; two states, $K = 10$; no channel state information.

Figures 9.9 and 9.10 illustrate the analogous results to Figs. 9.6 and 9.7 for the case where no channel state information is available.[10] Clearly, the lack of channel state information produces a noticeable degradation in system performance. To assess this additional degradation (at least for the Rayleigh channel) quantitatively, we now derive asymptotic results analogous to (9.45′), (9.46′), and (9.48′) for the no-channel state information case. In particular, we use the asymptotic (large argument) expansion for erfc x, namely,

$$\text{erfc } x \cong \frac{\exp(-x^2)}{\sqrt{\pi} x}\left[1 - \frac{1}{2x^2}\right] \tag{9.57}$$

in which case (9.53) simplifies to

$$\overline{D^{4\lambda\beta\rho}}\Big|_{D=Z} \cong \frac{1}{2\left[\lambda\beta\overline{E_b}/N_0\right]^2}. \tag{9.58}$$

---

[10] For simplicity of presentation, we have chosen not to illustrate the results for the values of optimum asymmetry determined from the no-fading analysis, since we have already made the point that asymmetry should be optimized in the fading environment.

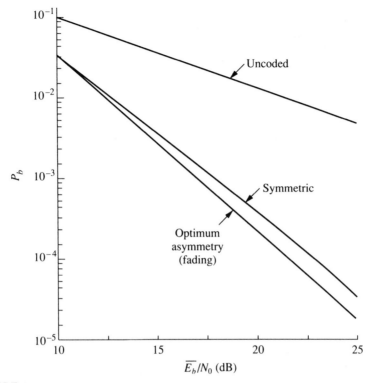

FIGURE 9.10 Bit error probability performance versus E_b/N_0 for rate 1/2 trellis-coded QPSK in the presence of Rayleigh fading; two states, no channel state information.

By using the appropriate values of β in (9.58), the branch gains of Fig. 9.5 become

$$a = \frac{1}{4\left(2\lambda\overline{E_b}/N_0\right)^2} \exp\left(4\lambda^2\overline{E_b}/N_0\right)$$

$$b = \frac{1}{4\left(\dfrac{2\lambda\overline{E_b}/N_0}{1+\alpha}\right)^2} \exp\left(\frac{4\lambda^2\overline{E_b}/N_0}{1+\alpha}\right) \qquad (9.59)$$

$$c = \frac{1}{4\left(\dfrac{2\alpha\lambda\overline{E_b}/N_0}{1+\alpha}\right)^2} \exp\left(\frac{4\alpha\lambda^2\overline{E_b}/N_0}{1+\alpha}\right).$$

Substituting (9.59) into the transfer function of (9.34a) and performing the differentiation required in (9.31) gives the approximate upper bound on P_b (valid for large $\overline{E}_b/N_0$):

$$P_b \leq \min_\lambda \frac{\exp\left(\frac{4(1+2\alpha)}{1+\alpha}\lambda^2\overline{E}_b/N_0\right)}{4(2\lambda\overline{E}_b/N_0)^4\left(\frac{\alpha}{1+\alpha}\right)^2} \quad (9.60)$$

$$\times \left\{1 - 0.5\left(\frac{2\lambda\overline{E}_b/N_0}{1+\alpha}\right)^{-2} \exp\left(\frac{4}{1+\alpha}\lambda^2\frac{\overline{E}_b}{N_0}\right)\right\}^{-2}.$$

Performing the minimization over λ required in (9.60) (actually, we minimize only the numerator, since the denominator has little effect on this operation) gives

$$\lambda_{\text{opt}} = \left[2\left(\frac{1+2\alpha}{1+\alpha}\right)\frac{\overline{E}_b}{N_0}\right]^{-1/2} \quad (9.61)$$

which when substituted in (9.60) yields

$$P_b \lesssim \left(\frac{1+2\alpha}{\alpha}\right)^2 \frac{e^2}{16(\overline{E}_b/N_0)^2}. \quad (9.62)$$

Finally, the desired asymptotic results are

$$P_b \lesssim \frac{9e^2}{16(\overline{E}_b/N_0)^2} \quad \text{(symmetric)}$$

$$P_b \lesssim \left(\frac{2\overline{E}_b/N_0 - \ln 3}{\overline{E}_b/N_0 - \ln 3}\right)^2 \frac{e^2}{16(\overline{E}_b/N_0)^2} \quad \text{(optimum asymmetry—no fading)}$$

$$P_b \lesssim \frac{e^2}{4(\overline{E}_b/N_0)^2} \quad \text{(optimum asymmetry—fading)}.$$

(9.63)

□

EXAMPLE 9.2

This example [10] illustrates the application of the theory to a multiple trellis-coded modulation (MTCM). The exposition of this example will be brief in comparison with Example 9.1 and merely serve to highlight the differences between the performance characterization of conventional and multiple trellis coding on the slow-fading channel.

Consider a rate 2/4 multiple ($k = 2$) trellis-coded QPSK system with a two-state trellis diagram as in Fig. 9.11. The corresponding state diagram is illustrated in Fig. 9.12 and the equivalent superstate diagram is shown in Fig. 9.13. In Fig. 9.12, the branches are labeled with the input bit and output QPSK symbol *pairs* that cause that particular transition, whereas in Fig. 9.13, the branches are once again labeled with gains of the form given in (9.32) or (9.33) as appropriate.

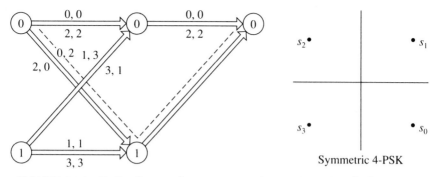

FIGURE 9.11 Trellis diagram for rate 1/2 multiple trellis-coded QPSK; $k = 2$.

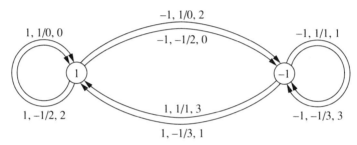

FIGURE 9.12 State diagram for rate 1/2 multiple ($k = 2$) trellis-coded QPSK.

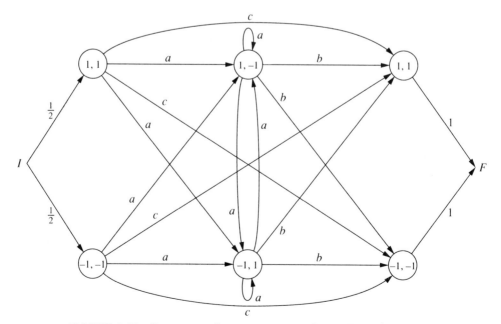

FIGURE 9.13 Superstate diagram corresponding to Fig. 9.12.

In the absence of fading, the transfer function of Fig. 9.13 is easily computed as

$$T(D, I) = 2c + \frac{4ab}{1 - 2a} \tag{9.64a}$$

or

$$T(D, I) = \frac{(2I + 2I^2 + I^3)D^8 - (I^2 + I^3)D^{12}}{1 - (I + I^2)D^4} \tag{9.64b}$$

where, in accordance with (9.32),

$$a = \frac{1}{2}(I + I^2)D^4 D^0$$

$$b = \frac{1}{2}(1 + I)D^2 D^2 \tag{9.65}$$

$$c = \frac{I}{2}D^4 D^4.$$

Substituting (9.34b) into (9.31) and ignoring the overbar for the no-fading case, we obtain the upper bound on P_b given by

$$P_b \leq \frac{1}{2}Z^8 \left(\frac{9 - 8Z^4 + 4Z^8}{(1 - 2Z^4)^2} \right). \tag{9.66}$$

Ideal Channel State Information

Analogous to the branch label gain replacements made in an earlier section to convert the no-fading superstate transition diagram into one suitable for the Rician slow-fading channel, we arrive at the following upper bound on P_b:

$$P_b \leq \frac{1}{2} \frac{(1 + \xi_2 Z^{\zeta_2})\xi_2^2 Z^{2\zeta_2} + 4(2 - \xi_2 Z^{\zeta_2})\xi_2 \xi_1^2 Z^{\zeta_2 + 2\zeta_1} - (5 - 4\xi_2 Z^{\zeta_2})\xi_2^3 Z^{3\zeta_2}}{(1 - 2\xi_2 Z^{\zeta_2})^2} \tag{9.67}$$

where ξ_i and ζ_i, $i = 1, 2$, are as given in (9.41), now with $\beta_1 = \beta_2 = 4$. Figure 9.14 is a plot of the upper bound in (9.67) as a function of $\overline{E_b}/N_0$ for a Rician parameter $K = 10$.

For the Rayleigh channel, (9.41) simplifies to $\xi_i = (\beta_i \gamma)^{-1}$ and $\zeta_i = 0$, $i = 1, 2$, in which case (9.67) becomes

$$P_b \leq \frac{\left(2 + \dfrac{\overline{E_b}}{N_0}\right)\left(1 + \dfrac{\overline{E_b}}{N_0}\right)\left(1 + \dfrac{\overline{E_b}}{2N_0}\right)^2 + 4\left(1 + 2\dfrac{\overline{E_b}}{N_0}\right)\left(1 + \dfrac{\overline{E_b}}{N_0}\right)^2 - \left(1 + 5\dfrac{\overline{E_b}}{N_0}\right)\left(1 + \dfrac{\overline{E_b}}{2N_0}\right)^2}{2\left(\dfrac{\overline{E_b}}{N_0} - 1\right)^2\left(1 + \dfrac{\overline{E_b}}{N_0}\right)^2\left(1 + \dfrac{\overline{E_b}}{2N_0}\right)^2}.$$

$$\tag{9.68}$$

This result is illustrated graphically in Fig. 9.15.

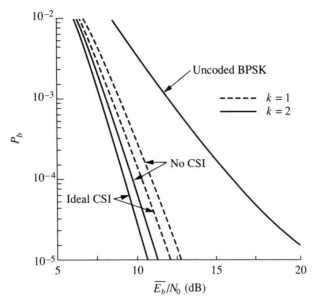

FIGURE 9.14 Average bit error probability performance versus $\overline{E_b}/N_0$ for conventional ($k = 1$) rate 1/2 trellis-coded QPSK and multiple ($k = 2$) rate 2/4 trellis-coded QPSK in the presence of Rician fading with and without ideal channel state information; two states, $K = 10$.

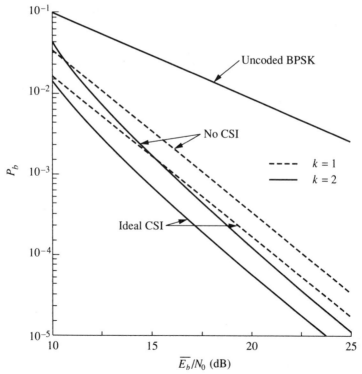

FIGURE 9.15 Average bit error probability performance versus $\overline{E_b}/N_0$ for conventional ($k = 1$) rate 1/2 trellis-coded QPSK and multiple ($k = 2$) rate 2/4 trellis-coded QPSK in the presence of Rayleigh fading with and without ideal channel state information; two states, $K = 0$.

For large $\overline{E_b}/N_0$, (9.68) is approximated by

$$P_b \leq \frac{0.5}{(\overline{E_b}/N_0)^2}. \qquad (9.69)$$

When compared with (9.45′) for the equivalent multiplicity one (conventional) trellis code, we observe that the multiple trellis code offers improved performance on the fading channel. Even when compared with the conventional trellis code with optimized asymmetry [equation (9.48′)], the multiple trellis code performs better.

No Channel State Information

Here we proceed as in the previous section, that is, we evaluate $T(D, I)$ of (9.64a) with D^β replaced by $D^{-4\lambda^2\beta}\overline{D^{4\lambda\beta\rho}}$ in each factor of (9.65) and $\beta = 2$ or 4 as appropriate. Using $\overline{D^{4\lambda\beta\rho}}$ of (9.51) or (9.53) in the resulting $\overline{T}(D, I)|_{D=Z}$ and performing the differentiation required in (9.31) gives the upper bound

$$P_b \leq$$
$$\min_{\lambda \geq 0} \frac{1}{2} \frac{(1 + \xi_2 Z^{-16\lambda^2})\xi_2^2 Z^{-32\lambda^2} + 4(2 - \xi_2 Z^{-16\lambda^2})\xi_2 \xi_1^2 Z^{-32\lambda^2} - (5 - 4\xi_2 Z^{-16\lambda^2})\xi_2^3 Z^{-48\lambda^2}}{(1 - 2\xi_2 Z^{-16\lambda^2})^2}$$

$$(9.70)$$

where ξ_i, $i = 1, 2$, is given in (9.55) or (9.56), respectively.

Superimposed on the results of Figs. 9.14 and 9.15 are the upper bounds computed from (9.70) together with (9.55) or (9.56), respectively. A comparison between the two reveals the amount to be gained by employing CSI. Also indicated on these figures are dashed-line curves depicting the performance results corresponding to Example 9.1. The difference between the dashed and solid curves indicates the relative gain of a multiplicity $k = 2$ system. Higher values of multiplicity (at least up to $k = 4$ for this example) will yield further gain yet.

Finally, for purpose of comparison, the corresponding upper bound on the performance of uncoded BPSK (same bandwidth as in the trellis-coded system) in the presence of Rician and Rayleigh fading is also illustrated in Figs. 9.14 and 9.15. The analytical results for these curves are given by (9.49) and (9.50), respectively.

□

9.1.5 Simulation Results

In this section, we describe and present the results of a software simulation of the system block diagram of Fig. 9.1. The development of a simulation has a manyfold purpose. First, it can be used to "verify" the theoretical results obtained in Section 9.1.4, keeping in mind that the simulation is indicative of the exact system performance, whereas the theoretical bit error rate expressions are upper bounds. Second, when the number of states in the trellis diagram becomes large, determining the state transition diagram and its associated transfer function is a tedious task; in such cases, simulation is the more expedient approach. Finally, system degradations due to the finite size of interleaving and decoder buffer are analytically intractable, particularly when coupled with that caused by the "noisy" carrier demodulation reference[11] produced by the pilot tone extractor. Hence to predict true system performance corresponding to the real-world environment, one must again turn to simulation. In the next paragraph we expand upon the last of these issues.

The block interleaver of Fig. 9.1 can be regarded as a buffer with d rows that represent the *depth* of interleaving and s columns that represent the *span* of interleaving. Thus the size of the interleaver (in symbols) is $d \times s$. Data are written into the buffer in successive rows and read out of the buffer (the order in which they are transmitted over the channel) in columns. At the receiver, the block deinterleaver performs the reverse operation; that is, the received soft quantized symbols are written into the buffer in successive columns and read out in rows. In practice, the interleaving depth should be chosen on the order of the maximum fade duration anticipated, which for the mobile satellite application, for example, depends on the Doppler frequency, or equivalently, the vehicle speed. The smaller the Doppler frequency, the longer the fade duration, and vice versa. The interleaving span should be chosen on the order of the decoder buffer size. When this is done, the performance degradation (relative to that for the analytically tractable assumption of infinite interleaving depth and buffer size) will be inversely proportional to the product of interleaving size and Doppler frequency.

When pilot tones are used for coherent demodulation (as suggested in Fig. 9.1), performance will degrade directly proportional to Doppler frequency. The reason for this is that the bandpass filter(s) used in the pilot tone extractor to isolate the pilot tone(s) from the modulation must have bandwidth sufficiently wide to include the Doppler shift. Thus the larger the Doppler, the wider the bandwidth of the filter(s) and hence the "noisier" the extracted modulation reference. Assuming infinite interleaving and decoder buffer size, one can, in principle, use the same analytical approach as discussed previously to derive upper bounds on the bit error probability in the presence of the noisy carrier reference. In Chapter 11 we discuss this computation in the absence of fading. However, for the case where both noisy demodulation references and fading are present, such evaluations become quite tedious. Thus a simulation is preferable.

EXAMPLE 9.3

The first example simulated is identical to Example 9.1 and is considered here for the purpose of "verifying" the analytical results. In the simulation, the interleaving size was chosen equal to 512 QPSK symbols (or, equivalently, 512 input bits), which for all practical purposes, approximates infinite interleaving. A Doppler frequency of 100 Hz was chosen, which at the data rate of 4800 bit/s selected for the simulation [7] makes the channel rapidly varying enough for the fading to be assumed independent from symbol to symbol (see Appendix A for a more de-

---

[11] We discuss the performance of trellis codes in the presence of noisy carrier demodulation references in greater detail in Chapter 11.

tailed discussion of the relation between Doppler frequency and the statistics of the multipath fading induced by it). These two assumptions provide a memoryless channel as assumed in the analysis (see Fig. 9.2). Furthermore, the buffer size was chosen equal to 32 bits, which approximates the assumption of an infinite bit buffer.

Numerical simulation results for this example are superimposed as dashed lines on the analytical results of Figs. 9.6 and 9.7 corresponding, respectively, to the cases of Rician and Rayleigh fading with ideal channel state information. The discrepancy between the solid (theoretical) and dashed curves reflects the looseness of the upper bounds, but the relative behavior of the analytical curves compares well with that of the simulation results. □

EXAMPLE 9.4

As a second example, we consider a trellis code of sufficient complexity as to warrant analytical results intractable. In particular, a rate 2/3, 16-state trellis code combined with symmetric 8-PSK modulation was simulated.[12] Such a code was proposed for NASA's Mobile Satellite Experiment (MSAT-X) [11], whose objective it was to transmit 4800 to 9600 bit/s of digitally encoded speech over a 5-kHz UHF channel with a bit error rate of 10^{-3}. In applications such as this where speech is to be communicated, there exists a limitation on the total allowable delay between the speaker and the listener. At the foregoing data rate, this delay is on the order of 60 ms and imposes a constraint on the interleaving and decoder buffer sizes. For the delay constraint above, the size of the block interleaver and deinterleaver was chosen equal to 128 8-PSK symbols (or 256 input bits). With the interleaving size chosen above, the interleaving depth was optimized by computer simulation and found to be equal to 16 symbols. Thus the interleaving span was 128/16 = 8 symbols. It should be noted that at low Doppler frequencies such as 20 Hz (corresponding to a vehicle speed of 15 mph at UHF), one can indeed have fade durations much longer than 16 symbols. In this case, an interleaving size of 128 symbols is not sufficient, and as we shall see shortly, a significant performance penalty occurs. Once again, this could not be predicted by analytical means—thus the importance of the simulations. Finally, with the delay constraint above imposed, the decoder buffer size was optimized through simulation and found to be 32 symbols (or 64 bits).

Figures 9.16 and 9.17 illustrate the results of the simulation for perfect carrier and time synchronization and no intersymbol interference (ISI). In particular, Fig. 9.16 assumes a fixed Doppler frequency of 100 Hz (corresponding to a vehicle speed of 75 mph at UHF), a Rician parameter $K = 10$, and various interleaving and CSI options. Also shown as a reference point is the performance of uncoded QPSK (same bandwidth as rate 2/3 trellis-coded 8-PSK) at the chosen error rate of 10^{-3}. Figure 9.17 shows the effect of Doppler frequency on system performance for the case of ideal CSI and the same interleaving parameters as in Fig. 9.16.

---

[12] It was shown in [9] that for this case, the additional coding gain produced by the addition of asymmetry in the modulation is small, and thus we have chosen to ignore it.

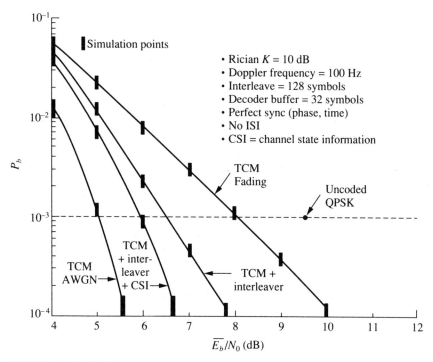

FIGURE 9.16 Performance of rate 2/3, 16-state code over fading channel with/without CSI, with/without interleaving.

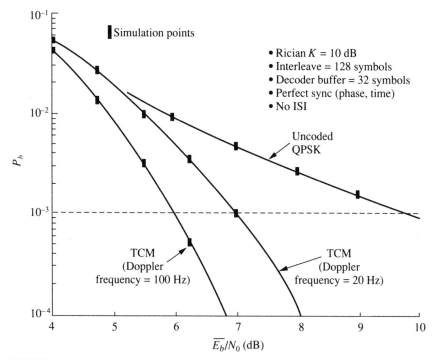

FIGURE 9.17 Performance of rate 2/3, 16-state trellis-coded 8-PSK modulation (TCM) over Rician fading channel with CSI and interleaving.

Table 9.1 summarizes the above results by tabulating the required $\overline{E_b}/N_0$ at a bit error rate of 10^{-3} for each case and also for Rician fading parameters of 5 and 7 dB. From these numerical results, one can assess the coding gain (reduction in required $\overline{E_b}/N_0$ relative to uncoded QPSK) achieved in each case. For example, without interleaving and CSI, transmitting 2/3 coded 8-PSK over the $K = 10$ Rician channel produces a coding gain of 1.6 dB. When 128-symbol interleaving is added, the coding gain is increased to 3.1 dB. If, in addition to interleaving, one was to provide ideal CSI, another 0.5-dB coding gain is achieved, which brings the total gain to 3.6 dB in this particular environment. As is obvious from Fig. 9.16, this coding gain would be greater at lower bit error rates. When the Doppler is decreased from 100 Hz to 20 Hz, Fig. 9.17 and Table 9.1 show a reduction in coding gain of 1 dB or more due to the limitation on the size of the interleaver.

As mentioned previously, all of the results in Figs. 9.16 and 9.17 and Table 9.1 are for the case of perfect carrier synchronization. To examine the degradation due to noisy carrier references, the previous simulation was modified to include a technique referred to as *dual pilot tone calibration* (DTCT) [12], wherein two tones of equal power are inserted symmetrically at the edges of the data spectrum for the purpose of coherent demodulation. When the total power of the two pilots was chosen to be 7 dB below that of the data (this ratio has been shown to be optimum [12]), it was found by simulation that the noisy carrier reference produced by the dual pilot tone extractor resulted in a 2- to 3-dB degradation in performance, depending on the value of the Rician parameter K. This rather large degradation is caused by the fact that (1) the bandwidth of the pilot tone bandpass filters has to be chosen wide enough to accommodate the maximum Doppler frequency of 100 Hz, thus allowing more noise than necessary at lower Doppler values, and (2) some of the total transmitted power has to be allocated to the two pilot tones, thus robbing the data-bearing signal. Simulation results with the DTCT technique are shown in Fig. 9.18 for Rayleigh and various Rician channels under the assumption of ideal CSI, 128-symbol interleaving, and a Doppler frequency of 20 Hz. □

TABLE 9.1 Summary of Results.[a]

Type of Modulation	Fading K (dB)	Doppler Hz	Block Interleave, 128 Symbols	Channel State Information (CSI)	Required Bit SNR (dB) at BER = 10^{-3}
QPSK	10	100	No	No	9.6
TCM	10	100	No	No	8.6
TCM	10	100	Yes	No	6.6
TCM	10	100	Yes	Yes	6.0
TCM	10	20	Yes	Yes	7.0
QPSK	7	100	No	No	12.0
TCM	7	100	Yes	Yes	7.0
TCM	7	20	Yes	Yes	8.5
QPSK	5	100	No	No	15.0
TCM	5	100	Yes	Yes	7.8
TCM	5	20	Yes	Yes	10.5

[a] Decoder buffer size = 32 symbols.

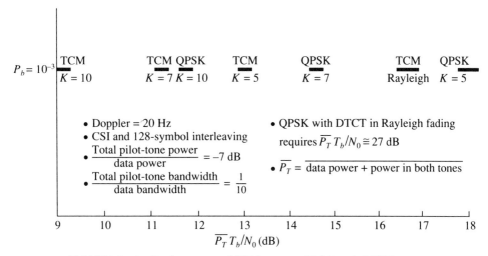

FIGURE 9.18 Performance of TCM versus QPSK with DTCT.

9.2 Differentially Coherent Detection of Trellis-Coded *M*-PSK on a Slow-Fading Rician Channel

In Section 9.1, we saw that the requirement for coherent detection of trellis-coded *M*-PSK on the fading channel resulted in a potentially large (2- to 3-dB) SNR degradation caused by the noisy carrier reference signals. In such situations, it is reasonable to consider the possibility of using trellis-coded multilevel *differential phase-shift-keying* (*M*-DPSK), with the hope that the performance penalty associated with the differential detection will not exceed that due to the noisy carrier demodulation reference in the coherent system. If this is indeed true (as we shall soon see is possible), the *M*-DPSK system has a decided implementation advantage over the coherent *M*-PSK one in that a means for extracting a carrier demodulation reference does not have to be provided. Once again the results presented here will be obtained using a combination of analysis and simulation. Also, as in Section 9.1, we consider only the case where interleaving/deinterleaving is employed to combat the fading.

9.2.1 System Model

Analogous to Fig. 9.1, the simplified block diagram of the system is illustrated in Fig. 9.19. As can easily be seen, the two systems are quite similar in many ways, the primary differences being that (1) in Fig. 9.19 the mapped trellis-encoded *M*-PSK signals are differentially encoded before being modulated onto the RF carrier for transmission over the channel, and (2) the received faded, noise-corrupted signal is differentially detected prior to deinterleaving. As such, no pilot tone system is required. Thus, if CSI is to be provided, it must be obtained from envelope detection of the received signal.

In selecting a decoding metric, a trade-off exists between simplicity of implementation and the optimality associated with the degree to which the metric matches the differential detector output statistics. For the case of uncoded *M*-DPSK, a

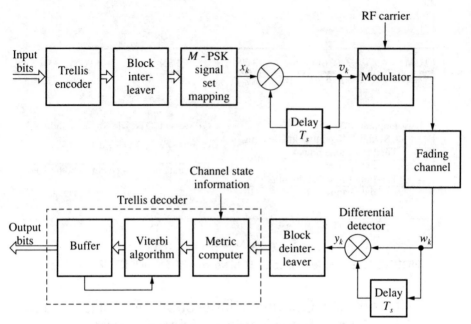

FIGURE 9.19 Block diagram of the trellis-coded M-DPSK system.

metric based on minimizing the distance between the received and transmitted signal vectors (equivalent to assuming Gaussian statistics as in the coherent M-PSK analysis of Section 9.1) is optimum in the sense of a minimum probability of error test. For the coded case, the optimum (maximum-likelihood) metric depends on the joint two-dimensional (amplitude and phase) statistics of a sequence of receptions. Although we consider both the maximum-likelihood (truly matched) and the much simpler to implement (and analyze) Gaussian metric, we only present system performance results for the latter.

9.2.2 Analysis Model

The basic analysis model for the block diagram of Fig. 9.19 is once again given by Figure 9.2. The box labeled "encoder," which is actually the combination of the trellis encoder and the mapping function, outputs the sequence **x** whose kth element, as before, represents the transmitted M-PSK symbol in the kth transmission interval. Before transmission over the channel, the sequence **x** is differentially encoded, producing the sequence **v**. In phasor notation, v_k and v_{k+1} can be written as

$$v_k = e^{j\phi_k}$$
$$v_{k+1} = v_k x_{k+1} = e^{j(\phi_k + \Delta\phi_{k+1})} = e^{j\phi_{k+1}} \qquad (9.71)$$

where

$$x_k = e^{j\Delta\phi_k} \qquad (9.72)$$

is the phasor representation of the M-PSK symbol $\Delta\phi_k$ assigned by the mapper in the kth transmission interval.

Corresponding to **x**, the channel outputs the sequence **y** where the $(k+1)$st element y_{k+1}, representing the output in the $(k+1)$st transmission interval, is given by

$$y_{k+1} = w_k^* w_{k+1}$$
$$= (X_k^* e^{-j\phi_k} + N_k^*)(X_k e^{j(\phi_k + \Delta\phi_{k+1})} + N_{k+1}) \quad (9.73)$$
$$= X_k^* X_{k+1} e^{j\Delta\phi_{k+1}} + \text{noise terms}.$$

Here N_k and N_{k+1} are samples of a stationary, complex Gaussian noise process $N(t)$ that represents the additive thermal noise at the receiver front end, and X_k, X_{k+1} are samples of a normalized, stationary, complex Gaussian noise process $X(t)$ [independent of $N(t)$] that represents the Rician fading. In phasor form,

$$X_k = \rho_k e^{j\phi_k} \quad (9.74)$$

where ρ_k has the pdf given in (9.1).

Finally, the first two moments of the random variables X_k and N_k are given by

$$E\{X_k\} = \eta \quad E\{|X_k|^2\} - |\eta|^2 = \sigma_x^2 \quad \sigma_x^2 + |\eta|^2 = 1$$
$$E\{N_k\} = 0 \quad E\{N_j N_k^*\} = \left(\frac{N_0}{E_s}\right)\delta_{jk} \quad \delta_{jk} = \begin{cases} 0, & j \neq k \\ 1, & j = k. \end{cases} \quad (9.75)$$

Since here again we shall assume infinite depth interleaving and deinterleaving in the analysis so that the coding channel is memoryless, the products $\rho_k \rho_{k+1}$ are independent random variables, and hence the joint channel probabilities again satisfy (9.3), and if CSI is available, also (9.4).

9.2.3 The Maximum-Likelihood Metric for Trellis-Coded *M*-DPSK

The maximum-likelihood metric of (9.6) requires obtaining the marginal pdf $p(y_n | x_n, z_n)$ when CSI is available or $p(y_n | x_n)$ in the absence of CSI. For *M*-DPSK with perfect CSI (i.e., $x_n = \rho_n$), the characteristic function method is used in Appendix 9B to derive $p(y_n | x_n, z_n)$ with the result

$$p_{Y_1, Y_2}(y_1, y_2) = \int_0^\infty \frac{R}{2\pi\left(1 + \left(\frac{N_0}{2E_s}\right)^2 R^2\right)} \exp\left\{-\frac{\rho_{k+1}^2 R^2\left(\frac{N_0}{2E_s}\right)}{\left(1 + \left(\frac{N_0}{2E_s}\right)^2 R^2\right)}\right\}$$

$$\times J_0\left(R\left|y_{k+1} - \frac{\rho_{k+1}^2 x_{k+1}}{1 + \left(\frac{N_0}{2E_s}\right)^2 R^2}\right|\right) dR$$

(9.76)

where $J_0(x)$ is the zero-order Bessel function of the first kind. The logarithm of (9.75) gives the maximum likelihood *branch* metric $m(y_n | x_n, z_n)$, which is quite complicated to implement. Furthermore, theoretical analysis of the bit error probability of the system in Fig. 9.19 with such a metric is difficult, if not impossible.

For this reason, we turn our attention now to the much simpler Gaussian metric for which bounds on bit error probability can be readily computed. The approach taken is analogous to that given in [13] and [14].

9.2.4 Upper Bound on Pairwise Error Probability

Here we set out to evaluate the upper bound on pairwise error probability as given by (9.8) for the metric that is optimum (maximum likelihood) for the additive Gaussian noise channel. We begin by treating the case when CSI is absent.

No channel state information

When no CSI is provided, the Gaussian metric takes the form of (9.23). Substituting (9.73) into (9.23) and again recalling that for M-PSK, $|x|^2 = |\hat{x}|^2 = 1$ independent of n, the difference of the metrics required in (9.8) becomes

$$m(y_n, \hat{x}_n) - m(y_n, x_n) = -|w_{n-1}^* w_n - \hat{x}_n|^2 + |w_{n-1}^* w_n - x_n|^2$$
$$= w_{n-1}^* w_n (\hat{x}_n - x_n) + w_{n-1} w_n^* (x_n - \hat{x}_n) \tag{9.77}$$

which can be conveniently written in the matrix form

$$m(y_n, \hat{x}_n) - m(y_n, x_n) = \mathbf{W}_n^{*T} (\hat{\mathbf{A}}_n - \mathbf{A}_n) \mathbf{W}_n \tag{9.78}$$

where

$$\mathbf{W}_n = \begin{bmatrix} w_{n-1} \\ w_n \end{bmatrix} \quad \hat{\mathbf{A}}_n = \begin{bmatrix} 0 & \hat{x}_n^* \\ \hat{x}_n & 0 \end{bmatrix} \quad \mathbf{A}_n = \begin{bmatrix} 0 & x_n^* \\ x_n & 0 \end{bmatrix} \tag{9.79}$$

and T denotes the transpose operation.

From (9.73) and (9.74), we have

$$w_{n-1}^* w_n = \rho_{n-1} \rho_n e^{j\Delta\phi_n} + \text{noise terms}. \tag{9.80}$$

Assuming that the fading is slowly varying enough that $\rho_{n-1} = \rho_n$, then substituting (9.78) into (9.8) gives

$$P(\mathbf{x} \to \hat{\mathbf{x}}) \le \prod_{n \in \eta} E \left\{ \exp\left(\lambda \mathbf{W}_n^{*T} (\hat{\mathbf{A}}_n - \mathbf{A}_n) \mathbf{W}_n \right) \mid x_n \right\} \tag{9.81}$$

where the expectation is over the additive noise and the fading process.

The expectation required in (9.81) was evaluated originally by Stein [15] and later by Johnston [13] in connection with the analysis of a block-coded M-DPSK system. In particular, for any $n \in \eta$,

$$E\left\{ \exp\left(\lambda \mathbf{W}_n^{*T} (\hat{\mathbf{A}}_n - \mathbf{A}_n) \mathbf{W}_n \right) \mid x_n \right\} = \frac{\exp\left\{ \lambda \boldsymbol{\mu}_n^{*T} \mathbf{F}_n (\mathbf{I} - 2\lambda \mathbf{R}_n^* \mathbf{F}_n)^{-1} \boldsymbol{\mu}_n \right\}}{\det(\mathbf{I} - 2\lambda \mathbf{R}_n^* \mathbf{F}_n)}$$

$$\tag{9.82}$$

where **I** is the identity matrix and[13]

$$\mathbf{F}_n \triangleq \hat{\mathbf{A}}_n - \mathbf{A}_n = \begin{bmatrix} 0 & (\hat{x}_n - x_n)^* \\ (\hat{x}_n - x_n) & 0 \end{bmatrix}$$

$$\boldsymbol{\mu}_n = E\{\mathbf{W}_n \mid x_n\} = \begin{bmatrix} E\{w_{n-1} \mid x_n\} \\ E\{w_n \mid x_n\} \end{bmatrix} \triangleq \begin{bmatrix} \mu_{n-1} \\ \mu_n \end{bmatrix}$$

$$\mathbf{R}_n = \frac{1}{2} E\{(\mathbf{W}_n - \boldsymbol{\mu}_n)^* (\mathbf{W}_n - \boldsymbol{\mu}_n)^T \mid x_n\}$$

$$= \frac{1}{2} \begin{bmatrix} E\{|w_{n-1} - \mu_{n-1}|^2\} & E\{(w_{n-1} - \mu_{n-1})^*(w_n - \mu_n)\} \\ E\{(w_{n-1} - \mu_{n-1})(w_n - \mu_n)^*\} & E\{|w_n - \mu_n|^2\} \end{bmatrix}.$$

(9.83)

Also it should be noted that (9.82) is valid only when $\det(\mathbf{I} - 2\lambda \mathbf{R}_n^* \mathbf{F}_n) > 0$. After much manipulation, (9.82) evaluates to

$$E\left\{\exp\left(\lambda \mathbf{W}_n^{*T}(\hat{\mathbf{A}}_n - \mathbf{A}_n)\mathbf{W}_n\right) \mid x_n, \rho_n\right\}$$

$$= \frac{\exp\left\{2\lambda \frac{E_s}{N_0} \rho_n^2 \left[\frac{\lambda |\hat{x}_n - x_n|^2 + \mathrm{Re}\{x_n(\hat{x}_n - x_n)^*\}}{1 - \lambda^2 |\hat{x}_n - x_n|^2}\right]\right\}}{1 - \lambda^2 |\hat{x}_n - x_n|^2}$$

(9.84)

where we have renormalized λ by dividing it by $2N_0$ and also indicated the conditional dependence on ρ_n.

One further simplification of (9.84) is possible for constant-envelope signal sets such as M-DPSK. In particular, it can easily be shown that

$$|\hat{x}_n - x_n|^2 = -2\mathrm{Re}\{x_n(\hat{x}_n - x_n)^*\}.$$

(9.85)

Substituting (9.85) into (9.84) and then into (9.81) gives the desired result for the conditional (on ρ_n) pairwise error probability bound:

$$P(\mathbf{x} \to \hat{\mathbf{x}} \mid \boldsymbol{\rho}) \leq \prod_{n \in \eta} \frac{\exp\left\{-\frac{\lambda \frac{E_s}{N_0} \rho_n^2 |\hat{x}_n - x_n|^2 (1 - 2\lambda)}{1 - \lambda^2 |\hat{x}_n - x_n|^2}\right\}}{1 - \lambda^2 |\hat{x}_n - x_n|^2}.$$

(9.86)

---

[13] Note that the results to be obtained from (9.82) and (9.83), namely, (9.86)–(9.88) depend only on $|x_n - \hat{x}_n|^2$ rather than on x_n and $\hat{x}_n$ themselves. As such, one can reverse x_n and $\hat{x}_n$ in the definition of $\mathbf{F}_n$ [i.e., define $\mathbf{F}_n = \mathbf{A}_n - \hat{\mathbf{A}}_n$ with no change in (9.86)–(9.88)].

Since we have assumed that the fading is constant over two symbol intervals (i.e., $\rho_{n-1}\rho_n$ has been replaced by ρ_n^2, and that the interleaving and deinterleaving makes the ρ_n's independent), the average over ρ in (9.81) can, insofar as the upper bound is concerned, be computed as the product of the averages. Averaging each term in the product of (9.86) over the pdf in (9.1) gives

$$E_{\rho_n}\left\{\frac{\exp\left\{-\dfrac{\lambda\dfrac{E_s}{N_0}\rho_n^2|\hat{x}_n - x_n|^2(1 - 2\lambda)}{1 - \lambda^2|\hat{x}_n - x_n|^2}\right\}}{1 - \lambda^2|\hat{x}_n - x_n|^2}\right\}$$

$$= \frac{1 + K}{1 + K + |\hat{x}_n - x_n|^2\left[\lambda\dfrac{\overline{E_s}}{N_0}(1 - 2\lambda) - \lambda^2(1 + K)\right]}$$

$$\times \exp\left\{-K\frac{\lambda\dfrac{\overline{E_s}}{N_0}|\hat{x}_n - x_n|^2(1 - 2\lambda)}{1 + K + |\hat{x}_n - x_n|^2\left[\lambda\dfrac{\overline{E_s}}{N_0}(1 - 2\lambda) - \lambda^2(1 + K)\right]}\right\}.$$

(9.87)

For the Rayleigh case ($K = 0$), (9.87) simplifies to

$$E_{\rho_n}\left\{\frac{\exp\left\{-\dfrac{\lambda\dfrac{E_s}{N_0}\rho_n^2|\hat{x}_n - x_n|^2(1 - 2\lambda)}{1 - \lambda^2|\hat{x}_n - x_n|^2}\right\}}{1 - \lambda^2|\hat{x}_n - x_n|^2}\right\}$$

(9.88)

$$= \frac{1}{1 + |\hat{x}_n - x_n|^2\left[\lambda\dfrac{\overline{E_s}}{N_0}(1 - 2\lambda) - \lambda^2(1 + K)\right]}.$$

The result in (9.87) cannot be optimized over λ independent of the index n. Thus for the Rician case, we first must compute the pairwise error probability (or better yet, the average bit error probability) and then optimize over the Chernoff parameter. On the other hand, the result in (9.88) can be optimized over λ independent of n. In particular, we wish to choose λ to maximize the term in brackets in the denominator of (9.88). Differentiating this expression with respect to λ and equating the result to zero gives the optimum Chernoff parameter

$$\lambda_{\text{opt}} = \frac{\overline{E_s}/2N_0}{1 + 2\overline{E_s}/N_0}. \tag{9.89}$$

Substituting (9.89) in (9.88) and simplifying gives the upper bound on pairwise error probability for the Rayleigh channel:

$$P(\mathbf{x} \to \hat{\mathbf{x}}) \leq \prod_{n \in \eta} \frac{1 - \nu^2}{1 - \nu^2 \left(\left| \frac{\hat{x}_n - x_n}{2} \right|^2 \right)} \tag{9.90}$$

where

$$\nu = \frac{\dfrac{\overline{E_s}}{N_0}}{1 + \dfrac{\overline{E_s}}{N_0}}. \tag{9.91}$$

Ideal channel state information

For the case where the receiver has absolute knowledge of the fading amplitude ρ in each symbol interval, then, analogous to (9.9), the Gaussian decoding metric becomes

$$m(y_n, x_n; z_n) = -|y_n - \rho_{n-1}\rho_n x_n|^2. \tag{9.92}$$

Following steps similar to (9.77)–(9.85), the conditional pairwise error probability analogous to (9.86) is given by

$$P(\mathbf{x} \to \hat{\mathbf{x}} | \boldsymbol{\rho}) \leq \prod_{n \in \eta} \frac{\exp\left\{ -\dfrac{\dfrac{1}{2}\lambda \dfrac{E_s}{N_0} \rho_n^4 |\hat{x}_n - x_n|^2 (1 - 2\lambda\rho_n^2)}{1 - \lambda^2 \rho_n^4 |\hat{x}_n - x_n|^2} \right\}}{1 - \lambda^2 \rho_n^4 |\hat{x}_n - x_n|^2}. \tag{9.93}$$

Unfortunately, even for the Rayleigh case, the average of (9.92) over the pdf of ρ cannot be accomplished in closed form, much less the minimization over the Chernoff parameter. More serious than this, however, is the fact that the condition on the determinant of $\mathbf{I} - 2\lambda \mathbf{R}_n^* \mathbf{F}_n$ being greater than zero is not satisfied for all values of ρ_n in the region $(0, \infty)$. As such, the denominator of any term on the right-hand side of (9.93) is not always positive, and thus averaging (9.93) over a Rayleigh or Rician pdf is not valid. In view of this, the upper bounding approach discussed thus far must be abandoned for the ideal CSI case. Later in this section when we illustrate the application of the theory with some examples, we shall have a few more words to say about the ideal CSI case.

9.2.5 Upper Bound on Average Bit Error Probability

When it comes to computing average bit error probability, the transfer function bound is still the preferred approach. Indeed, (9.31) is still valid, the only question being what to use for $\overline{D^{\delta^2}}$ in the average trellis branch gains of (9.33). Comparing (9.29) with (9.86) averaged over ρ, we immediately realize that (9.87) and (9.88) are the appropriate expressions.

EXAMPLE 9.5

So that we may compare the difference in performance between coherent and differentially coherent detection, we again consider Example 9.1, namely, rate 1/2, trellis-coded QPSK with a two-state trellis. Based on the previous discussion, for differential detection in the presence of Rician fading, the transfer function of (9.34) applies with, however, a, b, and c now defined by

$$a = \frac{I}{2}\xi_1 D^{\zeta_1} \qquad b = \frac{I}{2}\xi_2 D^{\zeta_2} \qquad c = \frac{1}{2}\xi_3 D^{\zeta_3} \qquad (9.94)$$

where

$$\xi_i = \frac{1 + K}{1 + K + \beta_i \left[\lambda \frac{\overline{E_b}}{N_0}(1 - 2\lambda) - \lambda^2(1 + K)\right]}$$

$$\zeta_i = \frac{4\beta_i K \lambda (1 - 2\lambda)}{1 + K + \beta_i \left[\lambda \frac{\overline{E_b}}{N_0}(1 - 2\lambda) - \lambda^2(1 + K)\right]} \qquad i = 1, 2, 3 \qquad (9.95)$$

with

$$\beta_1 = 4 \qquad \beta_2 = \frac{4}{1 + \alpha} \qquad \beta_3 = \frac{4\alpha}{1 + \alpha}. \qquad (9.96)$$

Once again performing the differentiation required in (9.31) and recalling that we must minimize over the Chernoff parameter λ as well as the asymmetry parameter α, we get the following result:

$$P_b \leq \min_{\lambda \geq 0} \min_{\alpha} \frac{\xi_1 \xi_3 Z^{\zeta_1 + \zeta_3}}{(1 - \xi_2 Z^{\zeta_2})^2}. \qquad (9.97)$$

Figure 9.20 is a plot of (9.97) versus $\overline{E_b}/N_0$ for a Rician parameter $K = 10$. Also shown are the results for the symmetric signal set [i.e., setting $\alpha = 1$ in (9.97)]. Finally, the upper bound on the performance of uncoded BPSK in the same environment,

$$P_b \leq \min_{\lambda} \xi_1 Z^{\zeta_1} \qquad (9.98)$$

is also shown. This result is a special case ($M = 2$) of the more general result for uncoded M-PSK:

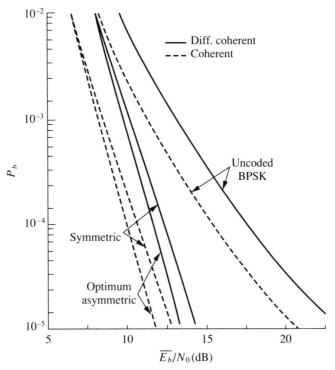

FIGURE 9.20 Bit error probability performance of rate 1/2 trellis-coded QPSK in the presence of Rician fading; two states, $K = 10$; no channel state information.

$$P_b \leq \min_\lambda \sum_{m=1}^{M-1} \xi_m Z^{\zeta_m} \quad (9.99)$$

where, rather than (9.96), $\beta_m = 4\sin^2 \pi m/M$; $m = 1, 2, \ldots, M$ is used in (9.95). The comparable results for coherent detection as obtained from Fig. 9.9 are superimposed on Fig. 9.20. We observe that over a large range of $\overline{E_b}/N_0$ values (where the curves are approximately straight lines), the results for differential detection track the coherent detection results with a fixed difference of about 1.5 dB for the coded cases and about 2 dB for the uncoded cases. Further analytical justification of this is given later on in the chapter when we discuss asymptotic behavior.

For the Rayleigh behavior, the previous results simplify somewhat. The appropriate a, b, and c for the transfer function of (9.34) are now

$$a = \frac{I}{2}(1 - \nu^2); \quad b = \frac{I}{2}\left(\frac{1 - \nu^2}{1 - \nu^2\left(\frac{\alpha}{1+\alpha}\right)}\right); \quad c = \frac{1}{2}\left(\frac{1 - \nu^2}{1 - \nu^2\left(\frac{1}{1+\alpha}\right)}\right).$$

(9.100)

Performing the differentiation required in (9.31) gives upon simplification

$$P_b \leq \left(\frac{1-\nu^2}{\nu^2}\right)^2 \frac{(1+\alpha-\nu^2\alpha)^2(1+\alpha)}{1+\alpha-\nu^2}. \tag{9.101}$$

For the symmetric QPSK signal set, (9.101) reduces to

$$P_b \leq 2\left(\frac{1-\nu^2}{\nu^2}\right)^2 (2-\nu^2). \tag{9.102}$$

On the other hand, optimizing (9.101) with respect to α produces the optimum asymmetry condition

$$\alpha_{opt} = -\left(1-\frac{3\nu^2}{4}\right) + \sqrt{\left(1-\frac{3\nu^2}{4}\right)^2 - \left(1-\nu^2 - \frac{\nu^2}{2(1-\nu^2)}\right)} \tag{9.103}$$

which can then be substituted into (9.101) to give the optimized bit error rate performance bound.

Illustrated in Fig. 9.21 is the upper bound of (9.101) together with (9.103). Also shown are the corresponding results for the coherent detection case as obtained from Fig. 9.10. Finally, the results for uncoded BPSK are superimposed. Although the

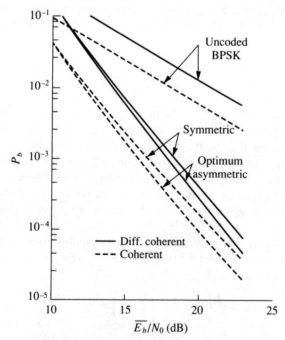

FIGURE 9.21 Bit error probability performance of rate 1/2 trellis-coded QPSK in the presence of Rayleigh fading; two states, $K = 0$; no channel state information.

performance of uncoded BPSK over a Rayleigh can be obtained in exact form, for fairness of comparison, the uncoded performance curves in Fig. 9.21 are upper bounds analogous to (9.98) and (9.99). In particular,

$$P_b \leq \sum_{m=1}^{M-1} \frac{1 - v^2}{1 - v^2 \cos^2 \frac{\pi m}{M}} \tag{9.104}$$

for differential detection and

$$P_b \leq \sum_{m=1}^{M-1} \frac{1}{1 - \left(\sin^2 \frac{\pi m}{M}\right)\left(\frac{\overline{E_s}}{N_0}\right)} \tag{9.105}$$

for coherent detection. For $M = 2$, (9.104) simplifies to

$$P_b \leq 1 - v^2 = \frac{1 + 2\overline{E_b}/N_0}{(1 + \overline{E_b}/N_0)^2} \tag{9.106}$$

whereas (9.105) becomes (9.50).

Once again, we observe from Fig. 9.21 that, over the range of $\overline{E_b}/N_0$ considered, the differential detection results track those for coherent detection with a fixed $\overline{E_b}/N_0$ difference of about 1.5 dB for the coded cases and about 3 dB for the uncoded case. Again this type of behavior will be expanded on later in the chapter.

□

EXAMPLE 9.6

As another example, consider a rate 2/3 trellis-coded asymmetric 8-PSK modulation again using a simple two-state trellis. The appropriate set partitioning is illustrated in Fig. 9.22, the trellis diagram in Fig. 9.23, and the superstate transition diagram in Fig. 9.24, the latter having a form similar to Fig. 9.13. The performance of this system in the absence of fading and with coherent detection was given originally in [9]. In particular, analogous to (9.64a), the transfer function is given by

$$T(D, I) = 2d + \frac{2(a_1 + a_2)c}{1 - 2b} \tag{9.107}$$

where

$$a_1 = \frac{1}{2}\left[D^{\delta_1^2}I^2 + D^{\delta_5^2}I\right] \quad c = \frac{1}{2}\left[D^{\delta_6^2}(I + 1)\right]$$

$$a_2 = \frac{1}{2}\left[D^{\delta_1^2}I + D^{\delta_5^2}I^2\right] \quad d = \frac{1}{2}\left[D^{\delta_4^2}I\right]. \tag{9.108}$$

$$b = \frac{1}{2}\left[D^{\delta_7^2}I + D^{\delta_3^2}I^2\right]$$

382 Ch. 9 / Analysis and Performance of TCM for Fading Channels

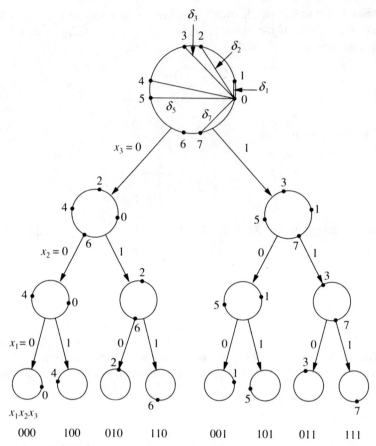

FIGURE 9.22 Set partitioning of asymmetric 8-PSK.

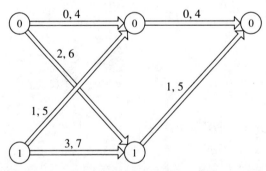

FIGURE 9.23 Two-state trellis diagram and signal assignment for 8-PSK.

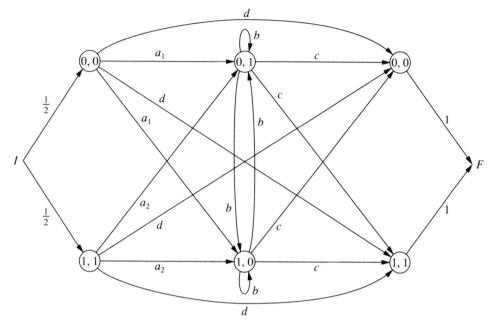

FIGURE 9.24 Superstate transition diagram for rate 2/3 trellis code.

The set of squared distances δ_j^2 from signal point 0 to signal point $j = 1, 2, \ldots, 7$ are given by

$$\delta_1^2 = 4\sin^2\frac{\phi}{2} = 2(1 - \cos\phi) \qquad \delta_5^2 = 4\sin^2\left(\frac{\pi}{2} - \frac{\phi}{2}\right) = 2(1 + \cos\phi)$$

$$\delta_2^2 = 2 \qquad \delta_6^2 = 2$$

$$\delta_3^2 = 4\sin^2\left(\frac{\pi}{4} + \frac{\phi}{2}\right) = 2(1 + \sin\phi) \qquad \delta_7^2 = 4\sin^2\left(\frac{\pi}{2} - \frac{\phi}{2}\right) = 2(1 - \sin\phi)$$

$$\delta_4^2 = 4.$$

(9.109)

Differentiating (9.107) in accordance with (9.31) (ignoring the overbar) gives the upper bound

$$P_b \leq \min_\phi \frac{1}{2}Z^{\delta_4^2} + \frac{Z^{\delta_5^2}\left(Z^{\delta_1^2} + Z^{\delta_5^2}\right)\left(2 - Z^{\delta_7^2}\right)}{\left(1 - Z^{\delta_7^2} - Z^{\delta_3^2}\right)^2}. \qquad (9.110)$$

For differential detection in the presence of Rician fading, we use (9.110) with $Z^{\delta_i^2}$ replaced by $\xi_i Z^{\zeta_i}$, $i = 1, 2, \ldots, 7$, where ξ_i and ζ_i are defined in (9.95) with $\overline{E_b}$ replaced by $\overline{E_s} = 2\overline{E_b}$ and the δ_i's are now defined by (9.109). Again performing the differentiation required in (9.31), we now get

$$P_b \leq \min_{\lambda \geq 0} \min_{\phi} \frac{1}{2}\xi_4 Z^{\zeta_4} + \frac{\xi_2 Z^{\zeta_2}(\xi_1 Z^{\zeta_1} + \xi_5 Z^{\zeta_5})(2 - \xi_7 Z^{\zeta_7})}{(1 - \xi_7 Z^{\zeta_7} - \xi_3 Z^{\zeta_3})^2}. \quad (9.111)$$

Figure 9.25 is a plot of (9.111) versus $\overline{E_b}/N_0$ for a Rician parameter $K = 10$. Also shown are the results for the symmetric signal set obtained by setting $\phi = \pi/4$ and the comparable results for coherent detection as determined from (9.111) with ξ_i and ζ_i defined, analogous to (9.51), by

$$\xi_i = e^{-K}\left[1 - \frac{1}{\sqrt{\pi}}\int_0^\pi \eta_i(\theta)\,\text{erfc}\left[\eta_i^2(\theta)\right]\,\text{erfc}\,\eta_i(\theta)\,d\theta\right]$$

$$\eta_i(\theta) = \frac{\lambda \delta_i^2(\overline{E_b}/N_0)}{\sqrt{1+K}} - \sqrt{K}\cos\theta \quad (9.112)$$

$$\zeta_i = -4\lambda^2\delta_i^2.$$

Finally, the upper bound on the performance of uncoded QPSK in the same environment is determined from (9.99) with $M = 4$.

Although not as obvious as in Fig. 9.20, we observe that except for the possibility of a proportionality constant, the bit error probability performances of the differentially coherent and coherent detection schemes approach each other asymp-

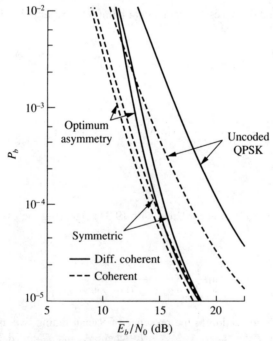

FIGURE 9.25 Bit error probability performance of rate 2/3 trellis-coded 8-PSK in the presence of Rician fading; two states, $K = 10$; no channel state information.

totically as $\overline{E_b}/N_0$ get sufficiently large. To see this better, Fig. 9.26 superimposes the symmetric results of Figs. 9.20 and 9.25 on a single grid, at the same time extending them to a broader range of $\overline{E_b}/N_0$. Similar results to Figs. 9.25 and 9.26 are plotted in Figs. 9.27 and 9.28 for the Rayleigh channel.

It is also interesting to observe from Figs. 9.26 and 9.28 that the rate of decrease of P_b with $\overline{E_b}/N_0$ is steeper for the rate 1/2 trellis-coded 4-DPSK (or 4-PSK) case than for the rate 2/3 trellis-coded 8-DPSK (or 8-PSK) case. This behavior will become apparent later in the chapter when we discuss the general asymptotic behavior of trellis codes on fading channels. ☐

9.2.6 Simulation Results

The simulation program that was used to obtain the results in Section 9.1.5 was modified to incorporate differential encoding/decoding and differential detection, and as such models the block diagram of Fig. 9.19. The example used for illustration is identical to Example 9.4, namely, rate 2/3-coded, 16-state 8-PSK with a 128 (8 × 16) symbol block interleaver. Figure 9.29 illustrates the results of the simulation for perfect Doppler tracking and time synchronization and no ISI. Additional assumptions are a Doppler frequency spread of 40 Hz, a fixed Rician parameter $K = 10$, and the two cases of no interleaving and limited interleaving, the latter in accordance with the above-mentioned interleaver size. Also shown are

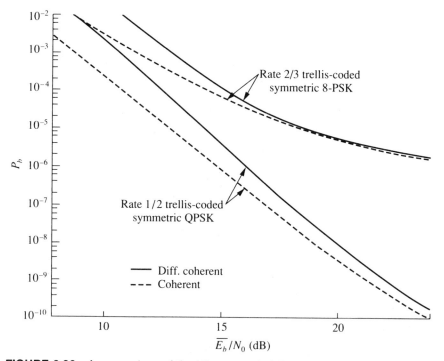

FIGURE 9.26 A comparison of the bit error probability performance of rate 1/2 trellis-coded symmetric QPSK and rate 2/3 trellis-coded symmetric 8-PSK in Rician fading; $K = 10$.

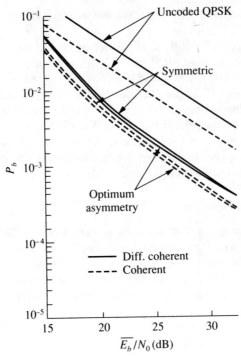

FIGURE 9.27 Bit error probability performance of rate 2/3 trellis-coded 8-PSK in the presence of Rayleigh fading; two states, $K = 10$; no channel state information.

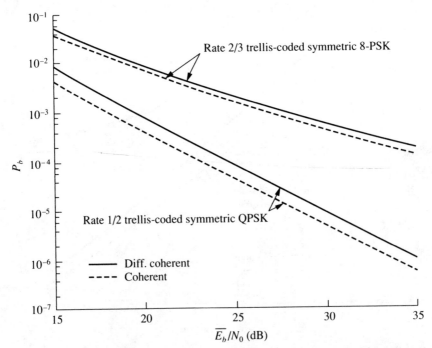

FIGURE 9.28 A comparison of the bit error probability performance of rate 1/2 trellis-coded symmetric QPSK and rate 2/3 trellis-coded symmetric 8-PSK in Rayleigh fading; $K = 0$.

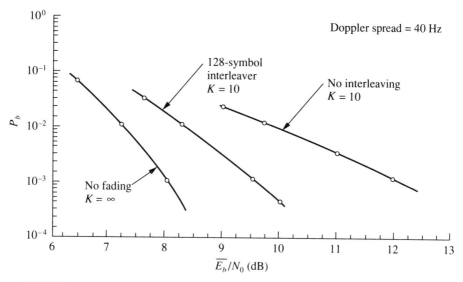

FIGURE 9.29 Simulation results for bit error rate performance of rate 2/3, 16-state trellis-coded 8-DPSK.

the corresponding results for no fading ($K = \infty$) corresponding to the AWGN. As in Fig. 9.16, we observe that a large performance penalty is paid when interleaving/deinterleaving is not employed. Quantitatively speaking, at a bit error rate of 10^{-3}, the performance with interleaving is 1.5 dB worse than that under ideal conditions (i.e., no fading). Without interleaving, we must pay an additional 2.5-dB penalty in average bit energy-to-noise ratio.

9.3 Differentially Coherent Detection of Trellis-Coded *M*-PSK on a Fast-Fading Rician Channel

In this section we consider the performance of trellis-coded *M*-DPSK over fast-fading channels. By *fast fading* we mean that the fading process, in particular the phase of the fading, varies rapidly over the symbol interval. In Section 9.2 the assumption was made that the fading phase process is constant over one or two symbol intervals. As such, the performance depended only on the first-order pdf of the fading envelope. When a fading phase variation exists over the symbol interval, as occurs, for example, in a mobile communications environment with high vehicle speeds, one must consider the second-order statistics (e.g., the statistical correlation function) of the fading process in evaluating system performance.

As before, we model the envelope of the fading process as a Rician pdf with parameter *K* [see equation (9.1)]. Based on this model, upper bounds on the bit error probability will be derived and the sensitivity of the performance to an increase in the speed of the mobile vehicle will be demonstrated as a function of *K*. In particular, as *K* decreases, the performance degradation due to the phase variation increases. All of the results [16] will again be obtained from a combination of analysis and computer simulation, the latter being used to verify the former and also reflect the practical limitation of finite interleaving and deinterleaving size.

9.3.1 Analysis Model

The analysis model for the system is still given in Fig. 9.19. Equations (9.71)–(9.75) characterize the various signal and noise processes in the block diagram, with one addition. The samples X_k and X_{k+1} that represent the complex fading process $X(t)$ in the kth and the $(k + 1)$st transmission intervals are now correlated as follows:

$$E\{(X_k - \eta)(X_{k+1} - \eta)^*\} = R_{X(t)}(T_s) - |\eta|^2 \triangleq \zeta \sigma_x^2 \qquad 0 \leq |\zeta| \leq 1$$

$$R_{X(t)}(T_s) = E\{X(t)X^*(t + T_s)\} = E\{X_k X_{k+1}^*\}$$

(9.113)

where $R_{X(t)}(\tau)$ is the autocorrelation function of the complex fading process $X(t)$. Note that the case $\zeta = 1$ corresponds to the slow-fading case treated previously. Later we consider specific functional forms for $R_{X(t)}(\tau)$ corresponding to various types of fast-fading channels.

9.3.2 Upper Bound on Pairwise Error Probability

As in Section 9.2.5, we consider only the simpler Gaussian metric for which an upper bound on pairwise error probability can be readily computed. The analysis approach for the no channel state information case follows exactly as in (9.77)–(9.83). Omitting the details, for the Rician channel, (9.82) evaluates to

$$E\left\{\exp\left(\lambda \mathbf{W}_n^{*T}(\hat{\mathbf{A}}_n - \mathbf{A}_n)\mathbf{W}_n\right) \mid x_n\right\}$$
$$= \frac{1}{\Delta} \exp\left\{-\frac{\lambda}{\Delta} \frac{E_s}{N_0} \frac{|\hat{x}_n - x_n|^2 K}{1 + K}\left[1 - 2\lambda\left(\frac{1}{1 + K}\frac{E_s}{N_0}(1 - \zeta) + 1\right)\right]\right\}$$

(9.114)

where

$$\Delta = 1 + |\hat{x}_n - x_n|^2 \left\{\lambda \frac{E_s}{N_0} \frac{1}{1 + K}(\zeta - 2\lambda) - \lambda^2\left[1 + (1 - \zeta^2)\left(\frac{E_s}{N_0} \frac{1}{1 + K}\right)^2\right]\right\}.$$

(9.115)

For the Rayleigh case ($K = 0$), (9.114) simplifies to

$$E\left\{\exp\left(\lambda \mathbf{W}_n^{*T}(\hat{\mathbf{A}}_n - \mathbf{A}_n)\mathbf{W}_n\right) \mid x_n\right\}$$
$$= \frac{1}{1 + |\hat{x}_n - x_n|^2 \left\{\lambda \frac{E_s}{N_0}(\zeta - 2\lambda) - \lambda^2\left[1 + (1 - \zeta^2)\left(\frac{E_s}{N_0}\right)^2\right]\right\}}.$$

(9.116)

9.3 Differentially Coherent Detection on a Fast-Fading Rician Channel

For the Rician case, we must again first compute the pairwise error probability by substituting (9.115) into (9.81) and then optimize over the Chernoff parameter. For the Rayleigh case, however, (9.116) can be optimized over the Chernoff parameter independent of the index n. Doing this results in

$$\lambda_{\text{opt}} = \frac{\zeta \dfrac{E_s}{N_0}}{4 \dfrac{E_s}{N_0} + 2\left[1 + \left(\dfrac{E_s}{N_0}\right)^2 (1-\zeta^2)\right]} \qquad (9.117)$$

which when substituted in (9.116) and then in (9.81) yields the pairwise error probability

$$P(\mathbf{x} \to \hat{\mathbf{x}}) \leq \prod_{n \in \eta} \frac{1}{1 + |\hat{x}_n - x_n|^2 \left\{\dfrac{\zeta^2}{4}\left(\dfrac{E_s}{N_0}\right)^2 \left[2\dfrac{E_s}{N_0} + 1 + (1-\zeta^2)\left(\dfrac{E_s}{N_0}\right)^2\right]^{-1}\right\}}. \qquad (9.118)$$

Note that for $\zeta = 1$, (9.118) agrees with (9.90) together with (9.91).

The asymptotic behavior of (9.118) is of significant interest. We shall consider two special cases of this behavior corresponding to $\zeta = 1$ (slow fading) and $\zeta \neq 1$.

Case I: $\zeta = 1$

$$P(\mathbf{x} \to \hat{\mathbf{x}}) \leq \prod_{n \in \eta} \frac{1}{1 + \dfrac{|\hat{x}_n - x_n|^2}{8} \dfrac{E_s}{N_0}}. \qquad (9.119a)$$

Case II: $\zeta \neq 1$

$$P(\mathbf{x} \to \hat{\mathbf{x}}) \leq \prod_{n \in \eta} \frac{1}{1 + \dfrac{|\hat{x}_n - x_n|^2}{4} \dfrac{\zeta^2}{(1-\zeta^2)}}. \qquad (9.119b)$$

Note that (9.119b) *is independent of signal-to-noise ratio!* This has the interpretation that regardless of how much we increase the signal-to-noise ratio, we get a nonzero error probability, which is referred to as an "error floor."

9.3.3 Upper Bound on Average Bit Error Probability

Once again we shall use the transfer function bound approach taken in the previous sections. In particular, (9.31) is valid with $\overline{D^{\delta^2}}$ in the average trellis branch gains of (9.33) replaced by (9.114) for the Rician case and (9.116) or (9.118) for the Rayleigh case.

EXAMPLE 9.7

The example used here for illustration of the application of the theory is that in Example 9.1, namely, rate 1/2 trellis-coded QPSK with a two-state trellis. The trellis diagram with the appropriate M-PSK symbol assignments is illustrated in Fig. 9.4

and the corresponding superstate transition diagram in Fig. 9.5. The performance of this system in the absence of fading and with coherent detection (QPSK) is described by the transfer function bound in (9.34a). Based on the discussion for differential detection in the presence of fast Rayleigh fading, this transfer function still applies with, however, a, b, and c defined by

$$a = \frac{1}{2}\left(\frac{1}{1+4\gamma}\right) \qquad b = \frac{1}{2}\left(\frac{1}{1+2\gamma}\right) \qquad c = \frac{1}{2}\left(\frac{1}{1+2\gamma}\right) \qquad (9.120)$$

where

$$\gamma = \frac{\zeta^2\left(\dfrac{\overline{E_b}}{N_0}\right)^2}{4\left[2\dfrac{\overline{E_b}}{N_0} + 1 + \left(\dfrac{\overline{E_b}}{N_0}\right)^2(1-\zeta)^2\right]}. \qquad (9.121)$$

Performing the differentiation required in (9.31) gives, upon simplification,

$$P_b \leq \frac{1+2\gamma}{4\gamma^2(1+4\gamma)} \qquad (9.122)$$

which is the required upper bound on bit error probability.

The previous results can also be used to compute an upper bound on bit error probability performance of an uncoded DPSK system (same bandwidth as rate 1/2 trellis-coded DQPSK). In particular, we obtain

$$P_b \leq \frac{1}{1+4\gamma}. \qquad (9.123)$$

We observe that when $\zeta \neq 1$, both the coded and uncoded systems exhibit the error floor mentioned previously, which for this example is given by

$$P_b \leq \frac{2(1-\zeta^2)(2-\zeta^2)}{\zeta^4} \quad \text{(coded)} \qquad (9.124)$$

$$P_b \leq 1 - \zeta^2 \quad \text{(uncoded)}. \qquad \square$$

9.3.4 Characterization of the Autocorrelation and Power Spectral Density of the Fading Process

The appropriate value of ζ (and hence γ) to be used in the previous results depends on the type of autocorrelation function (or equivalently, power spectral density) used to characterize the fading process. Mason [17] has tabulated such results for various types of fading processes of practical interest. These results are repeated here in Table 9.2, where for convenience, we have normalized the autocorrelation function and power spectral density such that $\sigma_x(0) = 1$. In the table, B_d denotes the fading bandwidth. For example, for a land mobile fading channel B_d is equal to the Doppler frequency shift $f_d = (v/c)f_c$, where v is the velocity of the mobile, c is the speed of light, and f_c is the carrier frequency.

For the land mobile channel, Figs. 9.30 and 9.31 illustrate the upper bounds on bit error probability performance versus $\overline{E_b}/N_0$ (in dB) for the rate 1/2 trellis-coded

9.3 Differentially Coherent Detection on a Fast-Fading Rician Channel

TABLE 9.2

Type of Fading Spectrum	Normalized Spectrum	ζ
Rectangular	$(2B_d)^{-1};\ |f| \leq B_d$	$\dfrac{\sin(2\pi B_d T_s)}{2\pi B_d T_s}$
Gaussian	$\exp\left[-\left(\dfrac{f}{B_d}\right)^2\right]\left(\sqrt{\pi}B_d\right)^{-1}$	$\exp[-(\pi B_d T_s)^2]$
Land mobile	$[\pi^2(f^2 - B_d^2)]^{-1/2};\ |f| \leq B_d$	$J_0(2\pi B_d T_s)$
First-order Butterworth	$\left[\pi B_d\left(1 + \left(\dfrac{f}{B_d}\right)^2\right)\right]^{-1}$	$\exp(-2\pi|B_d T_s|)$
Second-order Butterworth	$\left[1 + 16\left(\dfrac{f}{B_d}\right)^4\right]^{-1}$	$\exp\left(-\dfrac{\pi|B_d T_s|}{\sqrt{2}}\right)$ $\times\left[\cos\left(\dfrac{\pi B_d T_s}{\sqrt{2}}\right) + \sin\left(\dfrac{\pi|B_d T_s|}{\sqrt{2}}\right)\right]$

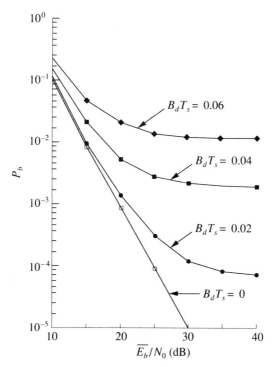

FIGURE 9.30 Bit error probability performance of rate 1/2 trellis-coded DQPSK in the presence of fast Rayleigh fading; two states.

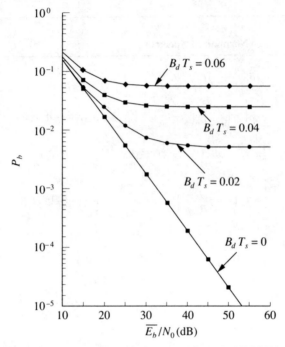

FIGURE 9.31 Bit error probability performance of uncoded DPSK in the presence of fast Rayleigh fading.

DQPSK and uncoded DPSK systems as given by (9.122) and (9.123), respectively, and with $B_d T_s$ as a parameter. Figure 9.32 illustrates the superposition of these results for two values of $B_d T_s$. We observe that even for very small values of $B_d T_s$, the bit error probability performance exhibits a serious degradation relative to the slow-fading case ($B_d T_s = 0$), illustrating the extreme sensitivity of this performance on the Rayleigh fading channel to the assumption made on the variation of the fading process.

9.3.5 Simulation Results

The computer simulation program used to obtain the results in Section 9.2.6 was modified to model the fast-fading channel. As before, because of practical limitations, the interleaver size in the simulation was 32 × 16 bits, whereas the analysis assumes infinite interleaving. Also, to approximate a Rayleigh channel, the Rician factor, K, was set equal to -30 dB.

The simulation results are superimposed on Fig. 9.32 where the shape of the points matches those along the analytical curves. We observe that the simulation results follow quite closely the trend of the analytical results, keeping in mind that the latter are upper bounds. A closer *absolute* agreement between the two could be had by including a factor of 1/2 in the Chernoff bound as discussed in

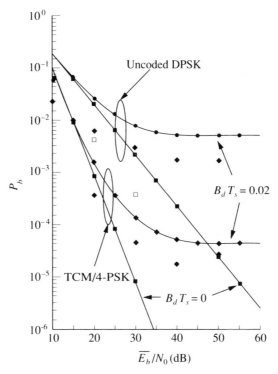

FIGURE 9.32 A comparison of the bit error probability performance of rate 1/2 trellis-coded DQPSK (two states) with uncoded DPSK in the presence of fast Rayleigh fading.

[18, Appendix 4B]. The simulation results for the trellis-coded case with $B_d T_s = 0$ are not indicated, since due to the finite interleaver size, they would yield worse performance than for the case $B_d T_s = 0.02$.

9.4 Asymptotic Results

In many previous sections we have alluded to the asymptotic behavior (large $\overline{E_b}/N_0$) of TCM over fading channels, particularly in regard to several of the examples given for illustration. Here we formalize the results in terms of a simple upper bound that applies for all cases and identifies the key parameters that affect the design of trellis codes for such channels [19]. The latter is the subject of discussion in Chapter 10.

To begin the discussion, recall from Section 9.1.3 that for coherent detection of trellis-coded M-PSK with ideal CSI, the upper bound on pairwise error probability for the Rayleigh channel was shown [see equation (9.22)] to be inversely proportional to (1) the *product* of the squared Euclidean distances along the error event path and (2) a power of $\overline{E_s}/N_0$ equal to the *length* of the error event path (i.e., the number of elements in the set η). This is in direct contrast to the AWGN channel, where the comparable upper bound is an exponential function of the negative *sum* of these same squared Euclidean distances multiplied by $E_s/4N_0$. For the Rician channel, one can obtain an analogous behavior to (9.22) for high SNR. In particular, for large $\overline{E}_s/N_0$, (9.17) simplifies to

$$P(\mathbf{x} \to \hat{\mathbf{x}}) \leq \prod_{n \in \eta} \frac{1}{\left(\dfrac{\overline{E}_s}{N_0}\right) \dfrac{|\hat{x}_n - x_n|^2}{4(1 + K)}} e^{-K} = \frac{4^{L_\eta}\left[(1 + K)e^{-K}\right]^{L_\eta}}{\left(\dfrac{\overline{E}_s}{N_0}\right)^{L_\eta} \prod_{n \in \eta} |\hat{x}_n - x_n|^2} \qquad (9.125)$$

where L_η is the "length" of the error event path corresponding to $\hat{\mathbf{x}}$ (i.e., the number of elements in the set η). Thus it is natural to wonder whether this type of asymptotic behavior can be demonstrated for other detection schemes (e.g., differential detection) and conditions of CSI. Furthermore, how does one relate the asymptotic behavior of the pairwise error probability to that of the average bit error probability?

For the case of coherent detection with no CSI, we have from (9.27) and (9.28) that the unconditional pairwise error probability is given by

$$P(\mathbf{x} \to \hat{\mathbf{x}}) \leq \prod_{n \in \eta} E_{\rho_n}\left\{\exp\left[-\frac{\overline{E}_s}{N_0}\lambda(\rho_n - \lambda)|x_n - \hat{x}_n|^2\right]\right\}$$

$$= \prod_{n \in \eta} \exp\left[-\frac{\overline{E}_s}{N_0}\lambda^2|x_n - \hat{x}_n|^2\right] E_{\rho_n}\left\{\exp\left[-\frac{\overline{E}_s}{N_0}\lambda\rho_n|x_n - \hat{x}_n|^2\right]\right\}.$$

(9.126)

Performing the expectation over the Rician pdf of (9.1) gives

$$P(\mathbf{x} \to \hat{\mathbf{x}}) \leq \prod_{n \in \eta} \exp\left[-\frac{\overline{E}_s}{N_0}\lambda^2|x_n - \hat{x}_n|^2\right]$$

$$\times e^{-K}\left[1 - \frac{1}{\sqrt{\pi}}\int_0^\pi \eta(\theta)\exp\left[\eta^2(\theta)\right]\text{erfc }\eta(\theta)\,d\theta\right]$$

(9.127)

where, analogous to (9.51),

$$\eta(\theta) = \frac{\lambda|x_n - \hat{x}_n|^2\left(\overline{E}_s/2N_0\right)}{\sqrt{1 + K}} - \sqrt{K}\cos\theta. \qquad (9.128)$$

For sufficiently large $\overline{E}_s/N_0$, the first term of (9.128) dominates, in which case, $\eta(\theta)$ becomes independent of θ and the evaluation of the integral becomes trivial.

Thus approximating (9.128) by its first term allows (9.127) to be simplified to

$$P(\mathbf{x} \to \hat{\mathbf{x}}) \leq \prod_{n \in \eta} \exp\left(\lambda^2 \frac{\overline{E_s}}{N_0} |x_n - \hat{x}_n|^2\right)$$

$$\times e^{-K} \left[1 - \sqrt{\pi} \lambda \frac{|x_n - \hat{x}_n|^2}{\sqrt{1+K}} \left(\frac{\overline{E_s}}{2N_0}\right) \right.$$

$$\left. \times \exp\left\{ \left[\lambda \frac{|x_n - \hat{x}_n|^2}{\sqrt{1+K}} \left(\frac{\overline{E_s}}{2N_0}\right) \right]^2 \right\} \operatorname{erfc}\left(\lambda \frac{|x_n - \hat{x}_n|^2}{\sqrt{1+K}} \left(\frac{\overline{E_s}}{2N_0}\right) \right) \right].$$

(9.129)

Furthermore, since for large $\overline{E_s}/N_0$, we can use the asymptotic expansion for the complementary error function given by (9.57), then, using this relation in (9.129) gives the further simplification

$$P(\mathbf{x} \to \hat{\mathbf{x}}) \leq \prod_{n \in \eta} \frac{\exp\left(\lambda^2 \frac{\overline{E_s}}{N_0} |x_n - \hat{x}_n|^2\right)(1+K)e^{-K}}{2\lambda^2 |x_n - \hat{x}_n|^4 \left(\frac{\overline{E_s}}{2N_0}\right)^2}$$

$$= \frac{\exp\left(\lambda^2 \frac{\overline{E_s}}{N_0} \sum_{n \in \eta} |x_n - \hat{x}_n|^2\right)(1+K)^{L_\eta} e^{-L_\eta K}}{2^{L_\eta} \lambda^{2L_\eta} \left(\frac{\overline{E_s}}{2N_0}\right)^{2L_\eta} \prod_{n \in \eta} |x_n - \hat{x}_n|^4}.$$

(9.130)

The result in (9.130) can be optimized over the Chernoff parameter. Performing this optimization gives

$$\lambda^2_{\text{opt}} = \frac{L_\eta}{\frac{\overline{E_s}}{N_0} \sum_{n \in \eta} |x_n - \hat{x}_n|^2}$$

(9.131)

which, when substituted in (9.130), gives the tightest upper bound on pairwise error probability,

$$P(\mathbf{x} \to \hat{\mathbf{x}}) \leq \frac{\left(\frac{2e}{L_\eta}\right)^{L_\eta}}{\left(\frac{\overline{E_s}}{N_0}\right)^{L_\eta}} \left(\prod_{n \in \eta} \frac{|x_n - \hat{x}_n|^2}{\left(\sum_{n \in \eta} |x_n - \hat{x}_n|^2\right)^{1/2}} \right)^{-2} \left[(1+K)e^{-K}\right]^{L_\eta}.$$

(9.132)

Finally, for the case of differentially coherent detection of M-PSK with no CSI, the pairwise error probability upper bound is obtained from (9.86) together with (9.87) for the Rician channel or (9.90) together with (9.91) for the Rayleigh channel. For high SNR, the optimum Chernoff parameter of (9.89) simplifies to

$$\lambda_{\text{opt}} \cong \tfrac{1}{4} \qquad (9.133)$$

Although (9.133) is not the optimum value of λ for (9.86), we shall use it nevertheless (resulting in a looser upper bound), to arrive at a result in a desirable form. Thus substituting (9.133) into (9.86) gives for the Rician channel

$$P(\mathbf{x} \to \hat{\mathbf{x}}) \le \prod_{n \in \eta} \frac{1+K}{1 + K + \dfrac{|x_n - \hat{x}_n|^2}{16}\left[2\dfrac{\overline{E_s}}{N_0} - (1+K)\right]}$$

$$\times \exp\left\{ -\frac{K \dfrac{\overline{E_s}}{8 N_0} |x_n - \hat{x}_n|^2}{1 + K + \dfrac{|x_n - \hat{x}_n|^2}{16}\left[2\dfrac{\overline{E_s}}{N_0} - (1+K)\right]} \right\} \qquad (9.134)$$

$$= \exp\left(-\frac{\overline{E_s}}{4N_0} d^2 \right)$$

where, analogous to (9.19),

$$d^2 = \sum_{n \in \eta} \frac{\tfrac{1}{2}|x_n - \hat{x}_n|^2 K}{1 + K + \dfrac{|x_n - \hat{x}_n|^2}{16}\left[2\dfrac{\overline{E_s}}{N_0} - (1+K)\right]} \qquad (9.135)$$

$$+ \left(\frac{\overline{E_s}}{4N_0}\right)^{-1} \ln\left(\frac{1 + K + \dfrac{|x_n - \hat{x}_n|^2}{16}\left[2\dfrac{\overline{E_s}}{N_0} - (1+K)\right]}{1+K} \right)$$

is a distance metric for $\overline{E_s}/N_0 > (1+K)/2$.

For sufficiently high SNR, (9.134) can be further approximated by

$$P(\mathbf{x} \to \hat{\mathbf{x}}) \le \prod_{n \in \eta} \frac{1}{\left(\dfrac{\overline{E_s}}{N_0}\right) \dfrac{|\hat{x}_n - x_n|^2}{8(1+K)}} e^{-K} = \frac{8^{L_\eta}[(1+K)e^{-K}]^{L_\eta}}{\left(\dfrac{\overline{E_s}}{N_0}\right)^{L_\eta} \prod_{n \in \eta} |\hat{x}_n - x_n|^2}. \qquad (9.136)$$

To translate the results of (9.125), (9.132), and (9.136) into asymptotic upper bounds on P_b, we start with the well-known relation between pairwise and bit error probability,

$$P_b \le \sum_{\mathbf{x}, \hat{\mathbf{x}} \in \mathscr{C}} a(\mathbf{x}, \hat{\mathbf{x}}) p(\mathbf{x}) P(\mathbf{x} \to \hat{\mathbf{x}}) \qquad (9.137)$$

where $a(\mathbf{x}, \hat{\mathbf{x}})$ is the number of bit errors that occur when the sequence $\mathbf{x}$ is transmitted and the sequence $\hat{\mathbf{x}} \ne \mathbf{x}$ is chosen by the decoder, $p(\mathbf{x})$ is the a priori probability of transmitting **x**, and $\mathscr{C}$ is the set of all coded sequences. This bound has already been evaluated many times throughout this chapter by the transfer function approach as described by (9.31). For our purpose here, however, (9.137) is the desired form. Substituting (9.125), (9.132), or (9.136) into (9.137) gives an asymptotic upper bound on average bit error probability for TCM transmitted over Rician channels using, respectively, coherent detection with ideal CSI, coherent detection with no CSI, and differentially coherent detection with no CSI.

To identify the important considerations for such a design in a fading environment (to be discussed in more detail in Chapter 10), we first observe that the upper bound of (9.137) will be dominated by the pairwise error probability term in the summation that has the slowest rate of descent with $\overline{E_s}/N_0$. This in turn corresponds to the error event path with the smallest number of elements in η. We refer to this path as the *shortest error event path* and define it more formally as the error event path with the smallest number of nonzero distances between itself and the correct path. We also define the length, L, of the shortest error event path by the number of nonzero pairwise distances between the symbols along its branches and those along the correct path.[14] It is to be emphasized that pairwise distance refers to Euclidean distance between corresponding *symbols* on the pair of paths being compared.

In terms of the foregoing, we see that asymptotically with high SNR, the average bit error probability is approximately given by[15]

$$P_b \cong \frac{1}{b} C \left(\frac{(1+K)e^{-K}}{\overline{E_s}/N_0} \right)^L \qquad \overline{E_s}/N_0 \gg K \qquad (9.138)$$

where C is a constant that depends on the distance structure of the code, that is,

$$C = \begin{cases} 4^L \left[\sum_{n \in \eta} |x_n - \hat{x}_n|^2 \right]^{-1} & \text{(coherent detection with ideal CSI)} \\ \left(\frac{2e}{L} \right)^L \left(\prod_{n \in \eta} \frac{|x_n - \hat{x}_n|^2}{\left(\sum_{n \in \eta} |x_n - \hat{x}_n|^2 \right)^{1/2}} \right)^{-2} & \text{(coherent detection with no CSI)} \\ 8^L \left[\sum_{n \in \eta} |x_n - \hat{x}_n|^2 \right]^{-1} & \text{(differentially coherent detection with no CSI)}. \end{cases}$$

(9.139)

---

[14] Equivalently, we mean the Hamming distance between the M-ary symbols on the error event path and those on the correct path.

[15] The approximation in (9.138) stems from the fact that we consider only a single term in (9.137), that due to the shortest (in length) error event path. Also, for simplicity, we shall ignore the *number* of such paths in the computation of C. As such (9.138) also represents a strict lower bound on P_b.

For now, the important point to be observed in (9.138) is that P_b varies inversely with $(\overline{E_s}/N_0)^L$ and thus L has the notion of *diversity*. This diversity effect was also observed by Wilson and Leung in [20] for the special case of Rayleigh fading channels. Clearly, the primary goal for designing a trellis code to give best performance on the Rician fading channel should be to maximize L. A secondary goal would be to minimize C. These considerations are explored in detail in Chapter 10, where a formal procedure is described for optimally designing trellis codes for transmission over fading channels.

For conventional trellis coding, wherein each branch in the trellis corresponds to a single M-PSK output channel symbol, the shortest error event path is that error event path with the fewest number of branches having nonzero distance from the correct path. For most cases this also corresponds to the shortest length (in branches) error event, and thus L is just the number of branches on this path.

For multiple trellis coding, wherein each branch in the trellis corresponds to more than one M-PSK output channel symbol, the "length" of the shortest error event path is always equal to or greater than the number of branches along the shortest error event path. In view of (9.138), the possibility of a value of L greater than the length (in branches) of the shortest error event path is significant and what affords multiple trellis coding the opportunity of improving trellis coding performance on the fading channel. A simple example of this comment, which is explored in a more general context in Chapter 10, pertains to trellis diagrams with parallel paths between states. This occurs whenever 2^b (i.e., the number of possible transitions from a given state) exceeds the number of states of the trellis diagram. In such cases, with conventional trellis coding, the minimum distance error event path is often the parallel path (i.e., the shortest error event path is of length one branch), and thus $L = 1$. With MTCM, we have the option of still having a trellis diagram with parallel paths, yet because of the multiplicity, we can have more than one nonzero pairwise Euclidean distance along that path, hence the opportunity of achieving a value of L greater than 1.

9.4.1 An Example

Consider the rate 1/2 trellis-coded QPSK modulation with two-state diagram corresponding to Example 9.1. Corresponding to the state diagram of Fig. 9.4, the shortest error event path is of length 2 branches, and both of these have nonzero distance with respect to the branches of the correct path (assumed to be the all-zeros path). Thus the length L of the shortest error event path is equal to 2, or equivalently, the code has a diversity equal to 2.

For coherent detection with no CSI, we need to compute the product of the squared branch distances normalized by the square root of their sum in accordance with C of (9.139). From Fig. 9.4, the square of this ratio is easily computed as

$$\left(\prod_{n\in\eta}\frac{|x_n-\hat{x}_n|^2}{\left(\sum_{n\in\eta}|x_n-\hat{x}_n|^2\right)^{1/2}}\right)^{-2} = \left[\left(\frac{4}{\sqrt{4+2}}\right)\left(\frac{2}{\sqrt{4+2}}\right)\right]^{-2} = \frac{9}{16}.$$

(9.140)

Thus letting $L_\eta = 2$ in (9.132), $\overline{E_s} = \overline{E_b}$, and substituting (9.140) into this same equation gives the upper bound on pairwise error probability

$$P(\mathbf{x} \to \hat{\mathbf{x}}) \leq \frac{9e^2 e^{-2K}(1+K)^2}{16\left(\dfrac{\overline{E_b}}{N_0}\right)^2} \qquad (9.141)$$

which, for large $\overline{E_b}/N_0$, is also approximately equal to the upper bound on bit error probability. Letting $K = 0$ gives the identical result as that of the first equation in (9.63).

For coherent detection with CSI, we need to compute the product of the distances in accordance with C of (9.139). For the shortest error event path this product is easily computed as

$$\prod_{n \in \eta} |x_n - \hat{x}_n|^2 = 4 \times 2 = 8. \qquad (9.142)$$

Thus keeping only the term in (9.137) corresponding to the shortest error event path, we get

$$P_b \cong \frac{2e^{-2K}(1+K)^2}{\left(\dfrac{\overline{E_b}}{N_0}\right)^2} \qquad (9.143)$$

which for $K = 0$ agrees with (9.45′).

Finally, for differentially coherent detection with no CSI, (9.139) also requires calculation of the product of squared branch distances. Using (9.142) and again keeping only the term in (9.137) corresponding to this path, we get

$$P_b \cong \frac{8e^{-2K}(1+K)^2}{\left(\dfrac{\overline{E_b}}{N_0}\right)^2} \qquad (9.144)$$

which for $K = 0$ agrees with (9.102) together with (9.91) when $\overline{E_s}/N_0$ is large.

9.4.2 No Interleaving/Deinterleaving

It is interesting to consider the effect of interleaving/deinterleaving on the diversity parameter L in (9.138). If no interleaving/deinterleaving is employed, the assumption that the fading is independent from symbol to symbol is no longer valid. In fact, if the fading is sufficiently slow as to be constant over the duration of a number of symbols equal to the minimum-distance error event path, then for coherent detection with a Gaussian metric, the average bit error probability is asymptotically upper bounded by

$$P_b \lesssim C_1 E_\rho \left\{ \exp\left(-\rho^2 d_{\text{free}}^2 \frac{\overline{E_s}}{4N_0}\right) \right\} \qquad (9.145)$$

where C_1 is a constant and d_{free}^2 is the squared free distance of the code, that is,

$$d_{\text{free}}^2 = \min_{\mathbf{x},\hat{\mathbf{x}}} \sum_{n \in \eta} |x_n - \hat{x}_n|^2. \tag{9.146}$$

Performing the average over the Rician pdf as required in (9.145) gives

$$P_b \cong C_1 \frac{1+K}{1+K+C_2} \exp\left(-K \frac{C_2}{1+K+C_2}\right) \quad C_2 = d_{\text{free}}^2 \left(\frac{\overline{E_s}}{4N_0}\right) \tag{9.147}$$

which can be represented for large $\overline{E_s}/N_0$ by

$$P_b \cong C \frac{1+K}{d_{\text{free}}^2 \left(\dfrac{\overline{E_s}}{N_0}\right)} e^{-K} \tag{9.148}$$

where $C = 4C_1$.

For differentially coherent detection with a Gaussian metric, the analogous result to (9.145) is obtained from (9.86) and is given by

$$P_b \lesssim \min_{\lambda \geq 0} C_1' E_\rho \left\{ \exp\left[-\lambda \rho^2 \left(\frac{\overline{E_s}}{N_0}\right) \sum_{n \in \eta} \frac{|x_n - \hat{x}_n|^2 (1 - 2\lambda)}{1 - \lambda^2 |x_n - \hat{x}_n|^2}\right]\right\}$$
$$\times \left[\prod_{n \in \eta} (1 - \lambda^2 |x_n - \hat{x}_n|^2)\right]^{-1} \tag{9.149}$$

where η corresponds to the dominant error event path. Again performing the average over the fading pdf gives

$$P_b \lesssim \min_{\lambda \geq 0} C_1' \frac{1+K}{1+K+C_2} \exp\left(-K \frac{C_2}{1+K+C_2}\right)$$
$$\times \left[\prod_{n \in \eta} (1 - \lambda^2 |x_n - \hat{x}_n|^2)\right]^{-1} \tag{9.150}$$
$$C_2 = \lambda \left(\frac{\overline{E_s}}{N_0}\right) \sum_{n \in \eta} \frac{|x_n - \hat{x}_n|^2 (1 - 2\lambda)}{1 - \lambda^2 |x_n - \hat{x}_n|^2}.$$

For large $\overline{E_s}/N_0$, (9.150) simplifies to

$$P_b \lesssim \min_\lambda \frac{C_1'(1+K)e^{-K}}{C_2 \prod_{n \in \eta}(1 - \lambda^2 |x_n - \hat{x}_n|^2)}$$

$$= \min_\lambda \frac{C_1'(1+K)e^{-K}}{\lambda(1-\lambda)\left(\dfrac{\overline{E_s}}{N_0}\right)\sum_{n\in\eta}|x_n - \hat{x}_n|^2 \prod_{\substack{k\in\eta \\ k\neq n}}(1 - \lambda^2|x_k - \hat{x}_k|^2)} \tag{9.151}$$

which, when optimized over the Chernoff parameter, can be put in the form of (9.148).

Thus for either coherent or differentially coherent detection, comparing (9.148) with (9.138), we observe that with no interleaving/deinterleaving, *independent of the trellis code*, the asymptotic steepest rate of descent of P_b with $\overline{E_s}/N_0$ is inverse linear (i.e., has diversity equal to 1).

9.5 Further Discussion

There are a number of other publications in the open literature that deal with the performance of TCM on fading channels and thus deserve mention. Edbauer [21] has considered the performance of a trellis-coded 8-DPSK with differentially coherent detection over the fading channel. In his paper, he describes a modem complete with digital signal processing algorithms suitable for use in a Nyquist system. Also included are AFC and symbol timing based on tentative symbol decisions of the soft-decision Viterbi decoder. Numerical performance results are obtained by computer simulation for Rician factors of $K = 7$ dB or 10 dB and four- or eight-state trellises.

Lodge and Moher [22] exploit the ability of TCM schemes to provide *time* diversity (a time-division multiplex of trellis encoders) in a fading environment. In particular, the trade-off between time diversity and Euclidean distance of trellis-coded M-PSK and QAM is examined for the mobile satellite channel. Results are obtained by computer simulation for a variety of different code designs having eight and sixteen states with both uniform and nonuniform symbol spacing (i.e., modulation asymmetry).

More recently, Schlegel and Costello [23, 24] have performed work similar to that of Divsalar and Simon [7, 19] by examining the performance of TCM over fading channels using Chernoff bounding techniques. Indeed, many of the results they obtain agree with results given in this chapter. Also considered in their work are design criteria for constructing trellis codes suitable for transmission over fading channels, a subject we shall treat in detail in Chapter 10.

For more general fading channels (e.g., a Rician channel with log normal shadowing of the line-of-sight component), McKay et al. [25] have obtained analytical bounds on the average error probability performance of trellis-coded coherent PSK. Such a model includes the effect of foliage attenuation or blockage on a mobile satellite channel. Also included in their work were Monte Carlo simulation results to support the theory. Prior to that, McLane et al. [2] had examined via simulation the performance of rate 2/3 trellis-coded 8-PSK and 8-DPSK on a fast-fading log normal shadowed channel.

The work of Crepeau [26] is of considerable interest in that it clearly identifies the need to accurately model the fading channel statistics. The particular case considered by Crepeau is the Nakagami-m fading model [27], which is a generalization of the Rayleigh fading model and which characterizes fading and scintillation channels. Despite the fact that Crepeau considers block (as opposed to trellis) codes in his work, the important point is that he shows that *both* the channel and code contribute diversity, with the effective diversity being the product of the two. One must be careful, however, in applying his results to the Rician channel despite the fact that according to Nakagami [27] there is a close fit between the m-distribution and the Rician distribution when K and m are related by

$$K = \frac{\sqrt{m^2 - m}}{m - \sqrt{m^2 - m}}. \tag{9.152}$$

In particular, as we have already seen, the Rician channel does not contribute any diversity on its own. Thus despite the fact that the average bit error probability results for the Nakagami-m channel and the Rician channel fit closely for small $\overline{E_b}/N_0$, for large values of this parameter (i.e., asymptotic performance), the departure between the two can be considerable. Crepeau does indeed point this out and his numerical illustrations support the validity of this statement. Perhaps the simplest demonstration of this point is afforded by examining the asymptotic average bit error probability performance of uncoded noncoherent binary FSK on the two channels. In particular, for large $\overline{E_b}/N_0$, we obtain from [26] the following results:

$$P_b = \begin{cases} \dfrac{2^{m-1} m^m}{\left(\dfrac{\overline{E_b}}{N_0}\right)^m} & \text{Nakagami-}m \\[2ex] \left(\dfrac{1+K}{\dfrac{\overline{E_b}}{N_0}}\right) e^{-K} & \text{Rice} \end{cases} \tag{9.153}$$

which clearly illustrates the above point.

Finally, as this book was going to press, the authors became aware of the work of Cavers and Ho [28], who, for the case of Rayleigh fading, were able to derive an *exact* analytical expression for pairwise error probability. Furthermore, they showed that their exact result could be expressed as the upper bound found by Divsalar and Simon [19] (e.g., (9.125) with $K = 0$), multiplied by a correction factor whose value depends on the poles of the two-sided Laplace transform of the pdf of the decision variable. For an eight-state rate 2/3 trellis-coded 8-PSK, they were able to show a difference of 3.6 dB between the exact result and the upper bound, which is indeed significant. Furthermore, Cavers and Ho were able to extend their approach to the trellis-coded M-DPSK case. Although the authors recognize and acknowledge the significance of this work, they were unable to include the details of it, since at the time this book went to press, reference [28] was as yet unpublished.

APPENDIX 9A: PROOF THAT d WHOSE SQUARE IS DEFINED IN (9.19) SATISFIES THE CONDITIONS FOR A DISTANCE METRIC

Theorem Let

$$\delta_0^2(x, y) = |x - y|^2 \tag{9A.1}$$

be the usual distance metric on the complex numbers. Then for $K \geq 0$ and $\gamma > 0$,

$$d_n(x, y) = \left(\frac{K \delta_0^2(x, y)}{1 + K + \delta_0^2(x, y)} + \frac{1}{\gamma} \ln\left(\frac{1 + K + \delta_0^2(x, y)}{1 + K} \right) \right)^{1/2} \quad (9A.2)$$

is a distance metric.

Proof Letting

$$\delta^2(x, y) = \frac{\gamma}{1 + K} \delta_0^2(x, y) \quad (9A.3)$$

we can rewrite (9A.2) as

$$d_n(x, y) = \frac{1}{\sqrt{\gamma}} \left(\frac{K \delta^2(x, y)}{1 + \delta^2(x, y)} + \ln\left(1 + \delta^2(x, y)\right) \right)^{1/2}. \quad (9A.4)$$

Since multiplication of a metric by a positive constant does not change its metric status, it is sufficient to show that

$$d(x, y) = \left(\frac{K \delta^2(x, y)}{1 + \delta^2(x, y)} + \ln\left(1 + \delta^2(x, y)\right) \right)^{1/2} \quad (9A.5)$$

is a metric.

Consider a function $\phi(t)$ defined over the domain $\mathcal{R}^+$ (the set of positive real numbers) and taking on values over $\mathcal{R}^+$. Let $\phi(t)$ have the following four properties:

1. $\phi(0) > 0$
2. $\phi(t) > 0$
3. $\phi'(t) > 0$
4. $\phi''(t) > 0$.

$\quad (9A.6)$

We shall now show that $\phi(\delta(x, y))$ is a metric. We observe that (9A.6) implies that $\phi(t)$ is a monotonically increasing function with decreasing slope as t increases. From conditions 1 to 3 of (9A.6), we have that (see Fig. 9A.1)

$$\phi(t_3) < \phi(t_1 + t_2) \quad \text{for } t_3 < t_1 + t_2. \quad (9A.7)$$

Further imposing condition 4 on (9A.6) results in

$$\frac{\phi(t_1 + t_2) - \phi(t_1)}{t_2} < \phi'(t_1)$$

$$\frac{\phi(t_2)}{t_2} > \phi'(t_2) > \phi'(t_1) \quad \text{for } t_1 > t_2.$$

$\quad (9A.8)$

Combining (9A.7) and (9A.8), we have that

$$\phi(t_3) < \phi(t_1) + \phi(t_2). \quad (9A.9)$$

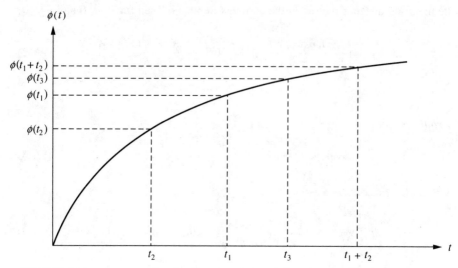

FIGURE 9A1. An example of a function that satisfies the conditions for a metric.

Letting

$$t_1 = \delta(x, y)$$
$$t_2 = \delta(y, z) \quad\quad (9A.10)$$
$$t_3 = \delta(x, z)$$

then, from the triangular inequality for δ,

$$\delta(x, z) \ge \delta(x, y) + \delta(y, z) \quad\quad (9A.11)$$

and (9A.9), we see that $\phi(\delta)$ satisfies the triangular inequality

$$\phi(\delta(x, z)) \ge \phi(\delta(x, y)) + \phi(\delta(y, z)) \quad\quad (9A.12)$$

and is thus a metric.

It remains to show that $\phi(\delta(x, y)) = d(x, y)$ satisfies the conditions of (9A.6). Letting

$$\phi(t) = \left[\frac{Kt^2}{1 + t^2} + \ln(1 + t^2) \right]^{1/2} \quad\quad (9A.13)$$

we immediately observe that conditions 1 and 2 are satisfied. The first derivative of $\phi(t)$ is given by

$$\phi'(t) = \frac{t(1 + K + t^2)}{\phi(t)(1 + t^2)^2} \quad\quad (9A.14)$$

which is obviously greater than zero for all t greater than zero. (Condition 3 is satisfied.)

Finally, the second derivative of $\phi(t)$ is

$$\phi''(t) = \frac{1 + K - 3Kt^2 - t^4}{(1 + t^2)^3} - \frac{t^2(1 + K + t^2)^2}{\phi^2(t)(1 + t^2)^4}. \quad (9\text{A}.15)$$

Since $\ln(1 + t^2) \leq t^2$, we have from (9A.13) that

$$\phi^2(t) \leq \frac{t^2(1 + K + t^2)}{1 + t^2}. \quad (9\text{A}.16)$$

Combining (9A.15) and (9A.16) gives

$$\phi''(t) \leq \frac{(3K + 1)t^2 + t^4}{(1 + t^2)^3} \quad (9\text{A}.17)$$

which obviously satisfies condition 4.

A simple application of the Cauchy–Schwarz inequality shows that

$$d = \left(\sum_{n \in \eta} d_n^2(x, y) \right)^{1/2} \quad (9\text{A}.18)$$

is also a metric, which is the desired result. $\qquad \square$

APPENDIX 9B: DERIVATION OF THE MAXIMUM-LIKELIHOOD BRANCH METRIC FOR TRELLIS-CODED *M*-DPSK WITH IDEAL CHANNEL STATE INFORMATION

Referring to (9.73), we wrote w_k^* and w_{k+1} in rectangular complex notation as

$$\begin{aligned} w_k^* &= X_k^* e^{-j\phi_k} + N_k^* = V_1 + jV_2 \\ w_{k+1} &= X_{k+1} e^{j(\phi_k + \Delta\phi_{k+1})} + N_{k+1} = U_1 + jU_2 \end{aligned} \quad (9\text{B}.1)$$

where U_1, U_2 are independent Gaussian random variables as are V_1, V_2. Furthermore, U_1 and U_2 are independent of V_1 and V_2. The detector output, y_{k+1}, in the $(k + 1)$st transmission interval is given by the product of w_k^* and w_{k+1}, which, in view of (9B.1), can be written as

$$y_{k+1} = w_k^* w_{k+1} = (U_1 V_1 - U_2 V_2) + j(U_1 V_2 + U_2 V_1) = Y_1 + jY_2. \quad (9\text{B}.2)$$

The joint characteristic function of Y_1 and Y_2 is

$$\Phi_{Y_1, Y_2}(\omega_1, \omega_2) = E\left\{ e^{j(\omega_1 Y_1 + \omega_2 Y_2)} \right\} \quad (9\text{B}.3\text{a})$$

$$= E_\mathbf{V}\left\{ E_\mathbf{U}\left\{ e^{j[(\omega_1 V_1 + \omega_2 V_2)U_1 + (\omega_2 V_1 - \omega_1 V_2)U_2]} \Big|_{V_1, V_2} \right\} \right\}$$

which because of the independence between U_1 and U_2 becomes

$$\Phi_{Y_1,Y_2}(\omega_1,\omega_2) = E_\mathbf{V}\left\{E_{U_1}\left\{\left.e^{j[(\omega_1 V_1+\omega_2 V_2)U_1]}\right|_{V_1,V_2}\right\} \times E_{U_2}\left\{\left.e^{j[(\omega_2 V_1-\omega_1 V_2)U_2]}\right|_{V_1,V_2}\right\}\right\}$$

(9B.3b)

Since for a Gaussian random variable X with mean $\overline{X}$ and variance σ_X^2, we have

$$\phi_X(\omega) = E\{e^{j\omega X}\} = e^{j\overline{X}\omega - \sigma_X^2 \omega^2/2}$$

(9B.4)

then applying (9B.4) to each expectation in the product inside the braces of (9B.3b) gives upon simplification

$$E_{U_1}\left\{\left.e^{j[(\omega_1 V_1+\omega_2 V_2)U_1]}\right|_{V_1,V_2}\right\} \times E_{U_2}\left\{\left.e^{j[(\omega_2 V_1-\omega_1 V_2)U_2]}\right|_{V_1,V_2}\right\}$$

$$= e^{j(\omega_1 \overline{U}_1+\omega_2 \overline{U}_2)V_1 - \sigma^2(\omega_1^2+\omega_2^2)V_1^2/2} \times e^{j(\omega_2 \overline{U}_1-\omega_2 \overline{U}_2)V_2 - \sigma^2(\omega_1^2+\omega_2^2)V_2^2/2}$$

(9B.5)

where σ^2 denotes the variance of U_1 and U_2 (or V_1 and V_2).

Finally, using the relation

$$E\left\{e^{j\alpha X-\sigma^2\beta X^2/2}\right\} = \frac{1}{\sqrt{1+\sigma^4\beta}}\exp\left\{\frac{2j\alpha\overline{X} - \sigma^2(\beta\overline{X}^2 + \alpha^2)}{2(1+\sigma^4\beta)}\right\}$$

(9B.6)

to evaluate the expectations over V_1 and V_2 required in (9B.3b), we get after much simplification

$$\phi_{Y_1,Y_2}(\omega_1,\omega_2) = \frac{1}{1+\sigma^4(\omega_1^2+\omega_2^2)}\exp\left\{\frac{2jf_1(\omega_1,\omega_2) - \sigma^2 f_2(\omega_1,\omega_2)}{2[1+\sigma^4(\omega_1^2+\omega_2^2)]}\right\}$$

(9B.7a)

where

$$f_1(\omega_1,\omega_2) = \left(\overline{U_1 V_1} - \overline{U_2 V_2}\right)\omega_1 + \left(\overline{U_1 V_2} + \overline{U_2 V_1}\right)\omega_2$$

$$f_2(\omega_1,\omega_2) = \left(\omega_1^2+\omega_2^2\right)\left(\overline{U}_1^2 + \overline{U}_2^2 + \overline{V}_1^2 + \overline{V}_2^2\right).$$

(9B.7b)

The next step is to perform the inverse Fourier transform of (9B.7a), thereby obtaining the joint probability density function of Y_1 and Y_2. In particular,

$$p_{Y_1,Y_2}(y_1,y_2) = \frac{1}{(2\pi)^2}\int_{-\infty}^{\infty}\int_{-\infty}^{\infty}\phi_{Y_1,Y_2}(\omega_1,\omega_2)e^{-j(\omega_1 y_1+\omega_2 y_2)}d\omega_1\,d\omega_2.$$

(9B.8)

Making the change of variables

$$\omega_1 = R\cos\theta \qquad \omega_2 = R\sin\theta$$

(9B.9)

Appendix 9B / Derivation of the Maximum-Likelihood Branch Metric

and performing the integrations gives, after some simplification,

$$p_{Y_1,Y_2}(y_1, y_2) = \frac{1}{(2\pi)^2} \int_0^\infty W(R) J_0\left(\frac{R\sqrt{A^2(R) + B^2(R)}}{1 + \sigma^4 R^2}\right) dR \qquad (9B.10)$$

where

$$W(R) = \frac{R}{2\pi(1 + \sigma^4 R^2)} \exp\left\{-\frac{\sigma^2 R^2(\overline{U_1}^2 + \overline{U_2}^2 + \overline{V_1}^2 + \overline{V_2}^2)}{2(1 + \sigma^4 R^2)}\right\}$$

$$A(R) = \overline{U_1}\,\overline{V_1} - \overline{U_2}\,\overline{V_2} - y_1(1 + \sigma^4 R^2) \qquad (9B.11)$$

$$B(R) = \overline{U_1}\,\overline{V_2} + \overline{U_2}\,\overline{V_1} - y_2(1 + \sigma^4 R^2).$$

Finally, recognizing from (9.74) and (9B.1) that

$$\overline{U_1}\,\overline{V_1} - \overline{U_2}\,\overline{V_2} = \rho_{k+1}^2 \cos \Delta\phi_{k+1}$$

$$\overline{U_1}\,\overline{V_2} + \overline{U_2}\,\overline{V_1} = \rho_{k+1}^2 \sin \Delta\phi_{k+1} \qquad (9B.12)$$

$$\overline{U_1}^2 + \overline{U_2}^2 + \overline{V_1}^2 + \overline{V_2}^2 = 2\rho_{k+1}^2$$

where we have made the assumption that ρ_k is constant over two symbol intervals [i.e., $\rho_k \rho_{k+1} = (\rho_{k+1})^2$], then (9B.10) together with (9B.11) becomes

$$p_{Y_1,Y_2}(y_1, y_2)$$

$$= \int_0^\infty \frac{R}{2\pi(1 + \sigma^4 R^2)} \exp\left\{-\frac{\rho_{k+1}^2 R^2 \sigma^2}{(1 + \sigma^4 R^2)}\right\}$$

$$\times J_0\left(R\left[\left(y_1 - \frac{\rho_{k+1}^2 \cos \Delta\phi_{k+1}}{1 + \sigma^4 R^2}\right)^2 + \left(y_2 - \frac{\rho_{k+1}^2 \sin \Delta\phi_{k+1}}{1 + \sigma^4 R^2}\right)^2\right]^{1/2}\right) dR.$$

(9B.13)

If we evaluate the variance σ^2 from the noise statistics in (9.75), we find that $\sigma^2 = N_0/2E_s$. Also, recognizing from (9.72) that $\text{Re}\{x_{k+1}\} = \cos \Delta\phi_{k+1}$ and $\text{Im}\{x_{k+1}\} = \sin \Delta\phi_{k+1}$, then (9B.13) can be put in the final desired form,

$$p_{Y_1,Y_2}(y_1, y_2) = \int_0^\infty \frac{R}{2\pi\left(1 + \left(\frac{N_0}{2E_s}\right)^2 R^2\right)} \exp\left\{-\frac{\rho_{k+1}^2 R^2 \left(\frac{N_0}{2E_s}\right)}{\left(1 + \left(\frac{N_0}{2E_s}\right)^2 R^2\right)}\right\}$$

$$\times J_0\left(R\left|y_{k+1} - \frac{\rho_{k+1}^2 x_{k+1}}{1 + \left(\frac{N_0}{2E_s}\right)^2 R^2}\right|\right) dR$$

(9B.14)

which agrees with (9.76).

REFERENCES

1. S. O. RICE, "Statistical properties of a sine wave plus random noise," *Bell Syst. Tech. J.*, Vol. 27, pp. 109–157, 1948.
2. P. J. MCLANE, P. H. WITTKE, P. K.-M. HO, and C. LOO, "PSK and DPSK trellis codes for fast fading, shadowed mobile satellite communication channels," *IEEE Trans. Commun.*, Vol. COM-36, No. 11, pp. 1242–1246; Nov. 1988. See also ICC '87 Conf. Rec., Seattle, Wash., June 1987.
3. J. MCGEEHAN and A. BATEMAN, "Phase-lock transparent tone-in-band (TTIB): A new spectrum configuration particularly suited to the transmission of data over SSB mobile radio networks," *IEEE Trans. Commun.*, Vol. COM-32, No. 1, pp. 81–87, Jan. 1984.
4. F. DAVARIAN, "Mobile digital communications via tone calibration," *IEEE Trans. Veh. Technol.*, Vol. VT-36, No. 2, pp. 55–62, May 1987.
5. G. C. CLARK, JR., and J. B. CAIN, *Error-Correction Coding for Digital Communications*, Plenum Press, New York, N.Y. 1981.
6. J. HAGENAUER, "Viterbi decoding of convolutional codes for fading- and burst-channels," *1980 International Zurich Seminar on Digital Communications*, Zurich, Switzerland, pp. G2.1–G2.7, Mar. 1980.
7. D. DIVSALAR and M. K. SIMON, "Trellis-coded modulation for 4800 to 9600 bps transmission over a fading satellite channel," *JPL Publication 86-8 (MSAT-X Report 129)*, Pasadena, Calif., June 1, 1986. See also *IEEE J. Select. Areas Commun.*, Vol. SAC-5, No. 2, pp. 162–175, Feb. 1987.
8. D. DIVSALAR, "Performance of mismatched receivers on bandlimited channels," Ph.D. dissertation, University of California, Los Angeles, Calif., 1978.
9. D. DIVSALAR, M. K. SIMON, and J. H. YUEN, "Trellis coding with asymmetric modulations," *IEEE Trans. Commun.*, Vol. COM-35, No. 2, pp. 130–141, Feb. 1987. See also "Combined trellis coding with asymmetric MPSK modulation," *JPL Publication 85-24 (MSAT-X Report 109)*, Pasadena, Calif., May 1, 1985.
10. M. K. SIMON and D. DIVSALAR, "Multiple trellis-coded modulation (MTCM) performance on a fading mobile satellite channel," *Globecom'87 Conf. Rec.*, Tokyo, Japan, pp. 43.8.1–43.8.6, Nov. 1987.
11. W. RAFFERTY, K. DESSOUKY, and M. SUE, "NASA's mobile satellite development program," *Proc. Mobile Satellite Conference*, Pasadena, Calif., pp. 11–22, May 3–5, 1988.
12. M. K. SIMON, "Dual pilot tone calibration technique (DTCT)," *IEEE Trans. Veh. Technol.*, Vol. VT-35, No. 2, pp. 63–70, May 1986.
13. D. A. JOHNSTON and S. K. JONES, "Spectrally efficient communication via fading channels using coded multilevel DPSK," *IEEE Trans. Commun.*, Vol. COM-29, No. 3, pp. 276–284, Mar. 1981.
14. M. K. SIMON and D. DIVSALAR, "The performance of trellis-coded multilevel DPSK on a fading mobile satellite channel," *JPL Publication 87-8 (MSAT-X Report 144)*, Pasadena, Calif., June 1, 1987. See also *IEEE Trans. Veh. Technol.*, Vol. 37, No. 2, pp. 78–91, May 1988.
15. M. SCHWARTZ, W. R. BENNETT, and S. STEIN, *Communication Systems and Techniques*, McGraw-Hill, New York, N.Y., 1966.

16. D. DIVSALAR and M. K. SIMON, "Performance of trellis-coded MDPSK on fast fading channels," *ICC'89 Conference Rec.*, Boston, Mass., pp. 9.1.1–9.1.7, June 11–14, 1989.
17. L. J. MASON, "Error probability evaluation for systems employing differential detection in a Rician fast fading environment and Gaussian noise," *IEEE Trans. Commun.*, Vol. COM-35, No. 1, pp. 39–46, Jan. 1, 1987.
18. M. K. SIMON, J. K. OMURA, R. A. SCHOLTZ, and B. K. LEVITT, *Spread Spectrum Communications*, Vol. 1, Computer Science Press, Rockville, MD, 1985.
19. D. DIVSALAR, and M. K. SIMON, "The design of trellis codes for fading channels," *JPL Publication 87-39 (MSAT-X Report 147)*, Pasadena, Calif., Nov. 1, 1987. See also *IEEE Trans. Commun.*, Vol. 36, No. 9, pp. 1004–1012, Sept. 1988.
20. S. G. WILSON and Y. S. LEUNG, "Trellis-coded phase modulation on Rayleigh channels," *ICC'87 Conference Rec.*, Seattle Wash., pp. 21.3.1–21.3.5, June 7–10, 1987.
21. F. EDBAUER, "Coded 8-DPSK modulation with differentially coherent detection — An efficient modulation scheme for fading channels," *Globecom'87 Conf. Rec.*, Tokyo, Japan, pp. 42.2.1–42.2.4, Nov. 1987. See also "Performance of interleaved trellis-coded differential 8-PSK modulation over fading channels," *IEEE J. Select. Areas Commun.*, Vol. 7, No. 9, pp. 1340–1346, Dec. 1989.
22. J. H. LODGE and M. L. MOHER, "Time diversity for mobile satellite channels using trellis-coded modulation," *Globecom'87 Conf. Rec.*, Tokyo, Japan, pp. 8.7.1–8.7.5, Nov. 1987. See also "TCMP—A modulation and coding strategy for Rician fading channels," *IEEE J. Select. Areas Commun.*, Vol. 7, No. 9, pp. 1347–1355, Dec. 1989.
23. C. SCHLEGEL, "Bandwidth efficient coding for non-Gaussian channels," Ph.D. Dissertation, University of Notre Dame, Notre Dame, Ind., 1988.
24. C. SCHLEGEL and D. COSTELLO, JR. "Bandwidth efficient coding on a fading channel," *Fifth International Conference on Systems Engineering*, Dayton, Ohio, Sept. 1987. See also "Bandwidth efficient coding for fading channels: code construction and performance analysis," *IEEE J. Select. Areas Commun.*, Vol. 7, No. 9, pp. 1356–1368, Dec. 1989.
25. R. G. MCKAY, P. J. MCLANE, and E. BIGLIERI, "Analytical performance bounds on average bit error probability for trellis-coded PSK transmitted over fading channels," *ICC'89 Conf. Rec.*, Boston, Mass., pp. 9.2.1–9.2.7, June 11–14, 1989.
26. P. CREPEAU, "Coding performance on generalized fading channels," *MILCOM'88 Conf. Rec.*, San Diego, Calif., pp. 15.1.1–15.1.7, Oct. 23–26, 1988.
27. M. NAKAGAMI, "The m-distribution—A general formula of intensity distribution of fading," in *Statistical Methods in Radio Wave Propagation*, W. C. Hoffman, Ed., Pergamon Press, Elmsford, N.Y., 1960.
28. J. CAVERS and P. HO, "Analysis of the error performance of trellis-coded modulations in Rayleigh fading channels," *IEEE Trans. Commun.*, to appear. Also presented at the *International Symposium on Information Theory (ISIT'90)*, San Diego, Calif., Jan. 14–19, 1990.

CHAPTER 10

Design of TCM for Fading Channels

In Chapter 9, we considered the performance of conventional and multiple trellis-coded M-PSK in a Rician fading environment characteristic of the mobile satellite channel. Results were given for both the case of coherent and differentially coherent detection with and without the use of channel state information (CSI). The primary emphasis in most of the chapter was on the degradation in performance produced by the fading, with little attention paid to the optimality of the code itself. In fact, several of the examples used for illustration were for trellis codes designed to be optimum on the additive white Gaussian noise (AWGN) channel.

Toward the end of the chapter we began to look more carefully into the properties of trellis-coded modulation (TCM) that enter into the various expressions for average bit error probability performance on the fading channel, especially at large SNR. In particular, it was shown that whereas maximizing free Euclidean distance (d_{free}) is the appropriate optimum design criterion on the AWGN, over Rician fading channels with interleaving/deinterleaving, the asymptotic (high SNR) performance of TCM is dominated by several other factors, depending on the value of the Rician parameter K. Specifically, for small values of K (the channel tends toward Rayleigh), the primary design criteria become (1) the *length* (in M-PSK symbols)[1] of the shortest error event path, and (2) the *product* of branch distances along that path, with d_{free} a secondary consideration. Thus, at low values of K, the longer the shortest error event path is and the larger the product of the branch distances along that path is, the better the code will perform *even though d_{free} does not achieve its optimum value over the AWGN!* As K increases, the significance of these primary and secondary considerations shift relative to one another until K reaches infinity (AWGN), in which case optimum performance is once again achieved by a trellis code designed solely to maximize d_{free}.

In this chapter we focus on implementing the foregoing considerations into a formal design procedure for constructing trellis codes to give optimum performance on the fading channel. This construction procedure, which deviates considerably from the Ungerboeck set-partitioning method described in Chapter 3, applies to both conventional and multiple trellis codes; however, considerable emphasis will be placed on the latter since it is indeed the fading channel that fully exploits the diversity potential offered by MTCM. We begin our discussion with an explanation

---

[1] Here, again, we mean the Hamming distance between the M-ary symbols on the shortest error event path and those on the correct path.

of this statement along the lines of that following (9.140) but in more detail and with emphasis on the design of the code.

10.1 Multiple Trellis-Code Design for Fading Channels

Recall that with conventional trellis coding (i.e., one symbol per trellis branch), the length L of the shortest error event path is equal to the number of trellis branches along that path. Equivalently, if we assume that the all-zeros path in the trellis diagram represents the transmitted sequence, L is the number of branches in the shortest length path to which a nonzero M-PSK symbol is associated. Since a trellis diagram with parallel paths is constrained to have a shortest error event of length one branch, we immediately have $L = 1$ (i.e., the average bit error probability asymptotically varies *inverse linearly* with $\overline{E_s}/N_0$). Thus we conclude that for conventional trellis coding on the fading channel, from an error probability performance standpoint it is undesirable to design the code to have parallel paths in its trellis diagram. Unfortunately, however, for a conventional rate $m/(m + 1)$ trellis code, when 2^m exceeds the number of states, one is forced into a trellis with parallel paths. Thus, in these instances, there is no choice but to accept an inverse linear asymptotic performance on the fading channel.

When multiple trellis coding is employed, we regain the option of designing a trellis diagram with parallel paths yet are still able to achieve an asymptotic performance on the fading channel, which varies inversely with $\overline{E_s}/N_0$ at a rate faster than linear. The reason behind this lies in the fact that even if there exist parallel paths in the trellis, it is now possible to have more than one M-PSK symbol with nonzero Euclidean distance associated with an error event of length one branch. In fact, even if the multiplicity k is equal to just two, as long as all of the pairs of M-PSK symbols assigned to the parallel paths are not alike in either of the two symbols positions (i.e., they both represent nonzero Euclidean distances), the pairwise error probability associated with that error event path will vary inversely with the *square* of $\overline{E_s}/N_0$. Thus the primary objective for good multiple trellis-code design on the fading channel is to *maximize the number of symbols with nonzero Euclidean distance along the error event path of shortest length* (smallest Hamming distance). A secondary objective is to minimize the constant C in (9.139) which, depending on the detection scheme (i.e., coherent or differentially coherent), requires maximizing the product of the squared branch distances or the product of the squared branch distances, each normalized by the square root of its sum along this shortest-length path [see (9.140)].

The simplest way of illustrating the foregoing considerations is with an example. Consider a rate 2/3 conventional trellis-coded 8-PSK optimally designed for the AWGN. The appropriate trellis diagram for a two-state trellis is illustrated in Fig. 9.23. Since for a rate 2/3 code there are $2^2 = 4$ possible transitions from each state to the next state, a conventional two-state trellis must have two parallel paths between states. In Fig. 9.23 these parallel paths are labeled with the M-PSK output symbol transmitted over the channel when that particular transition occurs.

An upper bound on the average bit error probability performance of this conventional TCM scheme used on the Rician fading channel with coherent detection at the receiver is obtained from (9.110) and the results of Section 9.1. Because of the existence of parallel paths in Fig. 9.23, this performance will asymptotically (for sufficiently large $\overline{E}_s/N_0$) vary inversely with $\overline{E}_s/N_0$. In particular, since, for the parallel paths $|x_n - \hat{x}_n|^2 = 4$, then, for example, for coherent detection with ideal CSI, (9.138) together with (9.139) yields

$$P_b \cong \frac{1}{2} \frac{1+K}{\left(\dfrac{\overline{E}_s}{N_0}\right)} e^{-K} = \frac{1}{4} \frac{1+K}{\left(\dfrac{\overline{E}_b}{N_0}\right)} e^{-K} \qquad \frac{\overline{E}_s}{N_0} \gg K \qquad (10.1)$$

where we have also noted that for a rate 2/3 code, $E_s = 2E_b$.

Now consider the rate 4/6, two-state multiple ($k = 2$) trellis code, with trellis diagram illustrated in Fig. 10.1. For this code, $b = 4$, $s = 6$, and thus there are $2^b = 16$ possible paths leaving each state. Since there are only two states, each transition between states has eight parallel paths. The sets of 8-PSK symbol pairs for these transitions are illustrated directly on the branches of the trellis diagram and correspond to the signal points in the 8-PSK signal constellation as shown. The construction of these sets[2] is given in (10.2).

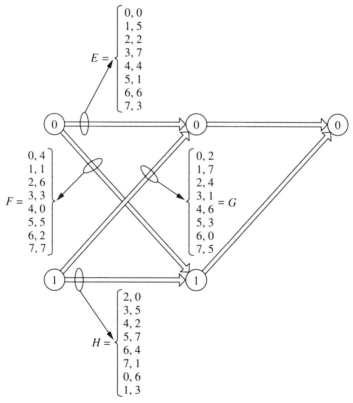

FIGURE 10.1 Trellis diagram for multiple ($k = 2$) rate 2/3-coded 8-PSK; two states.

---

[2] The formalism behind this construction will come later in this chapter.

$$A = \begin{bmatrix} 0 & 0 \\ 4 & 4 \end{bmatrix} \quad B = A + [2 \ 2] = \begin{bmatrix} 2 & 2 \\ 6 & 6 \end{bmatrix} \quad C = A \cup B = \begin{bmatrix} 0 & 0 \\ 2 & 2 \\ 4 & 4 \\ 6 & 6 \end{bmatrix}$$

$$D = C + [1 \ 5] = \begin{bmatrix} 1 & 5 \\ 3 & 7 \\ 5 & 1 \\ 7 & 3 \end{bmatrix} \quad E = C \cup D = \begin{bmatrix} 0 & 4 \\ 1 & 1 \\ 2 & 6 \\ 3 & 3 \\ 4 & 0 \\ 5 & 5 \\ 6 & 2 \\ 7 & 7 \end{bmatrix}$$

$$F = E + [0 \ 4] = \begin{bmatrix} 0 & 4 \\ 1 & 1 \\ 2 & 6 \\ 3 & 3 \\ 4 & 0 \\ 5 & 5 \\ 6 & 2 \\ 7 & 7 \end{bmatrix} \quad G = E + [0 \ 2] = \begin{bmatrix} 0 & 2 \\ 1 & 7 \\ 2 & 4 \\ 3 & 1 \\ 4 & 6 \\ 5 & 3 \\ 6 & 0 \\ 7 & 5 \end{bmatrix}$$

$$H = E + [2 \ 0] = \begin{bmatrix} 2 & 0 \\ 3 & 5 \\ 4 & 2 \\ 5 & 7 \\ 6 & 4 \\ 7 & 1 \\ 0 & 6 \\ 1 & 3 \end{bmatrix}.$$

(10.2)

First, we note that all of the parallel paths have a distinct pair of 8-PSK symbols that differ from each other in *both* symbol positions. Thus, insofar as single branch error events are concerned, the number of 8-PSK symbols with nonzero Euclidean distance from the correct path is two. Second, for an error event of length two branches, there are at least two out of the possible four 8-PSK symbols that have nonzero Euclidean distance from the correct path. This is true for each of the 64 such possible paths. Finally, then, in accordance with the previous definition of L, the length of the shortest error event path is two; that is, the asymptotic average bit error probability performance of this coded modulation scheme will vary inversely with the *square* of $\overline{E_s}/N_0$, as desired. As a specific demonstration of this result, consider again the case of coherent detection with ideal CSI. If we arbitrarily take the all-zeros path as being the correct one, then, for the two-branch error event, the one parallel path in set F that differs by one symbol from the correct path is [0 4] (or [4 0]), which has squared Euclidean distance 4. Similarly, the one parallel path in set G that differs by one symbol from the correct path is [0 2] (or [6 0]), which has squared Euclidean distance 2. For the one-branch paths in parallel with [0 0], the smallest squared Euclidean distances[3] occur for path [1 5], that is, $4\sin^2(\pi/8)$ and $4\sin^2(5\pi/8)$ whose product is less than $(4)(2) = 8$. Thus the dominant term in (9.137) will correspond to the one-branch error event (i.e., a parallel path). Thus (9.138) becomes

$$P_b \cong \frac{1}{4}\left(\frac{4(1+K)}{\overline{E_s/N_0}}e^{-K}\right)^2 \frac{1}{\left(4\sin^2\left(\frac{\pi}{8}\right)\right)\left(4\sin^2\left(\frac{5\pi}{8}\right)\right)} = \frac{1}{2}\frac{(1+K)^2}{(\overline{E_b/N_0})^2}e^{-2K}.$$

(10.3)

It goes without saying that the signal sets assigned to the trellis of Fig. 10.1 will not produce optimum performance on the AWGN channel. We now investigate the extent to which that performance is degraded relative to the optimum assignment for the AWGN illustrated in Fig. 7.5(a). Since on the AWGN channel asymptotic bit error probability performance is measured by the free distance of the trellis code, we shall now compare this quantity for the trellis diagrams of Figs. 10.1 and 7.5(a).

From Table 7.1, we find that $d_{\text{free}}^2 = 3.172$ for the trellis code of Fig. 7.5(a). The minimum Euclidean distance path for the trellis code of Fig. 10.1 also has length two branches. Then, since the minimum squared Euclidean distance between sets E and F and between sets E and G is $2[4\sin^2(\pi/8)] = 1.1715$, we have that $d_{\text{free}}^2 = 2(1.1715) = 2.343$ or a penalty of $10\log_{10}(3.172/2.343) = 1.315\,\text{dB}$.

As a second example, consider a four-state, rate 4/5 multiple ($k = 2$) trellis code whose five output symbols are mapped into one QPSK symbol and one 8-PSK symbol. This is an example of the general form of MTCM described by (7.1). As mentioned in Chapter 7, the advantage of such a hybrid MTCM scheme over one whose multiple output symbols *all* come from the same alphabet is that the former is much less sensitive to carrier synchronization errors at the receiver. In particular, one can derive the carrier reference necessary for demodulation from only the received QPSK symbols, which have greater distance between them than the 8-PSK symbols. Potentially, then, one can obtain an overall improvement in average system bit error probability performance relative to a four-state, rate 4/6 coded 8-PSK system also of multiplicity $k = 2$ and throughput 2 bits/s/Hz, despite the fact that the latter would perform better in an ideal (perfect carrier synchronization) environment.

Figure 10.2 is an illustration of the trellis for the foregoing hybrid scheme. Since $b = 4$, there are $2^b = 16$ paths emanating from each node. Thus with four states, we assign four parallel paths between nodes and the trellis is fully connected. The construction of the signal sets for assignment to the branches of this trellis that produce optimum performance on the fading channel is given in (10.4).

---

[3] As in previous chapters, we again assume a unit-energy PSK when referring to Euclidean distances.

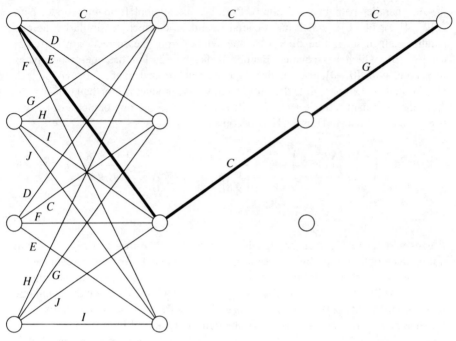

FIGURE 10.2 Four-state trellis for rate 4/5 hybrid QPSK/8-PSK multiple TCM.

$$A = \begin{bmatrix} 0 & 0 \\ 4 & 4 \end{bmatrix} \qquad B = \begin{bmatrix} 0 & 0 \\ 4 & 4 \end{bmatrix} + [2 \quad 2] = \begin{bmatrix} 2 & 2 \\ 6 & 6 \end{bmatrix}$$

$$C = A \cup B = \begin{bmatrix} 0 & 0 \\ 2 & 2 \\ 4 & 4 \\ 6 & 6 \end{bmatrix} \qquad D = C + [0 \quad 4] = \begin{bmatrix} 0 & 4 \\ 2 & 6 \\ 4 & 0 \\ 6 & 2 \end{bmatrix}$$

$$E = C + [0 \quad 2] = \begin{bmatrix} 0 & 2 \\ 2 & 4 \\ 4 & 6 \\ 6 & 0 \end{bmatrix} \qquad F = C + [0 \quad 6] = \begin{bmatrix} 0 & 6 \\ 2 & 0 \\ 4 & 2 \\ 6 & 4 \end{bmatrix}$$

$$G = C + [0 \quad 1] = \begin{bmatrix} 0 & 1 \\ 2 & 3 \\ 4 & 5 \\ 6 & 7 \end{bmatrix} \qquad H = D + [0 \quad 1] = \begin{bmatrix} 0 & 5 \\ 2 & 7 \\ 4 & 1 \\ 6 & 3 \end{bmatrix}$$

$$I = E + [0 \quad 1] = \begin{bmatrix} 0 & 3 \\ 2 & 5 \\ 4 & 7 \\ 6 & 1 \end{bmatrix} \qquad J = F + [0 \quad 1] = \begin{bmatrix} 0 & 7 \\ 2 & 1 \\ 4 & 3 \\ 6 & 5 \end{bmatrix}.$$

(10.4)

For the set assignment above, the error event path with the shortest length is the parallel path. Since for each set of parallel paths, both symbol positions represent distinct assignments (i.e., nonzero Euclidean distance), the length of this one branch path is $L = 2$ symbols and, from (9.138), the asymptotic behavior of the average bit error probability on the Rician fading channel varies as the inverse *square* of $\overline{E_s}/N_0$. The minimum squared distance between the parallel paths is 4. However, the three-branch error event path with signal set assignments E, C, and G (see Fig. 10.2) has a smaller squared distance, equal to $2 + 0 + 4\sin^2(\pi/8) = 2.586$, and thus we have $d_{\text{free}}^2 = 2.586$.

The conventional four-state, rate 2/3 trellis-coded 8-PSK found by Ungerboeck [1] for the AWGN (see also Section 3.6) had two parallel paths between nodes (i.e., a half-connected trellis). For his scheme, the parallel paths represented the minimum-distance error event and it was found that $d_{\text{free}}^2 = 4$. However, because of the absence of multiplicity, if this code were used on the fading channel, it would have an asymptotic error probability performance that varied only inverse *linearly* with $\overline{E_s}/N_0$. Thus with the hybrid MTCM scheme above, we obtain a performance on the *ideal* AWGN channel inferior to that of the equivalent Ungerboeck code but with much improved performance on the fading channel and perhaps equivalent performance in the presence of imperfect carrier synchronization.

10.2 Set Partitioning for Multiple Trellis-Coded *M*-PSK on the Fading Channel

With the previous examples as a basis, we now describe a set-partitioning method for the design of multiple trellis-coded *M*-PSK to achieve optimum asymptotic performance on the Rician fading channel. Two different approaches [2, 3] will be described, depending on whether the method proceeds from *root* to *leaf* of the tree (as in Ungerboeck's work) or from *leaf* to *root*. For multiplicity $k = 2$, the former approach is simpler and thus we shall describe that one first [2].

10.2.1 The First Approach

In [4], Ungerboeck presented a set-partitioning method for multidimensional trellis coding on AWGN channels. The method, which makes use of k-fold (in the MTCM case, k represents the multiplicity) Cartesian products of the sets found in Ungerboeck's original set-partitioning method for conventional ($k = 1$) trellis codes [1], is in essence the k-dimensional generalization of the latter. Since as we have already observed, the criteria for designing optimum trellis codes on the fading channel are quite different from that for the AWGN channel (i.e., maximize d_{free}), one might anticipate that the set-partitioning method would also be significantly different from that discussed in [4]. Indeed such is the case, the only commonality between the two being that we start the procedure with a k-fold Cartesian product of the complete *M*-PSK signal set. The remainder of the procedure, along with the motivation for it, is described in what follows. As stated above, we focus our attention first on the multiplicity $k = 2$ case. Later we consider the more general case of arbitrary multiplicity, k, whereupon the procedure will instead proceed from *leaf* to *root* of the tree.

Let A_0 denote the complete *M*-PSK signal set (i.e., signal points $0, 1, 2, \ldots, M - 1$) and $A_0 \otimes A_0$ denote a twofold Cartesian product of A_0 with itself. Thus an element of the set $A_0 \otimes A_0$ is a 2-tuple whose first and second symbols are each chosen from the set A_0. The first step is to partition $A_0 \otimes A_0$ into M signal sets defined by the ordered Cartesian product[4] $A_0 \otimes B_i$; $i = 0, 1, 2, \ldots, M - 1$, where the jth element ($j = 0, 1, 2, \ldots, M - 1$) of B_i is defined by $nj + i$ and the addition is performed modulo M. Thus the jth 2-tuple from the product $A_0 \otimes B_i$ is the ordered pair ($j, nj + i$). The selection of the odd integer multiplier n is the key to the set-partitioning method. Before presenting the relation whose solution provides the desired value(s) of n, we discuss what the first partitioning step is trying to accomplish.

The first partitioning step accomplishes two purposes. First, it guarantees that within any of the M partitions, each of the two symbol positions has distinct elements. That is, for a 2-tuple within a partitioned set, the Euclidean distance of each of the two symbols from the corresponding symbols in any other 2-tuple within the same set is nonzero. We recall that this is the desired property from the standpoint of maximizing the length of the shortest error event path. Stated another way, if the shortest error path is of length one branch (i.e., parallel paths exist in the trellis and have the smallest Euclidean distance from the correct path), the length L of this path is guaranteed to have value 2 and the error probability performance on the fading channel will vary as the inverse square of $\overline{E}_s/N_0$.

The second purpose accomplished is that the minimum Euclidean distance *product* between 2-tuples within a partitioned set (i.e., the minimum of the product of the distances between corresponding symbol positions of all pairs of 2-tuples) is maximized. To determine the value of this distance, we observe that the set $B_i + 1$ is merely a cyclic shift of the set B_i (i.e., a clockwise rotation of the corresponding signal points by an angle $2\pi/M$). Thus since the squared Euclidean distance between a pair of 2-tuples is the sum of the squared Euclidean distances between corresponding symbols in the 2-tuples, the set partitioning above guarantees that the *intra*distance structure of all of the partitions $A_0 \otimes B_i$ is *identical*. Thus it is sufficient to study the distance structure of $A_0 \otimes B_0$, henceforth called the *generating set*. For this set, the product of the squared distances between the ith 2-tuple and the jth 2-tuple is

$$\prod d_{ij}^2 = \left(4\sin^2\left(\frac{(j-i)\pi}{M}\right)\right)\left(4\sin^2\left(\frac{n(j-i)\pi}{M}\right)\right). \tag{10.5}$$

Thus based on the requirement above, we wish to choose n such that the minimum of $\prod d_{ij}^2$ over all pairs of 2-tuples in $A_0 \otimes B_0$ is maximized. Making use of the symmetry properties of the M-PSK signaling set around the circle, we can write the above as follows. If we let n^* denote the desired value(s) of n, then n^* has the maximin solution

$$f(n^*) = \max_{n^*=1,3,5,\ldots,M/2-1} \min_{m=1,3,5,\ldots,M/2-1} 16\sin^2\left(\frac{m\pi}{M}\right)\sin^2\left(\frac{n^*m\pi}{M}\right) \tag{10.6}$$

which has the equivalent vector form

$$g(n^*) = \max_{n^*=1,3,5,\ldots,M/2-1} \min_{m=1,3,5,\ldots,M/2-1} \left|z^m - 1\right|^2 \left|z^{n^*m} - 1\right|^2 \tag{10.7}$$

---

[4] By ordered Cartesian product we mean the concatenation of corresponding elements in the two sets forming the product.

where $z = \exp(j2\pi/M)$ represents a unit vector with phase equal to that between adjacent points in the signal constellation. For $M = 2$, we have the degenerate solution $n^* = 1$.

Note that the additive inverse(s) of n^* (i.e., $M - n^*$) is (are) also valid solutions. This conclusion is easily derived by substituting $z^{M-n} = \exp[j2\pi(M-n)/M] = z^{-n}$ for z^n in (10.7) and observing that the equation is unchanged. Table 10.1 gives the solution of (10.7) for $M = 4, 8, 16, 32,$ and 64.

The sets obtained by this first partition are illustrated below for the case $M = 8$ and $n^* = 3$, which is the single solution of (10.7)[5]:

$$A_0 \otimes B_0 = \begin{bmatrix} 0 & 0 \\ 1 & 3 \\ 2 & 6 \\ 3 & 1 \\ 4 & 4 \\ 5 & 7 \\ 6 & 2 \\ 7 & 5 \end{bmatrix} \quad A_0 \otimes B_1 = \begin{bmatrix} 0 & 1 \\ 1 & 4 \\ 2 & 7 \\ 3 & 2 \\ 4 & 5 \\ 5 & 0 \\ 6 & 3 \\ 7 & 6 \end{bmatrix}$$

$$A_0 \otimes B_2 = \begin{bmatrix} 0 & 2 \\ 1 & 5 \\ 2 & 0 \\ 3 & 3 \\ 4 & 6 \\ 5 & 1 \\ 6 & 4 \\ 7 & 7 \end{bmatrix} \quad A_0 \otimes B_3 = \begin{bmatrix} 0 & 3 \\ 1 & 6 \\ 2 & 1 \\ 3 & 4 \\ 4 & 7 \\ 5 & 2 \\ 6 & 5 \\ 7 & 0 \end{bmatrix}$$

$$A_0 \otimes B_4 = \begin{bmatrix} 0 & 4 \\ 1 & 7 \\ 2 & 2 \\ 3 & 5 \\ 4 & 0 \\ 5 & 3 \\ 6 & 6 \\ 7 & 1 \end{bmatrix} \quad A_0 \otimes B_5 = \begin{bmatrix} 0 & 5 \\ 1 & 0 \\ 2 & 3 \\ 3 & 6 \\ 4 & 1 \\ 5 & 4 \\ 6 & 7 \\ 7 & 2 \end{bmatrix}$$

$$A_0 \otimes B_6 = \begin{bmatrix} 0 & 6 \\ 1 & 1 \\ 2 & 4 \\ 3 & 7 \\ 4 & 2 \\ 5 & 5 \\ 6 & 0 \\ 7 & 3 \end{bmatrix} \quad A_0 \otimes B_7 = \begin{bmatrix} 0 & 7 \\ 1 & 2 \\ 2 & 5 \\ 3 & 0 \\ 4 & 3 \\ 5 & 6 \\ 6 & 1 \\ 7 & 4 \end{bmatrix}.$$

(10.8)

---
[5] Note that the additive inverse $n^* = 5$ could also have been used to generate (10.8).

TABLE 10.1
Solutions of (10.7) for Various
Various Values of M; $k = 2$.

M	n^*	$g(n^*)$
2	1	16
4	1	4
8	3	2
16	7	0.343
32	7, 9	0.062
64	19, 27	0.025

Note that sets $A_0 \otimes B_0$, $A_0 \otimes B_2$, $A_0 \otimes B_4$, $A_0 \otimes B_6$ of (10.8), which have the largest distance between them (i.e., the largest *inter*distance), are identical, respectively, to sets E, G, F, and H in Fig. 10.1, where only four sets of eight elements each were needed and $n^* = 5$ rather than $n^* = 3$ was used. Equivalently, one could have employed sets $A_0 \otimes B_1$, $A_0 \otimes B_3$, $A_0 \otimes B_5$, and $A_0 \otimes B_7$ in Fig. 10.1.

If one were to follow tradition, the second step in the set-partitioning procedure would be to partition each of the M sets $A_0 \otimes B_i$, $i = 0, 1, 2, \ldots, M - 1$, as in (10.8), for example, into two sets $C_0 \otimes D_{i0}$ and $C_1 \otimes D_{i1}$, with the first containing the even elements ($j = 0, 2, 4, \ldots, M - 2$) and the second containing the odd elements ($j = 1, 3, 5, \ldots, M - 1$). Although it is true that for each of these partitioned sets, the elements in each of the two symbol positions would still be distinct, unfortunately, it is not always true that these sets have the minimum Euclidean product distance between 2-tuples maximized. Thus we immediately conclude that the appropriate method to generate the sets on the second level of partition does not necessarily follow a tree structure.

After a little thought it becomes obvious that one should partition in such a way that the resulting sets (of dimensionality $M/2$) should have an *intra*distance product structure equal to that which would be achieved by a first-level partitioning in accordance with (10.7) with, however, M replaced by $M/2$. Interestingly enough, this second level of set partitioning can still be achieved by an odd–even split of a first-level partitioning like that described previously; however, the value of n^* used to generate the sets on this first level should be that corresponding to the solution of (10.7) with M replaced by $M/2$ (or its additive inverse). Note that if one of the solutions of (10.7) (including the additive inverse) is equal to one of the solutions of (10.7) (again including the additive inverse) when M is replaced by $M/2$, then indeed the first two levels of set partitioning follow a *tree structure*. By inspection of Table 10.1 we observe that when $M = 4, 8$, and 32, there exists a value of n^* common to these values of M and the corresponding values of $M/2$. Thus for $M = 4, 8$, and 32, the first two levels of set partitioning follow a tree structure, whereas for $M = 16$, they do not.

As an example of the second level of partitioning, the sets that result from the partitioning of the sets in (10.8) are as follows.

10.2 / Set Partitioning for Multiple Trellis-Coded M-PSK on the Fading Channel

$$C_0 \otimes D_{00} = \begin{bmatrix} 0 & 0 \\ 2 & 6 \\ 4 & 4 \\ 6 & 2 \end{bmatrix} \quad C_1 \otimes D_{01} = \begin{bmatrix} 1 & 3 \\ 3 & 1 \\ 5 & 7 \\ 7 & 5 \end{bmatrix}$$

$$C_0 \otimes D_{10} = \begin{bmatrix} 0 & 1 \\ 2 & 7 \\ 4 & 5 \\ 6 & 3 \end{bmatrix} \quad C_1 \otimes D_{11} = \begin{bmatrix} 1 & 4 \\ 3 & 2 \\ 5 & 0 \\ 7 & 6 \end{bmatrix}$$

$$C_0 \otimes D_{20} = \begin{bmatrix} 0 & 2 \\ 2 & 0 \\ 4 & 6 \\ 6 & 4 \end{bmatrix} \quad C_1 \otimes D_{21} = \begin{bmatrix} 1 & 5 \\ 3 & 3 \\ 5 & 1 \\ 7 & 7 \end{bmatrix}$$

$$C_0 \otimes D_{30} = \begin{bmatrix} 0 & 3 \\ 2 & 1 \\ 4 & 7 \\ 6 & 5 \end{bmatrix} \quad C_1 \otimes D_{31} = \begin{bmatrix} 1 & 6 \\ 3 & 4 \\ 5 & 2 \\ 7 & 0 \end{bmatrix}$$

$$C_0 \otimes D_{40} = \begin{bmatrix} 0 & 4 \\ 2 & 2 \\ 4 & 0 \\ 6 & 6 \end{bmatrix} \quad C_1 \otimes D_{41} = \begin{bmatrix} 1 & 7 \\ 3 & 5 \\ 5 & 3 \\ 7 & 1 \end{bmatrix}$$

$$C_0 \otimes D_{50} = \begin{bmatrix} 0 & 5 \\ 2 & 3 \\ 4 & 1 \\ 6 & 7 \end{bmatrix} \quad C_1 \otimes D_{51} = \begin{bmatrix} 1 & 0 \\ 3 & 6 \\ 5 & 4 \\ 7 & 2 \end{bmatrix}$$

$$C_0 \otimes D_{60} = \begin{bmatrix} 0 & 6 \\ 2 & 4 \\ 4 & 2 \\ 6 & 0 \end{bmatrix} \quad C_1 \otimes D_{61} = \begin{bmatrix} 1 & 1 \\ 3 & 7 \\ 5 & 5 \\ 7 & 3 \end{bmatrix}$$

$$C_0 \otimes D_{70} = \begin{bmatrix} 0 & 7 \\ 2 & 5 \\ 4 & 3 \\ 6 & 1 \end{bmatrix} \quad C_1 \otimes D_{71} = \begin{bmatrix} 1 & 2 \\ 3 & 0 \\ 5 & 6 \\ 7 & 4 \end{bmatrix}$$

(10.9)

The corresponding tree structure is illustrated in Fig. 10.3.

The third and succeeding steps are identical in construction to the second step; namely, we partition each set on the present level into two sets containing the alternate rows with the sets for the present level determined by a value of n^* computed from (10.7) with M successively replaced by $M/4$, $M/8$, and so on.

To extend the previous procedure to higher multiplicity of order $k \geq 2$, we can simply form the $k/2$-fold ordered Cartesian product of all the sets on a given partition level created by the procedure for $k = 2$. The result of this procedure

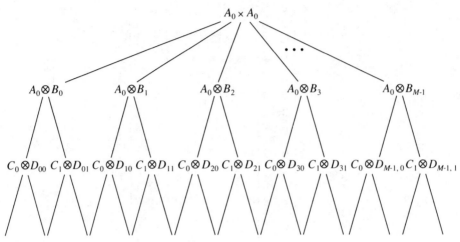

FIGURE 10.3 Set-partitioning method for multiple ($k = 2$) trellis codes on the fading channel.

is illustrated in Fig. 10.4 for $k = 4$. If the number of sets required to satisfy the trellis is less than the number of sets generated on a particular partition level, one would choose those that having largest *inter*distance, as was done in the example of Fig. 10.1. Also, as for the $k = 2$ case, the sets formed by this generalized set-partitioning procedure will all have distinct elements in any of the k symbol positions. Thus the length of a one-branch error event path will have value k, and hence the asymptotic bit error rate performance of such a trellis code on the fading channel will vary inversely as $(\overline{E_s/N_0})^n$ with $n \geq k$. Thus we have reinforced the notion introduced in Chapter 9 that insofar as the rate of decay of average bit error probability with $\overline{E_s/N_0}$ is concerned, incorporating multiplicity in the design of the trellis code has a similar effect to using *diversity*. Another procedure for $k > 2$ would be to generalize (10.7) to

$$g(\mathbf{n}^*) = \max_{\mathbf{n}^*} \min_{m=1,2,\ldots,N/2-1} |z^m - 1|^2 \prod_{i=1}^{j-1} |z^{n_i m} - 1|^2 \qquad (10.10)$$

where $\mathbf{n}^* = (n_1^*, n_2^*, \ldots, n_{j-1}^*)$ whose elements take on values from the set of odd integers $1, 3, 5, \ldots, N/2 - 1$. For $j = k$ and $N = M$, the maximin solution $\mathbf{n}^*$ can be used to produce all of the necessary sets on any level of partition. As we shall see shortly, the second approach to the design of trellis codes for fading channels is also based on the solutions to (10.10).

EXAMPLE 10.1 FOUR-STATE RATE 4/6 TRELLIS-CODED 8-PSK

Consider a four-state rate 4/6 trellis-coded 8-PSK system designed for optimum performance on the Rician fading channel. The trellis diagram appears in Fig. 10.2, where the signal point sets assigned to the branches are derived from the

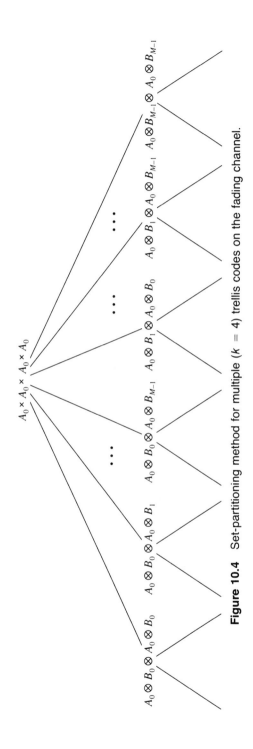

Figure 10.4 Set-partitioning method for multiple ($k = 4$) trellis codes on the fading channel.

previous procedure and are given by

$$C = \begin{bmatrix} 0 & 0 \\ 2 & 2 \\ 4 & 4 \\ 6 & 6 \end{bmatrix} \quad D = \begin{bmatrix} 1 & 5 \\ 3 & 7 \\ 5 & 1 \\ 7 & 3 \end{bmatrix} \quad E = \begin{bmatrix} 0 & 4 \\ 2 & 6 \\ 4 & 0 \\ 6 & 2 \end{bmatrix} \quad F = \begin{bmatrix} 1 & 1 \\ 3 & 3 \\ 5 & 5 \\ 7 & 7 \end{bmatrix}$$

$$G = \begin{bmatrix} 0 & 2 \\ 2 & 4 \\ 4 & 6 \\ 6 & 0 \end{bmatrix} \quad H = \begin{bmatrix} 1 & 7 \\ 3 & 1 \\ 5 & 3 \\ 7 & 5 \end{bmatrix} \quad I = \begin{bmatrix} 0 & 6 \\ 2 & 0 \\ 4 & 2 \\ 6 & 4 \end{bmatrix} \quad J = \begin{bmatrix} 1 & 3 \\ 3 & 5 \\ 5 & 7 \\ 7 & 1 \end{bmatrix}.$$

(10.11)

For this assignment, each set has a minimum squared *intra*distance equal to $4 + 4 = 8$, which represents the minimum squared Euclidean distance between parallel paths. Each of these one-branch paths when viewed as an error event path has a length $L = 2$ with respect to any of the other paths in parallel with it. Every other error event path (consisting of two or more branches) has a length L greater than 2 regardless of which path is chosen as the correct path. Thus the dominant term in the asymptotic bit error probability expression corresponds once again to the parallel paths. Since the minimum squared *intra*distance for each of the two symbols in any of these 2-tuples is 4, then analogous to (10.3), which describes the performance of the same scheme using only a two-state trellis, we get

$$P_b \cong \frac{1}{4} \left(\frac{4(1+K)}{\overline{E_s/N_0}} e^{-K} \right)^2 \frac{1}{(4)(4)} = \frac{1}{16} \frac{(1+K)^2}{(\overline{E_b/N_0})^2} e^{-2K} \quad (10.12)$$

(i.e., a gain of 4.5 dB in SNR). □

EXAMPLE 10.2 TWO-STATE RATE 4/12 TRELLIS-CODED 8-PSK

This is an example of a multiplicity 4 trellis code optimally designed for the Rician fading channel. Since four 8-PSK symbols are transmitted over the channel for each 4 bits into the encoder, the throughput of the code is 1 bit/s/Hz. The state diagram is as in Fig. 10.1, where E, F, G, and H are chosen as those sets that have the largest *inter*distance in the construction of Fig. 10.4. Thus we have

$$E = A_0 \otimes B_0 \otimes A_0 \otimes B_0 = \begin{bmatrix} 0 & 0 & 0 & 0 \\ 1 & 5 & 1 & 5 \\ 2 & 2 & 2 & 2 \\ 3 & 7 & 3 & 7 \\ 4 & 4 & 4 & 4 \\ 5 & 1 & 5 & 1 \\ 6 & 6 & 6 & 6 \\ 7 & 3 & 7 & 3 \end{bmatrix}$$

$$F = A_0 \otimes B_4 \otimes A_0 \otimes B_4 = \begin{bmatrix} 0 & 4 & 0 & 4 \\ 1 & 1 & 1 & 1 \\ 2 & 6 & 2 & 6 \\ 3 & 3 & 3 & 3 \\ 4 & 0 & 4 & 0 \\ 5 & 5 & 5 & 5 \\ 6 & 2 & 6 & 2 \\ 7 & 7 & 7 & 7 \end{bmatrix}$$

$$G = A_0 \otimes B_2 \otimes A_0 \otimes B_2 = \begin{bmatrix} 0 & 2 & 0 & 2 \\ 1 & 7 & 1 & 7 \\ 2 & 4 & 2 & 4 \\ 3 & 1 & 3 & 1 \\ 4 & 6 & 4 & 6 \\ 5 & 3 & 5 & 3 \\ 6 & 0 & 6 & 0 \\ 7 & 5 & 7 & 5 \end{bmatrix}$$

$$H = A_0 \otimes B_6 \otimes A_0 \otimes B_6 = \begin{bmatrix} 0 & 6 & 0 & 6 \\ 1 & 3 & 1 & 3 \\ 2 & 0 & 2 & 0 \\ 3 & 5 & 3 & 5 \\ 4 & 2 & 4 & 2 \\ 5 & 7 & 5 & 7 \\ 6 & 4 & 6 & 4 \\ 7 & 1 & 7 & 1 \end{bmatrix}$$

(10.13)

For this code, all sets have squared Euclidean *intra*distance equal to 8. The asymptotic average bit error probability for coherent detection with ideal CSI is computed analogous to (10.3) and is given by

$$P_b \cong \frac{1}{4} \left(\frac{4(1+K)}{\overline{E_s/N_0}} e^{-K} \right)^4 \frac{1}{\left(4\sin^2\left(\frac{\pi}{8}\right)\right)^2 \left(4\sin^2\left(\frac{5\pi}{8}\right)\right)^2} = \frac{(1+K)^4}{(\overline{E_b/N_0})^4} e^{-4K}.$$

(10.14)

Note that because the multiplicity is equal to 4, the average bit error probability varies inversely with $(\overline{E_b/N_0})^4$, where now $\overline{E_s} = \overline{E_b}$. □

EXAMPLE 10.3 FOUR-STATE RATE 5/6 TRELLIS-CODED 8-PSK

This is an example of a multiplicity 2 trellis code with noninteger throughput (i.e., 2.5) optimally designed for the Rician fading channel. The trellis diagram is as in

Fig. 10.1 with the following set assignments:

$$
\begin{aligned}
C &= A_0 \otimes B_0 & D &= A_0 \otimes B_2 \\
E &= A_0 \otimes B_4 & F &= A_0 \otimes B_6 \\
G &= A_0 \otimes B_1 & H &= A_0 \otimes B_3 \\
I &= A_0 \otimes B_5 & J &= A_0 \otimes B_7
\end{aligned}
\quad (10.15)
$$

where the sets $A_0 \otimes B_i$, $i = 0, 1, 2, \ldots, 7$, are as in (10.8). Again, we remind the reader that $n^* = 5$ rather than $n^* = 3$ could have been used to generate these sets. By construction, the parallel paths in each of the sets above have length $L = 2$. Also, the minimum squared Euclidean distance product for these parallel paths is $\prod d^2 = [4\sin^2(\pi/8)] \times [4\sin^2(5\pi/8)] = 2$. If we examine all of the two-branch error event paths, we find that the shortest length of these paths is also $L = 2$. The minimum squared Euclidean distance product for these two branch paths is $[4\sin^2 2(\pi/8)] \times [4\sin^2(\pi/8)] = 1.172$, which is smaller than 2 and thus dominates the asymptotic error probability performance. In particular, for coherent detection with ideal CSI, we have that

$$
P_b \cong \frac{1}{5}\left(\frac{4(1+K)}{\overline{E_s}/N_0}e^{-K}\right)^2 \frac{1}{1.172} = 0.437\frac{(1+K)^2}{(\overline{E_b}/N_0)^2}e^{-2K} \quad (10.16)
$$

where $\overline{E_s} = 2.5\overline{E_b}$.

The squared free distance of this code is determined by the error event path of length three branches indicated in Fig. 10.1 and is given by $d_{\text{free}}^2 = 2[4\sin^2(\pi/8)] + 4\sin^2(\pi/8) = 1.757$. The equivalent code optimized for the AWGN channel [5] has $d_{\text{free}}^2 = 2$ but only $L = 1$. □

EXAMPLE 10.4 EIGHT-STATE RATE 3/6 TRELLIS-CODED 8-PSK

This example is another one with noninteger throughput (i.e., 1.5). Also, since there are $2^3 = 8$ branches emanating from each node and the trellis has eight states, there are no parallel paths and the trellis is fully connected. The trellis diagram is illustrated in Fig. 10.5 with the following set assignments:

$$
A^{(0)} = A_0 \otimes B_0 \qquad B^{(0)} = A_0 \otimes B_4. \quad (10.17)
$$

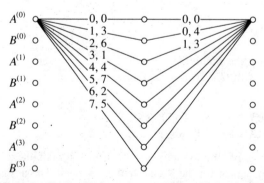

FIGURE 10.5 Trellis diagram for rate 3/6 trellis-coded 8-PSK.

10.2 / Set Partitioning for Multiple Trellis-Coded M-PSK on the Fading Channel

Also, $A^{(i)}$ and $B^{(i)}$; $i = 1, 2, 3$, are cyclic shifts of $A^{(0)}$ and $B^{(0)}$, respectively, by i rows. This code achieves a minimum diversity $L = 3$ corresponding to an error event path of length two branches. Also, the minimum product of squared Euclidean distances is given by $\prod d^2 = [4\sin^2(\pi/8)] \times [4\sin^2(5\pi/8)] \times 4 = 8$, which corresponds to the error event path (relative to the all-zeros path) with 8-PSK symbols $(1, 3)$ and $(0, 4)$ along its branches. Finally, then, the bit error probability is asymptotically approximated by

$$P_b \cong \frac{1}{3}\left(\frac{4(1+K)}{\overline{E_s}/N_0}e^{-K}\right)^3 \frac{1}{8} = 0.79\frac{(1+K)^3}{(\overline{E_b}/N_0)^3}e^{-3K} \qquad (10.18)$$

where $\overline{E_s}/N_0 = 1.5\overline{E_b}/N_0$. The squared free distance is determined by the two-branch path with 8-PSK symbol assignments $(3, 1)$ and $(1, 7)$ and is given by $d_{\text{free}}^2 = 4\sin^2(3\pi/8) + 4\sin^2(\pi/8) = 5.172$.

Another interesting generalization of this example is as follows. When there are no parallel paths in the trellis, as is true here, it may be desirable to go to a larger modulation constellation (e.g., 16-PSK rather than 8-PSK) to achieve an increase in diversity. To demonstrate this idea, consider the following set assignment to the trellis diagram of Fig. 10.5. First construct sets $A_0 \otimes B_0$ and $A_0 \otimes B_4$ with $n^* = 3$ and 16-PSK modulation. Next, partition these sets in accordance with Fig. 10.4. Now choose the sets in Fig. 10.5 as

$$A^{(0)} = C_0 \otimes D_{00} = \begin{bmatrix} 0 & 0 \\ 2 & 6 \\ 4 & 12 \\ 6 & 2 \\ 8 & 8 \\ 10 & 14 \\ 12 & 4 \\ 14 & 10 \end{bmatrix} \qquad B^{(0)} = C_0 \otimes D_{41} = \begin{bmatrix} 1 & 7 \\ 3 & 13 \\ 5 & 3 \\ 7 & 9 \\ 9 & 15 \\ 11 & 5 \\ 13 & 11 \\ 15 & 1 \end{bmatrix} \qquad (10.19)$$

together with the appropriate cyclic shifts as before. We note that with this assignment, all of the two-branch error event paths differ from the correct path in four 16-PSK symbols, and thus the diversity is now $L = 4$. Unfortunately, the minimum product of squared Euclidean distances will be reduced to the value $\prod d^2 = [4\sin^2(2\pi/16)] \times [4\sin^2(10\pi/16)] \times [4\sin^2(\pi/16)] \times [4\sin^2(9\pi/16)] = 1.172$, and thus the choice between 8-PSK and 16-PSK modulations depends on the value of $\overline{E_b}/N_0$ at which one is operating. □

10.2.2 The Second Approach

We have observed previously that conventional trellis codes (e.g., the original Ungerboeck codes [1] introduced in Chapter 3) that have parallel paths can only achieve unit diversity (i.e., they must have $L = 1$). On the other hand, trellis codes with multiplicity k and parallel paths can, with proper design, achieve a diversity $L = k$. To see under what circumstances the above can be achieved, we proceed as follows.

As discussed in Chapter 7, a multiple trellis code with b input bits entering and k M-PSK symbols leaving the encoder per trellis branch has a throughput

$r = b/k$ bits/s/Hz. In order for the code to have redundancy, we must have $b + 1 \le s$, or, equivalently from (7.3) with $M_i = M$ for all i,

$$M \ge 2^{r+1/k}. \tag{10.20}$$

To design a code with $L = k$ and *the smallest number of states*, we should use M parallel paths per state transition.[6] Since the total number of branches emanating from any node in the trellis diagram is 2^b, then if ν is the encoder memory, we would have

$$2^b = M 2^\nu \tag{10.21a}$$

or, equivalently, a number of states given by

$$2^\nu = \frac{2^{rk}}{M}. \tag{10.21b}$$

Note from (10.21b) that by increasing M, we can decrease the required number of states even to the limit of *one state* ($M = 2^{rk}$), in which case we have a ($k \log_2 M$, b) block code designed for the fading channel. The price paid for choosing M large enough to give the minimum number of required states is an increase in the constant C in (9.139), which leads to a degradation in performance. However, as mentioned above, for fading channels, the achievement of a large diversity is of primary importance and overshadows the necessity of achieving the minimum value of C. Finally, we remind the reader that for the AWGN, Ungerboeck has demonstrated, based on curves of channel capacity, that little performance is gained by increasing M beyond 2^{b+1}. From the above we see that this is not necessarily true for the fading channel.

To design a trellis code with the foregoing properties, we need to construct $2^{b+1}/M$ distinct sets each with M k-tuples of M-PSK symbols ($2^b/M$ sets for the branches emanating from the *even*-numbered states and a similar number for the branches emanating from the *odd*-numbered states). To obtain $2^{b+1}/M$ *distinct* sets, we can choose $m = \log_2 M$ such that $m = \lfloor (b+1)/2 \rfloor$ where, as in previous chapters, $\lfloor x \rfloor$ denotes the smallest integer greater than or equal to x. After choosing M as above for the given value of $b = rk$, the k-tuples of M-PSK symbols should be assigned to the parallel paths such that the maximum of C is minimized over all error events and no error event should result in a value of $L < k$. For even values of k, we propose the following construction[7] for the sets A_i, $i = 0, 1, 2, \ldots, (2^{b+1})/M - 1$, each containing M k-tuples:

$$A_i = \{(n, N_1^* n, \ldots, N_{\frac{k}{2}-1}^* n, N_{\frac{k}{2}}^* n + m_i, N_{\frac{k}{2}+1}^* n + M_1^* m_i, \ldots, N_{k-1}^* n + M_{\frac{k}{2}-1}^* m_i)$$

$$n = 0, 1, 2, \ldots, M-1\} \quad i = 0, 1, 2, \ldots, \frac{2^{b+1}}{M}$$

$$\tag{10.22}$$

---

[6] Note that we cannot use more than M parallel paths per transition, since the code design procedure described in [2] requires that each symbol position in the code set assigned to a group of parallel paths have *distinct* elements chosen from the symbol set of size M.

[7] This is a generalization of the set construction procedure described in Section 10.2.1 for the case of $k = 2$.

where $m_i = iM^2/2^{b+1}$ and all additions and multiplications are performed modulo M. For even values of i, we assign the sets to the branches emanating from the first state and for odd values of i, we assign the sets to the branches emanating from the second state. The set assignments for branches emanating from the remainder of the even and odd states are simply permutations of the set assignments for the first and second states.

The integers N_i^* and M_i^* used in (10.22) are determined as follows: Let $\mathbf{n}^* = (n_1^*, n_2^*, \ldots, n_{j-1}^*)$ be the solution of (10.10) where the elements of the vector $\mathbf{n}^*$ take on values from the set of odd integers $1, 3, 5, \ldots, N/2 - 1$. Then, for $j = k$ and $N = M$, we have $N_i^* = n_i^*$, while for $j = k/2$ and $N = 2^{b+1}/M$, we have $M_i^* = n_i^*$. Note that the additive inverses (i.e., $N - n_i^*$) are also valid solutions of (10.10), and hence N_i^* and M_i^* can also be determined from these values. Also note that for $k = 2$, there are no solutions for M_i^* and the values of $N_i^* = n_i^*$ agree with those found from the solution of (10.7).

The optimum values[8] $n_1^*, n_2^*, \ldots, n_{k-1}^*$ for $k = 3, 4$ and $M = 2^m$, $m = 1, 2, 3, \ldots, 6$, are given in Tables 10.2 and 10.3. Also indicated in the tables are the corresponding values of maximin squared distance product, $g(\mathbf{n}^*)$ as computed from (10.10).

TABLE 10.2 Solutions of (10.10) for Various Values of M; $k = 3$.

M	$n_1^*\ n_2^*$					$g(\mathbf{n}^*)$
2	1	1					64
4	1	1					8
8	1	3,	3	3			1.1716
16	3	5,	3	7,	5	7	0.05198
32	7	9,	7	15,	9	15	8.9×10^{-2}
64	11	27,| 15	29,	17	19	2.86×10^{-2}

TABLE 10.3 Solutions of (10.10) for Various Values of M; $k = 4$.

M	$n_1^*\ n_2^*\ n_3^*$	$g(\mathbf{n}^*)$
2	1 1 1	256
4	1 1 1	16
8	1 3 3	4
16	3 5 7	2
32	7 9 15	0.1177
64	11 17 19, 11 23 27, 15 27 29, 19 25 29	4.56×10^{-2}

[8] Other solutions are possible for $\mathbf{n}^*$; however, these solutions would merely correspond to permutations of the elements in any solution presented in Tables 10.2 and 10.3 and thus would yield the identical distance product. Hence, for simplicity, we have listed only those solutions that are distinctly different from one another. For example, for $k = 3$ and $M = 32$, $n_1^* = 9$ and $n_2^* = 7$ is also a possible solution. However, this is simply a permutation of the solution $n_1^* = 7$ and $n_2^* = 9$ that is reported in Table 10.2.

EXAMPLE 10.5

Let $r = 2$ and $k = 2$. Hence $b = 4$ and we choose

$$M = 2^{\lfloor \frac{b+1}{2} \rfloor} = 2^3 = 8. \qquad (10.23)$$

The number of states, $N_s = 2^\nu$, is determined from (10.21) as $N_s = 2$. There are $2^{b+1}/M = 4$ signal sets to be assigned to the branches of the trellis. These sets are given by

$$A_i = \{(n, N_1^* n + m_i), n = 0, 1, 2, \ldots, 7\} \qquad m_i = 0, 2, 4, 6 \qquad (10.24)$$

where from Table 10.1, $N_1^* = 3$ or its additive inverse $8 - 3 = 5$. The resulting code is illustrated in Fig. 10.1, where $E = A_0, F = A_2, G = A_1$, and $H = A_3$. Note that this code design is identical to that obtained from the first approach for Example 10.1. □

EXAMPLE 10.6

Let $r = 1$ and $k = 4$. Here $b = 4$, once again $M = 8$, and the number of states is $N_s = 2$. Again there are four signal sets required for constructing the trellis code,

$$A_i = \{(n, N_1^* n, N_2^* n + m_i, N_3^* n + M_1^* m_i), n = 0, 1, 2, \ldots, 7\} \qquad m_i = 0, 2, 4, 6.$$

$$(10.25)$$

From Table 10.3, we have $N_1^* = 1, N_2^* = 3, N_3^* = 3$, and $M_1^* = 1$. The state diagram for the code is the same as that illustrated in Fig. 10.1, where the A_i's are now given by (10.25). □

EXAMPLE 10.7

Let $r = 2$ and $k = 4$. Here $b = 8$ but now $M = 32$ and the number of states is $N_s = 8$. The signal sets are determined from

$$A_i = \{(n, N_1^* n, N_2^* n + m_i, N_3^* n + M_1^* m_i), n = 0, 1, 2, \ldots, 31\}$$

$$m_i = 0, 2, 4, \ldots, 14$$

$$(10.26)$$

where from Table 10.3, $N_1^* = 7, N_2^* = 9, N_3^* = 15$, and $M_1^* = 7$. The state diagram for the code can be obtained by assigning $A_0, A_2, \ldots, A_{14}$ to the even-numbered states and $A_1, A_3, \ldots, A_{15}$ to the odd-numbered states, together with the required permutations. □

10.3 Design of Ungerboeck-Type Codes (Unit Multiplicity) for Fading Channels

Under certain conditions it is possible to reduce a multiple trellis code with parallel paths, such as those discussed above, to an Ungerboeck-type code (multiplicity equal to 1) with no parallel paths but an increased number of states. To see this, we first note that in order for an Ungerboeck-type code with rate $b/b+1$ and no parallel paths to achieve diversity L, the smallest number of states required is

$$2^\nu = 2^{b(L-1)}. \tag{10.27}$$

For example, for $b = 2$, the minimum number of states required to achieve $L = 5$ would be $N_s = 2^\nu = 256$.

To construct such codes, we note that any $k = 1$ code can be converted into an equivalent trellis code with multiplicity $k > 1$. Unfortunately, the reverse operation is not, in general, true; that is, a trellis code with multiplicity $k > 1$ cannot always be converted to an equivalent $k = 1$ trellis code. However, by imposing certain constraints on the signal sets, it is possible to construct a procedure for designing a multiple trellis code that *always* has an equivalent $k = 1$ code. In what follows, we describe this procedure briefly.

To design an Ungerboeck-type code with diversity L, we first design a multiple trellis code with $k = L$. The generator set for such a code consists of $M/2 = 2^b$ L-tuples and is in the form

$$A_0 = \{(\pi_0(n), \pi_1(n), \ldots, \pi_{L-1}(n)), \; n = 0, 1, \ldots, 2^b - 1\}. \tag{10.28}$$

Here $\pi_i(n)$, $n = 0, 1, \ldots, 2^b - 1$, is the ith column of A_0 and contains a permutation of 2^b M-PSK symbols. In particular, $\pi_{2i+1}(n)$, $n = 0, 1, \ldots, 2^b - 1$, should contain an equal number of even and odd M-PSK symbols and $\pi_{2i}(n)$, $n = 0, 1, \ldots, 2^b - 1$, should contain only even M-PSK symbols. Furthermore, $\pi_0(n) = 2n$, $n = 0, 1, \ldots, 2^b - 1$. The $\pi_i(n)$'s should be chosen to maximize the minimum product of distances analogous to (10.10) and guarantee that there is no path with $L < k$ in the trellis and no smaller product of distances. The other set assignments for the multiple trellis code can be obtained from the generator set in (10.28) by a procedure analogous to (10.22).

The set assignments for the equivalent $k = 1$ code are obtained from A_0 as follows. Define a set of $L - 1$ indices $k_1, k_2, \ldots, k_{L-1}$, each of which ranges over the integers $0, 1, 2, \ldots, 2^b - 1$. For any choice of these indices, define the index N by

$$N = \sum_{i=1}^{L-1} 2^{b(L-i-1)} k_{L-i} \tag{10.29}$$

which ranges over the values $0, 1, \ldots, 2^{b(L-1)} - 1$ and corresponding signal set of 1-tuples

$$A_N = \left\{ \sum_{i=0}^{L-1} \pi_i(k_i), \; k_0 = 0, 1, \ldots, 2^b - 1 \right\}. \tag{10.30}$$

Since, from (10.27), there are $N_s = 2^{b(L-1)}$ states, the equivalent $k = 1$ trellis code is defined by assigning the set A_N to the transitions emanating from state N.

As an illustration of the foregoing procedure, Table 10.4 shows the $\pi_i(n)$'s required to design 4-, 16-, and 64-state Ungerboeck-type codes for $b = 2$. Interestingly, the 16-state code derived here for the fading channel is equivalent to

TABLE 10.4 Required $\pi_i(n)$ Values for Construction of Ungerboeck-Type Codes.

Diversity, L	N_s	$\pi_1(n)$	$\pi_2(n)$	$\pi_3(n)$	Minimum Squared Distance Product
2	4	n			1.17
3	16	$n(2n^2 + 3n + 4)$	$n(2n^2 + 7n + 3)$		4.69
4	64	$n(2n^2 + 3n + 4)$	$n(2n^2 + 7n + 3)$	$n(4n + 1)$	8.00

the original Ungerboeck code of the same number of states that was designed for the AWGN channel. This is merely a coincidence, and one should not conclude that codes designed in accordance with (10.30) will be the same as those optimized for the AWGN.

10.4 Comparison of Error Probability Performance with Computational Cutoff Rate

The throughput performance of trellis codes can be compared with the computational cutoff rate R_0 of the coding channel to determine the efficiency of the code design. In Chapter 7 we made such a comparison for the AWGN. Here we make an analogous comparison for the Rician fading channel using codes that are optimally designed for that channel.

From (9.139), an approximate (asymptotically approached at high SNR) expression for the first error event probability on the Rayleigh channel is given by[9]

$$P_e \cong \frac{1}{g(\mathbf{n}*)} \left(\frac{4}{\overline{E_s/N_0}} \right)^L N(L) \qquad (10.31)$$

where $N(L)$ is the number of error event paths at Hamming distance L from the all-zeros path. This can be simplified (at a slight expense in tightness) still further by ignoring $N(L)$. Then, for a given P_e (e.g., 10^{-6}), one can readily compute the required $\overline{E_s/N_0}$ for any particular trellis code with given values of b, k, and M once we determine $g(\mathbf{n}*)$ and L. In Fig. 10.6 we illustrate this procedure by superimposing on the R_0 curves the communication efficiency of various multiple trellis codes designed for the Rayleigh channel. Many of these examples correspond to cases discussed previously in this chapter. The points having $N_s = L = k = 1$ correspond to uncoded M-PSK modulation.

---

[9] Here $g(\mathbf{n}*)$ represents the minimum product of distances, which is not necessarily achieved by the parallel paths.

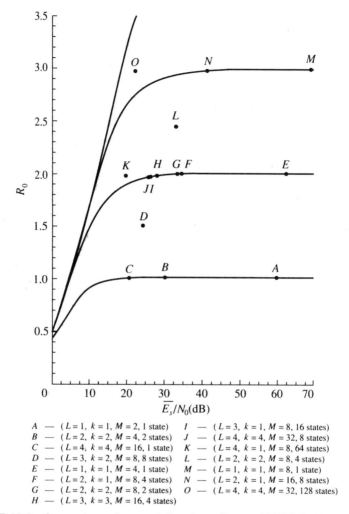

FIGURE 10.6 Comparison of computational cutoff rate of *M*-PSK with throughput performance of trellis-coded *M*-PSK over Rayleigh channel.

10.5 Simulation Results

Illustrated in Fig. 10.7 are computer simulation results for the bit error probability performance of the $L = 2$, $k = 2$, $M = 8$, $N_s = 2$ code (code G in Fig. 10.6) transmitted over a Rayleigh channel and differentially detected at the receiver. The interleaver is a block interleaver with 32 rows and 16 columns. For comparison, simulation results are presented for the $r = 2$, $k = 2$, $M = 8$, $N_s = 2$ code (see Chapter 7) designed for the AWGN. Figure 10.8 illustrates simulation results for an $L = 3$, $k = 1$, $M = 8$, $N_s = 16$ code (code I in Fig. 10.6) transmitted over a Rician ($K = 10$) channel along with measurement results from [6]. Here the interleaver has 16 rows and 8 columns and a decoder buffer size of 32

434 Ch. 10 / Design of TCM for Fading Channels

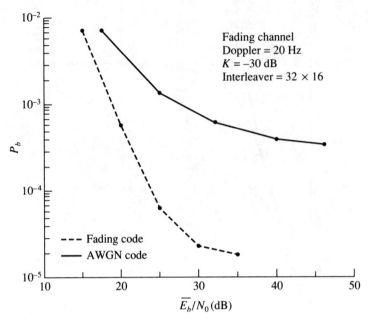

FIGURE 10.7 Comparison of bit error rate of code G with optimum equivalent AWGN code.

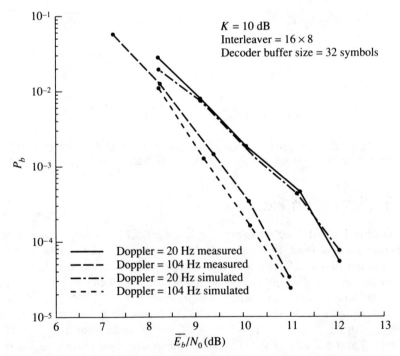

FIGURE 10.8 Performance of TCM/8-DPSK over Rician fading channel.

10.6 Further Discussion

In [7], Wilson and Leung proposed a conventional four-state rate 2/3 8-PSK code for use on Rayleigh fading channels. The code has a trellis diagram as in Fig. 10.2 where the sets C, D, E, F, G, H, I, and J now correspond to single 8-PSK symbols in accordance with

$$\begin{array}{llll} C = 0 & D = 4 & E = 2 & F = 6 \\ G = 1 & H = 5 & I = 3 & J = 7. \end{array} \quad (10.32)$$

The shortest error event path corresponding to the minimum squared Euclidean distance product has length three branches and is identical to that indicated in Fig. 10.2. Again since the middle branch has a symbol assignment identical to the all-zeros path, the diversity of this code is only $L = 2$. Furthermore, the minimum squared distance product is

$$\prod d^2 = 4\left(\sin^2 2\left(\frac{\pi}{8}\right)\right) \times \left(4\sin^2\left(\frac{\pi}{8}\right)\right) = 1.172. \quad (10.33)$$

On the AWGN, this code achieves a minimum squared Euclidean distance $d^2_{\min} = 2.586$ that is 1.9 dB inferior to the optimum code four-state found by Ungerboeck for this channel [1].

Most recently, Jamali and Le-Ngoc [8] introduced a set of design rules that allowed them to further improve this code. In particular, using the same trellis structure as in Fig. 10.2, they proposed the following ordered symbol assignment to the trellis branches emanating from states 1, 2, 3, and 4:

State	8-PSK Symbols
1	$CDEF$
2	$GHJI$ (or $IJGH$)
3	$FEDC$
4	$IJHG$ (or $HGJI$)

where C, D, E, F, G, H, I, and J are as in (10.32). In doing this, they still achieved a code with diversity $L = 2$; however, the shortest error event path corresponding to the minimum squared distance product was now of length two branches and the value of this product *doubled* relative to (10.33), that is,

$$\prod d^2 = \left(4\sin^2 4(\pi/8)\right) \times \left(4\sin^2(\pi/8)\right) = 2.344. \quad (10.34)$$

This implies an asymptotic improvement in performance on the Rayleigh fading channel of $10\log\sqrt{2} = 1.5$ dB. Furthermore, on the AWGN this code achieves a minimum squared Euclidean distance $d^2_{\min} = 3.172$, which is now only 1 dB inferior to Ungerboeck's optimum four-state code for this channel.

REFERENCES

1. G. UNGERBOECK, "Channel coding with multilevel/phase signals," *IEEE Trans. Inf. Theory*, Vol. IT-28, No. 1, pp. 55–67, Jan. 1982.
2. D. DIVSALAR and M. K. SIMON, "The design of trellis codes for fading channels," *JPL Publication 87-39 (MSAT-X Report 147)*, Pasadena, Calif., Nov. 1, 1987. See also *IEEE Trans. Commun.*, Vol. 36, No. 9, pp. 1004–1021, Sept. 1988.
3. D. DIVSALAR, M. K. SIMON, and T. JEDREY, "Trellis coding techniques for mobile communications," *MILCOM'88 Conf. Rec.*, Oct. 23–26, 1988, San Diego, Calif., pp. 35.3.1–35.3.7.
4. G. UNGERBOECK, "Trellis-coded modulation with redundant signal sets. Part I: Introduction; Part II: State of the art," *IEEE Commun. Mag.*, Vol. 25, No. 2, pp. 5–21, Feb. 1987.
5. S. G. WILSON, "Rate 5/6 trellis-coded 8-PSK," *IEEE Trans. Commun.*, Vol. COM-34, No. 18, pp. 1045–1049, Oct. 1986.
6. N. LAY, T. JEDREY, D. DIVSALAR, C. CHEETHAM, and D. BLACK, "Modem and terminal processor testing on the JPL fading channel simulator," *MSAT-X Q.*, No. 15, Apr. 1988.
7. S. G. WILSON and Y. S. LEUNG, "Trellis-coded phase modulation on Rayleigh channels," *ICC'87 Conf. Rec.*, June 7–10, 1987, Seattle, Wash., pp. 21.3.1–21.3.5.
8. S. H. JAMALI and T. LE-NGOC, "A new 4-state 8PSK TCM scheme for fast fading, shadowed mobile radio channels," *IEEE Trans. Veh. Technol.*, to appear.

CHAPTER 11

Analysis and Design of TCM for Other Practical Channels

11.1 Intersymbol Interference Channels

Communication channels are characterized as occupying a portion of a spectral band of frequencies. The portion of the spectral band where attenuation is low is the channel passband and is taken to have a given spectral width. Two examples can be found in Figs. 1.10 and 1.24 of Chapter 1. Intersymbol interference occurs when the energy in a modulation symbol of length T is dispersed over a time interval longer than T. Since a new modulation symbol is transmitted every T seconds, these symbols will interfere. An example for a low-pass filter is given in Fig. 1.10.

So far we have presented a time-domain definition of intersymbol interference. In the frequency-domain interpretation this interference occurs when the transmitted spectrum is wider than the channel passband. Recall that the conditions for distortionless transmission are a flat gain and a linear phase over the significant frequencies of an input signal [1, pp. 324 ff.]. Thus, if a transmitted spectrum (e.g., like the one in Fig. 1.11) is wider than the channel passband, the input spectrum will be altered. Thus pulse distortion will occur, causing energy from the modulation interval to spill over into many modulation intervals: The result is intersymbol interference.

In the receiver, a filter is used to attempt to realize the goal of distortionless transmission for the channel on an end-to-end basis. Thus this filter attempts to flatten the overall transfer function magnitude and linearize its phase. This is attempted over the selected passband of the transmitted signal. Such a receiver filter is called an *equalizer*, a name taken from the same function a filter would perform in analog transmission (see, e.g., [1, pp. 360 ff.]).

The question is: Over what frequencies should the overall channel have a flat spectrum? The answer is supplied by the Nyquist theory of data transmission [1–4]. Nyquist showed that error-free transmission in noiseless channels was possible if the symbol rate, R_s, did not exceed $2B$, where B is the one-sided channel bandwidth. Perfect timing synchronization was assumed. With just the smallest timing error it was found that performance would seriously degrade. Thus one chooses R_s from 20 to 100% less than $2B$. The parameter B is also known as the minimum bandwidth to support the symbol rate R_s.

Actually, TCM was invented for application to telephone channels. An average gain and delay characteristic for such a channel is shown in Fig. 11.1 [5]. The

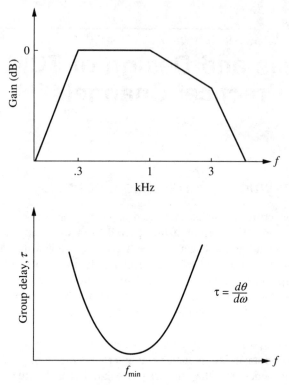

FIGURE 11.1 Basic telephone line transfer function.

equalizer in the receiver was assumed to make the overall channel essentially an AWGN channel. The TCM signals were then optimized for the AWGN channel. In this chapter we examine the sensitivity of this assumption. The traditional receiver in these applications is an equalizer followed by a Viterbi decoder, which works only on the code-state variables. That is, the assumption is that the equalizer transforms the intersymbol interference plus Gaussian noise channel into just the Gaussian noise channel [6]. At the end of this section we consider Viterbi receivers that work jointly on intersymbol interference states and code states. No consideration is given to designing TCM for intersymbol interference channels, as little has been done on this important problem.

11.1.1 Model

The basic transmission model is given in Fig. 11.2. We consider QAM modulation, and the basic transmitter for four-level QAM is shown in Fig. 11.3. This will serve as our reference uncoded system. The coded system is shown in Fig. 11.4; following [9, 10] we use independent trellis codes on each QAM rail.

In complex baseband notation the channel impulse response is

$$h(t) = h_I(t) + jh_Q(t). \tag{11.1}$$

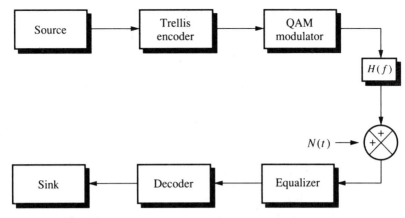

FIGURE 11.2 Block diagram for linear dispersive channel.

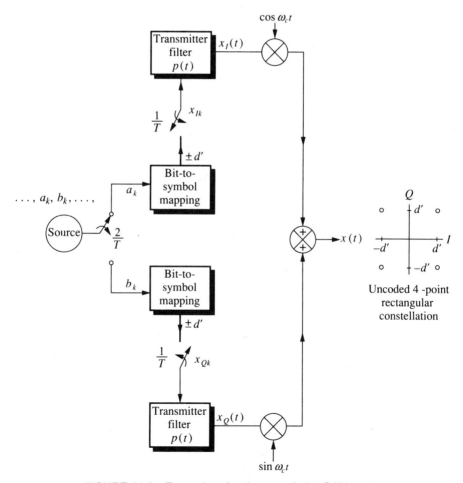

FIGURE 11.3 Transmitter for the uncoded 4-QAM system.

440 Ch. 11 / Analysis and Design of TCM for Other Practical Channels

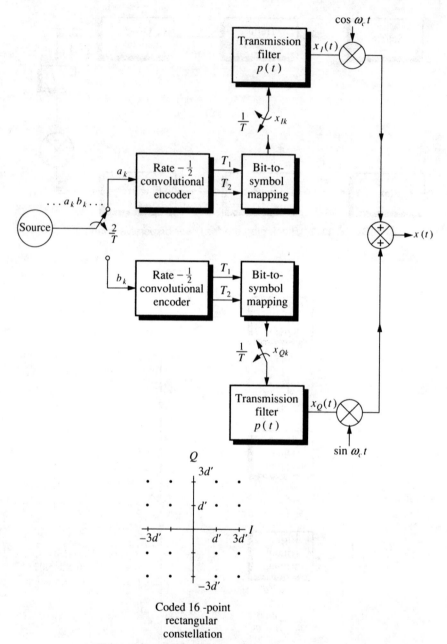

FIGURE 11.4 Transmitter for the coded 16-QAM system.

The sample values of a channel that we call channel A is shown in Fig. 11.5. Figure 11.6 shows three other channels, channels B, C, and D. Channel C contains a null at the band edge, which will prove to be a problem. In Figs. 11.5 and 11.6, symbol rate and factor-of-2 symbol rate samples are given. The length of the impulse responses is typical of high-frequency (HF) channels. Tabular data on

11.1 / Intersymbol Interference Channels 441

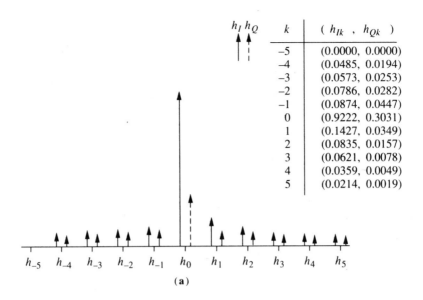

(a)

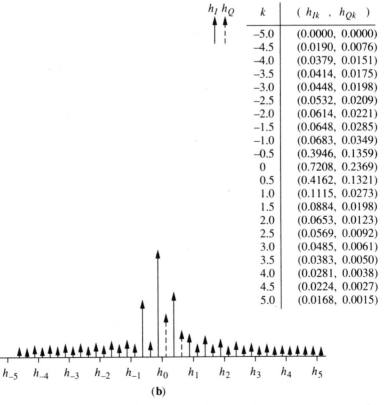

(b)

FIGURE 11.5 Channel A.

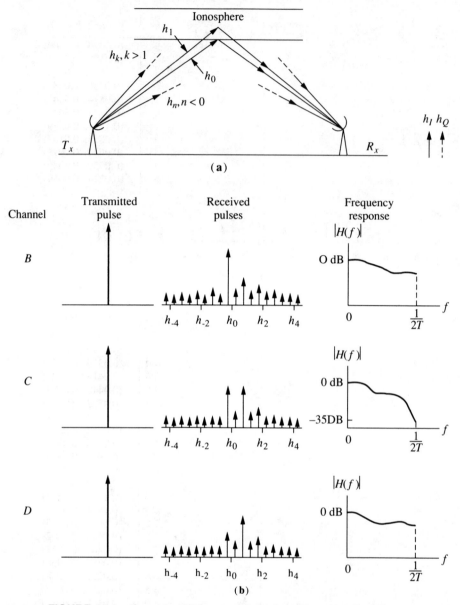

FIGURE 11.6 Example of the responses of some multipath channels.

these channels are given in Tables 11.1 and 11.2. The channel output is

$$y(t) = \sum_k x_k h(t - kT) + n(t) \tag{11.2}$$

$$n(t) = n_I(t) + jn_Q(t) \tag{11.3}$$

$$x_k = x_{Ik} + jx_{Qk} \tag{11.4}$$

TABLE 11.1 Symbol-Spaced Impulse Response of Channels B, C, and D.

	h_{Ik}, h_{Qk}		
k	Channel B	Channel C	Channel D
−5	(0.0000, 0.0000)	(0.0000, 0.0000)	(0.0000, 0.0000)
−4	(0.0410, 0.0109)	(0.0410, 0.0109)	(0.0410, 0.0109)
−3	(0.0495, 0.0123)	(0.0495, 0.0123)	(0.0495, 0.0123)
−2	(0.0672, 0.0170)	(0.0672, 0.0170)	(0.0672, 0.0170)
−1	(0.0919, 0.0235)	(0.0919, 0.0235)	(0.0919, 0.0235)
0	(0.7920, 0.1281)	(0.6278, 0.1000)	(0.3960, 0.0871)
1	(0.3960, 0.0871)	(0.6278, 0.1000)	(0.7920, 0.1281)
2	(0.2715, 0.0498)	(0.2715, 0.0498)	(0.2715, 0.0498)
3	(0.2291, 0.0414)	(0.2291, 0.0414)	(0.2291, 0.0414)
4	(0.1287, 0.0154)	(0.1287, 0.0154)	(0.1287, 0.0154)
5	(0.1032, 0.0119)	(0.1032, 0.0119)	(0.1032, 0.0119)

TABLE 11.2 $T/2$-Spaced Impulse Response of Channels B, C, and D.

	h_{Ik}, h_{Qk}		
k	Channel B	Channel C	Channel D
−5	(0.0000, 0.0000)	(0.0000, 0.0000)	(0.0000, 0.0000)
−4.5	(0.0290, 0.0077)	(0.0290, 0.0077)	(0.0290, 0.0077)
−4	(0.0290, 0.0077)	(0.0290, 0.0077)	(0.0290, 0.0077)
−3.5	(0.0350, 0.0087)	(0.0350, 0.0087)	(0.0350, 0.0087)
−3	(0.0350, 0.0087)	(0.0350, 0.0087)	(0.0350, 0.0087)
−2.5	(0.0410, 0.0109)	(0.0410, 0.0109)	(0.0410, 0.0109)
−2	(0.0475, 0.0120)	(0.0475, 0.0120)	(0.0475, 0.0120)
−1.5	(0.0530, 0.0138)	(0.0530, 0.0138)	(0.0530, 0.0138)
−1	(0.0650, 0.0166)	(0.0650, 0.0166)	(0.0650, 0.0166)
−0.5	(0.2100, 0.0387)	(0.2100, 0.0387)	(0.2100, 0.0387)
0	(0.5600, 0.0906)	(0.4439, 0.0707)	(0.2800, 0.0616)
0.5	(0.6100, 0.1225)	(0.4439, 0.0707)	(0.6100, 0.1225)
1	(0.2800, 0.0616)	(0.4439, 0.0707)	(0.5600, 0.0906)
1.5	(0.0970, 0.0095)	(0.4439, 0.0424)	(0.0970, 0.0095)
2	(0.1920, 0.0352)	(0.1920, 0.0352)	(0.1920, 0.0352)
2.5	(0.2100, 0.0387)	(0.2100, 0.0387)	(0.2100, 0.0387)
3	(0.1620, 0.0293)	(0.1620, 0.0293)	(0.1620, 0.0293)
3.5	(0.0605, 0.0063)	(0.0605, 0.0063)	(0.0605, 0.0063)
4	(0.0910, 0.0109)	(0.0910, 0.0109)	(0.0910, 0.0109)
4.5	(0.0970, 0.0095)	(0.0970, 0.0095)	(0.0970, 0.0095)
5	(0.0730, 0.0084)	(0.0730, 0.0084)	(0.0730, 0.0084)

where the noise terms $n_I(t)$ and $n_Q(t)$ are independent and Gaussian with zero mean and variance σ^2, and x_k is the complex data symbol. Note from Fig. 11.5 that significant intersymbol interference occurs as the symbol samples are nonzero over many symbol intervals.

11.1.2 LMS Equalization

Let the estimate of the data symbol in (11.4) be a quantization of z_k, where

$$z_k = \sum_j c_j y_{k-j} \tag{11.5}$$

with y_k the samples of $y(t)$ in (11.2). The LMS equalizer is defined by (11.5) and has the structure given in Fig. 11.7. Both T-spaced (i.e., symbol spaced) and $T/2$-spaced equalizers [7, 8] (i.e., a fractionally spaced equalizer) are given in Fig. 11.7. The role of the equalizer is to minimize the mean-square estimation error $|e_k|^2$, $e_k = x_k - z_k$. The tap-gain vector, c_k, for the linear equalizer that minimizes e_k can be computed using the stochastic gradient algorithm

$$c_{k+1} = c_k + \Delta e_k y_k^* \tag{11.6}$$

where the superscript * denotes complex conjugate. Here Δ is the step size for the stochastic gradient algorithm. Large Δ leads to fast but erroneous computation of c_k, whereas a low Δ leads to a slow but accurate calculation of c_k. Thus here a speed–accuracy trade-off exists. The adaptive aspects are not treated here, as our goal is merely to determine the least-mean-square error taps c_k for channels A to D. Adaptive aspects are given in [3] and [4].

As an example, consider channel A in Table 11.1. The result of computing (11.6) for 300 or so iterations for 16-QAM data, a periodic timing sequence (see [7, p. 1377]), and for an SNR of 20 dB, where

$$SNR = \frac{\sum_n |h_k|^2}{2\sigma^2}$$

gives the results shown in Table 11.3. The equalized sequence is given by (11.5) for the taps in Table 11.3 and no input noise. The result is shown in Table 11.4.

Important measures of the effectiveness of equalization are the signal-to-interference ratio (SIR), the output signal-to-noise ratio (SNR), and the signal-to-noise-plus-interference ratio (SNIR). These are defined by $SNR = |h_0|^2/2\sigma_0^2$,

$$\sigma_0^2 = \sigma^2 \sum_j |c_j|^2$$

$$SIR = \frac{|h_0|^2}{\sum_{i \neq 0} |h_i|^2}$$

$$SNIR = \frac{|h_0|^2}{2\sigma_0^2 + \sum_{i \neq 0} |h_i|^2}.$$

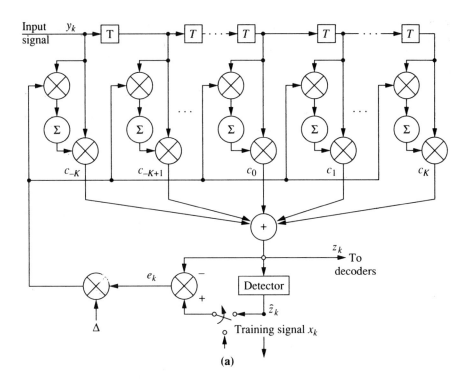

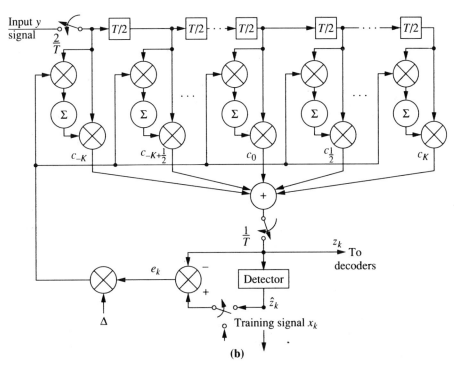

FIGURE 11.7 Structure of the (a) T-spaced and (b) $T/2$-spaced linear equalizers.

TABLE 11.3 Tap-Weight Coefficients of Linear Least-Mean-Square Equalizer on Channel A at 20 dB.

j	c_{Ij}	c_{Qj}	j	c_{Ij}	c_{Qj}
−10	−0.00042	0.00370	0	1.00673	−0.33066
−9	−0.00434	−0.00320	1	−0.12601	0.05121
−8	−0.00103	−0.00125	2	−0.04966	0.02502
−7	0.00198	−0.00141	3	−0.03541	0.02302
−6	0.00778	0.00090	4	−0.01675	0.00413
−5	0.01100	−0.00473	5	−0.00923	0.00528
−4	−0.03980	0.01138	6	0.00863	−0.00616
−3	−0.03974	0.01139	7	0.00120	−0.00263
−2	−0.05915	0.01919	8	−0.00084	0.00079
−1	−0.07756	0.00520	9	0.00223	0.00024
			10	−0.00188	0.00037

These data for an SNR of 20 dB are given in Table 11.5. Note that the equalizer is effective in all cases except channel C, which has a band-edge null. In all cases SNR is sacrificed to get the best SNIR. This is called *noise enhancement* [3, 4, 7].

As it contains an in-band null, channel C must be equalized using a nonlinear equalizer. We use the decision-feedback equalizer shown in Fig. 11.8 [1, 3, 4, 7]. Here decisions are fed back to cancel postcursor symbols in the equalized response. The precursor symbols are processed by a linear feedforward section. The decision feedback equalizer is effective for channel C.

TABLE 11.4 Equalized Response of Channel A at 20 dB Using Linear Least-Mean-Square Equalizer.

n	h'_{In}	h'_{Qn}	n	h'_{In}	h'_{Qn}
−15	0.00000	0.00000	0	0.98693	−0.00085
−14	−0.00012	0.00016	1	0.00023	−0.00203
−13	−0.00029	−0.00005	2	0.00247	−0.00037
−12	−0.00037	−0.00011	3	0.00031	0.00158
−11	−0.00040	−0.00022	4	0.00066	−0.00298
−10	−0.00174	0.00270	5	−0.00062	0.00042
−9	−0.00208	−0.00375	6	−0.00006	0.00003
−8	−0.00180	−0.00174	7	−0.00113	−0.00104
−7	−0.00132	−0.00124	8	−0.00194	0.00061
−6	−0.00121	0.00211	9	0.00165	0.00061
−5	−0.00254	−0.00351	10	−0.00139	−0.00030
−4	0.00372	−0.00097	11	0.00010	−0.00012
−3	0.00203	0.00234	12	−0.00003	0.00000
−2	0.00113	−0.00045	13	−0.00006	0.00004
−1	−0.00191	−0.00022	14	−0.00002	0.00001
			15	−0.00004	0.00000

TABLE 11.5 SIR and SNIR Measures (dB) before and after Linear, Symbol-Spaced Equalization at SNR = 20 dB.

Channel	SNR	SIR	SNIR
A			
Before	20.0	12.13	11.43
After	19.24	38.83	19.19
B			
Before	20.0	2.76	2.64
After	19.13	24.87	18.11
C			
Before	20.0	−1.57	−1.58
After	14.23	10.42	8.91
D			
Before	20.0	−6.98	−7.03
After	15.13	20.83	14.10

11.1.3 Trellis-Code Performance

We shall consider the bit error probability of the 16-QAM system illustrated in Fig. 11.4. Independent 4-AM codes are used on each rail of the transmitter. The 4-AM, four-state code is illustrated in Fig. 11.9 and the signal constellation is shown in Fig. 11.10.

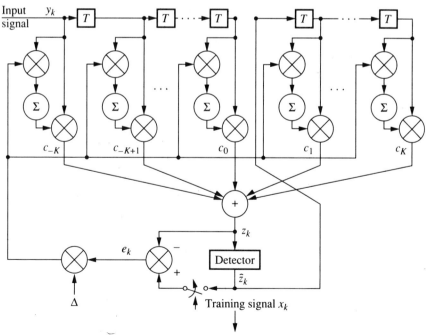

FIGURE 11.8 Decision feedback equalizer.

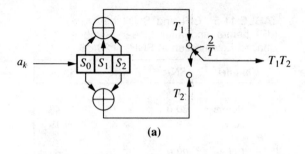

(a)

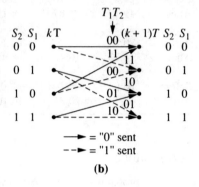

(b)

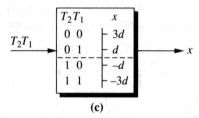

(c)

FIGURE 11.9 Rate 1/2, four-state, 4-AM trellis code.

An estimate of performance for the coded and uncoded systems is given by

$$P(e) \sim \frac{1}{2}\mathrm{erfc}\left\{\sqrt{\frac{d_{\text{free}}^2}{8\sigma_0^2}}\right\}$$

where $\sim$ means asymptotically, as $\sigma_0 \to 0$ with σ_0^2 is the equalizer output noise variance. Thus, for the coded system,

$$P(e) \sim \frac{1}{2}\mathrm{erfc}\left[\sqrt{\frac{(h'_{I0})^2 d_{\text{free}}^2}{\sum_j |c_j|^2 \, 8\sigma^2}}\right]$$

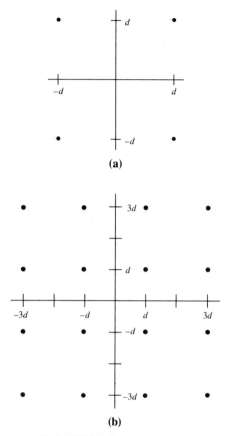

FIGURE 11.10 The uncoded 4-QAM signal constellation and the trellis-coded 16-QAM signal set.

where $h'_{10} = h_{10}/d$ and we have used the fact that

$$\sigma_0^2 = \sigma^2 \sum_j |c_j|^2$$

where σ^2 is the channel noise variance. Thus the gain relative to a perfectly equalized system with SNR, $d_{\text{free}}^2/4\sigma^2$, is

$$\text{gain} = 10 \log_{10} \left[\frac{(h'_{10})^2}{\sum_j |c_j|^2} \right] \text{ dB}.$$

For channel A the gain is -0.76 dB. Other gains relative to $d_{\text{free}}^2/4\sigma^2$ are tabulated in Table 11.6 for channels A to D and for both linear (T-spaced and T/2-spaced) and decision feedback equalizers. The error event performance of channels A and B are given in Figs. 11.11 and 11.12, respectively. This performance was obtained

TABLE 11.6 Estimated and Simulated Gain (dB) in SNR Relative to the Ideal for Channels $A, B, C,$ and D at Large SNR.[a]

Channel	LE Estimated	LE Simulated	DFE Estimated	DFE Simulated[b]	FSE Estimated	FSE Simulated
A						
Coded	−0.76	−0.9	−0.52	−0.7 (−0.5)	+1.3	+1.2
Uncoded	−3.36	−3.2	−3.12	—	−1.3	−1.4
B						
Coded	−3.7	−4.0	−2.5	−4.2 (−2.6)	−0.8	−0.9
Uncoded	−6.3	−6.7	−6.1	—	−3.4	−3.5
C						
Coded	−6.0	×	−3.8	× (−4.2)	−4.7	×
Uncoded	−8.6	×	−6.4	−6.8	−7.3	×
D						
Coded	−4.9	−5.2	−3.2	× (−3.3)	−1.2	−1.4
Uncoded	−7.5	−8.0	−5.8	−5.8	−3.8	−4.2

[a] ×, Very bad performance.
[b] Terms in parentheses indicate the genie's performance.

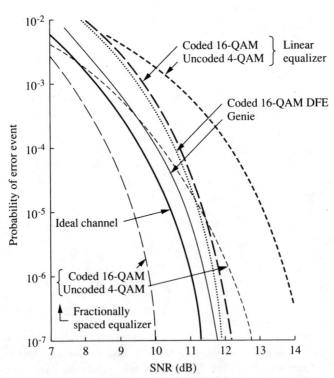

FIGURE 11.11 Error event performance for channel A.

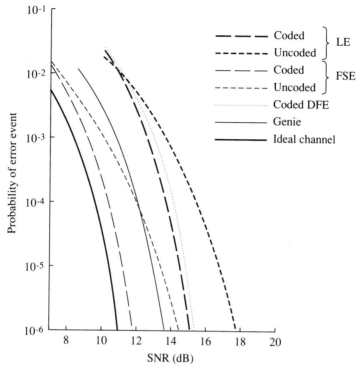

FIGURE 11.12 Error event performance for channel B.

through digital computer simulation [9]. The results given are similar to those given in [10], which considered voice-band channels. Note that the $T/2$ equalizer outperforms all receivers. This is because it approximates the performance of the matched filter plus T-spaced equalizer [11] plus Viterbi decoder, which is probably close to an optimum receiver structure. The genie-based DFE makes sure that all feedback decisions are correct. Recall that the feedback decisions are from the equalizer output, not the decoder output. This is because, in the high-frequency channel modeled here, the channel variations are fast enough that one cannot tolerate the delay introduced from the decoder. As such, the equalizer output represents an early decision for decision feedback.

The DFE performed poorly for channel C. This is unfortunate because this equalizer is needed here due to in-band nulls in the input spectrum to the equalizer. One technique to improve the situation is not to attempt to cancel all feedback terms. Rather, one should pass the job on to the decoder. A receiver and its trellis are shown in Fig. 11.13. Performance is shown in Fig. 11.14 and one observes that the DFE equalizer with $T/2$-spaced feedforward section comes within 1 dB in SNR of the genie-based receiver. The ideal response for channel C is $h = (1/\sqrt{2}, 1/\sqrt{2})$, that is, a partial response (PR) channel. The desired impulse response (DIR) receiver [12] maps the input sequence to the PR response and then decodes the result. Note from Fig. 11.14 that further performance gains are possible. This is called a *modified decision feedback equalizer* (MDFE).

452 Ch. 11 / Analysis and Design of TCM for Other Practical Channels

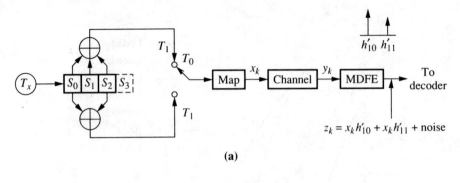

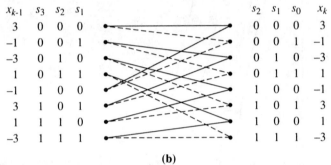

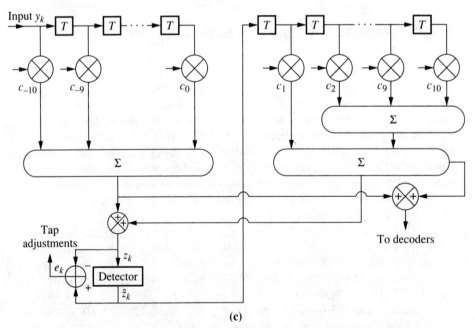

FIGURE 11.13 (a) Modified system and receiver for a one-dimensional case; (b) the trellis for the system depicted in (a); (c) modified decision feedback equalizer.

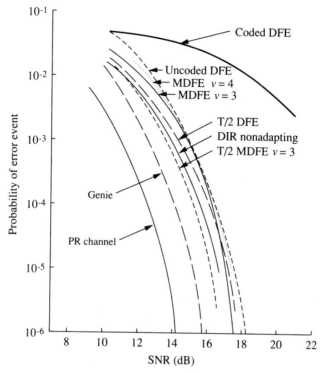

FIGURE 11.14 Error event performances of channel C with modified decoder.

11.2 Channels with Phase Offset

In the analyses developed so far, an implicit assumption was made that the reference signal used for demodulation is perfectly phase synchronized to the transmitted signal. As such, the performance results given have been for the case of ideal coherent detection. In a practical receiver, the coherent demodulation reference is derived from a carrier synchronization subsystem, for example, a type of phase-locked loop or Costas loop, resulting in a performance degradation due to the phase error between the received signal and the locally generated reference. Since this subsystem forms its demodulation reference from a noise-perturbed version of the transmitted signal, the phase error is a random process. The manner in which this process degrades the system error probability performance depends on the ratio of data rate to loop bandwidth (i.e., the rate of variation of the phase error over the data symbol interval). In many applications, this rate of variation is large (i.e., the data rate is much higher than the loop bandwidth), and thus the phase error can be considered constant (independent of time) over the data symbol time interval. This is the case we consider here.

In this section we discuss the foregoing degradation in signal-to-noise ratio (SNR) performance, which often is referred to as *radio* or *noisy reference loss*, and show how it can be reduced by employing interleaving/deinterleaving. Specific closed-form results are derived for both discrete and suppressed carrier systems, and the differences between the two are discussed and numerically illustrated. The case of convolutionally encoded BPSK systems is also considered as a special case of the more general results.

11.2.1 Upper Bound on the Average Bit Error Probability Performance of TCM

Perfect carrier phase synchronization

In [13] to [15] it was assumed that a TCM signal is transmitted over a perfectly phase-synchronized AWGN channel. An upper bound on the average bit error probability was obtained as

$$P_b \leq \sum_{\mathbf{x}} \sum_{\hat{\mathbf{x}} \in \mathscr{C}} a(\mathbf{x}, \hat{\mathbf{x}}) p(\mathbf{x}) P(\mathbf{x} \to \hat{\mathbf{x}}) \qquad (11.7)$$

where $a(\mathbf{x}, \hat{\mathbf{x}})$ is the number of bit errors that occur when the sequence $\mathbf{x}$ is transmitted and the sequence $\hat{\mathbf{x}} \neq \mathbf{x}$ is chosen by the decoder, $p(\mathbf{x})$ is the a priori probability of transmitting $\mathbf{x}$, and $\mathscr{C}$ is the set of all coded sequences. Also, in (11.7), $P(\mathbf{x} \to \hat{\mathbf{x}})$ represents the pairwise error probability, that is, the probability that the decoder chooses $\hat{\mathbf{x}}$ rather than the transmitted sequence $\mathbf{x}$. The upper bound of (11.7) is efficiently evaluated using the *transfer function bound* approach described in Chapter 4.

In general, evaluation of the pairwise error probability depends on the proposed decoding metric, the presence or absence of channel state information (CSI), and the type of detection used (i.e., coherent versus differentially coherent). For the case of interest here, coherent detection with no CSI and a Gaussian metric (optimum for the AWGN channel), the pairwise error probability is given by

$$P(\mathbf{x} \to \hat{\mathbf{x}}) \leq \exp\left[-\frac{E_s}{4N_0} d^2(\mathbf{x}, \hat{\mathbf{x}})\right] \qquad (11.8)$$

where

$$d^2(\mathbf{x}, \hat{\mathbf{x}}) = \sum_{n \in \eta} |x_n - \hat{x}_n|^2 = \sum_{n \in \eta} \delta_n^2 \qquad (11.9)$$

represents the sum of the squared Euclidean distances between the two symbol sequences $\mathbf{x}$ (the correct one) and $\hat{\mathbf{x}}$ (the incorrect one), and η is the set of all n for which $\hat{x}_n \neq x_n$. Also, in (11.9), E_s is the energy per output coded symbol and N_0 is the single-sided noise spectral density.

Imperfect carrier phase synchronization

When a carrier phase error $\phi(t)$ exists between the received signal and the locally generated demodulation reference, the result in (11.8) should be modified. In this subsection we derive an upper bound on the pairwise error probability for trellis-coded M-PSK with imperfect carrier phase reference. Let $\mathbf{y} = (y_1, y_2, \ldots, y_N)$ denote the received sequence when the normalized (to unit energy) sequence of

11.2 / Channels with Phase Offset

M-PSK symbols $\mathbf{x} = (x_1, x_2, \ldots, x_N)$ is transmitted. A *pairwise error* occurs if $\hat{\mathbf{x}} = (\hat{x}_1, \hat{x}_2, \ldots, \hat{x}_N) \neq \mathbf{x}$ is chosen by the receiver, which, if the receiver uses a distance metric to make this decision, implies that $\mathbf{y}$ is closer to $\hat{\mathbf{x}}$ than to $\mathbf{x}$, that is,

$$\sum_{n=1}^{N} |y_n - \hat{x}_n|^2 < \sum_{n=1}^{N} |y_n - x_n|^2. \tag{11.10}$$

Since M-PSK is a constant-envelope signaling set, we have $|\hat{x}_n|^2 = |x_n|^2 =$ constant and (11.10) reduces to

$$\sum_{n=1}^{N} \operatorname{Re}\{y_n \hat{x}_n^*\} > \sum_{n=1}^{N} \operatorname{Re}\{y_n x_n^*\}. \tag{11.11}$$

If we let n_n represent the additive noise in the nth signaling interval, and ϕ_n the phase shift introduced by imperfect carrier demodulation in that interval, then y_n and x_n are related by

$$y_n = x_n e^{j\phi_n} + n_n \qquad n = 1, 2, \ldots, N. \tag{11.12}$$

Substituting (11.12) into (11.11) and simplifying gives

$$\operatorname{Re}\left\{\sum_{n \in \eta} (\hat{x}_n - x_n)^* n_n\right\} > \operatorname{Re}\left\{\sum_{n \in \eta} x_n(x_n - \hat{x}_n)^* e^{j\phi_n}\right\} \tag{11.13}$$

where η is the set of all n such that $\hat{x}_n \neq x_n$.

Since for an AWGN channel, n_n is a complex Gaussian random variable whose real and imaginary components have variance

$$\mathbf{E}\{[\operatorname{Re}(n_n)]^2\} = \mathbf{E}\{[\operatorname{Im}(n_n)]^2\} = \sigma^2 \tag{11.14}$$

then

$$\operatorname{var}\left\{\operatorname{Re}\left\{\sum_{n \in \eta} (\hat{x}_n - x_n)^* n_n\right\}\right\} = \sigma^2 \sum_{n \in \eta} |x_n - \hat{x}_n|^2 \tag{11.15}$$

and the conditional pairwise error probability $P(\mathbf{x} \to \hat{\mathbf{x}} | \boldsymbol{\phi})$ is given by

$$\begin{aligned} P(\mathbf{x} \to \hat{\mathbf{x}} | \boldsymbol{\phi}) &= P\left\{\operatorname{Re}\left\{\sum_{n \in \eta}(\hat{x}_n - x_n)^* n_n\right\} > \operatorname{Re}\left\{\sum_{n \in \eta} x_n(x_n - \hat{x}_n)^* e^{j\phi_n}\right\}\right\} \\ &= Q\left(\frac{\operatorname{Re}\left\{\sum_{n \in \eta} x_n(x_n - \hat{x}_n)^* e^{j\phi_n}\right\}}{\sqrt{\sigma^2 \sum_{n \in \eta} |x_n - \hat{x}_n|^2}}\right) \end{aligned}$$

$$\tag{11.16}$$

where $\boldsymbol{\phi} = (\phi_1, \phi_2, \ldots, \phi_N)$ is the sequence of carrier phase errors and $Q(x)$ is the Gaussian integral, defined by

$$Q(x) = \frac{1}{\sqrt{2\pi}} \int_x^\infty \exp\left(-\frac{y^2}{2}\right) dy = \frac{1}{2} \operatorname{erfc}\left(\frac{x}{\sqrt{2}}\right). \tag{11.17}$$

To simplify (11.16), we proceed as follows. Since for constant-envelope signals

$$\begin{aligned} 2\operatorname{Re}\{x_n(x_n - \hat{x}_n)^*\} &= |x_n - \hat{x}_n|^2 \\ 2\operatorname{Im}\{x_n(x_n - \hat{x}_n)^*\} &= (x_n - \hat{x}_n)^*(x_n + \hat{x}_n) \end{aligned} \tag{11.18}$$

the numerator of the argument of the Gaussian integral in (11.16) becomes

$$\begin{aligned} \sum_{n\in\eta} \operatorname{Re}\{x_n(x_n - \hat{x}_n)^* e^{j\phi_n}\} &= \frac{1}{2}\sum_{n\in\eta} |x_n - \hat{x}_n|^2 \cos\phi_n \\ &\quad + j(x_n - \hat{x}_n)^*(x_n + \hat{x}_n)\sin\phi_n \\ &= \frac{1}{2}\sum_{n\in\eta} |x_n - \hat{x}_n|^2 \left(\cos\phi_n + j\frac{x_n + \hat{x}_n}{x_n - \hat{x}_n}\sin\phi_n\right). \end{aligned}$$

$$\tag{11.19}$$

Performing some further trigonometric simplification of (11.19) gives the desired result:

$$\sum_{n\in\eta} \operatorname{Re}\{x_n(x_n - \hat{x}_n)^* e^{j\phi_n}\} = \sum_{n\in\eta} |x_n - \hat{x}_n| \cos(\phi_n - \eta_n) \tag{11.20}$$

$$\eta_n = \tan^{-1}\alpha_n \qquad \alpha_n = \sqrt{\frac{4 - |x_n - \hat{x}_n|^2}{|x_n - \hat{x}_n|^2}} = \sqrt{\frac{4 - \delta_n^2}{\delta_n^2}}.$$

The argument of the Gaussian integral in (11.16) is in the form $a/\sqrt{b}$. For $a > 0$, we can upper bound this integral by[1]

$$Q\left(\frac{a}{\sqrt{b}}\right) \leq \frac{1}{2} e^{-a^2/2b} \tag{11.21}$$

Observe now that for any λ, we have $(a - 2\lambda b)^2 > 0$ or, equivalently,

$$\frac{a^2}{b} \geq 4\lambda a - 4\lambda^2 b. \tag{11.22}$$

Consequently, for any $a > 0$, we have

$$Q\left(\frac{a}{\sqrt{b}}\right) \leq \frac{1}{2} \exp\{-2\lambda[a - \lambda b]\}. \tag{11.23}$$

---

[1] Note that for perfect carrier demodulation (i.e., $\phi = 0$), we always have $a > 0$ [see (11.18)].

For $a < 0$, we must use the loose upper bound

$$Q\left(\frac{a}{\sqrt{b}}\right) = Q\left(-\frac{|a|}{\sqrt{b}}\right) = 1 - Q\left(\frac{|a|}{\sqrt{b}}\right) \leq 1. \tag{11.24}$$

Finally, using (11.20) together with (11.23) and (11.24) in (11.16) gives the desired upper bound on pairwise error probability as

$$P(\mathbf{x} \to \hat{\mathbf{x}} | \boldsymbol{\phi}; \lambda) \leq \begin{cases} \frac{1}{2} \exp\left\{-\frac{E_s}{4N_0} \sum_{n \in \eta} 4\lambda(\cos\phi_n + \alpha_n \sin\phi_n - \lambda)\delta_n^2\right\}; \\ \qquad\qquad\qquad\qquad \sum_{n \in \eta} \delta_n^2(\cos\phi_n + \alpha_n \sin\phi_n) > 0 \\ 1; \qquad\qquad\qquad \sum_{n \in \eta} \delta_n^2(\cos\phi_n + \alpha_n \sin\phi_n) \leq 0. \end{cases}$$

$$\tag{11.25}$$

In (11.25), we have also made use of the fact that for the unnormalized system, $1/2\sigma^2 = E_s/N_0$, where E_s is the symbol energy and N_0 the noise spectral density, and further replaced λ by the normalized quantity $\lambda\sigma^2$. Also, note that if (11.25) is minimized over λ, it is identically in the form of a Chernoff bound.

Discrete carrier (no interleaving)

Assuming the case where the data symbol rate $1/T$ is high compared to the loop bandwidth B_L, then $\phi(t)$ can be assumed constant (independent of time), say ϕ, over a number of symbols on the order of $1/B_L T$. Since the decoder has no knowledge of ϕ, the decoding metric can make no use of this information and as such is *mismatched* to the channel. Under these conditions, using the maximum-likelihood metric for a perfectly phase-synchronized AWGN, by setting $\phi_n = \phi$ for all n in (11.25), one obtains

$$P(\mathbf{x} \to \hat{\mathbf{x}} | \phi; \lambda) \leq \begin{cases} \frac{1}{2} \exp\left\{-\frac{E_s}{4N_0} \sum_{n \in \eta} 4\lambda(\cos\phi + \alpha_n \sin\phi - \lambda)\delta_n^2\right\}; \\ \qquad\qquad\qquad\qquad \sum_{n \in \eta} \delta_n^2(\cos\phi + \alpha_n \sin\phi) > 0 \\ 1; \qquad\qquad\qquad \sum_{n \in \eta} \delta_n^2(\cos\phi + \alpha_n \sin\phi) \leq 0 \end{cases}$$

$$\tag{11.26}$$

where

$$\alpha_n = \sqrt{\frac{4 - |x_n - \hat{x}_n|^2}{|x_n - \hat{x}_n|^2}} = \sqrt{\frac{4 - \delta_n^2}{\delta_n^2}} \tag{11.27}$$

and $\lambda \geq 0$ is a parameter to be optimized. Note that for $\phi = 0$, the optimum value of λ is independent of the summation index n. In particular, the expression

$4\lambda(1 - \lambda)$ is maximized by the value $\lambda = \frac{1}{2}$, which when substituted in (11.26) yields (11.8) as it should.

If we let $p(\phi)$ denote the probability density function (pdf) of the phase error ϕ, the average bit error probability is upper bounded by[2]

$$P_b \leq \sum_{\mathbf{x}} \sum_{\hat{\mathbf{x}} \in \mathscr{C}} a(\mathbf{x}, \hat{\mathbf{x}}) p(\mathbf{x}) \min_{\lambda} \mathbf{E}_\phi\{P(\mathbf{x} \to \hat{\mathbf{x}} | \phi; \lambda)\} \tag{11.28}$$

where $\mathbf{E}_\phi\{\cdot\}$ denotes averaging over the random variable ϕ.

To somewhat simplify notation, we introduce the Bhattacharyya parameter [16]

$$Z = \exp\left(-\frac{E_s}{4N_0}\right) \tag{11.29}$$

in which case

$$\mathbf{E}_\phi\{P(\mathbf{x} \to \hat{\mathbf{x}} | \phi; \lambda)\} \leq \frac{1}{2} \int_{\mathscr{R}} \left(\prod_{n \in \eta} Z^{\delta_n^2}\right)^{[4\lambda(\cos\phi - \lambda)]} \left(\prod_{n \in \eta} Z^{\alpha_n \delta_n^2}\right)^{[4\lambda \sin\phi]} p(\phi)\, d\phi$$

$$+ \int_{\overline{\mathscr{R}}} p(\phi)\, d\phi$$

$$\tag{11.30}$$

where $\mathscr{R}$ is the set of all ϕ in $(-\pi, \pi)$ for which

$$\sum_{n \in \eta} \delta_n^2 (\cos\phi + \alpha_n \sin\phi) > 0 \tag{11.31}$$

and $\overline{\mathscr{R}}$ is the complement of $\mathscr{R}$, that is, the remaining values of ϕ in $(-\pi, \pi)$ that do not satisfy (11.31). Defining

$$\phi_1 = \tan^{-1} \frac{\sum_{n \in \eta} \alpha_n \delta_n^2}{\sum_{n \in \eta} \delta_n^2} = \tan^{-1} \frac{\sum_{n \in \eta} \sqrt{\delta_n^2(4 - \delta_n^2)}}{\sum_{n \in \eta} \delta_n^2} \tag{11.32}$$

then $\mathscr{R}$ corresponds to the interval $0 \leq |\phi - \phi_1| \leq \pi/2$ and $\overline{\mathscr{R}}$ corresponds to the interval $\pi/2 \leq |\phi - \phi_1| \leq \pi$.

Discrete carrier (with interleaving)

Ordinarily, one thinks of using interleaving/deinterleaving to break up the effects of error bursts in coded communication systems. For example, in TCM systems operating in a multipath fading environment, it has been shown [14] that interleaving/deinterleaving is essential for good performance. To see how it may be applied in systems with noisy carrier phase reference, one can gain an intuitive notion by considering the $\cos\phi$ degradation factor as an "amplitude fade" whose duration is

---

[2] Later we present a tighter bound for this case by optimizing on λ prior to performing the expectation over ϕ.

on the order of $1/B_L T$ symbols. Thus if we break up this fade by interleaving to a depth on the order of $1/B_L T$, then, after deinterleaving, the degradation due to $\cos\phi$ and $\sin\phi$ will be essentially *independent* from symbol to symbol. From a mathematical standpoint, this is equivalent to the use of (11.25), but now the ϕ_n's are independent identically distributed (iid) random variables with pdf $p(\boldsymbol{\phi})$ and $\boldsymbol{\phi}$ refers to the vector whose components are the ϕ_n's. Therefore, we have

$$P(\mathbf{x} \to \hat{\mathbf{x}} \mid \boldsymbol{\phi}; \lambda) \le \begin{cases} \dfrac{1}{2} \exp\left\{ -\dfrac{E_s}{4N_0} \sum_{n\in\eta} 4\lambda(\cos\phi_n + \alpha_n \sin\phi_n - \lambda)\delta_n^2 \right\}; & \sum_{n\in\eta} \delta_n^2(\cos\phi_n + \alpha_n \sin\phi_n) > 0 \\ 1; & \sum_{n\in\eta} \delta_n^2(\cos\phi_n + \alpha \sin\phi_n) \le 0. \end{cases}$$

(11.33)

The expectation required in (11.28) now involves computation of multidimensional integrals over regions of $\boldsymbol{\phi}$ corresponding to the inequalities in (11.33). Since in these regions the intervals of integration per dimension are *dependent* on one another, the expectation required in (11.28) is extremely difficult to compute.

Thus, instead, we turn to a looser upper bound on conditional pairwise error probability that has the advantage of not having to separate the multidimensional integration required in (11.28) into two disjoint regions. Indeed, it is straightforward to see that the right-hand side of (11.33) is upper bounded by the exponential in its first line (without the factor of $\frac{1}{2}$) over the *entire* domain of $\boldsymbol{\phi}$ [i.e., $\{\phi_n \in (-\pi, \pi); n \in \eta\}$]. Hence

$$P(\mathbf{x} \to \hat{\mathbf{x}} \mid \boldsymbol{\phi}; \lambda) \le \exp\left\{ -\frac{E_s}{4N_0} \sum_{n\in\eta} 4\lambda(\cos\phi_n + \alpha_n \sin\phi_n - \lambda)\delta_n^2 \right\}$$

$$= \prod_{n\in\eta} \exp\left\{ -\frac{E_s}{N_0} \lambda(\cos\phi_n + \alpha_n \sin\phi_n - \lambda)\delta_n^2 \right\}$$

(11.34)

which is identically equal to the Chernoff bound. Now, substituting (11.34) into (11.28) gives the much simpler result

$$\mathbf{E}_{\boldsymbol{\phi}}\{P(\mathbf{x} \to \hat{\mathbf{x}} \mid \boldsymbol{\phi}; \lambda)\} \le \prod_{n\in\eta} \int_{-\pi}^{\pi} Z^{\delta_n^2[4\lambda(\cos\phi_n + \alpha_n \sin\phi_n - \lambda)]} p(\phi_n)\, d\phi_n. \quad (11.35)$$

Suppressed carrier (no interleaving)

When the carrier synchronization loop used to track the input phase is of the suppressed carrier type (e.g., a Costas loop), the previous results have to be modified somewhat, since the appropriate domain for ϕ is no longer $(-\pi, \pi)$. In fact, for suppressed carrier tracking of M-PSK with a Costas-type loop, and assuming perfect phase ambiguity resolution, ϕ takes on values only in the interval $(-\pi/M, \pi/M)$. Thus the regions $\mathcal{R}$ and $\overline{\mathcal{R}}$ required in (11.30) are reduced relative to those defined below (11.32), which assume that ϕ is allowed to take on values in $(-\pi, \pi)$. Specifically, $\mathcal{R}$ will now be the intersection of the intervals $0 \leq |\phi| \leq \pi/M$ and $0 \leq |\phi - \phi_1| \leq \pi/2$, where ϕ_1 is defined in (11.32). Similarly, $\overline{\mathcal{R}}$ is defined by the intersection of $\pi/M \leq |\phi| \leq \pi/2$ and $\pi/2 \leq |\phi - \phi_1| \leq \pi$. Here we wish to first show that for *any error event path*, the intersection of the intervals $0 \leq |\phi| \leq \pi/M$ and $0 \leq |\phi - \phi_1| \leq \pi/2$, where ϕ_1 is defined in (11.32), is indeed $0 \leq |\phi| \leq \pi/M$, which then defines the region $\mathcal{R}$ for the no-interleaving case. This is equivalent to showing that for *any error event path*, $\pi/2 - \phi_1 \geq \pi/M$. From (11.32), this inequality can be expressed as

$$\frac{\sum_{n \in \eta} \sqrt{\delta_n^2(4 - \delta_n^2)}}{\sum_{n \in \eta} \delta_n^2} \leq \cot \frac{\pi}{M} \tag{11.36}$$

or, equivalently,

$$\sum_{n \in \eta} \sqrt{\delta_n^2(4 - \delta_n^2)} \leq \sum_{n \in \eta} \frac{\delta_n^2}{\tan(\pi/M)}. \tag{11.37}$$

Inequality (11.37) will be satisfied if for each $n \in \eta$,

$$\delta_n^2(4 - \delta_n^2) \leq \frac{\delta_n^4}{\tan^2(\pi/M)} \tag{11.38}$$

or, equivalently,

$$\delta_n^2 \geq \frac{4\tan^2(\pi/M)}{1 + \tan^2(\pi/M)} = 4\sin^2 \frac{\pi}{M}. \tag{11.39}$$

However, for an M-PSK signaling set, the *smallest* squared Euclidean distance occurs between adjacent points in the constellation and has value $4\sin^2(\pi/M)$. Thus, (11.39) is satisfied for all $n \in \eta$. Therefore, the intersection of the intervals $0 \leq |\phi| \leq \pi/M$ and $0 \leq |\phi - \phi_1| \leq \pi/2$ is simply $0 \leq |\phi| \leq \pi/M$, which defines $\mathcal{R}$ and the intersection of $\pi/M \leq |\phi| \leq \pi/2$ and $\pi/2 \leq |\phi - \phi_1| \leq \pi$ is the null set that defines $\overline{\mathcal{R}}$. In short, for suppressed carrier tracking, the second integral in (11.30) disappears and the limits on the first integral become $(-\pi/M, \pi/M)$, that is,

$$\mathbf{E}_\phi\{P(\mathbf{x} \to \hat{\mathbf{x}} | \phi; \lambda)\} \leq \frac{1}{2} \int_{-\pi/M}^{\pi/M} \left(\prod_{n \in \eta} Z^{\delta_n^2} \right)^{[4\lambda(\cos\phi - \lambda)]} \left(\prod_{n \in \eta} Z^{\alpha_n \delta_n^2} \right)^{[4\lambda \sin\phi]} p(\phi) \, d\phi. \tag{11.40}$$

The significance of the second integral in (11.30) being equal to zero will soon be described in the context of a discussion on *irreducible error probability*.

Suppressed carrier (with interleaving)

For the interleaving case, first we show that for $0 \leq |\phi_n| \leq \pi/M$,

$$\sum_{n \in \eta} \delta_n^2 (\cos \phi_n + \alpha_n \sin \phi_n) \geq 0 \tag{11.41}$$

where, from (11.27),

$$\alpha_n = \sqrt{\frac{4 - \delta_n^2}{\delta_n^2}}. \tag{11.42}$$

Since if all the ϕ_n's are equal to $-\pi/M$, the left-hand side of (11.35) is most negative, then, equivalently, we must show that

$$\sum_{n \in \eta} \delta_n^2 \left(\cos \frac{\pi}{M} - \alpha_n \sin \frac{\pi}{M} \right) \geq 0. \tag{11.43}$$

The inequality in (11.43) is satisfied if each term in the sum is greater than or equal to zero. Thus we must show that

$$\cos \frac{\pi}{M} - \sqrt{\frac{4 - \delta_n^2}{\delta_n^2}} \sin \frac{\pi}{M} \geq 0 \tag{11.44}$$

or, equivalently,

$$\tan^2 \frac{\pi}{M} \leq \frac{\delta_n^2}{4 - \delta_n^2} \tag{11.45}$$

which is the same as (11.38), whose validity was established above. Once again assuming suppressed carrier tracking of M-PSK with a Costas-type loop and perfect phase ambiguity resolution, we get an inequality analogous to (11.35):[3]

$$E_\phi \{P(\mathbf{x} \to \hat{\mathbf{x}} \mid \boldsymbol{\phi}; \lambda)\} \leq \frac{1}{2} \prod_{n \in \eta} \left\{ \int_{-\pi/M}^{\pi/M} Z^{\delta_n^2 [4\lambda(\cos \phi_n + \alpha_n \sin \phi_n - \lambda)]} p(\phi_n) \, d\phi_n \right\}. \tag{11.46}$$

11.2.2 Carrier Synchronization Loop Statistical Model and Average Pairwise Error Probability Evaluation

To evaluate (11.28), using (11.30), (11.35), (11.40), or (11.46), we must specify the functional form of the pdf $p(\phi)$ of the modulo 2π-reduced phase error ϕ. For a discrete carrier synchronization loop of the phase-locked type, $p(\phi)$ is given

---

[3] Note that the factor of $\frac{1}{2}$ can be included here since for $0 \leq |\phi_n| \leq \pi/M$, $n \in \eta$, the condition on the first line of (11.33) is always satisfied and thus we need not use the looser upper bound of (11.34).

by the Tikhonov pdf [17],

$$p(\phi) = \begin{cases} \dfrac{\exp(\rho\cos\phi)}{2\pi I_0(\rho)} & |\phi| \le \pi \\ 0 & \text{otherwise} \end{cases} \qquad (11.47)$$

where ρ is the SNR in the loop bandwidth.

In order to allow evaluation of (11.28) in closed form, we recognize that for the case of no interleaving, (11.30) can be further upper bounded by using $(-\pi, \pi)$ instead of $\mathcal{R}$ in the first integral. Then, making this replacement, we get

$$\min_{\boldsymbol{\lambda}} \mathbf{E}_{\boldsymbol{\phi}}\{P(\mathbf{x} \to \hat{\mathbf{x}} | \boldsymbol{\phi}; \boldsymbol{\lambda})\} \le \min_{\boldsymbol{\lambda}} \left(\frac{1}{2}\exp\left\{ d^2(\mathbf{x},\hat{\mathbf{x}})\lambda^2 \frac{E_s}{N_0}\right\} \frac{I_0(\rho')}{I_0(\rho)} + I\right)$$

$$\rho' = \left(\left(\rho - d^2(\mathbf{x},\hat{\mathbf{x}})\lambda\frac{E_s}{N_0}\right)^2 + \left(d_1^2\lambda\frac{E_s}{N_0}\right)^2\right)^{1/2} \qquad d_1^2 = \sum_{n\in\eta}\alpha_n\delta_n^2$$

$$I = \frac{1}{2\pi I_0(\rho)}\left[\int_{\pi/2+\phi_1}^{\pi}\exp(\rho\cos\phi)\,d\phi + \int_{-\pi}^{-\pi/2+\phi_1}\exp(\rho\cos\phi)\,d\phi\right].$$

(11.48)

When (11.48) is substituted into (11.28), the term I will contribute an *irreducible error probability*; that is, the system will exhibit a nonzero error probability when ρ is held fixed and E_s/N_0 approaches infinity. An example of such a system might be one that employs a binary convolutionally encoded BPSK modulation and apportions a fixed amount of the total available input power to a discrete carrier component for the purpose of deriving a coherent carrier reference at the receiver. This case will be treated in detail shortly.

When interleaving is employed, (11.46) (minimized over λ) together with (11.47) becomes

$$\min_{\boldsymbol{\lambda}} \mathbf{E}_{\boldsymbol{\phi}}\{P(\mathbf{x} \to \hat{\mathbf{x}} | \boldsymbol{\phi}; \boldsymbol{\lambda})\} \le \min_{\boldsymbol{\lambda}} \left\{\prod_{n\in\eta}\exp\left\{\delta_n^2\lambda^2\frac{E_s}{N_0}\right\}\frac{I_0(\rho_n)}{I_0(\rho)}\right\}$$

$$= \min_{\boldsymbol{\lambda}} \left\{\exp\left\{d^2(\mathbf{x},\hat{\mathbf{x}})\lambda^2\frac{E_s}{N_0}\right\}\prod_{n\in\eta}\frac{I_0(\rho_n)}{I_0(\rho)}\right\}$$

$$\rho_n = \left[\left(\rho - \delta_n^2\lambda\frac{E_s}{N_0}\right)^2 + \left(\alpha_n\delta_n^2\lambda\frac{E_s}{N_0}\right)^2\right]^{1/2}.$$

(11.49)

For suppressed carrier tracking with an M-phase Costas loop, $p(\phi)$ again has a Tikhonov-type pdf, which is given by

$$p(\phi) = \begin{cases} \dfrac{\exp(\rho\cos M\phi)}{(2\pi/M)I_0(\rho)} & |\phi| \le \dfrac{\pi}{M} \\ 0 & \text{otherwise} \end{cases} \qquad (11.50)$$

Here ρ is the "effective" loop SNR, which includes the effects of S × S (signal times signal), S × N (signal times noise), and N × N (noise times noise) degradations, commonly referred to as *squaring loss* or more accurately, *M*th *power loss* [17]. Since suppressed carrier systems of this type derive their carrier demodulation reference from the data-bearing signal, the loop SNR, ρ, is directly proportional to E_s/N_0; thus there can be no irreducible error probability, since $\rho \to \infty$ when $E_s/N_0 \to \infty$. Furthermore, for perfect phase ambiguity resolution, we have previously shown that for no interleaving, the term I is identically zero, since the region $\overline{\mathcal{R}}$ is the empty set. Thus the average pairwise error probability results become

$$\min_\lambda \mathbf{E}_\phi \{P(\mathbf{x} \to \hat{\mathbf{x}} \mid \boldsymbol{\phi}; \lambda)\} \leq \min_\lambda \left\{ \frac{1}{2} \exp\left\{ \frac{d^2(\mathbf{x}, \hat{\mathbf{x}})\lambda^2 E_s}{N_0} \right\} \frac{f(\rho)}{I_0(\rho)} \right\}$$

$$f(\rho) = \frac{M}{2\pi} \int_{-\pi/M}^{\pi/M} \exp\left\{ \rho \cos M\phi - \lambda \frac{E_s}{N_0} \left(d^2(\mathbf{x}, \hat{\mathbf{x}}) \cos \phi + d_1^2 \sin \phi\right) \right\} d\phi$$

(11.51)

for no interleaving and

$$\min_\lambda \mathbf{E}_\phi \{P(\mathbf{x} \to \hat{\mathbf{x}} \mid \boldsymbol{\phi}; \lambda)\} \leq \min_\lambda \left\{ \prod_{n \in \eta} \exp\left\{ \delta_n^2 \lambda^2 \frac{E_s}{N_0} \right\} \frac{f_n(\rho)}{I_0(\rho)} \right\}$$

$$= \min_\lambda \left\{ \exp\left\{ d^2(\mathbf{x}, \hat{\mathbf{x}}) \lambda^2 \frac{E_s}{N_0} \right\} \prod_{n \in \eta} \frac{f_n(\rho)}{I_0(\rho)} \right\}$$

$$f_n(\rho) = \frac{M}{2\pi} \int_{-\pi/M}^{\pi/M} \exp\left\{ \rho \cos M\phi - \lambda \frac{E_s \delta_n^2}{N_0} (\cos \phi + \alpha_n \sin \phi) \right\} d\phi$$

(11.52)

for the case of interleaving.

In arriving at (11.48), (11.49), (11.51), and (11.52), we have assumed the "same type" of Chernoff bound in the sense that in all cases, the minimization over λ was performed *after* the averaging over ϕ. The principal reason for doing this is to allow comparison of performance with and without interleaving using bounds with "similar degrees of looseness." For the case of no interleaving, we can actually achieve a tighter bound than that given above by performing the minimization over λ on the conditional pairwise probability in (11.26). When this is done, we obtain

$$\lambda_{\text{opt}} = \frac{1}{2}[\cos \phi + \zeta(\mathbf{x}, \hat{\mathbf{x}}) \sin \phi] \qquad \zeta(\mathbf{x}, \hat{\mathbf{x}}) = \frac{d_1^2}{d^2(\mathbf{x}, \hat{\mathbf{x}})} = \frac{\sum_{n \in \eta} \sqrt{\delta_n^2(4 - \delta_n^2)}}{\sum_{n \in \eta} \delta_n^2}$$

(11.53)

and (11.26) becomes

$$P(\mathbf{x} \to \hat{\mathbf{x}}|(\phi)) \leq \begin{cases} \frac{1}{2}\exp\left\{-\frac{E_s}{4N_0}d^2(\mathbf{x},\hat{\mathbf{x}})(\cos\phi + \zeta(\mathbf{x},\hat{\mathbf{x}})\sin\phi)^2\right\}; & \sum_{n\in\eta}\delta_n^2(\cos\phi + \alpha_n\sin\phi) > 0 \\ 1; & \sum_{n\in\eta}\delta_n^2(\cos\phi + \alpha_n\sin\phi) \leq 0 \end{cases}$$

(11.54)

where $d^2(\mathbf{x},\hat{\mathbf{x}})$ is again defined in (11.9).[4]

Unfortunately, the integral of (11.54) over the pdf's of (11.47) and (11.50) cannot be obtained in closed form. Defining the integral

$$L(\beta; J) = \int_{\mathcal{R}} \exp\left\{-\frac{E_s d^2(\mathbf{x},\hat{\mathbf{x}})}{4N_0}[\cos\phi + \beta\sin\phi]^2\right\} \frac{\exp(\rho\cos J\phi)}{(2\pi/J)I_0(\rho)} d\phi$$

(11.55)

then, the average pairwise error probabilities are now as follows:

Discrete carrier

$$\mathbf{E}_\phi\left\{\min_\lambda P(\mathbf{x} \to \hat{\mathbf{x}}|\boldsymbol{\phi};\lambda)\right\} \leq \tfrac{1}{2}L(\zeta(\mathbf{x},\hat{\mathbf{x}}); 1) + I \qquad (11.56)$$

where I is defined in (11.48).

Suppressed carrier

$$\mathbf{E}_\phi\left\{\min_\lambda P(\mathbf{x} \to \hat{\mathbf{x}}|\boldsymbol{\phi};\lambda)\right\} \leq \tfrac{1}{2}L(\zeta(\mathbf{x},\hat{\mathbf{x}}); M) \qquad (11.57)$$

where the region $\mathcal{R}$ in the integral of (11.55) now corresponds to the interval $(-\pi/M, \pi/M)$.

Using (11.56) and (11.57) [rather than (11.48) and (11.51)] will result in a smaller improvement in performance due to interleaving/deinterleaving, since (11.56) and (11.57) result in a tighter bound on P_b (no interleaving).

11.2.3 The Case of Binary Convolutional Coded BPSK Modulation

Discrete carrier tracking

For BPSK modulation, there are only two points in the signal constellation and the squared Euclidean distance between them is equal to four (assuming a unit-energy signaling set; that is, the points are located opposite one another on a circle of unit radius). Substituting $\delta_n^2 = \delta^2 = 4$ into (11.26) results in $\alpha_n = 0$. Thus we see that α_n represents a "crosstalk" parameter, which reflects the degrading

---

[4] Note that (11.54) can be obtained directly by applying the bound of (11.21) to (11.16) together with (11.24).

effects of the quadrature modulation components (absent in BPSK) on the error probability performance. Substituting $\alpha_n = 0$ into (11.53), we obtain $\zeta(\mathbf{x}, \hat{\mathbf{x}}) = 0$ or $\lambda_{\text{opt}} = (\cos \phi)/2$, which when substituted in (11.54) yields the conditional pairwise error probability bound

$$P(\mathbf{x} \to \hat{\mathbf{x}} | \phi) \leq \begin{cases} \dfrac{1}{2} \exp\left\{ -\dfrac{E_s}{4N_0} d^2(\mathbf{x}, \hat{\mathbf{x}}) \cos^2 \phi \right\} & 0 \leq |\phi| \leq \dfrac{\pi}{2} \\ 1 & \dfrac{\pi}{2} \leq |\phi| \leq \pi \end{cases}$$

(11.58)

for the case of no interleaving. (Note that the regions $\mathcal{R}$ and $\overline{\mathcal{R}}$ are simply given by $0 \leq \phi \leq \pi/2$ and $\pi/2 \leq \phi \leq \pi$, respectively.) The corresponding average pairwise error probability (minimized over λ) becomes

$$\mathbf{E}_\phi \left\{ \min_\lambda P(\mathbf{x} \to \hat{\mathbf{x}} | \boldsymbol{\phi}; \lambda) \right\} \leq \tfrac{1}{2} L(0; 1) + I$$

$$I = \frac{1}{\pi I_0(\rho)} \int_{\pi/2}^{\pi} \exp(\rho \cos \phi) \, d\phi.$$

(11.59)

For the case of interleaving, even with $\alpha_n = 0$, it is still inconvenient, from the point of view of evaluation, to optimize over λ prior to averaging over ϕ_n. Nevertheless, we do get a considerable simplification of (11.49), namely,

$$\min_\lambda \mathbf{E}_\phi \left\{ P(\mathbf{x} \to \hat{\mathbf{x}} | \boldsymbol{\phi}; \lambda) \right\} \leq \min_\lambda \left\{ \left(\exp\left\{ 4\lambda^2 \frac{E_s}{N_0} \right\} \frac{I_0(\rho_0)}{I_0(\rho)} \right)^{d^H} \right\}$$

$$\rho_0 = \left| \rho - 4\lambda \frac{E_s}{N_0} \right|$$

(11.60)

where d^H is the Hamming distance of the error path corresponding to η.

To evaluate the upper bound on bit error probability [e.g., (11.28)], we use the transfer function bound approach [16], which for the ideal case of perfect carrier synchronization gives

$$P_b \leq \frac{1}{2} \frac{\partial}{\partial I} T(D, I) \Big|_{I=1, D=Z}$$

(11.61)

where $T(D, I)$ is the transfer function of the superstate diagram associated with the trellis diagram of the code (see Chapter 4). When noisy carrier synchronization references are present, the appropriate upper bound on bit error probability for the case of no interleaving becomes

$$P_b \leq \frac{1}{2} \left[2 \int_0^{\pi/2} \frac{\partial}{\partial I} T(D, I) \Big|_{I=1, D=Z(\phi)} p(\phi) \, d\phi + 2 \int_{\pi/2}^{\pi} p(\phi) \, d\phi \right]$$

$$= \int_0^{\pi/2} \frac{\partial}{\partial I} T(D, I) \Big|_{I=1, D=Z(\phi)} p(\phi) \, d\phi + I$$

(11.62)

where I is defined in (11.59) and from (11.58), the equivalent Bhattacharyya parameter becomes[5]

$$Z(\phi) = \exp\left\{-\frac{E_s}{N_0}\cos^2\phi\right\} \qquad 0 \le |\phi| \le \frac{\pi}{2}. \tag{11.63}$$

For the case of interleaving, we use (11.61) with Z defined in accordance with (11.60),

$$Z = \min_\lambda \left(\exp\left\{\frac{4\lambda^2 E_s}{N_0}\right\}\frac{I_0(\rho_0)}{I_0(\rho)}\right) \tag{11.64}$$

In arriving at (11.64), we have made use of the fact that for any d,

$$\min_\lambda\left\{\left(\exp\left\{4\lambda^2\frac{E_s}{N_0}\right\}\frac{I_0(\rho_0)}{I_0(\rho)}\right)^d\right\} = \left\{\min_\lambda\left(\exp\left\{4\lambda^2\frac{E_s}{N_0}\right\}\frac{I_0(\rho_0)}{I_0(\rho)}\right)\right\}^d. \tag{11.65}$$

Figures 11.15 to 11.17 illustrate the upper bound on bit error probability versus bit energy-to-noise ratio E_b/N_0 (E_b is related to E_s by $E_s = R_c E_b$, where R_c is

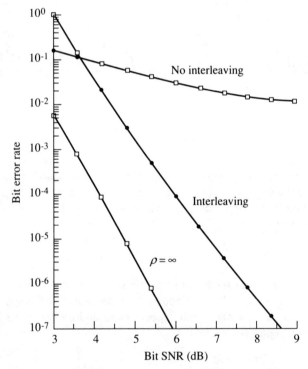

FIGURE 11.15 Upper bound on average bit error probability versus bit energy-to-noise ratio for rate 1/2, constraint length 7 convolutional code; loop SNR = 7 dB; discrete carrier.

---

[5] Note that for convolutional codes, $d^2(\mathbf{x}, \hat{\mathbf{x}})/4$ represents the Hamming distance between the two binary sequences $\mathbf{x}$ and $\hat{\mathbf{x}}$ that become the exponents (powers of D) in the transfer function $T(D, I)$.

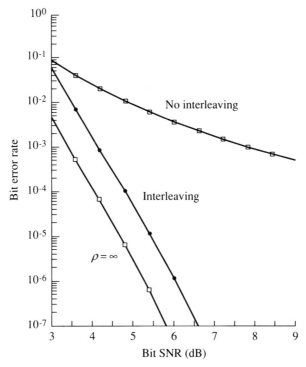

FIGURE 11.16 Upper bound on average bit error probability versus bit energy-to-noise ratio for rate 1/2, constraint length 7 convolutional code; loop SNR = 10 dB; discrete carrier.

the code rate) for a rate 1/2, constraint length 7 code, and various values of loop SNR, ρ. In these curves we have assumed carrier synchronization with a PLL (i.e., the discrete carrier case).[6] On each figure are plotted the results for the case of no interleaving, the case of interleaving, and the case of no radio loss (i.e., ideal carrier synchronization). The transfer function bound (truncated to 15 terms) on P_b for the code above is given in Table 11.7.

We observe from the results in Figs. 11.15 to 11.17 that even for large values of ρ (e.g., 13 dB), a substantial reduction of bit error rate is possible by using interleaving. Also, since the tighter bound was used for the interleaving case and the looser bound for the no-interleaving case, the performance improvement illustrated is somewhat pessimistic; that is, in reality, one will do even better than shown.

Suppressed carrier tracking

The upper bound on bit error probability for the no-interleaving case is, analogous to (11.62), given by

$$P_b \leq \int_0^{\pi/2} \frac{\partial}{\partial I} T(D, I)\Big|_{I=1, D=Z(\phi)} p(\phi) \, d\phi \qquad (11.66)$$

---

[6] Also, in the computation of (11.63), we have set the value of one-half the derivative of the transfer function evaluated at $I = 1$ to 1 whenever it would normally exceed 1. This is allowable, since the conditional error probability cannot exceed 1. Doing so results in a tighter bound.

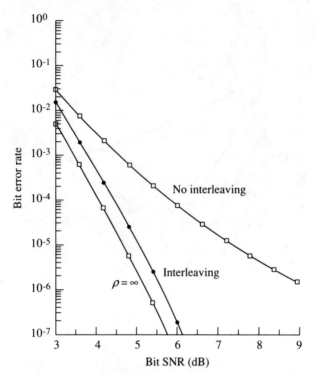

FIGURE 11.17 Upper bound on average bit error probability versus bit energy-to-noise ratio for rate 1/2, constraint length 7 convolutional code; loop SNR = 13 dB; discrete carrier.

TABLE 11.7 Transfer Function Bounds of $R_c = 1/2$, $K = 7$, Convolutional Code.

$$P_b \leq \frac{1}{2}\frac{\partial}{\partial I}T(Z,I)|_{I=1} = \frac{1}{2}\sum_{d=d_f}^{\infty}\beta_d Z^d$$

d	β_d
10	36
11	0
12	211
13	0
14	1404
15	0
16	11633
17	0
18	77433
19	0
20	502690
21	0
22	3322763
23	0
24	21292910

Source: F. Pollara, private communication.

with $p(\phi)$ as in (11.50) and $Z(\phi)$ as in (11.63). For the case of interleaving and the tighter upper bound, we again use (11.61). Z is now defined, analogous to (11.64), by

$$Z = \min_\lambda \left(\exp\left\{ 4\lambda^2 \frac{E_s}{N_0} \right\} \frac{f(\rho)}{I_0(\rho)} \right) \tag{11.67}$$

where $f(\rho)$ of (11.51) now simplifies to

$$f(\rho) = \frac{1}{\pi} \int_{-\pi/2}^{\pi/2} \exp\left\{ \rho \cos 2\phi - 4\lambda \frac{E_s}{N_0} \cos\phi \right\} d\phi. \tag{11.68}$$

By assuming a Costas loop with integrate-and-dump arm filters (matched filters), the equivalent loop SNR is given by [18]

$$\rho = \frac{1}{4}\left(\frac{S}{N_0 B_L}\right) S_L = \frac{1}{4}\left(\frac{E_b}{N_0}\right) \frac{1}{B_L T_b} S_L \qquad S_L = \frac{2(E_s/N_0)}{1 + 2(E_s/N_0)} \tag{11.69}$$

where S_L denotes the "squaring loss" associated with BPSK modulation. Figures 11.18 and 11.19 illustrate results analogous to Figs. 11.15 to 11.17 for the case

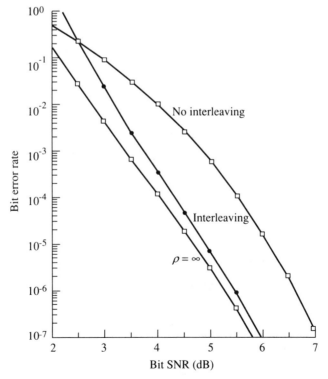

FIGURE 11.18 Upper bound on average bit error probability versus bit energy-to-noise ratio for rate 1/2, constraint length 7 convolutional code; suppressed carrier; $1/B_L I_b = 10$.

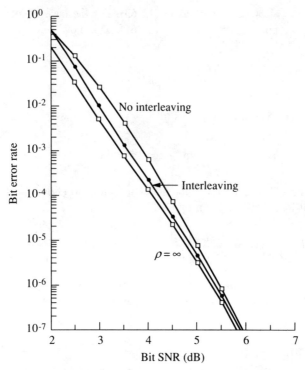

FIGURE 11.19 Upper bound on average bit error probability versus bit energy-to-noise ratio for rate 1/2, constraint length 7 convolutional code; suppressed carrier; $1/B_L T_b = 20$.

of suppressed carrier and various values of $B_L T_b$, where T_b is the bit duration. Since, as already mentioned, in suppressed carrier systems there is no irreducible error [since, from (11.69), $\rho \to \infty$ as $E_b/N_0 \to \infty$], the noisy reference losses are much smaller to begin with (i.e., no interleaving) than for the discrete carrier case. Thus, for sufficiently small $B_L T_b$, interleaving does not provide significant improvement.

11.2.4 A TCM Example

Consider a rate 1/2 trellis-coded QPSK using a simple two-state trellis. The code trellis structure with the appropriate QPSK symbol assignment is illustrated in Fig. 11.20, and the corresponding superstate transition diagram is shown in Fig. 11.21, where a, b, and c are branch label gains to be specified below. The transfer function of the superstate diagram is

$$T(D, I) = \frac{Iac}{1 - Ib}. \tag{11.70}$$

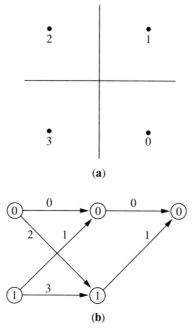

FIGURE 11.20 QPSK signal point constellation and trellis diagram showing QPSK signal assignments to branches.

No interleaving

For the case of no interleaving, we have

$$\left.\frac{\partial T(D, I)}{\partial I}\right|_{I=1, D=Z} = \frac{ac}{(1-b)^2} = \sum_{k=0}^{\infty} (k+1) a c b^k \quad (11.71)$$

where

$$a = Z^{16\lambda(\cos\phi - \lambda)}$$
$$b = c = Z^{8\lambda(\cos\phi + \sin\phi - \lambda)} \quad (11.72)$$

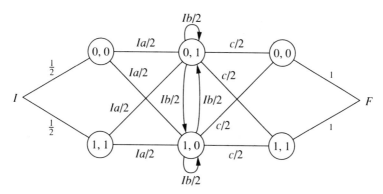

FIGURE 11.21 Superstate transition diagram for trellis diagram of Fig. 11.20.

and Z is the Bhattacharyya parameter for the ideal AWGN channel,

$$Z = \exp\left\{-\frac{E_s}{4N_0}\right\}. \tag{11.73}$$

Using (11.71) and the result of (11.54), we find that the upper bound on average bit error probability can be represented as

$$P_b \leq \sum_{k=0}^{\infty} (k+1) \int P_k(\phi) p(\phi) \, d\phi \tag{11.74}$$

where

$$P_k(\phi) = \begin{cases} \frac{1}{2} Z^{2(k+3)(\cos\phi + [(k+1)/(k+3)]\sin\phi)^2} & \phi - \tan^{-1}\frac{k+1}{k+3} < \frac{\pi}{2} \\ 1 & \phi - \tan^{-1}\frac{k+1}{k+3} \geq \frac{\pi}{2}. \end{cases} \tag{11.75}$$

For discrete carrier synchronization, $p(\phi)$ is given by (11.71), and for suppressed carrier tracking with a four-phase Costas loop, $p(\phi)$ is given by (11.50) with $M = 4$.

Interleaving

When interleaving is employed, then analogous to (11.71) we have

$$\left.\frac{\partial T(D, I)}{\partial I}\right|_{I=1, D=Z} = \frac{\bar{a}\bar{c}}{(1-\bar{b})^2} = \sum_{k=0}^{\infty} (k+1) \bar{a}\bar{c}\bar{b}^k \tag{11.76}$$

where

$$\bar{a} = \int Z^{16\lambda(\cos\phi - \lambda)} p(\phi) \, d\phi$$

$$\bar{b} = \bar{c} = \int Z^{8\lambda(\cos\phi + \sin\phi - \lambda)} p(\phi) \, d\phi. \tag{11.77}$$

For discrete carrier synchronization, (11.77) can be represented in closed form as

$$\bar{a} = \exp\left\{\frac{4\lambda^2 E_s}{N_0}\right\} \frac{I_0(\rho_0)}{I_0(\rho)}$$

$$\bar{b} = \exp\left\{\frac{2\lambda^2 E_s}{N_0}\right\} \frac{I_0\left(\sqrt{(\rho - 2\lambda E_s/N_0)^2 + (2\lambda E_s/N_0)^2}\right)}{I_0(\rho)} \tag{11.78}$$

where ρ_0 is defined in (11.60).

Upon using (11.76), an expression for the upper bound on average bit error probability, analogous to (11.74), is given by

$$P_b \leq \sum_{k=0}^{\infty} (k+1) \min_{\lambda} \gamma \bar{a}(\bar{b})^{k+1} \qquad (11.79)$$

where $\gamma = 1$ for a discrete carrier and $\gamma = \frac{1}{2}$ for a suppressed carrier.

The upper bounds of (11.74) and (11.79) are plotted in Figs. 11.22 and 11.23 for discrete carrier tracking with loop SNR values of $\rho = 13$ and 15 dB, respectively. In Figs. 11.24 and 11.25, the comparable results for the suppressed carrier case are illustrated. Here we have assumed a four-phase Costas loop with integrate-and-dump arm filters whose equivalent loop SNR is, analogous to (11.69),

$$\rho = \frac{1}{16}\left(\frac{S}{N_0 B_L}\right) S_L = \frac{1}{16}\left(\frac{E_b}{N_0}\right)\left(\frac{1}{B_L T_b}\right) S_L \qquad (11.80)$$

with S_L now denoting the "fourth power loss" and given by [17]

$$S_L = \left[1 + \frac{9}{2}\left(\frac{E_s}{N_0}\right)^{-1} + 6\left(\frac{E_s}{N_0}\right)^{-2} + \frac{3}{2}\left(\frac{E_s}{N_0}\right)^{-3}\right]^{-1}. \qquad (11.81)$$

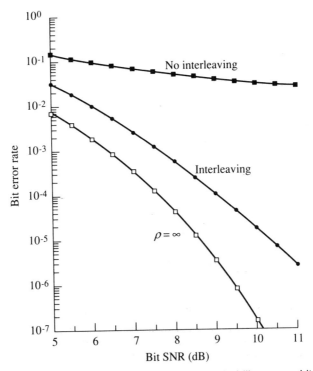

FIGURE 11.22 Upper bound on average bit error probability versus bit energy-to-noise ratio for rate 1/2, trellis-coded QPSK; two states; loop SNR = 13 dB; discrete carrier.

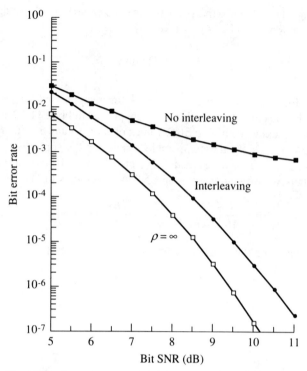

FIGURE 11.23 Upper bound on average bit error probability versus bit energy-to-noise ratio for rate 1/2, trellis-coded QPSK; two states; loop SNR = 15 dB; discrete carrier.

Also, in evaluating the numerical results, we have truncated the series in (11.74) and (11.79) to 15 terms.

11.2.5 Concluding Remarks

In this section we have shown that by interleaving the transmitted coded symbols in a trellis-coded system, the noisy reference loss can be reduced significantly. The amount of this reduction depends on the particular trellis code used and the region of operation of the system as characterized by such parameters as bit-error rate and loop SNR. In this section we have used a simple example (two-state, rate 1/2 trellis-coded QPSK) strictly for the purpose of illustrating the theoretical results. More complex trellis codes with a larger number of modulation levels and a larger number of states will show even more gain due to interleaving.

In general, whether or not coding and interleaving are employed, suppressed carrier systems have smaller radio losses than discrete carrier systems, since they are not subject to irreducible error probability. This is true despite the fact that for practical passive arm filters (e.g., *RC* filters) in the suppressed carrier tracking loop (Costas loop), one will experience larger squaring losses and thus larger radio losses than those shown here for active integrate-and-dump arm filters [18].

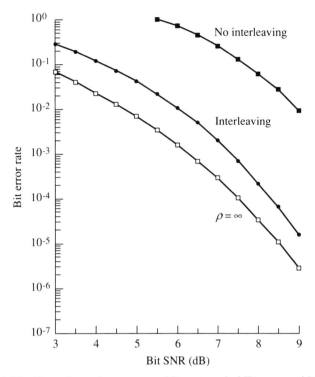

FIGURE 11.24 Upper bound on average bit error probability versus bit energy-to-noise ratio for rate 1/2, trellis-coded QPSK; two states; $1/B_L T_b = 10$; suppressed carrier.

Thus, if the radio loss is, without interleaving, small (as tends to be true in suppressed carrier systems), the use of interleaving cannot be of much additional help. Nevertheless, if the system can tolerate the delay associated with the interleaving/deinterleaving process, it is useful to include it in the system design, since it also helps to reduce other impairments of a bursty nature, such as intersymbol interference and fading. We also refer the reader to [19] to [21].

11.3 TCM over Satellite Channels

In this section we describe two implementations of TCM modems designed for transmission over satellite channels at 4800 bits/s and 64 kbits/s respectively. The development of modems for transmission at higher speeds over satellite channels is also under way. The description of a TCM modem for 120 Mbits/s operation is contained in [22]. The simulation of a TCM scheme for 140 Mbits/s transmission over 80-MHz transponders is described in [23]. Reference [24] describes a computational technique, based on the product-trellis method (see Chapter 4), to evaluate the error probability of TCM schemes used over nonlinear satellite channels.

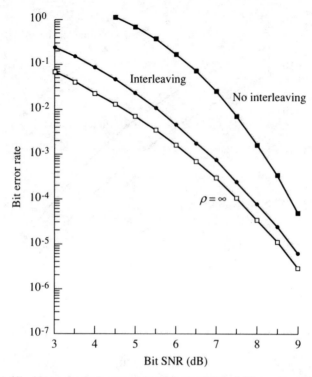

FIGURE 11.25 Upper bound on average bit error probability versus bit energy-to-noise ratio for rate 1/2, trellis-coded QPSK; two states; $1/B_L T_b = 20$; suppressed carrier.

11.3.1 A Modem for Land Mobile Satellite Communications

NASA, through the Jet Propulsion Laboratory, initiated a land mobile satellite experiment (MSAT-X) program with the goal of developing a mobile satellite service. In satellite-based land mobile communication systems, both bandwidth and power are limited resources. In fact, these systems employ frequency-division multiple access (FDMA) with a tight channel spacing, and the fraction of out-of-band power should be very small, to prevent interferences to adjacent channels. On the other hand, the satellite distance from earth, its power limitations, and the need for low-cost (and hence low-gain) mobile antennas put a serious limit on the power resources. Additionally, the fading environment of mobile communication further limits the power efficiency of the system. In a bandwidth- and power-limited environment, a bandwidth- and power-efficient coding/modulation scheme must be used. TCM offers an attractive scheme. In previous chapters we have described the performance and the design of TCM schemes for a fading channel. In this section we take a closer look at the modem developed for MSAT-X [25, 26].

Besides additive Gaussian noise, a number of additional sources of performance degradation must be taken into account to assess the merits of a proposed transmission scheme for mobile satellite channels. The most important among them are

- *Doppler shifts.* These are due to mobile vehicle motion. If differential detection is used, the information-bearing phase turns out to be shifted by an amount $2\pi f_d T_s$, where $1/T_s$ is the data symbol rate and f_d is the Doppler frequency spread. At L-band, vehicle speeds of 8 and 44 mph cause Doppler frequency spreads of 20 and 104 Hz, respectively.
- *Fading and shadowing.* The transmitted radio signal reaches the receiver through different paths caused by reflections from obstacles, yielding a signal whose components, having different phases and amplitudes, may either reinforce or cancel each other. Shadowing is caused by the obstruction of radio waves by buildings, trees, and hills.
- *Adjacent channel interference.* The 5 kHz mobile channel used for transmission operates in a channelized environment. As a result, signals suffer from interference from signals occupying adjacent channels.
- *Channel nonlinearities.* Primarily because of the high-power amplifier in the transmitter, operated at or near saturation for better power efficiency, the channel is inherently nonlinear.
- *Finite interleaving depth.* To break up the error bursts caused by amplitude fades of duration greater than symbol time, encoded symbols should be interleaved. Now, infinite interleaving provides a memoryless channel, but in practice the interleaving frame must be limited. In fact, for speech transmission the total coding/decoding delay must be kept below 60 ms in order not to cause perceptual annoyance. If the depth of interleaving cannot be larger than the maximum fade duration anticipated, this causes a performance degradation.

Frequency-division multiplexing was selected as the channel structure, with a bandwidth of 5 kHz per channel for a data transmission rate of 4800 bits. The required bit error rate is 10^{-3} at an average bit signal-to-noise ratio of 11 dB. The channel model assumed in the development of the modem is the Rician fading channel, with a coherent (desired) to noncoherent (scattered) power ratio $K = 10$ dB. Vegetative shadowing is also assumed to be present and to follow a log-normal distribution.

The modulator obtains packetized data at a rate of 4800 bits/s. A packet includes a packet identification word and a preamble and a postamble to be used for synchronization purposes. The data are encoded with a rate 2/3, 16-state trellis code with 8-PSK [27]. Every symbol carries two information bits, producing a rate of 2400 symbols per second. The TCM encoder is followed by a 128-symbol block interleaver for burst error protection. The interleaver output is differentially encoded and sent to the PSK modulator. The resulting signal is pulse shaped with a 100% rolloff root raised-cosine filter. The main reasons for choosing this particular pulse shape are to shape the power spectrum of the transmitted signal as well as to produce two intersymbol-interference (ISI) free points per symbol. The presence of these points is necessary for Doppler estimation and matched filtering at the demodulator [28].

The demodulator has a feedforward nature, which allows it to freewheel through deep fades and rapidly recover once the fade has ended. The received signal is filtered at the data bandwidth plus the maximum expected Doppler spread, and passed to the symbol-time-recovery and power/fade-detect circuits.

In the power/fade-detect algorithm, the in-phase and quadrature parts of the incoming signal are squared, summed, and low-pass filtered. The resulting signal is then thresholded. If the filter output exceeds the threshold, power (i.e., the start of a transmission) is detected. Further, if the filter output falls below a second threshold, a fade is detected, and the signal is used to reduce the effects of fading and to reduce the preamble false-alarm probabilities. In the symbol timing algorithm, the in-phase and quadrature parts of the incoming signal are squared, summed, and filtered through a bandpass filter centered at the symbol rate to produce a timing wave [29].

The Doppler estimator uses the two ISI-free points in each symbol period to derive a correction of the Doppler frequency shift [28]. After removal of the preamble and postamble, the data stream is deinterleaved and decoded by a 16-state Viterbi decoder. Experimental results are shown in Figs. 11.26 to 11.28, where the measured bit error probabilities are compared to simulation results. Perfect Doppler tracking and time synchronization are assumed in the simulation.

11.3.2 An SCPC Modem

In this section we describe a modem operating at 64 kbits/s developed to comply with INTELSAT specifications for single channel per carrier (SCPC) [30]. A

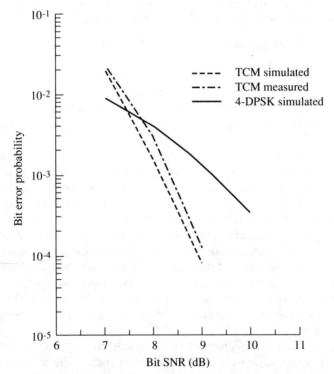

FIGURE 11.26 Bit error probability of TCM with differential 8-PSK and of Gray-coded differentially 4-PSK over the additive white Gaussian noise channel.

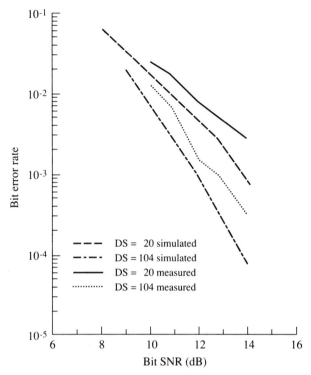

FIGURE 11.27 Bit error probability of TCM with differential 8-PSK over a Rician fading channel with $K = 5$ dB. DS denotes the Doppler spread in hertz.

four-state TCM scheme was used in conjunction with octonary PSK to transmit two information bits per waveform. The signal set, the four-state trellis, and the structure of the modulator are shown in Fig. 11.29. The asymptotic coding gain of this scheme is 3 dB over 4-PSK.

Figure 11.30 shows the sensitivity of this scheme to carrier-phase errors, in the form of the value of E_b/N_0 required to achieve an error probability 10^{-5} versus the carrier-phase offset $\Delta\Phi$. The figure also shows the sensitivity of an eight-state TCM scheme, which has an asymptotic coding gain of 3.6 dB. It is seen that for increasing $\Delta\Phi$ the eight-state scheme has a performance increasingly closer to that of the four-state scheme. In this situation it may be better to choose the four-state scheme because of its lower complexity. Carrier-phase tracking is performed by a decision-directed phase-locked loop that uses tentative decisions from the Viterbi decoder. Since the loop can lock at $\Delta\Phi = 0°$ and $180°$, differential encoding is needed to ensure data integrity under $180°$ phase shifts.

11.4 Trellis Codes for Partial Response Channels

In this section we present the results of Wolf and Ungerboeck [31] on the use of trellis codes on $1 - D$ and $1 + D$ channels (D represents unit time delay). The

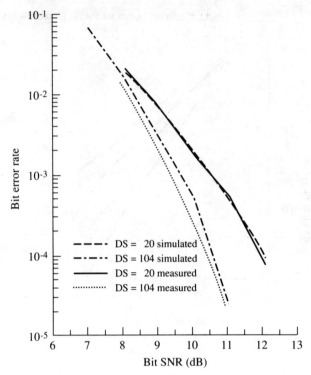

FIGURE 11.28 Bit error probability of TCM with differential 8-PSK over a Rician fading channel with $K = 10$ dB. DS denotes the Doppler spread in hertz.

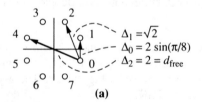

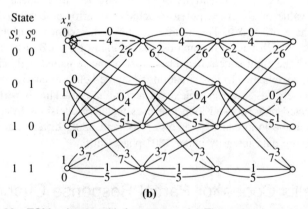

FIGURE 11.29 TCM scheme with four states. (a) The redundant signal set; (b) the trellis; and (c) structure of the modulator.

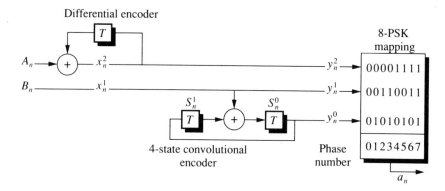

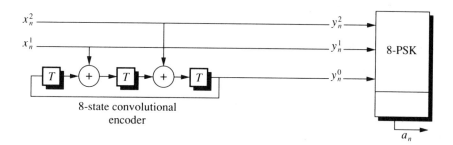

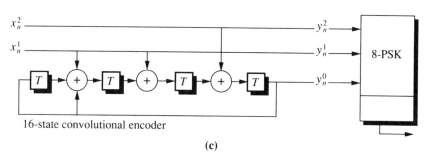

(c)

FIGURE 11.29 (continued)

$1 - D$ channel, called a *dicode channel*, introduces a spectral null at 0 Hz. This channel can be used as a very simple model for magnetic data recording devices. In this section we consider binary partial response channels [31]. The results are extended to M-ary partial response channels in [32].

To compare the performance of trellis-coded partial response signaling with the uncoded case, we should define a baseline uncoded system for such a comparison. Here we consider only the $1 - D$ channel. The approach taken here can easily be repeated for the $1 + D$ channel. The baseline system is shown in Fig. 11.31. In this figure the output samples of the partial response channel are

$$y_k = x_k - x_{k-1}. \tag{11.82}$$

482 Ch. 11 / Analysis and Design of TCM for Other Practical Channels

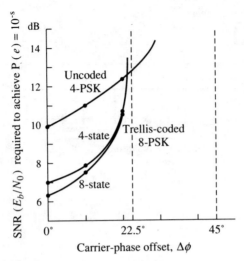

FIGURE 11.30 Sensitivity of uncoded 4-PSK and trellis-coded 8-PSK to carrier-phase offset. Reprinted with permission, ©IEEE 1986.

At the receiver, a sequence of independent Gaussian noise samples, the n_k's, with zero mean and variance $\sigma^2 = N_0/2E_b$, are added to the samples, y_k's. Then the received sample, z_k, is given by

$$z_k = y_k + n_k. \tag{11.83}$$

Each sample y_k depends on the present transmitted sample x_k and one past sample x_{k-1}. Therefore, the channel has unit memory. The equivalent finite state machine (FSM) for the combination of the transmitter and the channel is shown in Fig. 11.32. During the kth symbol time, the present state of the FSM is $s_k = x_{k-1}$ and the next state is $s_{k+1} = x_k$. The optimum maximum-likelihood (ML) sequence receiver for this FSM can be realized by a two-state Viterbi decoder [33, 34]. The optimum ML metric for the Viterbi decoder is a minimum Euclidean distance metric given by

$$M(\mathbf{y}, \mathbf{z}) = \sum_k \lambda(z_k, y_k) \tag{11.84}$$

where $\lambda(z_k, y_k) = (z_k - y_k)^2$ are the branch metrics. The trellis diagram for the

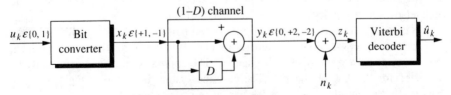

FIGURE 11.31 Baseline system.

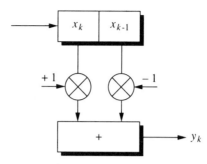

FIGURE 11.32 Finite state machine representing the combination of the transmitter and the channel.

Viterbi decoder is shown in Fig. 11.33. In this figure transitions to state "−1" correspond to the information input $u_k = 0$, and transitions to the state "1" correspond to information input $u_k = 1$. From Fig. 11.33 we see that the squared free Euclidean distance, d_{free}^2, is 8. Also from the same figure, the catastrophic nature of this type of channel can be seen (i.e., there are infinitely many error events with $d_{\text{free}}^2 = 8$). Moreover, the occurrence of unlimited runs of zero outputs can cause timing and gain-control problems. The bit error performance of this code can be found by using the superstate transfer function bound method (see Chapter 4). The superstate diagram for this system is shown in Fig. 11.34. In this figure the parameters are

$$a = \frac{I}{2}D^4$$

$$b = \frac{I}{2}$$

$$c = \frac{I}{2}D^{16}$$

$$d = \frac{1}{2}D^4$$

(11.85)

where D will take value Z, with

$$Z = e^{-E_b/4N_0}.$$

(11.86)

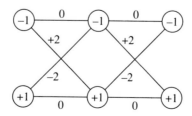

FIGURE 11.33 Trellis diagram for the Viterbi decoder.

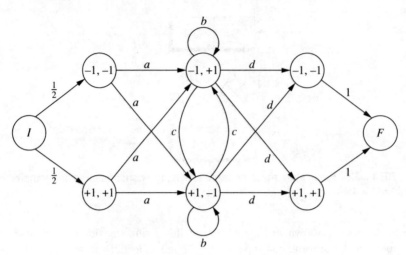

FIGURE 11.34 Superstate diagram.

The reduced version of the superstate diagram in Fig. 11.34 is shown in Fig. 11.35. From this figure we obtain the transfer function $T(D, I)$:

$$T(D, I) = \frac{4ad}{1 - (b + c)}. \tag{11.87}$$

Using (11.85) in (11.87), we get

$$T(D, I) = \frac{ID^8}{1 - (I/2)(1 + D^{16})}. \tag{11.88}$$

Finally, the upper bound on bit error probability is

$$P_b = Q\left(\sqrt{\frac{E_b\, d_{\text{free}}^2}{N_0\, 2}}\right) D^{-d_{\text{free}}^2} \left.\frac{\partial T(D, I)}{\partial I}\right|_{I=1,\ D=Z}$$

$$= \frac{4Q(\sqrt{4E_b/N_0})}{(1 - Z^{16})^2}. \tag{11.89}$$

We observe that although the channel has a catastrophic nature, the upper bound in (11.89) is finite. Also for paths with distance d_{free}, the average number of bit errors is 4. This baseline system has a 3-dB gain over the conventional partial response

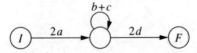

FIGURE 11.35 Reduced superstate diagram.

system with a detector that makes independent ternary decisions on each output sample z_k.

11.4.1 Trellis Codes for the Binary $(1 - D)$ Channel

For a binary alphabet (i.e., $x_i = +1$ or -1), the output y_i of the partial response $(1 - D)$ channel takes three values: $+2$, 0, and -2. The combination of trellis code and partial response channel can be regarded as an MTCM with b input bits and k output ternary symbols. As discussed previously, in order to obtain an MTCM, we should have

$$2^{b+1} \leq 3^k \tag{11.90}$$

where $b = rk$ and r is the throughput (i.e., the number of bits per ternary channel symbol). Unfortunately, due to the restrictions that $1 - D$ channels impose on the generation of k-tuple sequences, the number of possible sequences will be less than 3^k. In fact, the number of possible ternary k-tuples can be found from the following recursion (M. Shahshahani, private communication):

$$\begin{bmatrix} a_k \\ b_k \end{bmatrix} = \begin{bmatrix} 1 & \frac{4}{3} \\ \frac{3}{2} & 1 \end{bmatrix} \begin{bmatrix} a_{k-1} \\ b_{k-1} \end{bmatrix} \qquad k = 2, 3, \ldots \tag{11.91}$$

with

$$\begin{bmatrix} a_1 \\ b_1 \end{bmatrix} = \begin{bmatrix} 0 \\ 3 \end{bmatrix}. \tag{11.92}$$

Then the number of possible sequences of length k is $(a_k + b_k)$. For example:

$$\begin{array}{ll} k = 1: & a_1 + b_1 = 3 \\ k = 2: & a_2 + b_2 = 7 \\ k = 3: & a_3 + b_3 = 17 \\ k = 4: & a_4 + b_4 = 41. \end{array} \tag{11.93}$$

Therefore, for $k = 4$,

$$2^{4r+1} < 41 \tag{11.94}$$

or

$$r < \frac{\log_2 41 - 1}{4}. \tag{11.95}$$

Next we should select 2^{rk+1} sequences of k-tuples out of all available k-tuple ternary sequences, such that the minimum distance between the selected k-tuples is maximum. This procedure can be done by set partitioning. In assigning the selected k-tuples to the state transitions of the decoder trellis diagram, one should maximize d_{free}, and at the same time the run of consecutive zeros should be made as short as possible. Since the partial response $1 - D$ channel has memory 1, the number of states required at the decoder is generically twice the number of encoder states. Consider the following example.

EXAMPLE 11.1 [31]

Let $b = 3$ and $k = 4$; the minimum number of k-tuples required is 16, but in [31], 32 k-tuples have been chosen to get a good code that has a limited run of zeros. In designing the trellis code for the $(1 - D)$ channel, each state in the decoder trellis diagram is associated with polarity "+" or "−," which indicates the decoder's knowledge about the $(1 - D)$ channel state. The set partitioning is shown in Table 11.8, and the signal set assignment is shown in Fig. 11.36. The number of states for the decoder trellis is eight. Hence the number of states required for the encoder is four. The trellis shown in Fig. 11.36 has a minimum squared Euclidean distance equal to 16. Several such pairs of paths having $d_{free}^2 = 16$ are shown in Fig. 11.37. This code achieves a distance improvement of 3 dB over the baseline system, but with throughput $r = \frac{3}{4}$. From Fig. 11.37 we can observe that the longest run of consecutive zeros is 12. One such sequence of 12 consecutive zeros is shown in Fig. 11.38. □

TABLE 11.8 Sets and Subsets of Four Consecutive $(1 - D)$ Channel Outputs.

Set A: Starting State − Ending State −					Set C: Starting State + Ending State −				
Subset 0	0	0	0	0	Subset 8	0	0	0	−2
	2	−2	2	−2		−2	2	−2	0
Subset 1	0	2	0	−2	Subset 9	0	0	−2	0
	2	0	−2	0		−2	2	0	−2
Subset 2	0	2	−2	0	Subset 10	0	−2	0	0
	2	0	0	−2		−2	0	2	−2
Subset 3	0	0	2	−2	Subset 11	0	−2	2	−2
	2	−2	0	0		−2	0	0	0
Set B: Starting State − Ending State +					Set D: Starting State + Ending State +				
Subset 4	0	0	0	+2	Subset 12	0	0	0	0
	2	−2	2	0		−2	2	−2	2
Subset 5	0	0	2	0	Subset 13	0	−2	0	2
	2	−2	0	2		−2	0	2	0
Subset 6	0	2	0	0	Subset 14	0	−2	2	0
	2	0	−2	2		−2	0	0	2
Subset 7	0	2	−2	2	Subset 15	0	0	−2	2
	2	0	0	0		−2	2	0	0

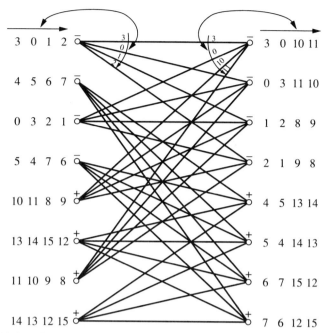

FIGURE 11.36 Decoder trellis for a rate 3/4 code with eight states and free Euclidean distance $d_{\text{free}} = \sqrt{16}$.

11.4.2 Convolutional Codes with Precoder for $(1 - D)$ Channels

Another method, which is a systematic construction of trellis codes for the partial response channels, was proposed in [31]. The block diagram for the system is shown in Fig. 11.39. In the transmitter, a rate $m/(m + 1)$ convolutional en-

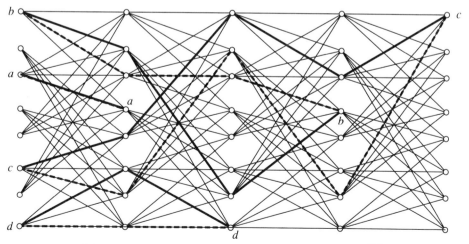

FIGURE 11.37 Pairs of paths with distance $d_{\text{free}} = \sqrt{16}$.

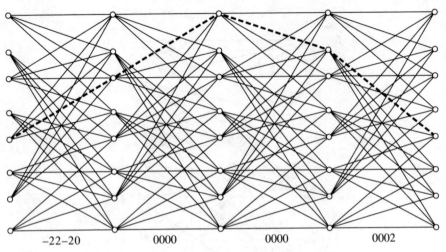

-22-20　　　　　0000　　　　　　0000　　　　　　0002

FIGURE 11.38 Longest sequence of 12 consecutive zeros.

coder with memory ν accepts an m-tuple of data bits $\mathbf{a}_j = (a_j^1, \ldots, a_j^m)$ and outputs a $(m + 1)$-tuple of coded bits $\mathbf{b}_j = (b_j^1, \ldots, b_j^{m+1})$. The coded bits are serially transmitted over the $(1-D)$ channel after being precoded. With the precoder present, the output of the $(1-D)$ channel is

$$y_i = \begin{cases} 0 & \text{for } b_i = 0 \\ \pm 2 & \text{for } b_i = 1. \end{cases} \tag{11.96}$$

The precoder for the $(1 - D)$ channel performs the function

$$b'_i = b'_{i-1} \oplus b_i \tag{11.97}$$

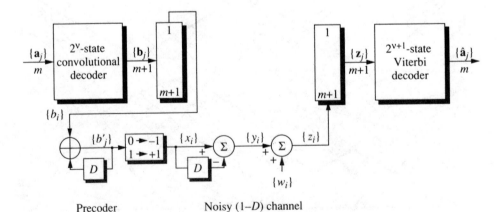

Precoder　　　　　　Noisy (1–D) channel

FIGURE 11.39 System block diagram.

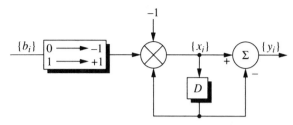

FIGURE 11.40 Equivalent representation of noiseless $(1 - D)$ channels with precoding.

where $\oplus$ denotes modulo-2 addition. Precoding does not increase the number of trellis states because of the equivalence of the signals stored simultaneously in the delay element of the precoder and the $(1 - D)$ channel. In Fig. 11.40 the equivalent representation of the precoder and $(1 - D)$ channel is shown.

In the receiver, the received noisy samples z_i's are grouped into $(m + 1)$-tuples from which a Viterbi decoder estimates the sequence of transmitted data m-tuples. As discussed previously, the precoded $(1 - D)$ channel, in general, increases the number of states in the trellis diagram for the combined encoder and channel, which can be called the *decoder trellis*, to $2^{\nu+1}$ states. In [35], some cases have been discussed where the encoder and decoder trellises require the same number of 2^{ν} states.

In [31], it has been shown that the minimum Euclidean distance of decoder trellises can be lower bounded in terms of the Hamming distance of the convolutional code as

$$d_{\text{free}}^2 \geq \begin{cases} 4d_{\text{free}}^{\text{H}} & d_{\text{free}}^{\text{H}} = \text{even} \\ 4(d_{\text{free}}^{\text{H}} + 1) & d_{\text{free}}^{\text{H}} = \text{odd} \end{cases} \quad (11.98)$$

In [31], several good codes have been found, which are summarized in Tables 11.9 to 11.11. In these tables $H^i(D), 1 < i < m + 1$ represent the parity check polynomials, which describe the convolutional codes found. In Fig. 11.41 the performance of these codes over the $(1 - D)$ channel is shown.

TABLE 11.9 $R_c = 1/2$ Codes for Use with Precoded $(1 - D)$ Channels.

	$\nu = 1$	$\nu = 2$	$\nu = 4$	$\nu = 6$
$d_{\text{free}}^{\text{H}}$	3	5	7	10
$H^1(D)$	01	101	10011	1011011
$H^2(D)$	11	111	11101	1111001
$2^{\nu+1}$	4	8	32	128
d_{free}	$\sqrt{16}$	$\sqrt{24}$	$\sqrt{32}$	$\sqrt{40}$

TABLE 11.10 $R_c = 2/3$ Codes for Use with Precoded $(1 - D)$ Channels.

	$v = 2$	$v = 4$	$v = 6$
$d_{\text{free}}^{\text{H}}$	3	5	7
$H^1(D)$	001	10011	1000101
$H^2(D)$	101	11001	1100101
$H^3(D)$	111	11101	1110011
2^{v+1}	8	32	128
d_{free}	$\sqrt{16}$	$\sqrt{24}$	$\sqrt{32}$

TABLE 11.11 $R_c = 3/4$ Codes for Use with Precoded $(1 - D)$ Channels.

	$v = 2$	$v = 5$	$v = 8$
$d_{\text{free}}^{\text{H}}$	3	5	7
$H^1(D)$	001	100101	100011001
$H^2(D)$	011	100111	101111101
$H^3(D)$	101	101001	110101111
$H^4(D)$	111	110011	111010111
2^{v+1}	8	64	512
d_{free}	$\sqrt{16}$	$\sqrt{24}$	$\sqrt{32}$

11.5 Trellis Coding for Optical Channels

In this section we consider the design of TCM schemes for a pulse-width-constrained (defined in Section 11.5.1) direct-detection optical channel. There are two types of optical channels: deep space and fiber optic. To achieve high data rates and a high level of energy efficiency over a deep-space link, very small optical pulse widths or, equivalently, very high bandwidths are required. Practical lasers cannot support very small pulse widths. Therefore, large-bandwidth systems are not implementable. Communication over optical fibers also encounters bandwidth limitations, due to the practical limit on the switching speed of lasers at high data rates.

Thus, since a pulse-width-constrained, direct-detection optical channel is band-limited for high-data-rate transmission, TCM is a natural choice for communication over such channels. In this section we follow the approach taken and the results

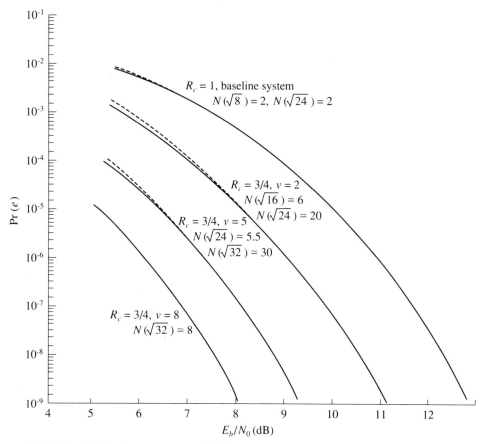

FIGURE 11.41 Error event probability versus signal-to-noise ratio for coded $(1 - D)$ channel systems using the $R_c = 3/4$ codes.

obtained in [36] and [37]. We consider signals that are on–off in nature and subject to a minimum pulse-width duration.

Consider a signal set of size M, given by

$$g_i(t) = \sqrt{\lambda_i(t)} \qquad i = 1, 2, \ldots, M, \quad 0 \leq t \leq T \tag{11.99}$$

where $\{\lambda_i(t), i = 1, 2, \ldots, M\}$ are the laser intensity rates at the receiver in photons per second, and

$$\lambda_0 \leq \lambda_i(t) \leq \lambda_s + \lambda_0. \tag{11.100}$$

In (11.100), λ_0 is the photon intensity of the dark current and λ_s is the received photon intensity in the absence of dark current and when the laser is fully on.

We normalize the signals $g_i(t)$ according to the following transformation:

$$\hat{g}_i\left(\frac{t}{T}\right) = \frac{g_i(t) - \sqrt{\lambda_0}}{\sqrt{\lambda_s + \lambda_0} - \sqrt{\lambda_0}} \qquad 0 \leq t \leq T. \tag{11.101}$$

In view of (11.99) and (11.100), we have

$$0 \le \hat{g}_i(t) \le 1 \qquad 0 \le t \le 1, \quad i = 1, 2, \ldots, M. \tag{11.102}$$

Let d_{ij} denote the distance between the signals $g_i(t)$ and $g_j(t)$:

$$d_{ij}^2 = \int_0^T [g_i(t) - g_j(t)]^2 \, dt. \tag{11.103}$$

Using (11.101), we can show that

$$d_{ij}^2 = \alpha \hat{d}_{ij}^2 \tag{11.104}$$

where

$$\alpha = T\left(\sqrt{\lambda_s + \lambda_0} - \sqrt{\lambda_0}\right)^2 \tag{11.105}$$

and $\hat{d}_{ij}$ is the distance between the normalized signals $\hat{g}_i(t)$ and $\hat{g}_j(t)$. The parameter α plays the same role as the signal-to-noise ratio in the AWGN channel. The cutoff rate for this channel with infinite bit quantization can be written as [38]

$$R_0 = -\log_2 \left\{ \min_q \sum_{i=1}^{M} \sum_{j=1}^{M} q(x_i) q(x_j) e^{-(\alpha/2)\hat{d}_{ij}^2} \right\} \tag{11.106}$$

where the x_i's are elements of the signal alphabet. The following upper bound on the error probability of this signal set is derived in [36]:

$$P_e \le \sum_{i=0}^{M-1} \sum_{\substack{j=0 \\ j \ne i}}^{M-1} e^{-(\alpha/2)\hat{d}_{ij}^2} \approx e^{-(\alpha/2)\hat{d}_{\min}^2}. \tag{11.107}$$

The approximation in (11.107) holds for large values of α, and $\hat{d}_{\min}$ represents the normalized minimum distance of the signal set according to (11.104).

It is apparent from (11.106) and (11.107) that the parameter α significantly affects both R_0 and P_e, which are the measures of performance of the degraded direct-detection optical communication channel. Both R_0 and P_e are obtained from the conditional probability of the channel, that is, the probability of observing n specific points during $[0, T]$, given that x_i is the transmitted signal;

$$P[\{N(t), 0 \le t \le T\} \mid x_i] = \begin{cases} \prod_{k=1}^{n} \lambda_i(t_k) e^{-\int_0^T \lambda_i(t) \, dt} & n \ge 1 \\ e^{-\int_0^T \lambda_i(t) \, dt} & n = 0. \end{cases} \tag{11.108}$$

Here $N(t)$ is a Poisson counting process, $\lambda_i(t)$ is the intensity function governing the process when the signal x_i is transmitted, and the t_k's are the locations of the individual photon counts, within the interval $[0, T]$.

11.5.1 Signal Sets with Amplitude and Pulse-Width Constraints

The signal sets considered in this section are of the on–off type; that is, the normalized signal $\hat{g}_i(t)$ takes on values 0 or 1, $0 < t < 1$; $i = 1, 2, \ldots, M$.

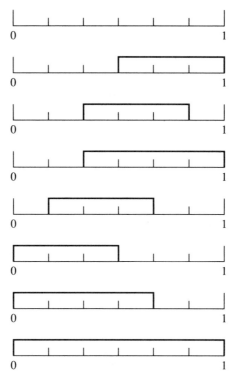

FIGURE 11.42 Signal set $(T/2, T/6)$, with normalized $T = 1$.

These signals are also pulse-width constrained, so that every nonzero pulse has a duration greater than or equal to β seconds. Furthermore, each nonzero signal consists of one pulse. The restriction to signals of the on–off type is due to the fact that it is difficult for practical lasers to produce multilevel pulses. An example of a signal set that satisfies the foregoing constraints is a signal set with $\beta = T/2$ and each interval of T seconds consists of six slots. We denote this signal set by $(T/2, T/6)$. The normalized version of this signal set with $T = 1$ and $\beta = \frac{1}{2}$ is shown in Fig. 11.42. For this set the normalized distance between two unequal signals is

$$\hat{d}_{ij}^2 = \int_0^1 (\hat{g}_i(t) - \hat{g}_j(t))^2 \, dt \qquad (11.109)$$
$$= |t_{e,i} - t_{e,j}| + |t_{b,i} - t_{b,j}|$$

where $t_{b,i}$ is the beginning time and $t_{e,i}$ is the ending time of the signal pulse $\hat{g}_i(t)$. Thus the square of the intersignal distance between two normalized, nonzero signals is the sum of the absolute differences of their beginning and ending times. This distance is sometimes called the *Manhattan distance*. Figure 11.43 demonstrates the Manhattan distance of two points in the two-dimensional plane. The distance between the zero signal and a nonzero signal is

$$\hat{d}_{i,0}^2 = t_{e,i} - t_{b,i}. \qquad (11.110)$$

$$p_1 = (x_1, y_1)$$
$$p_2 = (x_2, y_2)$$
$$d_m^2(p_1, p_2) = |x_1 - x_2| + |y_1 - y_2|$$

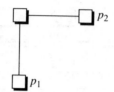

FIGURE 11.43 Manhattan distance.

Consider the mapping of the signal set into the plane, which assigns to each signal set the pair (x, y), where x is the beginning time of the signal pulse and y is the ending time of the signal pulse. The zero signal pulse is mapped to point $(0.5, 0.5)$. This mapping is an isometry relative to the Manhattan metrics on the signal set and the plane. We also note that since $\beta = \frac{1}{2}$, the beginning times must be in the first half of the interval and the ending times must be in the last half. Therefore, all signal points except the zero signal, which is at $(0.5, 0.5)$, lie on or within a right isosceles triangle located in the upper left-hand corner of the unit square. The mapping of signal set $(T/2, T/6)$ into a two-dimensional space with $\beta = \frac{1}{2}$, $M = 8$, and $T = 1$ is shown in Fig. 11.44.

11.5.2 Trellis-Coded Modulation for Optical Channels

In order to assess the potential coding gains and the number of signal levels required for coding, for a given throughput, one should compute R_0. The R_0 for

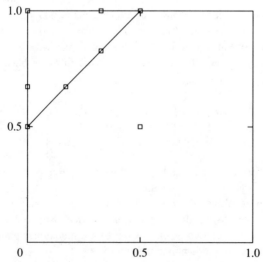

FIGURE 11.44 Mapping of signal set $(T/2, T/6)$ into a two-dimensional space, with $T = 1$.

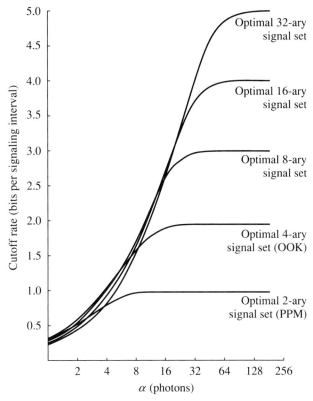

FIGURE 11.45 Cutoff rate of optimal signal sets for various sizes (M = 2, 4, 8, 16, and 32).

various signal levels are shown in Fig. 11.45. In this figure *optimum signal set* means those signal sets that maximize R_0. For example, consider a data throughput of 3 bits/signaling interval. If an uncoded 8-ary signal set is used, one can achieve $P_e = 10^{-6}$ with $\alpha = 111$ photons, and maintain a throughput of 3 bits/signaling interval. For the coded case, one should use a 16-ary signal set with a rate 3/4 coded system, as shown in Fig. 11.45. Only $\alpha = 19.4$ photons is required in this case. Hence the difference between the uncoded and coded systems is 7.5 dB in the required level of α. This example explains the motivation for using trellis codes over optical channels. To design a trellis-coded modulation scheme over an optical channel, one should choose an appropriate signal set, use the partitioning method of the signal set [39], and then choose a rate $m/m + 1$ convolutional code that preferably minimizes the bit error rate, or for large α, maximizes the minimum Manhattan distance.

In order to use the existing optimum codes for AWGN channels, in the case of optical channels, the signal sets are partitioned such that the distance-doubling relationship between the minimum distances at each level of the signal set-partitioning tree is preserved. To have an acceptable signal set with the distance-doubling property in the partitioning tree, one can consider the lattice $Z^2 = \{(x, y) \cdot x, y \text{ integer}\}$. To fit the points of the lattice Z^2 in the triangular acceptance region, one can scale up the sides of the triangular region. For example, in Fig. 11.44 for an 8-ary signal, one can enlarge the side of the triangle by a factor of 6. The scale-up factor should be chosen such that the required number of points can fit in the triangular acceptance region, as will be discussed shortly.

Lattices that produce sublattices with the distance-doubling property in partitioning were considered in Chapter 6 (see also [6] and [40]). Such lattices are Z^2, D_2, $2D_2$, and $4D_2$. These lattices are shown in Fig. 11.46. The coset decomposition of the Z^2 lattice by the D_2, $2D_2$, and $4D_2$ sublattices are shown in Fig. 11.47. The partitioning is shown by a tree with coset representatives at each node. To use the existing trellis codes, one should have a binary partitioning of the M-ary signal set. The coset decomposition of the Z^2 lattice does not provide a binary partitioning at every level. A binary decomposition can be obtained by trimming the tree to eliminate all but two paths leaving each node. The trimming should be done in such a way that the number of required signal points in each coset for the trellis code are preserved. Thus the partitioning of the signal set and the trimming of the tree should satisfy the desired distance-doubling property. It should also fulfill the requirement that the signal set at each node be partitioned into two smaller sets, each with exactly $\frac{1}{2}$ the number of signals in that node. For example, for $M = 16$, at the top of the tree, we need at least 16 points. Lowering the level by one reduces the number of required points by a factor of 2. The number of points that exist at each node of the tree are computed by placing the triangular acceptance region on each of the cosets at each node of the tree. The result of such a process should guarantee obtaining the population chain of $M, M/2, M/4, M/8, \ldots, 4, 2$. Consider the lattice Z^2 and its decomposition, which was shown in Fig. 11.47. We choose the triangular acceptance region with the smallest side size of 1, such that on or within this region we obtain at least $(M - 1)$ points from the top tree in Fig. 11.47. The number of points that can fit on or within a triangular acceptance region of a side size of L is $(L^2 + 3L - 2)/2$. Thus we should have

$$\frac{L^2 + 3L + 2}{2} \geq M - 1 \qquad (11.111)$$

or

$$L \geq \sqrt{2M - \frac{7}{4}} - \frac{3}{2}. \qquad (11.112)$$

The smallest admissible L should satisfy not only (11.112), but also all those properties required for partitioning the tree as described above. After choosing L, the scale-up factor will be $2L$. For example, let us consider the partitioning of the $M = 8$ signal set. Choosing $L = 3$ gives the smallest side size that satisfies (11.112). The process of partitioning is shown in Fig. 11.48. In this process we place the triangular acceptance region with side size 3 on the lattice Z^2 and on each subsequent coset of Fig. 11.47. Only those points of Fig. 11.47 that have been selected by the triangular acceptance region of side size of 3 have been

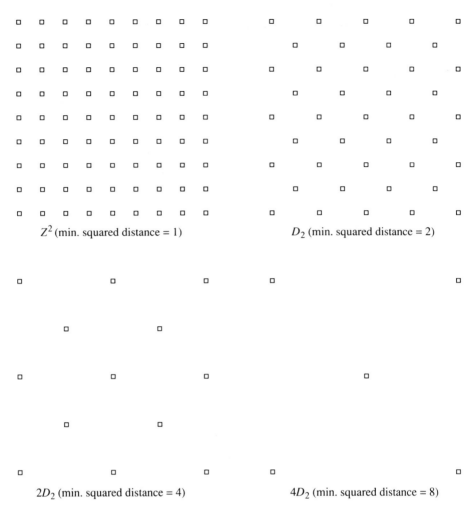

FIGURE 11.46 The Z^2, D_2, $2D_2$, and $4D_2$ two-dimensional lattices.

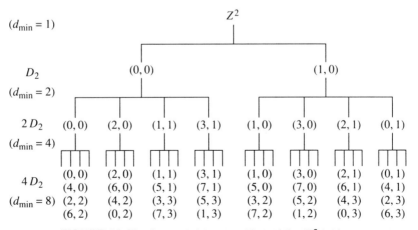

FIGURE 11.47 A coset decomposition of the Z^2 lattice.

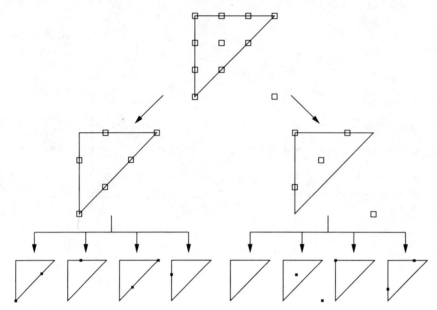

FIGURE 11.48 Signal set partitioning for $L = 3$ and $M = 8$.

shown in Fig. 11.48. The next step is trimming the signal cosets at each level to meet the population requirements. Note that the zero signal that is outside the triangular acceptance region is included in Fig. 11.48. Figure 11.49 shows the final result of this partitioning and trimming. Both Figs. 11.48 and 11.49 demonstrate the process, which guarantees obtaining the population chain of 8-4-2. The

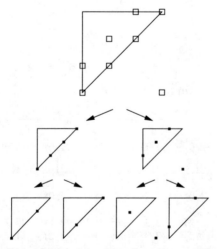

FIGURE 11.49 The final signal set resulting from the partitioning and trimming process for $L = 3$ and $M = 8$.

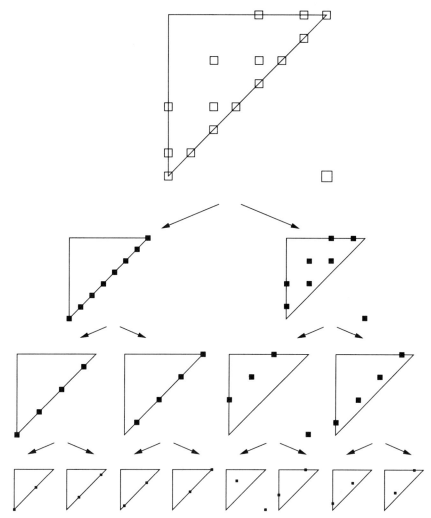

FIGURE 11.50 The final signal set resulting from the partitioning and trimming process for $L = 7$ and $M = 16$.

partitioning, in general, is not unique. The process for 16-ary and 32-ary signals is the same and the partitioning is shown in Figs. 11.50 and 11.51, respectively. Scale-up by factors of 14 and 24 is used for 16-ary and 32-ary signals, respectively.

The procedure for the signal set design, with the doubling minimum distance property at each node level of the partitioning tree, guarantees that the optimum trellis codes found for the AWGN channel can be used for the optical channel. Most trellis codes considered in this book are based on a set partitioning of the signal set with doubling minimum intersignal distance property at each level of the tree. The asymptotic performance gain of the trellis codes is defined as the

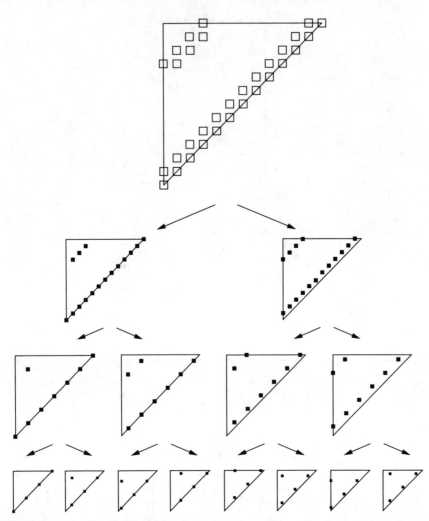

FIGURE 11.51 The final signal set resulting from the partitioning and trimming process for $L = 12$ and $M = 32$.

difference between the α's necessary to operate the coded system at the same error rate as the uncoded system. Then from (11.107) we have

$$e^{-(\alpha_c/2)\hat{d}^2_{\text{free}}} = e^{-(\alpha_u/2)\hat{d}^2_{\min}} \tag{11.113}$$

where α_c and α_u are the effective signal intensities for the coded and the uncoded systems, respectively, and $\hat{d}_{\text{free}}$ and $\hat{d}_{\min}$ are their respective minimum distances. From (11.113) the coding performance gain is defined by

$$G = 10\log_{10}\frac{\alpha_u}{\alpha_c} = 10\log_{10}\frac{\hat{d}^2_{\text{free}}}{\hat{d}^2_{\min}} \text{ dB}. \tag{11.114}$$

TABLE 11.12 Normalized Minimum Squared Distances of Partitioning of Sets ($\beta = 1/2$).

Partition Distance	8-ary	16-ary	32-ary
Δ_0	$\frac{1}{6}$	$\frac{1}{14}$	$\frac{1}{24}$
Δ_1	$\frac{2}{6}$	$\frac{2}{14}$	$\frac{2}{24}$
Δ_2	$\frac{4}{6}$	$\frac{4}{14}$	$\frac{4}{24}$
Δ_3	∞	$\frac{8}{14}$	$\frac{8}{24}$

The minimum squared distances of the set partitioning described earlier are shown in Table 11.12. In this table Δ_i, $i = 0, 1, 2, 3$, represents the minimum squared distance of the signal cosets at level i of the set-partitioning tree. Δ_0 is the minimum distance of the uncoded signal set. The computation of d_{free} is straightforward, since partitioning the tree for the cases we have considered is almost identical with most of the set partitioning used so far in this book for the design of TCM schemes. Actually, for coded 16-ary and 32-ary signals, one can simply copy the d_{free}'s from the designs obtained in previous chapters. For coded 8-ary signals, the sequence of Δ_i's does not quite double at each level. In this case d_{free} can be obtained by listing the minimum-distance pair of paths in each of the trellis diagrams and adding up the differences between these two paths. Table 11.13 contains $\hat{d}_{\text{free}}$'s for various trellis codes. From Tables 11.12 and 11.13 one derives the coding gains that are tabulated in Table 11.14.

TABLE 11.13 Normalized Squared Free Distances of TCM Schemes Based on Pulse-Width-Constrained Signals.

Number of States	$\hat{d}^2_{free}$		
	8-ary	16-ary	32-ary
4	$\frac{4}{6}$	$\frac{4}{14}$	$\frac{4}{24}$
8	$\frac{5}{6}$	$\frac{5}{14}$	$\frac{5}{24}$
16	$\frac{6}{6}$	$\frac{6}{14}$	$\frac{6}{24}$
32	$\frac{7}{6}$	$\frac{6}{14}$	$\frac{6}{24}$
64	$\frac{8}{6}$	$\frac{7}{14}$	$\frac{7}{24}$
128	$\frac{7}{6}$	$\frac{8}{14}$	$\frac{8}{24}$
256	$\frac{10}{6}$	$\frac{8}{14}$	$\frac{8}{24}$
512	—	$\frac{8}{14}$	$\frac{8}{24}$

TABLE 11.14 Coding Gains.

Number of States	Coding Gain (dB)		
	$G_{8/4}$ (2 bits/interval)	$G_{16/8}$ (3 bits/interval)	$G_{32/16}$ (4 bits/interval)
4	1.25	2.34	3.69
8	2.22	3.31	4.66
16	3.01	4.10	5.45
32	3.68	4.10	5.45
64	4.26	4.77	6.12
128	3.68	5.35	6.70
256	5.23	5.35	6.70
512	—	5.35	6.70

In closing this section, we refer the reader to the important recent contribution to the study of TCM on optical channels in [41].

11.6 TCM with Prescribed Convolutional Codes

In many applications it is desirable to have a multiple data rate capability for a given channel bandwidth allocation. TCM is ideally suited for this since it allows one to choose both the code rate and modulation alphabet size so that a single-channel symbol rate can correspond to a variety of different information bit rates. For example, for a channel symbol rate R_s, rate 1/2 coded QPSK, rate 2/3 coded 8-PSK, and rate 3/4 coded 16-PSK allow bit rates R_s, $2R_s$, and $3R_s$, respectively. The optimum design of such TCM schemes with varying complexity (e.g., 4 to 64 states) has been treated in great detail in several of the early chapters. For example, for a bit error rate equal to 10^{-5}, the optimum 64-state rate 1/2 coded QPSK, rate 2/3 coded 8-PSK, and rate 3/4 coded 16-PSK have E_b/N_0 and E_s/N_0 requirements on the AWGN as indicated in Table 11.15.

Unfortunately, to achieve the optimum code designs above in such a multirate environment requires implementing three distinct encoders, but even worse, *three different decoders with different trellis connectivities for each of the three data rate modes*. What would be desirable, then, would be a suboptimum scheme that unifies and simplifies the implementation with the hope of paying little, or even better, no performance penalty relative to the individual optimum designs.

Viterbi et al. [42, 43] have found a rather unique and cost-effective solution to the problem. Their approach is to implement the three modes using a *single* rate 1/2, constraint length 7 convolutional encoder/decoder with appropriate modifications of only the metric input sections. Since, as discussed in Chapter 3, a TCM scheme can be implemented as a suitable mapping of a combination of uncoded and convolutionally encoded bits, their suggestion is not far afield from what has been done before. The major contribution is that the convolutional encoder portion of the trellis encoder is now *fixed* for all three data rates and, moreover, selected

TABLE 11.15 E_b/N_0 and E_s/N_0 Requirements (in dB) with Trellis-Coded M-PSK for a Bit Error Probability of 10^{-5} and Constant Symbol Rate R_s.

Data Rate	Modulation	Code Rate	With Optimum 64-State Decoder		With Single 64-State Decoder	
			E_b/N_0	E_s/N_0	E_b/N_0	E_s/N_0
R_s	QPSK	1/2	4.5	4.5	4.5	4.5
$2R_s$	8-PSK	2/3	6.0	9.0	6.4	9.4
$3R_s$	16-PSK	3/4	9.6	14.4	9.6	14.4

as a standard item that has been around since the early 1970s [44] and for which VLSI implementations readily abound [45–47].

Since in the *optimum* design of a trellis code, the convolutional encoder portion will not necessarily turn out to be a conventional (standard) one, the configuration suggested by Viterbi et al. will generally be suboptimum. The interesting result, however, is that their suboptimum configuration yields a performance very close, if not equal, to that of the optimum one corresponding to each case. Such a pragmatic approach to coding applications whereby the "industry standard" convolutional code is used to achieve a multirate capability might possibly become the universal trellis coding standard.

Before going on to specific applications of the above, it is only fair to point out that the notion of unifying the implementation of convolutional encoder/decoders with different $m/(m-1)$ code rates into a single configuration goes back to a technique known as "puncturing," which was suggested over a decade ago. Punctured codes were first employed by Linkabit Corp. but first appeared in the literature in the work of Cain et al. [48] for $m = 2$ and 3, and later extended by Yasuda et al. [49] to arbitrary m. The idea there was to use the standard rate 1/2 convolutional encoder and achieve the higher rate by deleting a fraction of the symbols generated at its output. For the decoder, they again used the standard rate 1/2 configuration with the deleted symbols replaced by erasures.

11.6.1 Application to M-PSK Modulation

Figure 11.52 is an illustration of the scheme proposed in [43] for M-PSK modulation. The trellis codes generated by this method will be referred to here as "pragmatic" codes [43]. One of the two (for rate 2/3 coded 8-PSK) or three (for rate 3/4 coded 16-PSK) input bits is passed through a conventional rate 1/2, constraint length 7 convolutional encoder. The two output code symbols from the encoder define one of four M-ary symbols. For 8-PSK, these symbols are uniformly located in either the right or left half-plane, with the remaining (uncoded) input bit being used to choose between the two planes. For 16-PSK, the four M-ary signals are located in one of the four quadrants, and the remaining two (uncoded) input bits are used to choose the quadrant. In summary, for a rate 2/3 coded 8-PSK, one input bit remains uncoded and defines a half-plane, while the second bit is convolutionally encoded with a rate 1/2 code and the two output code symbols choose one of four 8-PSK symbols in

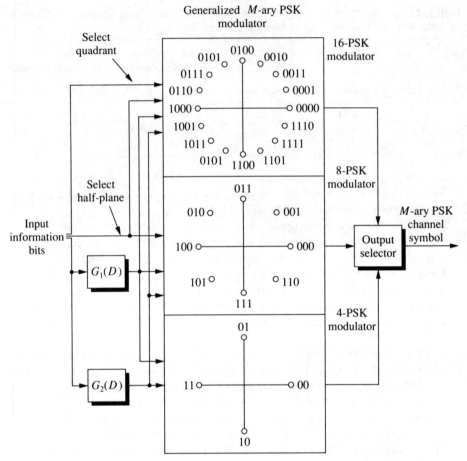

FIGURE 11.52 Generalized M-ary PSK modulator. (Reprinted with the permission of the authors of [43])

that plane. For a rate 3/4 code, two input bits remain uncoded and define a quadrant, while the remaining bit is again rate 1/2 convolutionally encoded and allows choice of one of four 16-PSK symbols in that quadrant.

The E_b/N_0 requirements for the single encoder/decoder 64-state (constraint length 7) implementation of Fig. 11.52 are given in Table 11.15 and compared with those of the optimum codes determined from Ungerboeck's search [39]. We observe that for operation at R_s and $3R_s$, we pay no penalty relative to the optimally designed 64-state trellis codes. At $2R_s$ (rate 2/3 coded 8-PSK), there is a mere 0.4-dB penalty. The reason for the penalty in the 8-PSK case stems from the fact that each state in the optimum code has four distinct input branches and four distinct output branches with no parallel paths. Thus d_{free} is determined by an error event path of length greater than 1. By limiting the code as in [44, 45] to incorporate a rate 1/2 convolutional encoder, we force the existence of two parallel paths per branch in the trellis and d_{free} is now limited by the minimum distance among the parallel paths, which has a smaller value.

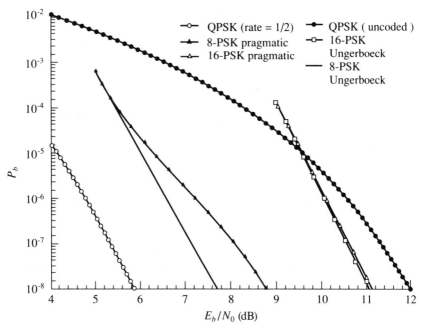

FIGURE 11.53 P_b performance comparison for some optimum and pragmatic trellis codes.

Figure 11.53, which plots P_b versus E_b/N_0, is a more complete picture of the comparison between the optimum and pragmatic trellis codes from which the values in Table 11.15 were extracted.

It is natural to ask whether this pragmatic approach to trellis coding for the AWGN channel is applicable to the fading channel. Unfortunately, the answer is "no," since uncoded bits in the implementation of Fig. 11.52 correspond to parallel paths in the trellis diagram and we have previously shown (see Chapters 9 and 10) that to achieve a diversity greater than 1 requires a trellis with no parallel paths.

Before turning our attention to M-AM and QAM modulations, we note that the scheme above for QPSK, 8-PSK, and 16-PSK modulations can be extended to apply to arbitrary M-PSK with $M = 2^{m+1}$ [43]. In particular, the lowest-order bit of the m input bits would be fed to the convolutional encoder whose two output bits choose one of four phases within a sector $(2\pi/2^{m-1}$ rad) according to the Gray-code rule:

$$\begin{aligned} 00 &\to 0 \text{ rad} \\ 00 &\to \pi/2^m \text{ rad} \\ 11 &\to 2\pi/2^m \text{ rad} \\ 10 &\to 3\pi/2^m \text{ rad.} \end{aligned} \tag{11.115}$$

The remaining $m - 1$ bits select the sector as follows. If j is the decimal equivalent of this sequence of $m - 1$ input bits, then the $(j + 1)$st sector is chosen when $0 \leq j \leq 2^{m-1} - 1$.

11.6.2 Application to M-AM and QAM Modulations

To apply the pragmatic approach to M-AM modulation, we merely perform a conformal mapping of the circular M-PSK constellation onto a segment of the real line according to the transformation $y = e^{jx}$, where y is an M-PSK symbol and x is the corresponding M-AM symbol. Thus the perimeter of the circle defines the range of the points in the M-AM constellation, and similarly, the arc length between adjacent M-PSK symbols now becomes the distance between M-AM points. As such, a symmetric M-PSK signaling set maps into a symmetric (about the origin) M-AM signaling set. Each of the $2^{m-1} = M/4$ sectors now becomes a segment of the real line of length, say, Δ, which is numbered from left to right according to the same decimal equivalent of the corresponding $m - 1$ uncoded bits as above. Each real line segment of length Δ contains four equally spaced points that are determined left to right in accordance with the convolutional encoder output sequences 00, 01, 11, and 10. This mapping produces M points equally spaced by $\Delta/4$ in the interval from $-(M/2 - 1)\Delta - \Delta/8$ to $(M/2 - 1)\Delta + \Delta/8$. Finally, the distance parameter Δ should be normalized by the square root of the average energy of the constellation,

$$\sqrt{E} = \sqrt{\frac{(M^2 - 1)\Delta^2}{192}}. \tag{11.116}$$

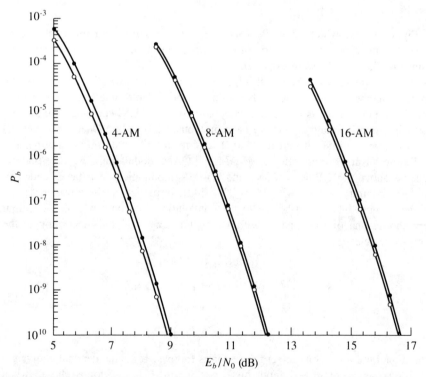

FIGURE 11.54 P_b performance of some pragmatic trellis codes with AM modulation.

Figure 11.54 illustrates the performance of rate 1/2 coded 4-AM, rate 2/3 coded 8-AM, and rate 3/4 coded 16-AM with 64 states using the pragmatic code structure above. Also indicated in the figure are the corresponding results for the 64-state Ungerboeck codes corresponding to each individual case. We observe that in all cases, the performance of the two are almost identical.

For the pragmatic trellis-coded QAM, Viterbi et al. [43] propose coding both dimensions (I and Q) with their own pragmatic trellis-coded M-AM as above resulting in a constellation with M^2 points conveying $2m$ bits/symbol. Although not necessarily optimum, surprisingly, this independent coding per dimension procedure yields a performance roughly similar to the more elaborate two-dimensional Ungerboeck codes discussed in Chapter 6 (see also [50]).

REFERENCES

1. S. BENEDETTO, E. BIGLIERI, and V. CASTELLANI, *Digital Transmission Theory*. Prentice-Hall, Englewood Cliffs, N.J., 1987.
2. R. W. LUCKY, J. SALZ, and E. J. WELDON, JR., *Principles of Data Communications*. McGraw-Hill, New York, N.Y., 1968.
3. J. G. PROAKIS, *Digital Communications*. McGraw-Hill, New York, N.Y., 1983.
4. A. E. LEE and D. G. MESSERSCHMIDT, *Digital Communication*. Kluwer, Boston, Hingham, Mass., 1988.
5. J. BINGHAM, *The Theory and Practice of Modem Design*. Wiley-Interscience, New York, N.Y., 1988.
6. G. D. FORNEY, JR., ET AL., "Efficient modulation for band-limited channels," *IEEE J. Select. Areas Commun.*, Vol. SAC-2, pp. 632–647, Sept. 1984.
7. S. H. QURESHI, "Adaptive equalization," *Proc. IEEE*, Vol. 73, Sept. 1985.
8. S. QURESHI and D. G. FORNEY, JR., "Performance and properties of a T/2 equalizer," *Proc. NTC'77*, pp. 11.1-1 to 11.1-9, 1977.
9. L.-N. WONG and P. J. MCLANE, "Performance of trellis codes for a class of equalized ISI channels," *IEEE Trans. Commun.*, Vol. 36, pp. 1330–1336, Dec. 1989.
10. H. K. THAPAR, "Real-time application of trellis coding to high-speed voiceband data transmission," *IEEE J. Select. Area Commun.*, Vol. SAC-2, pp. 648–658, Sept. 1984.
11. R. D. GITLIN and S. B. WEINSTEIN, "Fractionally spaced equalization: An improved digital transversal equalizer," *Bell Syst. Tech. J.*, Vol. 60, pp. 275–297, Feb. 1981.
12. D. D. FALCONER and F. R. MAGEE, JR., "Adaptive channel memory truncation for maximum likelihood sequence estimation," *Bell Syst. Tech. J.*, Vol. 52, pp. 1541–1562, Nov. 1973.
13. M. K. SIMON and D. DIVSALAR, "Combined trellis coding with asymmetric MPSK modulation," *JPL Publication 85-24 (MSAT-X Report 109)*, Pasadena, Calif., May 1, 1985; see also *IEEE Trans. Commun.*, Vol. COM-35, No. 2, pp. 130–141, Feb. 1987.
14. D. DIVSALAR and M. K. SIMON, "Trellis coded modulation for 4800–9600 bps transmission over a fading mobile satellite channel," *JPL Publication 86-8 (MSAT-X Report 129)*, Pasadena, Calif., June 1, 1986.

15. D. Divsalar and M. K. Simon, "Multiple trellis coded modulation (MTCM)," *JPL Publication 86-44 (MSAT-X Report 141)*, Pasadena, Calif., Nov. 15, 1986; see also *IEEE Trans. Commun.*, Vol. 36, No. 4, pp. 410–419, Apr. 1988.
16. A. J. Viterbi and J. K. Omura, *Principles of Digital Communication and Coding*. McGraw-Hill, New York, N.Y., 1979.
17. W. C. Lindsey and M. K. Simon, *Telecommunication Systems Engineering*. Prentice-Hall, Englewood Cliffs, N.J., 1973.
18. W. C. Lindsey and M. K. Simon, "Optimum performance of suppressed carrier receivers with Costas loop tracking," *IEEE Trans. Commun.*, Vol. COM-25, No. 2, pp. 215–227, Feb. 1977.
19. J. Hagenauer and C.-E. Sundberg, "Performance evaluation of trellis-coded 8-PSK with phase offset," *ICC'88 Conf. Rec.*, Philadelphia, Pa., pp. 23.4.1–23.4.7, June 1988.
20. H. Leib and S. Pasupathy, "Trellis-coded MPSK with reference phase errors," *IEEE Trans. Commun.*, Vol. COM-35, No. 9, pp. 888–900, Sept. 1987.
21. G. Kaplan and E. Zehavi, "Noise effects on M-ary PSK trellis codes," *IEEE International Symposium on Information Theory*, San Diego, Calif., p. 107, Jan. 14–19, 1990.
22. T. Fujino, Y. Moritani, M. Miyake, K. Murakami, Y. Sakato, and H. Shiino, "A 120 Mbit/s 8PSK modem with soft-decision Viterbi decoder," *7th International Conference on Digital Satellite Communications (ICDSC7)*, Munich, Germany, pp. 315–321, 1986.
23. R. J. F. Fang, "A coded 8-PSK system for 140-Mbit/s information rate transmission over 80-MHz nonlinear transponders," *7th International Conference on Digital Satellite Communications (ICDSC7)*, Munich, Germany, pp. 305–313, 1986.
24. E. Biglieri, "High-level modulation and coding for nonlinear satellite channels," *IEEE Trans. Commun.*, Vol. COM-32, No. 5, pp. 616–626, May 1984.
25. T. C. Jedrey, N. E. Lay, and W. Rafferty, "The design and performance of a modem for land mobile satellite communications," *Proc. IEE Conference on Satellite Systems for Mobile Communications and Navigation*, London, England. Oct. 17–19, 1988.
26. T. C. Jedrey, N. E. Lay, and W. Rafferty, "An all-digital 8-DPSK TCM modem for land mobile satellite communications," *Proc. IEEE International Conference on Acoustics, Speech, and Signal Processing*, New York, N.Y., Apr. 11–14, 1988.
27. D. Divsalar and M. K. Simon, "Trellis coded modulation for 4800–9600 bps transmission over a fading mobile satellite channel," *IEEE J. Select. Areas Commun.*, Vol. SAC-5, No. 2, pp. 162–175, Feb. 1987.
28. M. K. Simon and D. Divsalar, "Doppler-corrected differential detection of MPSK," *IEEE Trans. Commun.*, Vol. COM-37, No. 2, pp. 99–109, Feb. 1989.
29. L. E. Franks and J. P. Bubrouski, "Statistical properties of timing jitter in a PAM timing recovery scheme," *IEEE Trans. Commun.*, Vol. COM-22, pp. 913–920, July 1974.
30. G. Ungerboeck, J. Hagenauer, and T. Abdel-Nabi, "Coded 8-PSK experimental modem for the INTELSAT SCPC-system," *7th International Conference on*

Digital Satellite Communications (ICDSC7), Munich, Germany, pp. 299–304, 1986.
31. J. K. WOLF and G. UNGERBOECK, "Trellis coding for partial response channels," *IEEE Trans. Commun.*, Vol. COM-34, No. 8, Aug. 1986.
32. J. W. KETCHUM, "Performance of trellis codes for M-ary partial response," *Proc. Globecom'87*, Tokyo, Japan, Nov. 15–18, 1987.
33. H. KOBAYASHI, "Correlative level coding and maximum likelihood decoding," *IEEE Trans. Inf. Theory*, Vol. IT-17, pp. 586–594, Sept. 1971.
34. G. D. FORNEY, "Maximum likelihood sequence estimation of digital sequences in the presence of intersymbol interference," *IEEE Trans. Inf. Theory*, Vol. IT-18, pp. 363–378, May 1972.
35. E. ZEHAVI and J. K. WOLF, "On saving decoder states for some trellis codes and partial response channels," *IEEE Trans. Commun.*, Vol. 36, No. 2, Feb. 1988.
36. G. L. BECHTEL and J. W. MODESTINO, "Trellis coded modulation on the pulsewidth-constrained direct-detection optical channel," *IEEE Trans. Commun.*, to appear, 1990.
37. G. L. BECHTEL and J. W. MODESTINO, "Optimal pulsewidth-constrained signaling on the direct-detection optical channel," *IEEE Trans. Commun.*, to appear, 1990.
38. D. L. SNYDER and I. B. RHODES, "Some implications of the cutoff-rate criterion for coded direct-detection optical communication systems," *IEEE Trans. Inf. Theory*, Vol. IT-26, pp. 327–338, May 1980.
39. G. UNGERBOECK, "Channel coding with multilevel/phase signals," *IEEE Trans. Inf. Theory*, Vol. IT-28, No. 1, pp. 55–67, Jan. 1982.
40. A. R. CALDERBANK and N. J. A. SLOANE, "New trellis codes based on lattices and cosets," *IEEE Trans. Inf. Theory*, Vol. IT-33, No. 2, pp. 177–195, Mar. 1987.
41. C. N. GEORGHIADES, "Some implications for optical direct-detection channels," *IEEE Trans. Commun.*, Vol. COM-37, pp. 481–487, May 1989.
42. A. J. VITERBI, J. K. WOLF, and E. ZEHAVI, "Trellis-coded M-PSK modulation for highly efficient military satellite applications," *MILCOM'88 Conf. Rec.*, San Diego, Calif., pp. 35.2.1–35.2.5, Oct. 23–26, 1988.
43. A. J. VITERBI, E. ZEHAVI, R. PADOVANI, and J. K. WOLF, "A pragmatic approach to trellis-coded modulation," *IEEE Commun. Mag.*, Vol. 27, No. 7, pp. 11–19, July 1989.
44. J. A. HELLER and I. M. JACOBS, "Viterbi decoding for satellite and space communications," *IEEE Trans. Commun. Technol.*, Vol. COM-19, pp. 835–848, Oct. 1971.
45. "QUALCOMM announces single-chip $K = 7$ Viterbi decoder device," *IEEE Commun. Mag.*, Vol. 25, No. 4, pp. 75–78, Apr. 1987.
46. S. KUBOTA, S. KATO, T. ISHITANI, and M. NAGATANI, "High-speed and high-coding-rate Viterbi decoder VLSI design and performance of NUFEC," *IEEE International Symposium on Information Theory*, Kobe, Japan, June 1988.
47. H. SUZUKI, M. TAJIMA, and M. SHINAJA, "Viterbi decoder chip implemented with sub-parallel architecture," *IEEE International Symposium on Information Theory*, Kobe, Japan, June 1988.

48. J. B. Cain, G. C. Clark, Jr., and J. M. Geist, "Punctured convolutional codes of rate $(n-1)/n$ and simplified maximum likelihood decoding," *IEEE Trans. Inf. Theory*, Vol. IT-25, pp. 97–100, Jan. 1979.
49. Y. Yasuda, K. Kashusi, and Y. Hirata, "High-rate punctured convolutional codes for soft decision Viterbi decoding," *IEEE Trans. Commun.*, Vol. COM-32, pp. 315–319, Mar. 1984.
50. G. Ungerboeck, "Trellis-coded modulation with redundant signal sets. Part I: Introduction; Part II: State of the Art," *IEEE Commun. Mag.*, Vol. 25, pp. 5–21, Feb. 1987.

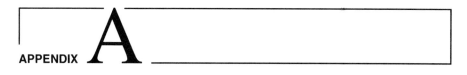

Fading Channel Models

One of the important considerations in the development of the trellis-coded modulation (TCM) schemes over fading channels is the effect of propagation on their performance. The material in this appendix covers the theoretical development of the Rician channel model (including antenna effects). For more details on the subject see [1] to [5].

A.1 The Theoretical Model

An RF signal at carrier frequency f_c (e.g., 1.5 GHz) with circular polarization that is transmitted by a satellite to a moving vehicle in a typical land environment exhibits extreme variation in both amplitude and apparent frequency. The fading effects are due to the random distribution of the electromagnetic field in space and originate from the motion of the vehicle. Here the electromagnetic field will be expressed by a linear superposition of plane waves of random phase, each of whose frequency is effected by a Doppler shift of the carrier frequency. This shift is due to the mobile velocity, carrier frequency, and the angle that the propagation vector makes with the velocity vector.

The power spectrum of the received signal depends on the density of the arrival angles of the plane waves and the mobile antenna directivity pattern. The instantaneous frequency of the signal received at the mobile has a random variation (i.e., a random frequency modulation).

We can identify three components in the received faded signal at the mobile antenna: the direct line-of-sight component, the specular component, and the diffuse component. The combined direct and specular components are usually referred to as the *coherent received component* and the diffuse component is referred to as the *noncoherent received component*. In this model it is assumed that only the coherent component can be attenuated by a lognormal-distributed shadowing process [6], due to the presence of trees, poles, buildings, and vegetation in the terrain. If the received coherent component is totally blocked, the noncoherent component dominates and the received signal envelope variation has a Rayleigh distribution, which will be discussed shortly.

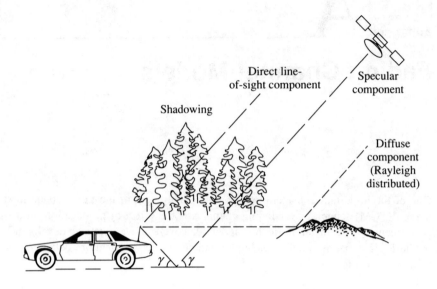

Complex received faded signal : $R(t) = m(t) * (R_{dir}(t) + R_{spec}(t)) + R_{dif}(t)$

$m(t)$: long-term signal fading (lognormal distributed)

FIGURE A.1

In Fig. A.1 the field-reflected diffuse component seen by the mobile can be represented as

$$e(t) = \sum_{n=1}^{N} e_n \cos\left[(\omega_c - \omega_n)t + \phi_n\right] \quad \text{(A.1)}$$

where

$$\omega_n = \frac{2\pi}{\lambda} v \cos \alpha_n. \quad \text{(A.2)}$$

In (A.1) and (A.2), ω_c is the angular carrier frequency, e_n the amplitude of the nth wave field, α_n the angle between the direction of the mobile movement and the nth incident wave, v the velocity of the mobile vehicle, λ the carrier wavelength, and ϕ_n a phase uniformly distributed between 0 and 2π. Note that the maximum Doppler frequency shift is

$$f_d = \frac{v}{\lambda} = \frac{v f_c}{c} \quad \text{(A.3)}$$

where f_c is the carrier frequency and c is the speed of light.

We note that for large N by the central limit theorem the field components are approximately Gaussian random processes and since the Doppler frequency shift f_d is much smaller than f_c, these components are narrowband. Furthermore, it has been assumed that for times that are short compared to the slow variations of the signal, the processes are wide-sense stationary and since they are Gaussian, they are stationary. Note that we can write

$$e(t) = N_c(t) \cos \omega_c t - N_s(t) \sin \omega_c t \tag{A.4}$$

where $N_c(t)$ and $N_s(t)$ are narrowband stationary independent Gaussian noise processes. We are assuming that the mean of the diffuse signal is zero and the variance is given by

$$\text{var}\{N_c(t)\} = \text{var}\{N_s(t)\} = \sigma^2. \tag{A.5}$$

Let the direct line-of-sight component be $A \cos[(\omega_c + \omega_d)t]$ and the specular component be $B \cos[(\omega_c + \omega_d)t + \phi_0]$, where $\omega_d = 2\pi f_d$ and ϕ_0 is some arbitrary phase. Define

$$\mu_c = A \cos \omega_d t + B \cos(\omega_d t + \phi_0) \tag{A.6}$$

and

$$\mu_s = A \sin \omega_d t + B \sin(\omega_d t + \phi_0). \tag{A.7}$$

Then the Rician parameter K is defined as the ratio of the direct and specular power to the diffuse power:

$$K = \frac{\mu_c^2 + \mu_s^2}{2\sigma^2} = \frac{A^2 + B^2 + 2AB \cos \phi_0}{2\sigma^2}. \tag{A.8}$$

We assume that the power of the fading signal is normalized to unity. Fix the time t and let

$$x = \mu_c + N_c(t) \tag{A.9}$$

and

$$y = \mu_s + N_s(t). \tag{A.10}$$

Next define

$$\rho(t) = \sqrt{x^2 + y^2} \tag{A.11}$$

$$\theta(t) = \tan^{-1} \frac{y}{x}. \tag{A.12}$$

It can be shown that the pdf's of $\rho(t)$ and $\theta(t)$ can be written as

$$p(\rho) = 2(K+1)\rho \exp\{-(K+1)\rho^2 - K\} I_0\left(2\rho \sqrt{K(1+K)}\right) \qquad \rho \geq 0 \tag{A.13}$$

for $E\{\rho^2\} = 1$ (i.e., power of fading process normalized to unity) and

$$p(\theta) = \frac{1}{2\pi} \exp\{-K\} + \frac{1}{2}\sqrt{\frac{K}{\pi}} \cos \theta e^{-K \sin^2 \theta}\left[1 + \text{erf}\left(\sqrt{K} \cos \theta\right)\right] \qquad |\theta| \leq \pi$$

$$\tag{A.14}$$

where $I_0(\cdot)$ is the zeroth-order modified Bessel function of the first kind. This is the Rician distribution. For the Rayleigh distribution we set $K = 0$, which means no line-of-sight and no specular component.

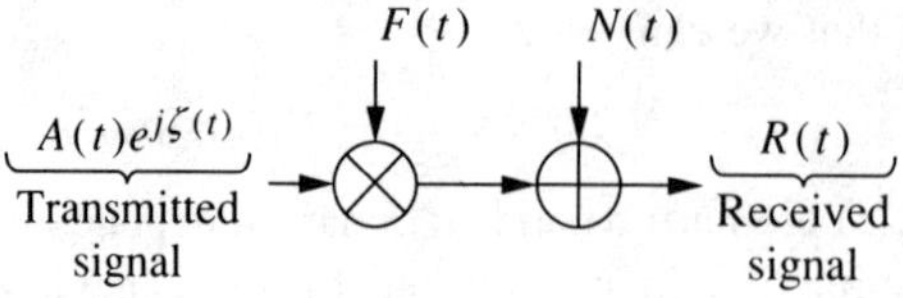

```
                    F(t)        N(t)
                     ↓           ↓
  A(t)e^{jζ(t)}    ⊗ ─────→ ⊕ ─────→   R(t)
  ─────────                            ─────────
  Transmitted                          Received
    signal                              signal

         F(t) =  m(t)e^{jω_d t}  +  n_d(t)e^{jψ(t)}
                 ─────────────      ──────────────
                  Long-term            Diffuse
                   signal              process
                   fading             (Rayleigh)
```

FIGURE A.2

The complex baseband model for the fading channel at any given time instant t is shown in Fig. A.2, where we assume that $\mathrm{Re}\{A(t)\exp([j\{\omega_c t + \zeta(t)\}]\}$ is the modulated transmitted signal and $N(t)$ is complex white Gaussian noise. Note that in Fig. A.2

$$F(t) = \rho(t)\exp\{j\theta(t)\} = [m(t)\mu_c + N_c(t)] + j[m(t)\mu_s + N_s(t)] \quad (A.15)$$

represents the fading process with envelope variation $\rho(t)$ and phase variation $\theta(t)$. In (A.15) $N_c(t)$ and $N_s(t)$ are independent colored Gaussian noise processes. Note that $m(t)$ in Fig. A.2 and (A.15) represents the log-normal shadowing process. In the absence of shadowing, $m(t)$ can be set to a constant; for the normalization given above, we set $m(t) = 1$.

In order to generate $F(t) = \rho(t)\exp[j\theta(t)]$, we should find the autocorrelation function or equivalently, the power spectral density of the signal $F(t)$. To find the power spectral density of $F(t)$, first we find the power spectral density of the diffuse signal,

$$N_c(t) + jN_s(t). \quad (A.16)$$

A.2 Power Spectrum of the Diffuse Signal

To a first-order approximation, we assume that all reflected diffuse waves are traveling in the horizontal plane. As shown in Fig. A.3, assume that the horizontal component of the boresight of the antenna makes an azimuthal angle η with the direction of movement of the mobile vehicle. Let the angle between the direction of a diffuse wave component to the mobile antenna and the direction of the mobile vehicle be γ, and let $G(\alpha)$ represent the gain function of the antenna in the direction of this diffuse wave component, where $\alpha = \gamma - \eta$.

We note that the power contribution to the received signal by diffuse waves arriving in the horizontal plane from the scattering of rough objects around the mobile vehicle to within an angle $d\gamma$ is proportional to

$$p(\gamma)G(\gamma - \eta)\,d\gamma \quad (A.17)$$

where $p(\gamma)$ is the angular density function of wave arrival and $G(\gamma - \eta)$ is the antenna gain pattern. Now if $S(f)$ is the power spectral density of the received diffuse signal, then the power contribution at Doppler frequency f within the range

A.2 / Power Spectrum of the Diffuse Signal

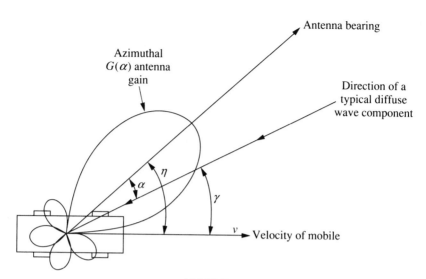

FIGURE A.3

of frequency df is

$$S(f)|df|. \tag{A.18}$$

But the Doppler frequency is

$$f = f_d \cos \gamma \tag{A.19}$$

where

$$f_d = v/\lambda \tag{A.20}$$

is the maximum Doppler frequency shift with v the mobile speed and λ the carrier wavelength.

Note that for given f_d, the two angles γ and $-\gamma$ can result in the same Doppler frequency. Therefore, the contribution of power at angles γ and $-\gamma$ within differential angle $d\gamma$ that results in Doppler frequency f is proportional to

$$k[p(\gamma)G(\gamma - \eta) + p(-\gamma)G(-\gamma - \eta)] \, d\gamma. \tag{A.21}$$

Equating (A.18) and (A.21), we get

$$S(f) = \frac{k[p(\gamma)G(\gamma - \eta) + p(-\gamma)G(-\gamma - \eta)]}{|df|} \tag{A.22}$$

where k is a constant. But

$$|df| = f_d|-\sin \gamma| \, d\gamma$$

$$= f_d \sqrt{1 - \left(\frac{f}{f_d}\right)^2} \, cf \, d\gamma \tag{A.23}$$

$$= \sqrt{f_d^2 - f^2} \, d\gamma.$$

Therefore,

$$S(f) = \frac{k[p(\gamma)G(\gamma - \eta) + p(-\gamma)G(-\gamma - \eta)]}{\sqrt{f_d^2 - f^2}} \quad (A.24)$$

where

$$\gamma = \left| \cos^{-1} \frac{f}{f_d} \right|. \quad (A.25)$$

The term $p(\gamma)$, which shows the intensity of diffuse waves from different directions, depends on various scenarios (shadowing, type of surrounding materials, roughness of materials, etc.).

For simplicity, assume that all angles of the incident waves are uniformly distributed; then

$$p(\gamma) = 1/2\pi. \quad (A.26)$$

Thus

$$S(f) = \frac{kG(\gamma - \eta)}{\pi \sqrt{f_d^2 - f^2}}. \quad (A.27)$$

Now consider the power spectrum that results from the use of a beam antenna. For simplicity, assume that the antenna pattern is uniform within angle β; (antenna beamwidth) in azimuthal (horizontal) plane. Then we identify two cases as follows.

Case 1:

This case corresponds to a situation when the antenna bearing angle satisfies (assuming $\beta \leq \pi$)

$$\frac{\beta}{2} \leq |\eta| \leq \pi - \frac{\beta}{2} \quad (A.28)$$

The case for $\beta/2 \leq \eta \leq \pi - \beta/2$ is shown in Fig. A.4. For $\beta/2 \leq \eta \leq \pi - \beta/2$

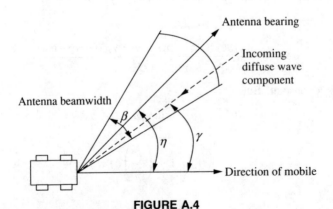

FIGURE A.4

A.2 / Power Spectrum of the Diffuse Signal

by assumption of having uniform antenna pattern, we should have

$$G(\gamma - \eta) = \begin{cases} G_0 & \text{if } |\gamma - \eta| < \dfrac{\beta}{2} \\ 0 & \text{otherwise} \end{cases} \quad (A.29)$$

and

$$G(-\gamma - \eta) = 0. \quad (A.30)$$

Let ξ denote the elevation angle. Figure A.5 shows the equivalent baseband power spectral density of the fading process, assuming that a single tone at carrier frequency f_c has been transmitted. In Fig. A.5 note that the carrier frequency is shifted by $f_d \cos\eta \cos\xi$ and, in general, is not inside the power spectral density of the diffuse process. If the power spectral density of the diffuse process at RF is desired, then we can simply shift the baseband power spectral density by f_c and $-f_c$. The power spectral density of the diffuse process at RF can be written as

$$S_{RF}(f) = \frac{1}{2}[S(f - f_c) + S(f + f_c)] \quad (A.31)$$

where the baseband power spectral density is

$$S(f) = \frac{kG_0}{2\pi \sqrt{f_d^2 - f^2}} \quad f_d \cos\left(\eta + \frac{\beta}{2}\right) < f < f_d \cos\left(\eta - \frac{\beta}{2}\right) \quad (A.32)$$

and it is shown in Fig. A.5. Figure A.6 presents a specific example of case 1 for $\eta = \pi/2$. The corresponding fading power spectral density is shown in Fig. A.7. Note that for this case there is no Doppler offset for the carrier frequency.

Case 2:

Here we consider the second case when $|\eta| < \beta/2$ or $|\eta| \leq \pi - \beta/2$. This case is shown in Fig A.8 and the corresponding fading power spectral density is shown in Fig A.9. In this case note that if an incoming wave with angle γ with respect

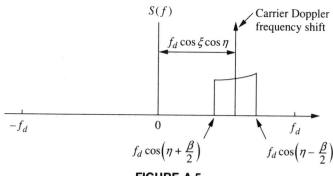

FIGURE A.5

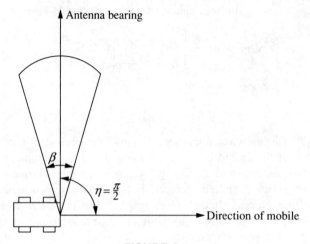

FIGURE A.6

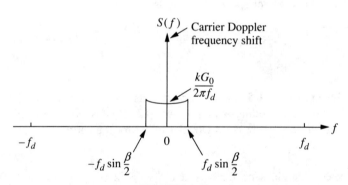

FIGURE A.7

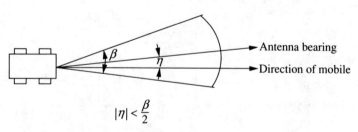

$|\eta| < \frac{\beta}{2}$

FIGURE A.8

A.2 / Power Spectrum of the Diffuse Signal

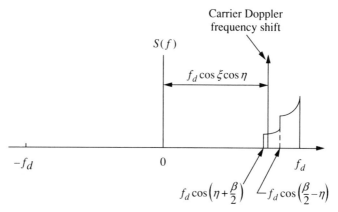

FIGURE A.9

to direction of mobile satisfies

$$\gamma < \left| \frac{\beta}{2} - \eta \right| \tag{A.33}$$

then both the γ ray and the $-\gamma$ ray will contribute to the power spectral density at frequency $f = f_d \cos \gamma$. However, if incoming waves possess an angle γ outside the above region, either the γ ray or the $-\gamma$ ray will contribute to the power spectral density. The baseband power spectral density, in general, can be written as

$$S(f) = \begin{cases} \dfrac{2kG_0}{2\pi \sqrt{f_d^2 - f^2}} & f_d \cos\left(\dfrac{\beta}{2} - \eta\right) < f < f_d \\[2ex] \dfrac{kG_0}{2\pi \sqrt{f_d^2 - f^2}} & f_d \cos\left(\eta + \dfrac{\beta}{2}\right) < f < f_d \cos\left(\dfrac{\beta}{2} - \eta\right) \end{cases} \tag{A.34}$$

Example for Case 2

If $\beta = 2\pi$, which is the case for an omnidirectional antenna, then the power spectral density is shown in Fig. A.10 with

$$S(f) = \frac{1}{\pi \sqrt{1 - (f/f_d)^2}} \qquad -f_d \leq f \leq f_d \tag{A.35}$$

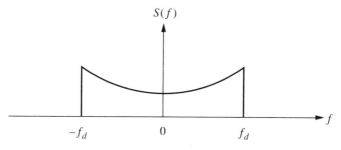

FIGURE A.10

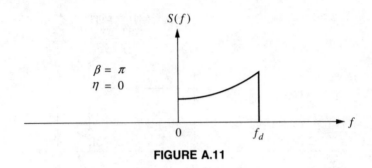

FIGURE A.11

which is normalized to unit fading power. The case of $\beta = \pi$ and $\eta = 0$ is shown in Fig. A.11.

A.3 Generation of Rician Fading Process

The Rayleigh fading process can be generated by passing a complex white Gaussian noise process through a filter with transfer function

$$H(f) = \sqrt{S(f)} \qquad (A.36)$$

where $S(f)$ is the power spectral density of the diffuse process. For example, the $S(f)$ given by (A.35) can be used in (A.36). The probability distribution of the envelope of the output process will have the Rayleigh density function. To generate the Rician process, a constant can be added to the Rayleigh process; then the envelope of the resulting process is distributed according to the Rician density function.

REFERENCES

1. W. C.-Y. LEE, *Mobile Communications Engineering*. McGraw-Hill, New York, N.Y., 1982.
2. R. S. KENNEDY, *Fading Dispersive Communication Channels*. Wiley-Interscience, New York, N.Y., 1969.
3. F. DAVARIAN, "Channel simulation to facilitate mobile-satellite communications research," *IEEE Trans. Commun.*, Vol. COM-35, No. 1, pp. 47–57, Jan. 1987.
4. D. DIVSALAR, "Software simulation of the LMSS propagation channel," *Proc. Progagation Workshop in Support of MSAT-X*, JPL, Pasadena, Calif., Jan. 30–31, 1985.
5. F. AMOROSO and W. W. JONES, "Modeling direct sequence psuedonoise (DSPN) signaling with directional antennas in the dense scatter mobile environment," *Proc. 38th IEEE Vehicular Technology Conference*, Philadelphia, Pa., June 15–17, 1988.
6. C. LOO, "A statistical model for a land mobile satellite link," *IEEE Trans. Veh. Technol.*, Vol. VT-34, pp. 122-127, Aug. 1985.

APPENDIX B

Computational Techniques for Transfer Functions

B.1 Transfer Function Bounds

In this appendix we describe how the superstate approach introduced in Chapter 4 can be used to evaluate upper bounds on bit or symbol error probability. Here we follow the approach and use the results in [1] to [5]. The modulator model used here is based on the equations

$$\mathbf{x}_n = f(\mathbf{u}_n, \sigma_n) \tag{B.1}$$

$$\sigma_{n+1} = g(\mathbf{u}_n, \sigma_n) \tag{B.2}$$

where σ_n is the modulator state at time n, $\mathbf{u}_n$ is the modular input (source output), and $\mathbf{x}_n$ is the modulator output (channel symbol).

To find the average bit error probability performance of the trellis (Viterbi) decoder, one must first find the pairwise error probability $P(\mathbf{x} \to \hat{\mathbf{x}})$ between the coded sequence $\mathbf{x} = \{\mathbf{x}_n\}$ and the estimated sequence $\hat{\mathbf{x}} = \{\hat{\mathbf{x}}_n\}$. Using the Bhattacharyya or Chernoff bound [2, 4], one obtains the following upper bound on the pairwise error probability:

$$P(\mathbf{x} \to \hat{\mathbf{x}}) \leq \prod_n Z^{\delta(\mathbf{U}_n, \mathbf{S}_n)}. \tag{B.3}$$

Here $\delta(\mathbf{U}_n, \mathbf{S}_n)$ is a function of supersymbol $\mathbf{U}_n$ and superstate $\mathbf{S}_n$, which are defined by

$$\mathbf{U}_n = (\mathbf{u}_n, \hat{\mathbf{u}}_n) \tag{B.4}$$

$$\mathbf{S}_n = (\sigma_n, \hat{\sigma}_n). \tag{B.5}$$

When $\hat{\sigma}_n = \sigma_n$ we define $\mathbf{S}_n$ as a good superstate; otherwise, it is a bad superstate. Here the state σ_n represents the transmitter state at the discrete time n and $\hat{\sigma}_n$ represents the decoder state at the same discrete time. In (B.3), Z is the Bhattacharyya parameter,

$$Z = \exp\left\{-\frac{1}{8\sigma^2}\right\} = \exp\left\{-\frac{E_s}{4N_0}\right\} \tag{B.6}$$

where E_s is the channel symbol energy and N_0 is one-sided power spectral density of the noise. The function $\delta(\mathbf{U}_n, \mathbf{S}_n)$ in (B.3) may depend on many factors, including the type of channel (e.g., AWGN, fading), the type of demodulation (coherent, differentially coherent, noncoherent), and the type of decoding metric, which may or may not include channel state information (CSI). The function $\delta(\mathbf{U}_n, \mathbf{S}_n)$ in general may also depend on the Chernoff parameter λ, which usually should be optimized after computation of the transfer function bound on the bit or symbol error probability. For example, for AWGN channels with coherent demodulation and minimum distance metric, we have

$$\delta(\mathbf{U}_n, \mathbf{S}_n) = \|f(\mathbf{u}_n, \sigma_n) - f(\hat{\mathbf{u}}_n, \hat{\sigma}_n)\|^2. \tag{B.7}$$

An upper bound on the average bit error probability can be obtained from

$$P_b \leq \sum_{\substack{\mathbf{x}, \hat{\mathbf{x}} \in \mathscr{C} \\ \mathbf{x} \neq \hat{\mathbf{x}}}} a(\mathbf{x}, \hat{\mathbf{x}}) p(\mathbf{x}) P(\mathbf{x} \to \hat{\mathbf{x}}) \tag{B.8}$$

where $a(\mathbf{x}, \hat{\mathbf{x}})$ is the number of bit errors that occur when the sequence $\mathbf{x}$ is transmitted and the sequence $\hat{\mathbf{x}} \neq \mathbf{x}$ is chosen by the decoder, $p(\mathbf{x})$ is the a priori probability of transmitting $\mathbf{x}$, and $\mathscr{C}$ is the set of all coded sequences. Equation (B.8) is also valid for symbol error probability when $a(\mathbf{x}, \hat{\mathbf{x}})$ is replaced by the number of input symbol errors. The upper bound given by (B.8) is efficiently evaluated using the transfer function bound approach.

The upper bound to (B.8) also has the representation

$$P_b \leq k_0 \sum_{i=1}^{\infty} a_i Z^{d_i^2} = \frac{k_0}{bN} \frac{\partial}{\partial I} T(D, I)\Big|_{I=1, D=Z} \tag{B.9}$$

where $d_1^2 < d_2^2 < d_3^2 < \cdots$, N is the number of trellis states and b is the number of information symbols associated with each branch. The distance d_1 is the free distance. For AWGN channels with coherent demodulation, d_1 is the free Euclidean distance d_{free}. The coefficient a_i in (B.9) represents the average number of bit errors for all error events with distance d_i. The coefficient k_0 in (B.9) is a factor that depends on the type of the channel and the type of the demodulation and the code structure. If Jacobs's conditions [6] are fulfilled, then $k_0 = \frac{1}{2}$. For AWGN with optimum coherent demodulation, k_0 is given by

$$k_0 = \frac{1}{2} \text{erfc}\left(\sqrt{\frac{E_s}{4N_0} d_{\text{free}}^2}\right) D^{-d_{\text{free}}^2} \tag{B.10}$$

which means that to compute k_0, we should have knowledge of d_{free}, the minimum distance. For all other cases we set $k_0 = 1$.

In (B.9) the function $T(D, I)$ is called the *generalized transfer function*. To compute $T(D, I)$ consider an $N^2 \times N^2$ superstate transition matrix $\mathbf{A}(D, I)$ with elements

$$a(\mathbf{S}_n, \mathbf{S}_{n+1}) = \begin{cases} \sum_{\mathbf{U}_n \in \mathscr{U}_n} 2^{-b} I^{W(\mathbf{U}_n)} Z^{\delta(\mathbf{U}_n, \mathbf{S}_n)} & \mathscr{U}_n \neq \emptyset \\ 0 & \text{otherwise} \end{cases} \tag{B.11}$$

where $\emptyset$ is the empty set and the set $\mathcal{U}_n$ is defined by

$$\mathcal{U}_n = \{(\mathbf{u}_n, \hat{\mathbf{u}}_n) | (\hat{\mathbf{u}}_n, \hat{\sigma}_n) \neq (\mathbf{u}_n, \sigma_n) \quad \mathbf{S}_{n+1} = (g(\mathbf{u}_n, \sigma_n), g(\hat{\mathbf{u}}_n, \hat{\sigma}_n))\}. \tag{B.12}$$

In (B.11), $W(\mathbf{U}_n)$ represents the Hamming distance between information bit sequences $\mathbf{u}_n$ and $\hat{\mathbf{u}}_n$.

For the symbol error probability computation we have

$$P_s \leq \frac{k_0}{N} \frac{\partial}{\partial I} T(D, I)\big|_{I=1, D=Z} \tag{B.13}$$

where we should redefine $W(\mathbf{U}_n)$ by

$$W(\mathbf{U}_n) = \begin{cases} 1 & \hat{\mathbf{u}}_n \neq \mathbf{u}_n \\ 0 & \text{otherwise.} \end{cases} \tag{B.14}$$

Note that in deriving (B.11), we have used (B.3) and the fact that $a(\mathbf{x}, \hat{\mathbf{x}})$ in (B.8) can be represented as

$$a(\mathbf{x}, \hat{\mathbf{x}}) = \frac{d}{dI} I^{\sum_n W(\mathbf{U}_n)} \big|_{I=1}. \tag{B.15}$$

Let us now consider the superstate transition matrix $\mathbf{A}(D, I)$. By suitably renumbering the pair states, $\mathbf{A}(D, I)$ can be partitioned as follows:

$$\mathbf{A}(D, I) = \begin{bmatrix} \mathbf{A}_{GG}(D, I) & \mathbf{A}_{GB}(D, I) \\ \mathbf{A}_{BG}(D, I) & \mathbf{A}_{BB}(D, I) \end{bmatrix} \tag{B.16}$$

where the $N \times N$ matrix $\mathbf{A}_{GG}(D, I)$ represents the transitions between "good" (correct) superstates, the $(N^2 - N) \times N$ matrix $\mathbf{A}_{BG}(D, I)$ represents the transitions from the "bad" (incorrect) superstates to the "good" (correct) superstates, the $N \times (N^2 - N)$ matrix $\mathbf{A}_{GB}(D, I)$ represents the transitions from the "good" (correct) superstates to the "bad" (incorrect) superstates, and finally, the $(N^2 - N) \times (N^2 - N)$ matrix $\mathbf{A}_{BB}(D, I)$ represents the transitions between the "bad" (incorrect) superstates, It can be shown that [1–3]

$$T(D, I) = \mathbf{1}^t [\mathbf{A}_{GG}(D, I) + \mathbf{A}_{GB}(D, I)[\mathbf{I} - \mathbf{A}_{BB}(D, I)]^{-1} \mathbf{A}_{BG}(D, I)] \mathbf{1} \tag{B.17}$$

where $\mathbf{1}$ denotes the column vector all of whose elements are 1, $\mathbf{I}$ denotes the identity matrix, and the superscript t represents the transpose operation. A sufficient condition for the existence of the inverse matrix in (B.17) is that the sum of the absolute values of the entries in each row (or column) of $\mathbf{A}(D, I)$ be less than 1.

Next, to obtain an upper bound on P_b one should take the derivative of $T(D, I)$ given in (B.17) with respect to the variable I and then set $I = 1$. Before taking the derivative, we can rewrite (B.17) as

$$T(D, I) = a(D, I) + \mathbf{b}^t(D, I)[\mathbf{I} - \mathbf{A}_{BB}(D, I)]^{-1} \mathbf{c}(D, I) \tag{B.18}$$

where

$$a(D, I) = \mathbf{1}'\mathbf{A}_{GG}(D, I)\mathbf{1} \tag{B.19}$$

$$\mathbf{b}(D, I) = \mathbf{1}'\mathbf{A}_{GB}(D, I) \tag{B.20}$$

$$\mathbf{c}(D, I) = \mathbf{A}_{BG}(D, I)\mathbf{1}. \tag{B.21}$$

Note that $a(D, I)$ is a scalar, while $\mathbf{b}(D, I)$ and $\mathbf{c}(D, I)$ are $(N^2 - N) \times 1$ vectors. After taking the derivative of (B.18) with respect to the variable I and setting $I = 1$, we obtain

$$\frac{\partial}{\partial I}T(D, I)\Big|_{I=1} = a'(D, 1) + \mathbf{b}'^t(D, 1)[\mathbf{I} - \mathbf{A}_{BB}(D, 1)]^{-1}\mathbf{c}(D, 1)$$

$$+ \mathbf{b}^t(D, 1)[\mathbf{I} - \mathbf{A}_{BB}(D, 1)]^{-1}\mathbf{c}'(D, 1)$$

$$+ \mathbf{b}^t(D, 1)[\mathbf{I} - \mathbf{A}_{BB}(D, 1)]^{-1}\mathbf{A}'_{BB}(D, 1)[\mathbf{I} - \mathbf{A}_{BB}(D, 1)]^{-1}\mathbf{c}(D, 1).$$

$$\tag{B.22}$$

In Equation (B.22) the prime denotes the derivative of a function with respect to the variable I.

B.2 Numerical Computation of Transfer Function Bounds

Consider the transfer function bound given by (B.18). The computation of $a(D, I)$, $\mathbf{b}(D, I)$, and $\mathbf{c}(D, I)$ in (B.18) is an easy task. Note that

$$[\mathbf{I} - \mathbf{A}_{BB}(D, I)]^{-1} = \sum_{n=0}^{\infty} \mathbf{A}_{BB}^n(D, I). \tag{B.23}$$

Therefore, (B.18) can be written as

$$T(D, I) = a(D, I) + \sum_{n=0}^{\infty} \mathbf{b}^t(D, I)\mathbf{A}_{BB}^n(D, I)\mathbf{c}(D, I). \tag{B.24}$$

To compute the summation in (B.24), let the $(N^2 - N) \times 1$ vector $\mathbf{z}_n$ be defined as

$$\mathbf{z}_n^t = \mathbf{b}^t(D, I)\mathbf{A}_{BB}^n(D, I). \tag{B.25}$$

The vector $\mathbf{z}_n$ can be found recursively [1, 5] from

$$\mathbf{z}_{n+1}^t = \mathbf{z}_n^t \mathbf{A}_{BB}(D, I) \tag{B.26}$$

with $\mathbf{z}_0 = \mathbf{b}(D, I)$. Then

$$T(D, I) = a(D, I) + \sum_{n=0}^{\infty} \mathbf{z}_n^t \mathbf{c}(D, I). \tag{B.27}$$

To compute (B.22), take the derivative of (B.27) with respect to I and set $I = 1$ to obtain

$$\frac{\partial}{\partial I} T(D, I) = a'(D, I) + \sum_{n=0}^{\infty} \frac{\partial \mathbf{z}_n^t}{\partial I} \mathbf{c}(D, I) + \sum_{n=0}^{\infty} \mathbf{z}_n^t \mathbf{c}'(D, I). \tag{B.28}$$

Differentiating (B.26) and substituting

$$\mathbf{y}_n^t = \frac{\partial \mathbf{z}_n^t}{\partial I} \tag{B.29}$$

we obtain

$$\mathbf{y}_{n+1}^t = \mathbf{y}_n^t \mathbf{A}_{BB}(D, 1) + \mathbf{z}_n^t \mathbf{A}'_{BB}(D, 1). \tag{B.30}$$

This says that $\mathbf{y}_n$ can be found recursively from (B.30) with $\mathbf{y}_0 = \mathbf{b}'(D, 1)$. Thus

$$P_b \leq \frac{k_0}{bN} \frac{\partial}{\partial I} T(D, I)\Big|_{I=1, D=Z} = a'(Z, 1) + \sum_{n=0}^{\infty} [\mathbf{y}_n^t \mathbf{c}(Z, 1) + \mathbf{z}_n^t \mathbf{c}'(Z, 1)]. \tag{B.31}$$

Obviously, to compute the right-hand side of (B.31), we should properly truncate the summation at some integer n_0. For example, n_0 might be chosen so that

$$\frac{\mathbf{y}_{n_0}^t \mathbf{c}(Z, 1) + \mathbf{z}_{n_0}^t \mathbf{c}'(Z, 1)}{\sum_{n=0}^{n_0} [\mathbf{y}_n^t \mathbf{c}(Z, 1) + \mathbf{z}_n^t \mathbf{c}'(Z, 1)]} < \epsilon \tag{B.32}$$

for some desired small ϵ.

Note that the $(N^2 - N) \times (N^2 - N)$ matrix $\mathbf{A}_{BB}(D, I)$ is very sparse, having no more than $2^{2b}/N_p^2$ nonzero elements for each row and column, where N_p represents the number of parallel paths joining two states. For example, in performing multiplication in (B.26), we should only multiply the nonzero column components of $\mathbf{A}_{BB}(D, I)$ by the corresponding components of the vector $\mathbf{z}_n$. By doing this, we can save tremendously in computational effort and storage requirements.

B.3 Computation of the Free Distance of a TCM Scheme

For some applications, to obtain a tight bound on P_b, we should compute k_0 given by (B.10) and use it in (B.9). But the computation of k_0 requires knowledge of the free distance d_{free}. Fortunately, one can obtain d_{free} from the computation of (B.18). It has been shown in [1] that

$$d_{\text{free}}^2 = \lim_{D \to 0} \log_2 \frac{T(2D, 1)}{T(D, 1)}. \tag{B.33}$$

Indeed, it can be shown that $\log_2[T(2D, 1)/T(D, 1)]$ decreases monotonically to the limit d_{free}^2 as $D \to 0$. Therefore,

$$d_{\text{free}}^2 \leq \log_2 \frac{T(2D, 1)}{T(D, 1)} \tag{B.34}$$

for all D. Furthermore, in [3] it is shown that a lower bound on d_{free}^2 is given by

$$d_{\text{free}}^2 \geq \frac{\log_2 T(D, 1)}{\log_2 D}. \tag{B.35}$$

In fact,

$$\frac{\log_2 T(D, 1)}{\log_2 D} = d_{\text{free}}^2 - \epsilon(D) \tag{B.36}$$

for $\epsilon(D) \geq 0$ and $\epsilon(D) \to 0$ monotonically as $D \to 0$. Thus, by taking a decreasing sequence of values of D, from $\log_2 T(2D, 1)/T(D, 1)$ and $\log_2 T(D, 1)/\log_2 D$, we obtain two sequences whose values are increasingly closer to each other and thus to d_{free}^2. Therefore, for very small values of D, we can obtain a very good approximation to d_{free}.

In summary, for computation of a tight upper bound on P_b or P_s, we can first compute (B.18), using the recursive technique of (B.27) to obtain d_{free}^2 from (B.33), then compute the upper bound on P_b or P_s given in (B.9) using (B.10) and the recursive technique of (B.31).

REFERENCES

1. D. DIVSALAR, "Performance of mismatched receivers on bandlimited channels," Ph.D. dissertation, University of California, Los Angeles, Calif., 1978, University Microfilms International.
2. A. J. VITERBI and J. K. OMURA, *Principles of Digital Communications and Coding*. McGraw-Hill, New York, N.Y., 1979.
3. E. BIGLIERI, "High-level modulation and coding for nonlinear satellite channels," *IEEE Trans. Commun.*, Vol. COM-32, No. 5, pp. 616–626, May 1984.
4. D. DIVSALAR, M. K. SIMON, and J. H. YUEN, "Trellis coding with asymmetric modulations," *IEEE Trans. Commun.*, Vol. COM-35, No. 2, pp. 130–144, Feb. 1987.
5. P. J. LEE, "A very efficient transfer function bounding technique on bit error rate for Viterbi decoded, rate 1/N convolutional codes," *Jet Propulsion Laboratory TDA Progress Report 42-79*, pp. 114–123, Nov. 1984.
6. I. M. JACOBS, "Probability of error bounds for binary transmission on the slowly fading Rician channel," *IEEE Trans. Inf. Theory*, Vol. IT-12, pp. 431–441, Oct. 1966.

APPENDIX C

Computer Programs: Design Technique

This appendix provides information on a PC-based, MS-DOS, computer program for the design technique given in Section 5.1. The programs in this appendix are available for the cost of a diskette, or by sending your own blank diskette ($3\frac{1}{2}$- or $5\frac{1}{4}$-inch size) to P. J. McLane. Two programs constitute the design package. One program, DMIN, computes the minimum distance of a trellis code. This is used to ensure that the specified code is not catastrophic. In the example it outputs the distance and length of the minimum distance error event. It also states if the distance for parallel transitions needs to be checked. The program is written in WATFOR. It is run by typing DMIN followed by pressing the RETURN key, at which time the user is requested to input the design data. Also, one is asked for the name of an output file; for instance, TEST.OUT.

Our other program is MATRIX. It solves for the Calderbank–Mazo form of a trellis code, that is, the solution to the matrix equation $BX = D$, as described in Section 5.1. It is run by typing MATRIX and then pressing the RETURN key. The user is then asked for the input data.

The remainder of the appendix considers two examples of DMIN followed by two examples for MATRIX. On the diskette both the source and object form of both programs are given.

DMIN

The program requires the following input:

- s = the number of states
- k, where 2^k = the number of transitions leaving a state or re-emerging at a state
- $\tilde{k}$, where $s^{k-\tilde{k}}$ = the number of parallel transitions
- D = the number of dimensions

The input file name and the output file name are also required. The input file provides the program with a description of the trellis.

EXAMPLE C.1 4-AM FOUR-STATE

The trellis and state assignment for this example are given in Fig. C.1.

Input file

```
 3
-1
 1
-3        Signal assignment beginning at the first state
-1
 3
-3
 1
 1
 3        Describes the interconnectivity of the trellis;
 1        denotes the uppermost transition form each state; for example,
 3            state 1 → state 1, state 2 → state 3
              state 3 → state 1, state 4 → state 3
```

EXAMPLE C.2 8-PSK FOUR-STATE

The trellis and state assignment for this example are given in Fig. C.2. The signal constellation is given in Fig. C.3. The data inputs for the input file are the (x, y) coordinates of the phase points in Fig. C.3.

Input file

```
       1,       0
      -1,       0
       0,       1
       0,      -1
   1/√2,    1/√2
```

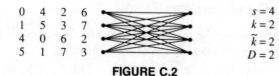

FIGURE C.2

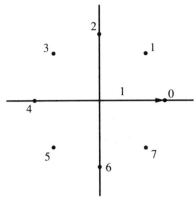

FIGURE C.3

$$\begin{array}{ll}
-1/\sqrt{2}, & -1/\sqrt{2} \\
-1/\sqrt{2}, & 1/\sqrt{2} \\
1/\sqrt{2}, & -1/\sqrt{2} \\
-1, & 0 \\
1, & 0 \\
0, & -1 \\
0, & 1 \\
-1/\sqrt{2}, & -1/\sqrt{2} \\
1/\sqrt{2}, & 1/\sqrt{2} \\
1/\sqrt{2}, & 1/\sqrt{2} \\
-1/\sqrt{2}, & 1/\sqrt{2} \\
1 & \\
1 & \\
1 & \\
1 & \\
\end{array}$$

□

MATRIX

- The program requires N, the number of input bits (the number of data bits + the number of memory bits).
- The program requires the names of the input file and output file. The input file contains the signal sent for each set of input bits from $0, 0, \ldots, 0$ to $1, 1, \ldots, 1$.

EXAMPLE C.3 4-AM FOUR-STATE

The trellis and state assignment for this example are given in Fig. C.4. The b_i follow from Fig. C.5. In the input file the signal assignment follows the binary triple (a_1, a_2, a_3) taken in natural order with $b_i = 1 - 2a_i$. For instance, the second signal assigned in the input file is -1, and this corresponds to $(1, 0, 0)$ or $b_1 = -1$ and $b_2 = b_3 = 1$.

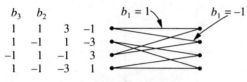

FIGURE C.4

Input file	Output file
3	$D(2) = 1$ $\therefore x = 2b_1b_3 + b_2$
-1	$D(13) = 2$
1	
-3	
-1	
3	
-3	
1	

□

EXAMPLE C.4 8-PSK FOUR-STATE

The state variables for this example are presented in Fig C.6 and the signal constellation is given in Fig C.7. The trellis is presented in Fig. C.8. In the input file the signal is assigned according to (a_1, a_2, a_3, a_4) taken in natural order with $b_i = 1 - 2a_i$. Note that 5 is the second signal assigned as it corresponds to $(1, 0, 0, 0)$ or $b_1 = -1$ and $b_2 = b_3 = b_4 = 1$.

Input file	Output file
1	$D(1) = -2$
5	$D(3) = -1$
-7	$D(24) = 4$
-3	
3	$\therefore x = 4b_2b_4 - 2b_1 - b_3$
7	
-5	
-1	
-7	
-3	
1	
5	
-5	
-1	
3	
7	

□

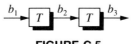

FIGURE C.5

FIGURE C.6

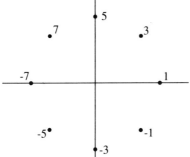

where
signal = $e^{i(x-1)\pi/8}$

FIGURE C.7

		(b_2, b_1)			
b_4	b_3	(1, 1)	(–1, 1)	(1, –1)	(–1, –1)
1	1	1	–7	5	–3
1	–1	3	–5	7	–1
–1	1	–7	1	–3	5
–1	–1	–5	3	–1	7

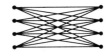

FIGURE C.8

```
C****************************************************************
C
        PROGRAM MATRIX
C
C Program identification
C =======================
C
C Given properly assigned channel signals, this program solves
C       the constant elements of D.  These constants determine
C       the terms involved in an analytic description of a
C       trellis-coded scheme.
C
C The signals are sequentially assigned in an input file (one
C       per line). The proper signal order is specified by a
C       fixed pattern of the Hadamard matrix. The signals must
C       be inputted according to the following sequence of input
C       bit combinations b1, ..., bn (transitions) :
C       0000...0, 1000...0, 0100...0, 1100...0, 0010...0, etc.
C
C For multi-dimensional signals, each dimension implies a
C       program execution with the input file containing the
C       proper coordinates.
C
C The program asks for the following input :
C       i) the value of N (N=k+v)
C       ii) the input data file name
C       iii) the output data file name
C
C Note : the real arrays must be dimensioned as follows
C       X(2**N), B(2**N,N), D(2**N - 1)
C
C WARNING : this program is designed for N < 10.
C
C****************************************************************
C
        REAL X(256), B(256,8), D(255)
        REAL E, PROD, SUM
C
C    X    : channel signal matrix
C    B    : Hadamard matrix reduced to its N prime columns
C    D    : matrix of constants
C    E    = +1 = element of B
C    PROD : product of prime column elements, element of the
C           specified column vector
C    SUM  : scalar being the dot product between X and a column
C           vector
C
        INTEGER N, I, J, K, L, M, Z, S, T, CPTR
        INTEGER TEMP, Y, CHECK
C
C    N = K + V : # of channel signals + # of memory bits
C    I         : loop-control variable (lcv) for each row of B,
C                and X
C    J         : lcv for each column of B, and for each row of D
C    K         : lcv for each prime column of B
C    M         : general lcv
C    L = 2**N  : # of rows of X (total # of trellis transitions)
C    Z         : # associated to a column vector and specifying
C                its combination of prime columns
C    S         : variable for specification of next Z
```

```
C     T          : variable for highest  possible digit values of
C                  specific digit positions
C     TEMP       : temporary  value  specifying the position of a
C                  digit
C     Y          : temporary number specifying a prime column
C     CPTR       : counter for # of digits of next Z
C     CHECK      : value of the last digit
C
      LOGICAL PASSED
C
C     PASSED : loop flag
C
      CHARACTER*12 INFILE,OUTFILE
C
      WRITE (6,*)
      WRITE (6,*)
      WRITE (6,5)
      READ (5,*) N
      WRITE (6,7)
      READ (5,9) INFILE
      WRITE (6,8)
      READ (5,9) OUTFILE
    5 FORMAT (' ENTER N              ---> ',$)
    7 FORMAT (' INPUT FILE NAME      ---> ',$)
    8 FORMAT (' OUTPUT FILE NAME     ---> ',$)
    9 FORMAT (A)
      OPEN (15,STATUS = 'OLD',FILE = INFILE)
      OPEN (8,STATUS = 'NEW',FILE = OUTFILE)
      L    = 2**N
      DO 10, I = 1, L
         READ(15, *) X(I)
   10 CONTINUE
C
C*****************************************************************
C
C Generation of the N prime columns
C =================================
C
      DO 20, K = 1, N
         DO 30, I = 1, L
C
            IF (MOD((I-1), 2**K) .GE. 2**K/2) THEN
               E = -1.0
            ELSE
               E = 1.0
            ENDIF
            B(I, K) = E
   30    CONTINUE
   20 CONTINUE
C
C*****************************************************************
C
      Z = 1
C
      DO 40, J = 1, L-1
         SUM = 0.0
         DO 50, I = 1, L
C
C              interpret Z as the combination of specific columns
C              ===================================================
C
```

```
              TEMP = Z
              PROD = 1.0
              DO 60, K = 1, N
                 IF (TEMP .NE. 0) THEN
                    Y = MOD(TEMP, 10)
                    TEMP = (TEMP - Y) / 10
                    PROD = PROD * B(I, Y)
                 ENDIF
   60         CONTINUE
C
              SUM = SUM + X(I) * PROD
   50      CONTINUE
C
           D(J) = SUM / FLOAT(L)
C
C          presentation of the constant elements of D
C          ==========================================
C
           IF (D(J) .NE. 0.0) THEN
              WRITE(8, *) 'D(', Z, ') = ', D(J)
              WRITE(8,*) ' '
           ENDIF
C
C define the combination of prime columns
C =======================================
C          for the next column vector
C          ==========================
C          e.g. for N = 3, we want 1, 2, 3, 12, 13, 23, and 123
C
C          basic incrementation
C          --------------------
C
           PASSED = .FALSE.
           T      = N + 1
           CPTR   = 0
           Z      = Z + 1
C          verify if the digits, starting at the end,
C          ------------------------------------------
C                    can be incremented
C                    ------------------
C
   70      CHECK = MOD(Z, 10)
           IF (CHECK .EQ. T) THEN
              PASSED = .TRUE.
              Z      = Z - CHECK
              CPTR   = CPTR + 1
              T      = T - 1
C
              IF (Z .NE. 0) THEN
                 Z = Z/10
                 Z = Z + 1
                 GOTO 70
              ENDIF
           ENDIF
C
C          define the next Z, given values of CPTR and present Z
C          -----------------------------------------------------
C
           IF (PASSED) THEN
              IF (Z .EQ. 0) THEN
                 S = 1
```

```
              ELSE
                  S = Z
              ENDIF
              Z    = 0
              TEMP = CPTR
              DO 80, M = 1, CPTR+1
                  Z    = S*(10**TEMP) + Z
                  S    = MOD(S+1, 10)
                  TEMP = TEMP - 1
    80        CONTINUE
          ENDIF
    40 CONTINUE
       CLOSE (15)
       CLOSE (8)
C
       STOP
       END
```

```
C*****************************************************************
C
       PROGRAM DMINIM
C
C PROGRAM IDENTIFICATION
C ======================
C
C Given properly assigned channel signals, this program finds
C     the minimum free Euclidean distance in a trellis code.
C
C The signals are sequentially assigned in an input file (one
C     per line). The proper signal order corresponds exactly
C     to the signal assignment of a trellis. Each dimension
C     is separated by a space.
C
C Following the signals, there are S additional entries.
C     Those entries specify the trellis structure (e.g. half-
C     connected, fully connected trellis). They are :
C     For each trellis state, the number of the upper trellis
C     state joined by a transition leaving the specified state.
C
C The program requires the following input
C     i) values of S(# of states), K, KT(K tilde), and
C                  D(# of dimensions).
C     ii) the input file name
C     iii) the output file name
C
C Note : the real variables should be dimensioned as follows
C          PT(S**2, 2**(2*KT), 2), CDT(S, S), TEMP(S, S),
C          X(S*2**K, D)
C
C *** Program input format of X to be specified according to
C     value of D
C
C*****************************************************************
C
       REAL  PT(64,16,2),  CDT(8,8), TEMP(8,8), X(256,4)
       REAL  DIST, MIN
       REAL  DMIN, TEMPO
C
```

```
C    PT     : Pair-state Table
C    CDT    : Cumulative Difference Table
C    TEMP   : temporary CDT
C    X      : matrix of channel signals
C    DIST   : cumulative  squared distance between signals  (due
C             to each coordinate)
C    MIN    : minimum squared distance  between any 2 signals of
C             two specific branches
C    DMIN   : minimum free Euclidean distance in a trellis code
C    TEMPO  : temporary new element value of CDT
C
      INTEGER I, J, L, M, N, R
      INTEGER S, K, KT, D
      INTEGER C1, C2, C3, SN, TD
      INTEGER P, Q, Y, Z
      INTEGER ROWPT, ROW, COL
C
C    I      : loop-control variable (lcv)  for each state  and for
C             each CDT row
C    J      : lcv for each state, for each CDT column, and each PT
C             column
C    L      : lcv for each branch per state
C    M      : lcv for each branch in the trellis
C    N      : lcv for each state
C    S      : # of states
C    K      : # of data bits
C    KT     : # of coded data bits
C    D      : # of dimensions
C    R      : lcv for each dimension
C    C1, C2, C3 : counters
C    SN     : number of present receiver state
C    TD     : trellis depth
C    P      : lcv for each parallel transition per state
C    Q      :                "
C    Y      : position index for reference channel signals
C    Z      : position index  for channel signals  compared to the
C             reference channel signals
C    ROWPT  : row number of PT
C    ROW    : row number of CDT and TEMP
C    COL    : column number of CDT and TEMP
C
      CHARACTER*12 INFILE,OUTFILE
C
C    INFILE  : input data file (device 16)
C    OUTFILE : output data file (device 8)
C
      LOGICAL FIRST, STILL
C
C    FIRST  : loop flag
C    STILL  : loop flag
C
C*****************************************************************
C
C General initialisation
C ======================
C
      WRITE (6,*)
      WRITE (6,*)
      WRITE (6,6)
      READ (5,*) S
      WRITE (6,7)
      READ (5,*) K
```

```
              WRITE (6,8)
              READ (5,*) KT
              WRITE (6,9)
              READ (5,*) D
              WRITE (6,11)
              READ (5,3) INFILE
              WRITE (6,12)
              READ (5,3) OUTFILE
              OPEN (16,FILE = INFILE,STATUS='OLD')
              OPEN (8,FILE = OUTFILE,STATUS='UNKNOWN')
3             FORMAT (A)
6             FORMAT (' ENTER S    (# OF STATES)        ==> ',$)
7             FORMAT (' ENTER K                         ==> ',$)
8             FORMAT (' ENTER KT (K TILDE)              ==> ',$)
9             FORMAT (' ENTER D    (# OF DIMENSIONS)    ==> ',$)
11            FORMAT (' ENTER INPUT FILE NAME    ==> ',$)
12            FORMAT (' ENTER OUTPUT FILE NAME   ==> ',$)
C
              Y   = 0
              Z   = 0
              DO 5, I = 1, S*2**K
                  READ(16, *) (X(I,J),J=1,D)
5             CONTINUE
C
              DO 40, I = 1, S
                 DO 50, J = 1, S
                    IF (I .EQ. J) THEN
                        CDT(I, J)    = 0.0
                        TEMP(I, J)   = 0.0
                    ELSE
                        CDT(I, J)    = 1000.0
                        TEMP(I, J)   = 1000.0
                    ENDIF
50              CONTINUE
40          CONTINUE
C
C****************************************************************
C
C Initialize Pair-state Table (PT)
C ================================
C
              TD = 0
              C1 = 0
              DO 10, N = 1, S
C
                  READ(16, *) C2
                  C2 = C2 - 1
                  C3 = 0
                  SN = 1
                  J  = 1
                  DO 20, M = 1, S * (2**KT)
C
                      DO 30, L = 1, 2**KT
                          MIN = 1000.0
                          DO 35, P =1, 2**(K-KT)
                              DO 38, Q = 1, 2**(K-KT)
                                  DIST = 0.0
                                  DO 39, R = 1, D
                                      DIST = DIST + (X(Y+P, R)
     1                                             - X(Z+Q, R))**2
39                                CONTINUE
```

```
C
                IF (DIST .LT. MIN) THEN
                   MIN = DIST
                ENDIF
38          CONTINUE
35        CONTINUE
C
          Z = Z + 2**(K-KT)
          IF (Z .GE. S*2**K) THEN
             Z = 0
             Y = Y + 2**(K-KT)
          ENDIF
C
          PT(C1*S + SN, C3*(2**KT) + L, 1) = MIN
          PT(C1*S + SN, C3*(2**KT) + L, 2) = C2*S + J
          J = J + 1
          IF (J .EQ. S+1) THEN
             J = 1
          ENDIF
30      CONTINUE
C
        SN = SN + 1
        IF (SN .EQ. S+1) THEN
           SN = 1
           C2 = C2 + 1
           C3 = C3 + 1
        ENDIF
20    CONTINUE
C
      C1 = C1 + 1
10  CONTINUE
C
C*****************************************************************
C
C Compute TEMP for the next trellis depth
C =======================================
C
      FIRST = .TRUE.
      DMIN = 1000.0
110   DO 70, I = 1, S
C
        DO 80, J = 1, S
          IF ((CDT(I, J) .NE. 0.0 .OR. FIRST) .AND.
     1                    CDT(I, J) .LT. 1000.0) THEN
C
            DO 90, L = 1, 2**(2*KT)
              ROWPT  = (I-1)*S + J
              TEMPO  = CDT(I, J) + PT(ROWPT, L, 1)
              ROW    = IFIX(PT(ROWPT, L, 2) - 1.0) / S + 1
              COL    = MOD(IFIX(PT(ROWPT, L, 2)), S)
              IF (COL .EQ. 0) THEN
                 COL = S
              ENDIF
C
              IF (TEMPO .LT. TEMP(ROW, COL) .OR.
     1                       TEMP(ROW, COL) .EQ. 0.0) THEN
                 TEMP(ROW, COL) = TEMPO
              ENDIF
C
              IF (ROW .EQ. COL .AND. TEMP(ROW, ROW) .GT. 0.0
     1                .AND. TEMP(ROW, ROW) .LT. DMIN) THEN
                 DMIN = TEMP(ROW, COL)
```

```
              ENDIF
C
90            CONTINUE
           ENDIF
80       CONTINUE
70    CONTINUE
C
C***************************************************************
C
C Update CDT from TEMP
C ====================
C
      STILL = .FALSE.
      FIRST = .FALSE.
      DO 120, I = 1, S
         DO 130, J = 1, S
            CDT(I, J) = TEMP(I, J)
            IF (I .EQ. J) THEN
               TEMP(I, J) = 0.0
            ELSE
               TEMP(I, J) = 1000.0
            ENDIF
            IF (CDT(I, J) .LT. DMIN) THEN
               STILL = .TRUE.
            ENDIF
130      CONTINUE
120   CONTINUE
C
C***************************************************************
C
C Presentation of results
C =======================
C
      IF (STILL) THEN
         TD = TD + 1
         IF (TD .GE. 100) THEN
            WRITE(8, *) 'The code appears to be catastrophic'
            WRITE(8, *) 'after a trellis depth of 100'
            WRITE(8, *)
         ELSE
            GOTO 110
         ENDIF
      ENDIF
      WRITE(8,500) DMIN
      WRITE(8,505) TD
      WRITE(8,*)
      WRITE(8,*) 'Now check for DMIN due to parallel transitions !'
500   FORMAT('Squared DMIN = ',F10.6,' for a trellis ',$)
505   FORMAT('depth of at most ',I4)
C
      CLOSE (16)
      CLOSE (8)
      STOP
      END
```

Index

Analytic description, TCM, 149
Anderson, J. B., 27
Antenna
 bearing, 518
 gain, 514–517
 omnidirectional antenna, 519
 pattern, 516
Asymmetric
 16-QAM, 176, 203
 8-PSK, 176, 187
 8-AM, 176
 4-PSK, 179
 16-PSK, 195
 4-AM, 201
 relative performance, 200, 204
Asymmetric modulation, trellis-coded, 174, 205
Asymmetric M-PSK, 221
Asymmetry angle, parameter (of M-PSK modulation), 176, 179, 353
 optimization of
  for AWGN channel, 178–200
  for slow Rician and Rayleigh fading, 356, 380
Asymmetry (of modulation), 174–205, 259
 CPM, 175
 M-AM, 201–203, 204
 M-PSK, 179–200
 QAM, 203, 205
Asymmetry translation, parameter (of M-AM modulation), 175–176
 optimization of (for AWGN channel), 201–203

Average information, 16
Azimuth angle, 514

Bandlimited Gaussian channel, 25
 channel capacity, 26
Bandwidth, 6, 69
 available, 69
 efficiency, 71, 246
 99% power bandwidth, 6
BCH codes
 binary, 48–52
 nonbinary, 52–56
Bellman, R., 59
BER performance: rotationally invariant TCM
 4/8 PSK, 337
 linear case, 336
 nonlinear case, 335
Bessel function, 513
Bhattacharyya parameter, 178, 287, 290, 351, 466, 521
Biglieri, E., 207
Binary erasure channel, 33
Binary symmetric channel (BSC), 3, 11, 33–40
 BSC for additive noise channel, 11
Binary tree, 16
Biorthogonal signals, 250
Bit error probability, 12
 of TCM in presence of phase error, 453–475
Block interleaving/deinterleaving, 344

541

Bound
  Bhattacharyya, 101, 202
  Blichfeldt, 143
  Chernoff, 114
  Hamming, 229
  Johnson, 138
  Levenshtein, 138
  union, 100
Branch metric, 87
Brändström, H., 207
Byte (8-bits), 55

Cain, J. B., 503
Calderbank, A. R., 79, 149–151, 217, 282
Calderbank–Mazo form for TCM, 149
Carrier phase synchronization error (jitter), 259, 293, 415, 454
  discrete carrier, 457–458
  suppressed carrier, 459
Cartesian set product, 16
Catastrophic
  code, 178, 183, 259, 262
  TCM, 113
Cavers, J., 402
Channel
  capacity, 18, 207
  complex baseband model, 438
  gain, 438
  group delay, 438
  errors, 3, 33–35
  passband, 437
  reliability function, 28
  state information (CSI), 343, 345–346
Channels with phase offset BPSK, 453
Chernoff bound (parameter), definition of, 347, 457, 459, 463, 522
Clark, A. P., 299
Code
  Hamming, 38, 48, 229
  Reed–Muller, 230
  Reed–Solomon, 54
  Role, 33, 52
Coding gain
  asymptotic (of TCM), 72, 99
Coherent demodulator, 295
  suppressed carrier 4-PSK case, 295
Coherent fading component, 511

Communication channel, 6
  additive white Gaussian noise channel (AWGN), 8
  flat fading channel, 8
  frequency selective fading channel, 8
  intersymbol interference (ISI) channel, 8
Communications receiver, 8
  integrate and dump, 12
  relation to channel decoder, 8
Complementary error function, erfc(x), 13
Computational cutoff rate, R_0, 27
  for 8-PSK, 29
Computer programs
  AM examples, 528, 529
  DMIN, 527
  MATRIX, 527
  PSK examples, 528, 530
Continuous-phase modulation
  decomposition of the modulator, 252–255
  definition, 240–242
  encoder, 252–255
  full-response, 241
  modulation index, 241
  partial-response, 241
  phase state, 242
  physical phase trellis, 251
  tilted phase trellis, 251
  time-variant trellis, 251–252
Convex ∪ function, 28
Convolutional codes
  elementary, 56–71
    finite state machine, 56
    trellis representation, 57
  noisy reference loss reduction by using interleaving, 453–475
  partial response channels, 487–491
  performance degradation due to phase error for, 464–470
  prescribed, combined with TCM (pragmatic approach), 502–507
    application to M-PSK, 503–505
    application to M-AM and QAM, 506–507
  punctured, 503
  weight distribution of rate 1/2, $k = 7$, 468
  $K = 7$, $R = 1/2$, in phase offset channel, 466–469

$1 - D$, partial response channel, 487–489
Conway, J. H., 209
Costas loop, 453, 459, 462, 469, 472
Costello, D., 401
CPFSK, 243
CPM; *see* Continuous-phase modulation
Crepeau, P., 401
Cutoff rate of optical channels, 495
Cyclic-group alphabets, 250

Decoder buffer, size of, 179, 346, 367
Deep space optical channel, 490
Demodulation
  of correlated symbols, 90
  maximum-likelihood, 88
  symbol-by-symbol, 89
Design procedure
  rotationally invariant TCM, 303
Design rules
  rotationally invariant TCM, 302
  TCM, 152
  Turgeon, T1, T2, 152
  Ungerboeck, U1, U2, U3, 152
Dicode channel, 481
Differential encoding, 297
Diffuse fading component, 343
Digital communications, 1
  receiver, 2
  structure, 1
  transmitter, 2
Directional antenna, 515–519
Direct-detection optical channel, 490
Discrete carrier, 458, 464, 472–473
Discrete memoryless channel (DMC), 19
Diversity, 358, 398, 422
Divsalar, D., 175, 401, 402
Doppler frequency shift, 343, 367–370, 385, 390, 477, 478, 512, 514
Dynamic programming (Bellman), 59

Edbauer, F., 401
Elia, M., 207
Elias, P., 34
Encoder
  channel encoder, 2, 33
  source encoder, 2

Energy efficiency, 71
Energy per bit, E_b, 14
Entropy, 2
  binary entropy function, $h(p)$, 17
  discrete source, 15
Equalizer, 437
  decision feedback, 444
  LMS equalizer, 444
  symbol spaced equalizer, 444
  $T/2$-spaced, 444
Error
  coefficient, 133
  event, 61, 73, 99–100
  probability, 1, 100, 111–112
  state diagram (convolutional codes), 61, 101–102
    branch gains, 61
    error, 61
    Hamming distance, 61
  weight matrix, 101, 126
Error-correcting code, 33
Error event path, 393–394, 397
  "length" of, defined (relation to diversity), 393–394, 397
  shortest, 397
Error floor, 343, 389, 460
Error performance of convolutionally coded BPSK over M-PSK, 464–470
Error performance of trellis-coded over interleaving for, 458, 459
Euclidean weight, 127
EXOR function, 34

Fading channel
  introduction, 171–173
  models, 511–520
  theoretical model, 511
Fair partition, 209, 223
Feedback realization TCM, 171–173
Feedforward realization TCM
  examples, 156, 159, 163, 166
  in relation to feedback form, 174
Fiber optic, 490
Field (from algebra), 41
  error locator field, 53
  extension field, 42
  Galois field, 40–48
Free distance, 72

544 Index

Free distance (*Continued*)
  computation, 125–131
  of TCM (computation of), 525–526
Frequency reuse, 212

Gallager, R. G., 24
Galois, E., 41
Gaussian random variable, 12
Generation: rotationally invariant TCM
  8-PSK example, 310
  16-PSK example, 314
  16-QAM example, 317
Gersho, A., 207
Ginzburg construction, 224–229
Gram–Schmidt orthogonalization, 95
Gray code, 5, 505
Group (from algebra), 40
Group alphabet
  generation, 220
  regular, 220
  separable, 220

Hadamard matrix, 85
Hadamard transform, 85, 283
Hamming, R. W., 33–38
Hamming distance, d_H, 3, 33, 178, 411, 412, 432, 465
Hamming spheres, 38
HF channel, 440
Hirakawa, S., v, 93
Ho, P., 402
Hybrid multiple trellis code, 293, 415

Imai, H., v, 93
Imperfect carrier phase synchronization, 454
Information rate, 69, 71
INTELSAT, 479
Interdistance, 209, 223, 231
Interleaving
  depth, 344, 367, 459, 477
  span, 367
Interleaving/deinterleaving, noisy reference loss reduction by, 453–475
Intradistance, 209, 223, 231
Irreducible error probability, 460–463

Isometry, 107–109
Iterated code, 34

Jamali, S. H., 435
Johnston, D. A., 374

Kotel'nikov, 70
Kuhn–Tucker theorem, 24
  verify channel capacity guess, 24

Land mobile channel, 390
Land mobile satellite channel, 476
Lattice
  Barnes–Wall, 212
  basis, 210
  cubic, 210
  definition, 209
  dual, 210
  equivalent, 210, 212
  generator matrix, 210
  Gosset, 212, 217
  kissing number, 210
  Leech, 212
  for optical channel, 496
  sublattice, 212, 216
  transformations, 210
Lawrence, V. B., 207
Lee, S. L., 207
Le-Ngoc, T., 435
Leung, Y. S., 398, 435
Lindell, G., 244
Line-of-sight (LOS) component, 344
Lodge, J. H., 401
Lognormal (shadowing) channel, 343, 511–512
Loop bandwidth, 453, 462, 469
Loop signal to noise ratio, 462–463

Manhattan distance, 493
Mason, L. J., 390
Massey, J. L., v, 93
Matrix
  Hadamard, 85
  orthogonal, 250
  permutation, 250

Matrix description, 36
  error-correction codes, 36–40
  generator matrix, block codes, 36–38
  generator matrix, convolutional codes, 58–59
  parity check matrix, block codes, 36–38
  parity check matrix, convolutional codes, 58–59
  parity check matrix, TCM, 170–173
  orthogonal, 151–152
Maximum entropy, 17
  continuous source, 18
  discrete source, 17
Maximum signal label (MSL), 153
Maximum signal value (MSV), 152–153
Mazo, J. E., 79, 149–151, 282
Measure of information
  conditional entropy, 20
  entropy, 2
  joint entropy, 20
Metric, 113
  additive, 113
Minimum polynomial, 49–53
  Peterson–Weldon (P-W) table, 50
Mismatched decoder, 457
Mobile satellite channel, 343
  Rician parameter of, 354
  speech delay of, 346
Mobile Satellite Experiment (MSAT-X) of NASA, 368
Modulation, 4
  amplitude modulation, AM, 154
  AM/PM modulation, 169
  baseband, 4
  binary phase shift keying (BPSK), 464
  carrier phase offset, M-PSK, 460
  duobinary MSK, 8
  minimum shift keying (MSK), 6
  quadrature amplitude modulation (QAM), 6, 163, 164
  quadrature phase shift keying (QPSK), 6, 470
  quaternary baseband, 5
  phase shift keying (PSK), 157, 160
  phase shift keying represented as AM, 157
  tamed frequency modulation (TFM), 8
Moher, M. L., 401
MSAT-X, 476

Mulligan, M. M., 128
Multidimensional modulation
  Chapter 6, 207
  2/4 PSK, 163
  M/N PSK, 167
  4/8 PSK, 322, 337
Multiple trellis-coded modulation (MTCM), 259–293
  analytical representation of, 282–286
  bit error probability performance of, 287–290
  complexity of, 292–293
  computational cutoff rate of, 260, 290–292
  generalized, 260, 268–282
    set mapping and minimum squared free distance, 273–282
    set-partitioning method, 269–273
  multiplicity of, 260
  transfer function of superstate diagram, 287
  two-state case, 261–268
    mapping procedure, 264–265
    minimum squared free distance, 265–268
Multiple trellis-coded M-PSK for fading Rician channels, 412–430
  computational cutoff rate of, 432–433
  design principles, 412–417
  set partitioning and error probability performance, 417–430
  generating set, defined, 418
  simulation results, 433–435

Nakagami-m fading channel model, 401–402
Natural mapping (Fig. 5.8), 155
Next distance, 126, 133
Nibble (4-bits), 55
Noisy reference loss, 453
Noncoherent fading component, 511
Nonlinear channels, 475, 477

Octal representation, $H(D)$, 172–173
Omnidirectional antenna, 519
  power spectral density of fading using, 519

Optical channels, 490
  lattices for, 496–500
  set partitioning and trimming for, 495–500
  TCM over, 490–502
  trellis coding for optical channel, 494
Orbit of a group, 250
Ordered Cartesian product, defined, 418
Orthogonal signals, 250
Orthonormal polynomials, 94

Packetized data, 477
Packing density, 141–142
Pair-state transition diagram, 181
Pairwise error probability, 100, 454, 461, 521
Parity check, 34
  parity check matrix, 38
  parity check polynomials, TCM, 169–173
Partial response channels, 479–491
  $1 - D$, dicode channel, 481
  precoder for, 487–489
  set-partitioning, 486–487
  TCM over, 485–487
  trellis-coded modulation, $1 - D$ channel, 485–487
  two-state Viterbi decider, 482–484
Perfect carrier phase synchronization, 454
Performance of TCM with discrete and suppressed carrier, 458–474
Permutation alphabets, 247
  selection of optimum initial vector, 247–249
Peterson, W. W., 50
Phase error, 453
Phase-locked loop, 453–461
  Costas loop, 453, 459, 461, 462, 469, 472
  decision directed, 479
  4th-power loop, 296
  4th-power loss, 473
  squaring loss, 463
Phase offset channel, 453
Photon intensity, 491
Pilot tone (dual) calibration techniques, 345, 367, 370
Pizzi, S. V., 244

Poisson counting probability density function, 492
Pollara, F., 468
Polynomial
  generator, 48
  minimum, 49–53
Pottie, G. J., 138
Power-fade detection, 478
Power spectrum
  of diffuse component of fading, 514
  diffuse signal, 514–517
  TCM, 145–147
Prabhu, V. K., 336
Precoded partial response channels, 487–489
Primitive element, 42
Probability of error, 14
Pulse-width-constrained optical channel, 490

Radio loss, 453, 475
Raised cosine filter, 477
Reduced state transition matrix, 199
Repetition code, 3, 34
Rician (Rayleigh) channel, 343, 511–520
  amplitude probability density of, 513
  fading generation for, 520
  phase probability density of, 513
Rician (Rayleigh) fast-fading channel, 387–393
  bandwidth, B_d, 390
  channel model, 388, 390–391
  differentially coherent detection of trellis-coded M-PSK over, 387–393
  pairwise error probability bound, 388–389
  simulation results, 388–389
  upper bound on bit error probability, 389–390
Rician parameter, K, definition of, 344, 513
Rician (Rayleigh) slow-fading channel, 344–387, 520
  channel model, 344
  coherent detection of trellis-coded M-PSK over, 344–371
  pairwise error probability bound, 346–350
  simulation results, 366–371

Index 547

transfer function of superstate transition diagram, 351
upper bound on bit error probability, 350–366
differentially coherent detection of trellis-coded M-PSK over, 371–387
maximum-likelihood metric, 373–374, 405–407
pairwise error probability bound, 374–377
simulation results, 385–387
theoretical model, 511–515
upper bound on bit error probability, 377–385
Rimoldi, B., 250
Ring (from algebra), 41
Rotational invariance
M-PSK modulation, 295
in TCM, introductory, 297–302
Rotationally invariant TCM
multidimensional cases, 322–329
nonlinear, multidimensional, 329–335
16-PSK example, 304
16-QAM example, 309
Run of consecutive zeros, 485–488

Satellite channels, 475
Saxena, R. C. P., 128
Schlegel, C., 401
Set partitioning, 77–78, 230
of asymmetric M-PSK, 343, 344
M-PSK, 177, 180
AM, 201
QAM, 203
Shadowing, 343, 344, 477
Shahshahani, M., 485
Shannon, C. E., 19, 26, 27, 28, 70
Shannon's two theorems, 26
channel coding theorem, 27
noiseless coding theorem, 27
Signal
constellation, 70
of asymmetric modulations, 176
energy, 70
label, 99, 225
multidimensional, 207
sinusoidal, 6
space, 70
Signal difference, 152

Simon, M. K., 401, 402
Simplex, 137
16-QAM modulation, 444
Single-channel per carrier (SCPC), 478
Slepian, D., 220, 249
Sliding block input bits, 153
Sloane, N. J. A., 209, 217
Source, 1
binary, 33
continuous, 17
discrete, 1
Specular fading component, 175, 343, 511
Sphere packing, 136, 138, 211
Squaring loss, 463, 469
Stark, H., 27
State of TCM encoder, 72
superstate, 523
Stein, S., 374
Sundberg, C.-E. W., 244
Super symbol, 178, 521
Superstate
diagram, 484
reduced, 484
(pair-state), 374
transfer function, 522–524
Suppressed carrier, 459–464, 467–470, 473, 475
Symmetric channels, 22
channel capacity, 21, 22
Symmetry of a TCM scheme, 105
Syndrome vector, 36

Tables for fields (algebra), 40–48
addition, 40
multiplication, 41
Tanner, R. M., 225
Taylor, D. P., 138
Throughput, 69
Transfer function bounds
computation, 112–113
of an error-state diagram, 62–64, 102
generalized, 522
computational techniques, 521–526
phase offset channel, 465
rotational invariant code, 336
Transition
adjacent, 77
parallel, 73, 77, 80

Trellis, 73
  product, 131
Turgeon, M., 152, 310
Tuteur, F., 27

UHF channel, 368
Unbalanced QPSK, 180
Ungerboeck, G., v, 77, 93, 149, 192, 194, 292, 298, 417, 427, 435
  code, 79
  representation, 79
Ungerboeck-coded asymmetric modulation, 174–205
  asymptotic coding gain of, 182–205
  bit error probability (upperbound) of, 178, 202
  design procedure, 179
  optimum-asymmetry, 175–205
  set-partitioning, 175–177
  simulation results, 183–184
  state transition matrix of, 180–199
  superstate transition diagram of, 181, 189, 202
  transfer function bound, 176–178
  trellis diagram, 180–201
Ungerboeck-coded M-PSK for Rician fading channels, 430–432
Ungerboeck design rules, U1→U3, 78, 152, 304, 310
Ungerboeck form for TCM, 150
Uniform TCM, 106–107, 126
Unlimited run of zero, 483
Upper bound on Q function, 456–457

Venn diagram, 39
  for decoding, 39
Viterbi algorithm, 59, 90
  complexity and storage requirements, 92
  decision depth, 60
  path metric, 60
  state metric, 60
  survivor paths, 60
  truncated, 93
Viterbi decoder, 59–61, 451, 482, 484
Volterra series, 94, 282

WATFOR, 527
Wei, L. -F., 232, 299
Wei construction, 232–233
Weight profile, 108, 118, 121
Weighted Euclidean distance, 348
Weldon, Jr., E. J., 50
Welti, G. R., 207
White Gaussian noise, 12
Wilson, S. G., 128, 207, 244, 398, 435
Wolf, J. K., 336, 337, 479

Yasuda, Y., 503
Yuen, J. H., 175

Zehavi, E., 336, 337
Zetterberg, L., 207
Zhu, Z. C., 299